TOPICS IN REMOTE SENSING

Archaeological prospecting and remote sensing

TOPICS IN REMOTE SENSING

Series editors
Garry Hunt and Michael Rycroft

Archaeological Prospecting and Remote Sensing

I. SCOLLAR
Rheinisches Amt für Bodendenkmalpflege

A. TABBAGH
Université Pierre et Marie Curie, Paris

A. HESSE
Centre de Recherches Géophysiques

I. HERZOG
Rheinisches Amt für Bodendenkmalpflege

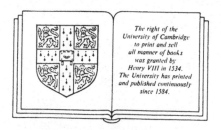

The right of the
University of Cambridge
to print and sell
all manner of books
was granted by
Henry VIII in 1534.
The University has printed
and published continuously
since 1584.

CAMBRIDGE UNIVERSITY PRESS
Cambridge
New York   Port Chester   Melbourne   Sydney

CAMBRIDGE UNIVERSITY PRESS
Cambridge, New York, Melbourne, Madrid, Cape Town, Singapore, São Paulo, Delhi

Cambridge University Press
The Edinburgh Building, Cambridge CB2 8RU, UK

Published in the United States of America by Cambridge University Press, New York

www.cambridge.org
Information on this title: www.cambridge.org/9780521115469

First published 1990
This digitally printed version 2009

A catalogue record for this publication is available from the British Library

ISBN 978-0-521-32050-4 hardback
ISBN 978-0-521-11546-9 paperback

CONTENTS

4   Archaeological image enhancement

PREFACE

There are a number of books which have been written by physicists for archaeologists describing advances in the application of physical methods to archaeological problems. There are a much larger number of books written by archaeologists describing the results obtained in the detection and mapping of archaeological sites through the application of remote sensing techniques. To our knowledge, there are no books written for those trained in the physical sciences which provide detailed information on the theory and practice of archaeological prospecting as remote sensing is called in Europe. This book is designed to fill that gap. It has been written by a team composed of an archaeologist, two geophysicists and an applied mathematician. Its goal is to present a systematic outline of the physical and mathematical principles lying at the roots of most of the currently employed methods and to describe how advances in the newly developed discipline of image processing can aid in the preparation and interpretation of results. It also aims at collecting the widely scattered literature of the field and summarising what the authors consider to be the most important contributions in one handy volume. We are most indebted to Cambridge University Press and especially to its science editor Dr Simon Mitton for giving us the opportunity to place this work before interested physical scientists. They will be aiding the archaeologists of the future, and it is also for those younger archaeologists whose training in mathematics and physical science permits them to profit from its content.

This book summarises the work of nearly four decades of research at the laboratories of the authors and elsewhere. It would not have been possible without the contributions of many collaborators and students. Some have now gone on to high positions in other branches of research. It has been a truly international effort, as can be judged from the three nationalities of the four authors, the fourth nationality of the editor and publisher and the many nationalities of their collaborators and students over the years.

The long term collaborators of the first author at Bonn, Aribert Lander, Bernd Weidner, Ian Graham and Waldemar Freund were responsible for much of what is descri'ed here. Irmela Herzog, the fourth author, has been with the first author since her high school days. Students from outside Germany who made important contributions while in Bonn were, in roughly chronological order: John Mudie, Ian Black, David Gubbins, David Bew, Ian Cunningham, Guy Abu Khalil, Mohamed El-Tahir, Gregory Tang, David Ray, Benjamin Santillan, Mark Dean, and Connie Bennett. Thomas S. Huang played a vital role during his year in Bonn as a Humboldt Prize winner in giving our fledgling image processing system wings while Karel Segeth has made repeated contributions on his visits.

The first author was guided as a student by O. G. S. Crawford, and at the instigation of Gerhard Bersu was asked to come to Germany three decades ago by the then director of the Rheinisches Landesmuseum Harald von Petrikovits, whose vision and patience made possible all that happened afterward. He had the benefit of discussing problems as a young researcher with Emile Thellier and Eugene Le Borgne, pioneers in the study of the magnetism of clays and soils which plays such a vital role in archaeological prospecting. Thellier's insight, guidance and material help during the early years contributed so much, and that is why the book is dedicated to his memory. Research at Bonn was financed by the Landschaftsverband Rheinland, the Deutsche Forschungsgemeinschaft, the Stiftung Volkswagenwerk, the Humboldt Stiftung, and the Deutsche Akademische Austausch Dienst.

At Garchy and Paris both the Centre Nationale de la Recherche Scientifique and the University of Paris 6 supported research over many years. The second and third author's principal collaborators were Alain Jolivet and Jeanne Tabbagh. Students who made important contributions were Vittorio Iliceto, Serge Renimel, Nguyen Van Hoa, Yannis Spahos, Christos Parchas, Marie-Claude Perisset, Gilles Boussuet and Michel Dabas. The third author was inspired to take up his line of research by Louis Cagniard and André Leroi-Gourhan, and the second author profited from this during their long-term collaboration.

The division of labour has been: I. Scollar conceived the book and wrote chapters 1, 3, 4, 7, and 8 and jointly authored chapters 2 and part of chapter 5. A. Tabbagh wrote chapters 9 and 10 and jointly authored chapter 2. A. Hesse wrote chapter 6, and I. Herzog wrote most of chapter 5. This division of labour reflects the principal research interests of each of us. Translation of chapters 5, 6, 9, and 10 from the original German and French versions was by the first author. Mrs Maureen Storey of Cambridge University Press edited this heterogeneous effort and has given the book unity of style as well as finding and correcting the many errors which accumulated during the long gestation period. For her sharp eye and great patience we are deeply grateful.

Irwin Scollar   Alain Tabbagh
Irmela Herzog   Albert Hesse
Bonn   Paris and Garchy

Emile Thellier
Louis Cagniard
in memoriam

1

Archaeological prospecting

1.1. Why prospecting?

Contemporary archaeology is very much concerned with the increasing mastery of nature by man. This mastery results in a transformation of the landscape through agriculture and construction. Earlier hunting and gathering economies left few visible traces in the environment as far as we know. In some remote arid parts of the world it is quite easy to see the remains left by human activity which sometimes survive for very long periods above or on the surface of the ground. But in heavily occupied and farmed regions most surface traces of the past have been obliterated. Traces remain in the ground just below the surface and these are usually studied through archaeological excavation. Since excavation is a tool of last resort, techniques for non-destructive investigation of sites have been developed during the last half-century which form the subject of this book.

To understand the development of earlier cultural and economic activity, it is necessary to obtain as much data as possible from a wide area under different natural conditions. For such purposes, large numbers of similar sites must be identified and a few typical examples excavated. Excavation is also tantamount to the destruction of the site and can be thought of as a non-repeatable experiment. It is also just about the most expensive way possible to move earth, since nearly all must be done by hand. Given limitations of money and time, excavation should be used only when a site is in imminent danger of destruction through modern activity. Sites excavated solely because of their scientific interest have become a rarity in most heavily populated regions of the planet. Therefore in these areas of high population density, it is of utmost importance to preserve as many sites as possible. Natural ecological advantage usually means a dense population relative to almost every historical period, so that the more interesting and varied sites are likely to be found in just those areas subject to greatest destruction.

The cheapest way to preserve archaeological sites is to arrange for cultivation to continue or for parks and green belts to be located over them. Since the fund of information about the past is dwindling at an alarming rate, there soon will be nothing left for future study if steps are not taken now with advanced techniques. Preserving archaeological sites underground in an undisturbed state requires giving sufficient notice to planning authorities. Monument protection services must therefore have as complete an inventory as possible of sites in their area in order to be able to intervene long before the bulldozers appear. This means that an information system has to be built up for features not ordinarily visible above ground. A major advantage of such a system is that geographical and historical studies are possible using the data, without the excavation of a large number of sites.

1.2. Methods

During the last 60 years, a number of highly ingenious methods for detecting and mapping otherwise invisible archaeological sites have been developed, mainly in northern Europe. The techniques have also been applied elsewhere in the world, but not on the same systematic and institutionalised basis, and without comparable financial resources. This book concentrates on techniques which have proven successful under north European climatic and soil conditions, but the laws of physics being universal, it is possible to apply these methods in other areas, provided that enough is known about the conditions.

Almost all the techniques were initially derived from methods used in geological geophysical prospecting and conventional aerial survey, but they have now grown into a separate discipline with its own specialties. Data acquisition is by low level oblique or vertical air photography and magnetic, thermal, electric and electromagnetic geophysical prospecting and it is these most frequently used methods that are the subject of this book.

The time honoured tradition of walking over the fields while searching from eye level for surface finds has not been completely abandoned, but in many areas there is no longer time for this. Seismic techniques, the mainstay of oil exploration, have found few applications in archaeology because of the difficulty of coupling a sufficient amount of very high frequency sonic energy to the ground which is required for the detection of objects close to the surface (1,2), although they have met with some success in finding small cavities in stone buildings. Similarly, gravity prospecting has been restricted to unusual structures like the Great Pyramid (5) and finding caves in Kaarst regions because the required mass differences are almost never encountered on archaeological sites.

1.3. Aims

The application of modern prospecting techniques to archaeology has produced an enormous mass of data stored either on film, on paper or in computer readable (digital) form which requires transformation to something which can be used by the archaeologist in his struggle with the planning authorities and the forces of destruction. Recent developments in the computer treatment of pictures lend themselves admirably to this purpose. Our ultimate aim is the generation of a visual geographic information system which can be used by anyone interested in the study or protection of the vestiges of the past in a given area. Processed data can also be used to plan unavoidable excavation so that the cost of earth moving is minimised.

Prospecting techniques provide raw data which must then be evaluated. The first stage is to present the data in a form which an archaeologist can understand, either by constructing a model of the physical phenomena thought to be responsible and changing this until the measured data are accounted for within a minimum error, or by using heuristics which provide some degree of separation of the components of the measurements which are due to archaeological sources from those of natural or modern origin. The next step is to construct a database from which the archaeologist can obtain a rough index of available information through which he can search either geographically, by type of site or by period. The information from the database and the evaluated measurements can then be treated by the methods of spatial statistics to analyse associations between sites of different types and periods, time or spatial trends, and the significance of observed geographical distributions. In this book, only the acquisition of measurements and the preparation of material for a database and for statistical analysis will be discussed, not the analysis or database techniques.

Two broad types of prospecting techniques exist. One may be called passive, the other, active. Passive techniques are those where the prospector does not introduce any change in the environment. He measures or records a physical phenomenon which, by its nature, reflects the presence of archaeological structures. The classical passive technique *par excellence* is aerial photography. Magnetic, thermal and gravity prospecting are also passive techniques. Active techniques in geophysics include electrical, electromagnetic, ground radar, seismic and induced polarisation methods, for in all of these, an external field is created locally and applied to the ground. Each approach has diverse applications and difficulties.

1.4. The archaeological site as a physical phenomenon

There are two kinds of remains left by man in the landscape; those which are vestiges of structures such as buildings, fortifications, industrial installations, roads, agricultural boundaries etc., and those which are small portable objects lying in the ground with no clear archaeological context. The latter can usually be found only by walking over the surface, observing chance cuttings produced by modern construction or consulting ancient texts and archives where relevant. Metal detectors can also be used for coins and small metallic objects when these are not far from the surface. However, only sites of the first class are capable of being detected by systematic prospecting techniques which can also produce a map of the remains.

1.4.1. The buried structure

When a structure is built, holes are usually dug in the ground, walls set in foundation trenches, materials foreign to the site are brought in, and the naturally developed ordering of the soil is disturbed. This disturbance is persistent. Without human intervention, soil develops over many thousands of years through weathering of rock and transport of fine mineral and organic debris by water and wind. A complex ecosystem with natural vegetation and burrowing animals completes the process. With the invention of agriculture at the beginning of the neolithic period, man began to perturb this natural sequence significantly for the first time. The first large artificial structures were raised as the population increased in response to a more dependable food supply, pits were needed for storage and waste disposal, ditches and banks for containment of recently domesticated animals, and houses for people. Before agriculture, tents or small sheds were sometimes constructed, but their remains are very difficult to spot. From remote periods, it is the caves used by man which, though undoubtedly exceptional, survive best and which have therefore received much attention, rather unjustly.

Land had to be cleared of forest and brush for agriculture and a prime technique was simply to set fire to the vegetation or to slash the bark of trees too large to cut, causing them to die and decay. Forest clearance through slash and burn agriculture caused a transformation of the minerals in the upper soil layer to which we will return later. The cleared areas were farmed, causing some reduction of the particle size of the soil minerals, and buildings were constructed by using newly developed and more efficient tools to dig holes in which posts for roof and wall supports were placed. Ditches were dug, surrounded by or containing banks on which a palisade may sometimes have been built to keep domestic animals in and perhaps other animals and men out. In later epochs, when stone cutting and brick making were mastered, massive structures were built of these more permanent materials. This meant that foundations had to be laid if the soil has any significant depth or compressibility.

Sites were abandoned after a time either because the land was exhausted, or the population was reduced by disease, or after they had been destroyed deliberately by enemies. The structures built with such effort then began to decay, a process assisted by the forces of gravity, wind and the weather. The remains vanished from the modern surface. The archaeologist of today seeks to reconstruct on paper the appearance of the remains of the past on the basis of the evidence left after hundreds of years of natural or deliberate decay and destruction. From this in turn he hopes to deduce something about the nature and development of the society which left such traces. For a site protection inventory, knowledge of the rough shapes of structures is often enough to enable a good guess to be made about dates and functions based on similar excavated examples. Most importantly, the true extent of the site must be known in order to protect it.

Buried remains differ from their surroundings in a number of physical properties which make their detection and mapping possible either from the surface or at a distance. For one particular type of structure, a ditched enclosure protected by a bank with a palisade, a sequence of events can be assumed as reconstructed from excavated or experimental examples: Natural forest cover was burned off repeatedly by fire. After cultivation for some time, clay minerals from the surface which had been subject to burning were incorporated in the uppermost layer. Then a ditch was dug, the material being thrown up to make a bank, with a palisade to protect a settlement. The earth from the ditch was mixed when thrown up in the bank. After the site was abandoned, this mixed surface material, together with the steep sides of the bank was washed back into the ditch which slowly filled up. The post holes of the palisade filled with surface material after the wood had rotted. The ditch was ploughed over at a later epoch when settlement was resumed and the bank was completely flattened with its material partly redistributed over the surface. The result is an inclusion under the modern plough layer of surface material mixed with that from lower layers. If walling was present, foreign or local stone would have been mixed with subsurface material and encapsulated under the surface.

Other types of structures produce different types of remains, but the general pattern of disturbance of the soil sequence and the incorporation of surface or foreign material at a greater depth than before characterises almost all of them. In the last two decades archaeologists have made full scale field experiments in a number of countries in order to see what really happens when a site is constructed and left to decay. These experiments, carried out under controlled conditions, will provide valuable knowledge in the future about many of the processes involved, about which we can only make intelligent guesses at the present time (3,6).

Construction and destruction often occur repeatedly on the same spot because of its natural advantages. This may leave a complex geometric palimpsest in the ground, often with many layers and sometimes forming an artificial mound. Such complex shapes are quite difficult objects for prospecting. When the archaeologist excavates, he looks for soil colour, texture and material differences which are inclusions of a perceptible nature in a natural context. In geophysical or aerial prospecting, one searches for a measurable physical difference between the inclusion and its surroundings and one attempts to map the differences to get a picture of the site. The differences may not be perceptible to the eye of the archaeologist. The problem of recognition of a significant archaeological pattern is especially severe when a site was occupied over a long period of time.

Fortunately archaeologists are accustomed to inferring the nature of quite complex buried remains based on fragmentary information. A number of them have decided to construct a series of experimental structures built according to the hypotheses usually employed in analysing the evidence from excavation data, and then allowing these to decay under normal weather conditions. The processes of decay have then to be studied at regular intervals thereafter. Such observations have created a somewhat more solid foundation for inferences from partial remains, but unfortunately there have been very few experiments of this type. These have been on only a limited few soils and have been running for relatively short times (4). Such experiments give us some idea of what happens mechanically to both soil and remains, but unfortunately no data have been amassed concerning magnetic and other non-visible properties.

One feature chosen for many of the experiments is the ditch and bank. Soils developed on chalk, volcanic ash, and loess have been tested. It is still too soon to see some of the changes in structure which archaeologists usually describe when making sections through ancient banks and ditches. In the case of chalk subsoil, the compact surface turf falls into the bottom of the ditch and the slope of the sides is reduced. The bank is somewhat compacted. The process is much more rapid when the ditches are constructed on the more mobile loess. Surface material is quickly incorporated into the bottom of the ditch, giving the so-called 'quick fill' which the archaeologist treasures, since it is nearly contemporary with construction. Afterwards, presuming that the site lies open for a time, the process of filling slows considerably, especially when a stabilising cover of vegetation establishes itself. The angles of ditch sides at which movement stabilises are characteristic of different mechanical soil properties and average local weather conditions with a given amount of rainfall available to wash the material to lower levels. The type of vegetation cover also plays an obvious role in this mechanical stabilisation process. In the following chapters, a process similar to that assumed in these experiments is presumed to be responsible for the burying of the remains. The assumptions are based on the existence of a temperate climate with adequate rainfall and are probably not very valid for sites in other parts of the world.

Notes

(1) Carabelli, E. 1966, A new tool for archaeological prospecting: the sonic spectroscope for the detection of cavities, *Prospezioni Archeologiche*, **1**, 25–36.
(2) Carabelli, E. 1968, Ricerca delle cavità superficiali con l'impiego di vibratori primi esperimenti, *Prospezioni Archeologiche*, **3**, 37–44.

(3) Jewell, P. A., Dimbleby, G. W. 1966, The experimental earthwork on Overton Down, *Proceedings of the Prehistoric Society*, **32**, 313–42.
(4) Küper, R., Löhr, H., Lüning, J., Stehli, P. 1974, Untersuchungen zur neolithischen Besiedlung der Aldenhovener Platte, *Bonner Jahrbücher*, **174**, 482–96.
(5) Lakashmanan, J., Montlucon, J. 1987, Microgravity probes the Great Pyramid, *Geophysics: The Leading Edge*, **6**, 10–16.
(6) Lüning, J. 1974, Das Experiment im Michelsberger Erdwerk in Mayen, *Archäologisches Korrespondenzblatt*, **4**, 125–32.

2

Soils and the effects of climate on prospecting

2.1. Introduction

The detection of archaeological structures is based on the measurement of a difference or contrast between the properties of the materials which constitute the structures and those of their environment. Therefore it is necessary to study the physical properties of soils upon which the various prospecting methods are based. Soil is, however, an extremely complex physical medium which can be viewed from various points of view: geologically as the product of the weathering of rocks by climate; agriculturally, as the medium which nourishes the vegetation; in civil engineering, as a support for structures; archaeologically, whereby parts of all these properties are the result of action by man in the past, whether voluntary or involuntary. All of these approaches lead to different interests depending on one's point of view, but it is desirable to define some general parameters which permit an understanding of physical properties.

2.2. Macroscopic characteristics

Physically, a soil appears to be composed of three phases: solid particles, water and the gas (mostly air) present in the pores between the particles. The solid particles or grains are of very different sizes. They come essentially from the parent rock, but they can also be formed in the soil itself. The grain size distribution is the principal factor which governs the majority of the physical and mechanical properties with the exception of the magnetic properties as discussed in chapter 7.

2.2.1. Granulometric analysis

Granulometric analysis consists of grouping the various soil grains in arbitrary size classes by their equivalent diameter. Since the particles are obviously not spheres, the equivalent diameter is defined for the largest particles as the minimum size of the holes in a sieve through which they can pass. For the finest particles, it is defined as the diameter of a particle of the same density which has an equivalent speed of sedimentation. The International Soil Science Society has adopted the following classification for the particles whose equivalent diameter is:

Clay:	$2\ \mu m$
Silt:	>2 and $<20\ \mu m$
Fine sand:	>20 and $<200\ \mu m$
Coarse sand:	$>200\ \mu m$ and <2 mm
Gravel:	>2 and <20 mm

Particle size distribution is determined by successive sieving for the coarser fractions when dry, followed by sedimentation techniques based on the settling rates of particles in a fluid which follow Stoke's law. A definite amount of the fluid is removed at a given depth with a pipette at specified time intervals, the samples then being dried and accurately weighed. This method has been used in analysis of soils from archaeological sites discovered through aerial photography which will be discussed shortly. Other methods are discussed in the soil physics literature.

The result of a granulometric analysis is displayed in the form of a cumulative percentage curve as a function of equivalent diameter, figure 2.1. The terminology based on granulometry and commonly used in agriculture is shown in figure 2.2. It may be noted that this classification is based on the dry weight of the sample and does not take into account any residues of organic material, the quantity of which is very variable and depends on the degree of preservation. When the percentage by weight of dry organic material is between 3 and 10%, the material is an organic mud, between 10 and 30% it is a peaty soil, and over 30% it is peat.

Organic matter also very much affects the properties of the surface soil layer. Structure, water penetration, and retention and aggregation of particles are regulated by the amount of decomposing organic matter present. The materials which have survived after very long periods of time combined with substances synthesised by soil bacteria make up

Fig. 2.1 The cumulative grain size distribution of a soil as a function of equivalent particle diameter. From ref. 3.

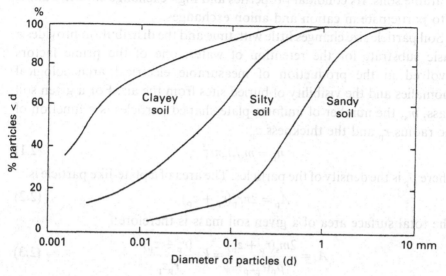

Fig. 2.2 The terminology of soils based on particle size distribution used in agriculture. From ref. 3.

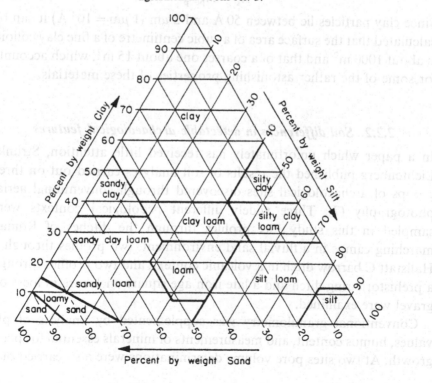

humus, a lignoproteinate which makes up from 0.1 to 5% of soil solids in arable soils. Its colloidal properties and high exchange capacity allow it to participate in cation and anion exchanges.

Soil particle size changes little with time and the distribution provides a basic substrate for the retention of water, one of the prime factors involved in the production of measurable electrical archaeological anomalies and the visibility of buried sites from the air. For a given soil mass, m_s, the number of uniform plate-shaped particles is a function of the radius r_p and the thickness z_p:

$$n_p = m_s/\rho_p \pi r_p^2 z_p \qquad (2.1)$$

where ρ_p is the density of the particles. The area of a plate-like particle is:

$$A_p = 2\pi r_p(r_p + z_p) \qquad (2.2)$$

The total surface area of a given soil mass is therefore:

$$A_s = \frac{2m_s(r_p + z_p)}{\rho_p(r_p z_p)} = k_p \frac{(r_p + z_p)}{r_p z_p} \qquad (2.3)$$

Thus the surface area is proportional to the sum of the size parameters. For spherical particles (2.3) reduces to:

$$A_s = 2k_p/r_p \qquad (2.4)$$

Since clay particles lie between 50 Å and 2 μm (1 μm $= 10^4$ Å) it can be calculated that the surface area of a cubic centimetre of a fine clay colloid is about 1000 m^2 and that of a coarser one about 15 m^2, which accounts for some of the rather astonishing properties of these materials.

2.2.2. *Soil differences in detectable archaeological features*

In a paper which unfortunately has received little attention, Strunk-Lichtenberg published the results of soil analyses carried out on three groups of archaeological sites discovered through conventional aerial photography (7). Three widely different pedological contexts were sampled in this study. Six profiles through the ditches of Roman marching camps in a fluvial sand environment, two profiles through a Hallstatt C barrow ditch in a volcanic context, and two profiles through a prehistoric ring ditch and a late iron age square enclosure in loess on gravel were examined.

Conventional granulometry was supplemented by analysis of pH values, humus content, and measurements of minerals essential for plant growth. At two sites pore volume determinations were also carried out.

This quantity was measured on samples, taken in rings to preserve soil structure, which were pressed into profiles and immediately sealed.

In a given volume of soil, there are solids, water, and air such that:

$$V = V_s + V_w + V_a \qquad (2.5)$$

where $(V - V_s)$ is the pore space and n the soil porosity:

$$n = (V - V_s)/V \qquad (2.6)$$

The water content is expressed for a given volume as:

$$s = V_w/V \qquad (2.7)$$

or:

$$s_s = V_w/(V - V_s) \qquad (2.8)$$

where s_s is the saturation ratio. Samples are dried in a desiccator and weighed. The samples are then saturated with water by placing them on continually wetted blotting paper in a closed container and allowing them to suck up as much water as possible. They are then weighed again. From this the water per unit volume can readily be calculated. An air picnometer is used to measure the volume of air filled pores directly. A description of the measurement technique is given by Taylor and Ashcroft (11).

Strunk-Lichtenberg's results for a Roman ditch in a sandy soil are shown in figure 2.3, and for loess on gravel in figure 2.4. Both show a considerably higher percentage of fine particulate material at greater depth than in the adjacent undisturbed soil. Characteristic is the mixing of the normally vertically structured distribution in the archaeological feature, so that although the total amount of fine substance may not be significantly greater, it is more uniform with depth. The pore and water volume determination confirms the granulometric results, showing that the archaeological features are capable of retaining a slight but significantly higher percentage of water at greater depth. Fine materials are capable of the adsorption of a greater number of hydrogen ions on their larger total surface area, and this is consistently reflected in slightly more acid pH values measured in the features as compared with the surrounding material. There are, of course, many other possible soil conditions the behaviour of which may differ from that of the three sites studied. From the results of electrical measurements it may be inferred, however, that the enrichment in fine particles is probably quite general, but more studies of samples from other sites would be useful.

2.2.3. Soil water

Soil water is manifest in four forms: the water present in the crystal
structure of clays, that present between platelets of certain types of clay
composed of thin flakes (Montmorillonite), water adsorbed on the
surface of the soil particles and free water. Only the last corresponds to

Fig. 2.3 (*a*) Air, water and solid substance against depth in a Roman ditch
and in the surrounding soil on riverine sands near Veen, Kreis Moers in
the Rhineland. (*b*) Percentages of sand, silt, and clay in the section. From
ref. 7.

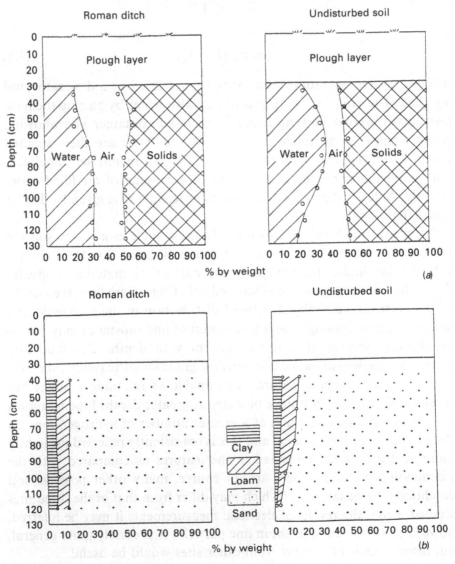

the usual meaning of humidity, which is measured by furnace drying at 105°C for 24 h or until constant weight is achieved. Other methods for measuring humidity are calibrated against this value. Free water can move in the soil either through gravity or via capillarity. The quantity of water which a soil can retain by capillarity is expressed with the help of a suction potential which takes into account the size and the shape of the pores, which the measurement of free water does not. The relative amount of water in an archaeological feature compared with its surroundings is of prime importance in several of the most important prospecting techniques. Water content is not constant in soil; it changes dynamically with the weather in an extremely complex way. Since it affects movements of solutes, there is a considerable time dependence in electrical conductivity as well. Hillel (3) and Taylor and Ashcroft (11) among others have devoted entire books to soil water and there is a vast literature on the subject which is obviously of capital importance for agriculture. The physics of water flow in unsaturated soils is extremely complex and poorly quantifiable since the water state and content change during the flow process. Usually numerical analytical techniques are employed, or rough approximations used.

Soil water flow takes place because there is a difference in potential and the rate of flow is proportional to the potential gradient and is influenced

Fig. 2.4 Percentages of gravel, sand, silt, and clay plotted against depth in a late Iron Age ditch and in the surrounding loess soil near Düren in the Rhineland. From ref. 7.

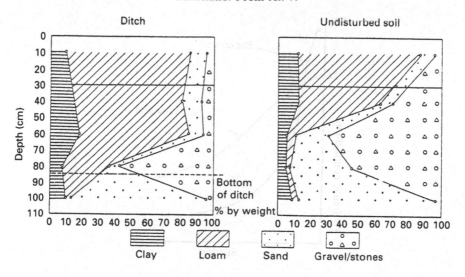

by the pores through which flow must take place. In unsaturated soils, water is subject to suction due to atmospheric pressure differences as well. This suction is attributable to capillary effects arising from the physical attachment of water to the surfaces of the finer particles. It is usually a much stronger force than that due to gravity. In a saturated soil, all of the pores which can be filled are filled and the conductivity of the water is at its highest. In an unsaturated soil, some of the larger pores contain air and the conductive cross-section decreases. The largest pores empty first and water flows only in the smaller ones. The transition from saturation to non-saturation is abrupt, and hydraulic conductivity drops by several orders of magnitude. Hydraulic conductivity is largest in coarse grained soils only when they are saturated. When unsaturated, finer grained soils may have higher conductivity at the same degree of suction. Thus flow persists for a longer time in clay soils. Figure 2.5 shows the dependence of hydraulic conductivity on suction of two types of soils and illustrates the phenomenon. It shows hysteresis and there is a difference when a soil is becoming wetter rather than drier. Empirical equations can be written to describe the process, but there is a fundamental mathematical model for the process. For the purposes of this book, a qualitative descriptive model suffices. A very detailed mathematical treatment of unsaturated flow is given by Hillel (3).

Fig. 2.5 The dependence of hydraulic conductivity on suction in two soil types (K_{s_1} = sandy soil saturated conductivity, K_{s_2} = clayey soil saturated conductivity). From ref. 3.

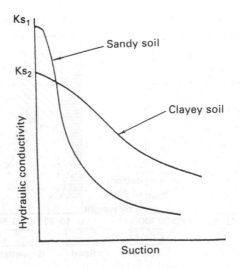

Water penetrates into soil after rainfall via the flow mechanism, and assuming that the soil is sufficiently deep, the profile shows a wet zone above a dry zone. The water is redistributed with time because the suction gradient between the wet and dry zones changes and the initially-wet zone loses water. The wetting front advances more slowly and a diagram like that of figure 2.6 shows the progress in an idealised case over a two week period. The kind of soil influences the process as shown in the lower part of the figure.

Fig. 2.6 Advance of the wetting front in a soil over a two week time interval ($t_0 = 0$ days, $t_1 = 1$ day, $t_2 = 4$ days, $t_3 = 14$ days after irrigation). From ref. 3.

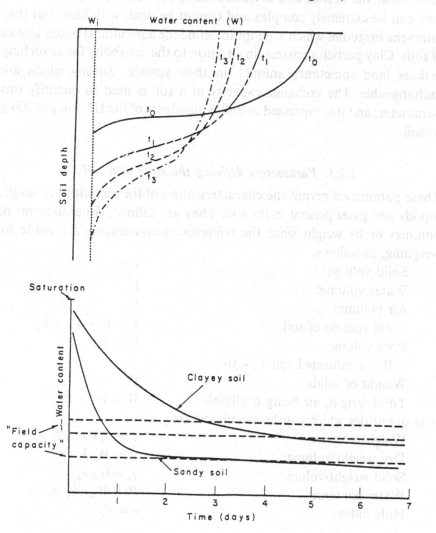

Drying of soil is a far more complex process, being influenced not only by the properties of the material, but also by the state of the overlying vegetation, the temperature, and the wind, the whole being lumped together under the name of evapotranspiration. The drying process will be described in chapter 3 on aerial photography because the interaction with soil colour and plant growth is of primary importance for this technique.

2.2.4. *Ions in the soil*

Free soil water is capable of containing various quantities of dissolved salts. Near the oceans and in deserts salty soils are found. The role of ions can be extremely complex and cannot be dealt with here, but they intervene in studies which attempt to define the agricultural potentialities of soils. Clay particles possess, in addition to the capability for adsorbing cations (and sometimes anions) on their surface, cations which are exchangeable. The exchange capacity of a soil is used to quantify this parameter, and it is expressed in milliequivalents of the H^+ ion per 100 g of soil.

2.2.5. *Parameters defining the state of a soil*

These parameters permit the characterisation of the quantities of solids, liquids and gases present in the soil. They are defined, either in terms of volumes or by weight since the reference measurements are made by weighing, as follows:

Solid volume: V_s

Water volume: V_w

Air volume: V_a

Total volume of soil: $V = V_s + V_w + V_a$

Pore volume: $V_v = V_w + V_a$

(for a saturated soil $V_a^* = 0$)

Weight of solids: W_s

Total weight, air being negligible: $W = W_s + W_w$

The quantities which can be calculated are:

Weight/volume: $\gamma = W/V$ in kN m^{-3}

Dry weight/volume: $\gamma_d = W_s/V$

Solid weight/volume: $\gamma_s = W_s/V_s$

Water content: $W = W_w/W_s$ in %

Hole index: $e = V_v/V_s$

Porosity: $n = V_v/V$
Degree of saturation: $s = V_w/V_v$

Also used are the apparent bulk density, the ratio of the weight of a given volume of soil to the weight of the same volume of water and the dry density which is usually between 2.3 and 2.6 (see also chapter 7).

2.3. The electrical conductivity of soils

Conductivity σ is defined by Ohm's Law $\mathbf{i} = \sigma \mathbf{E}$ as the proportionality factor between the current density $\mathbf{i}$ and the electric field $\mathbf{E}$. Its value may range widely between metals, the best conductors, and insulators. Conduction in soils is electrolytic, based on the displacement of ions in interstitial water. It is therefore aided by the presence of dissolved salts and the presence of water. Since it is a function of the mobility of the ions, conduction will depend on their size and on the viscosity of the water. Therefore, conductivity will increase with temperature because the quantity of dissolved salts rises with temperature and above all because the viscosity of water drops. The law which relates the conductivity of an electrolyte to temperature is approximately linear at ambient temperatures. For a solution of sodium chloride the coefficient of proportionality is 0.022 per degree Celsius. This value is about the same for other ions and one can write:

$$\sigma(\theta) = \sigma(20°C)[1 + 0.022(\theta - 20)] \tag{2.9}$$

where θ is the temperature in degrees Celsius. However soils are triphase materials in which the behaviour of the clays is very complex and it is difficult to give other empirical measurements of conductivity or its reciprocal, resistivity, in soils.

Nevertheless, for a saturated environment without clay, and neglecting the conductivity of the soil grains, Archie's Law allows us to relate the resistivity ρ to the resistivity of the interstitial water ρ_w and to the porosity:

$$\rho/\rho_w = an^{-m} \tag{2.10}$$

where a and m are empirical coefficients with a between 0.5 and 2 and m between 1.5 and 2.5. For non-saturated soils the empirical relationship between the resistivity and the volume humidity (5) is:

$$1/\rho = 1/\rho_w + (a\theta^2 + b\theta) + 1/\rho_s \tag{2.11}$$

This includes the surface resistivity ρ_s of the solid grains due to ions trapped in the adsorbed surface water, with the coefficients a and b

Table 2.1. *Resistivities of soils developed on various materials*

Soils developed on	Resistivity in Ωm
Granite rocks	5000+
Granitic or metamorphic rocks	300–3000
Sand and gravel	40–500
Calcareous rocks	40–200
Fine soils with high silt and clay content	20–50
Loess	20–40
Clays	5–25
Salt sands	1–10

depending on geometric parameters of the solid phase. The order of magnitude and the dispersion in resistivities which may be encountered in soils are shown in table 2.1 which indicates that the fundamental parameter is the grain size distribution if salt soils are considered an exception.

Electrical resistivity is thus a property which differentiates clearly between fine and coarse materials, and it is particularly suitable for the detection of stone structures in a fine soil or earth structures in a gravel soil as discussed in chapter 6.

2.3.1. Dielectric permittivity of soils

The dielectric permittivity of a body is the ratio between the electric induction **D** and the electric field **E**. It expresses the aptitude of molecules or of a crystal lattice to acquire electrical polarisation. One may write:

$$\mathbf{D} = \varepsilon\mathbf{E} = \varepsilon_r\varepsilon_0\mathbf{E}, \qquad \varepsilon_0 = 1/36\pi10^9 \tag{2.12}$$

where ε_0 is the permittivity of air and ε_r is the relative permittivity in SIu. Permittivity plays no role in electrical prospecting and only intervenes in electromagnetic prospecting at very high frequencies (see chapter 9). Since the absorption rises rapidly at ultra high frequencies, there is only a limited range between 100 MHz and 1 GHz for which it is of interest in archaeological prospecting. In this range as at lower frequencies, the dominant fact is the great difference which exists between the relative permittivity ε_r of free water, *ca.* 81, that of solid grains $3 < \varepsilon_r < 7$, and that of air, $\varepsilon_r = 1$. This difference makes the permittivity the physical property which is most sensitive to humidity. Between 100 MHz and1 GHz, permittivity must be considered as a complex quantity

$\varepsilon_r = \varepsilon'_r - i\varepsilon''_r$. *In situ* measurements on samples can be carried out by time domain reflectometry with a bifilar or coaxial line (1,4) or in the frequency domain with a coaxial line (6). The results obtained show an independence of frequency, and a linear relationship with humidity (figure 2.7) depending slightly on the kind of soil. For humidities less than 40% the relationship is $\varepsilon'_r = ac^{bw}$, where c and w are empirical constants. In

Fig. 2.7 Variation of the average value of permittivity ε'_r as a function of humidity for six different soils. (*a*), (*b*), (*c*) Soils at Grignon, Noyen and Garchy, respectively. (*d*), (*e*), (*f*) Sands of the Landes, Seine and Loire, respectively. From ref. 6.

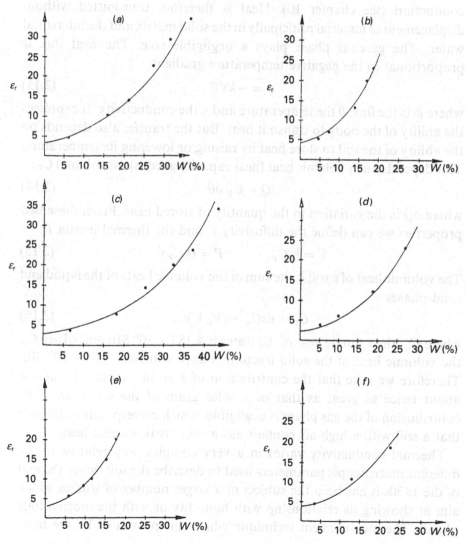

this expression, the coefficient b is clearly correlated with the grain size distribution, and it increases with an increasing content of coarse particles (6). The values are between 0.05 and 0.11, whereas the variation in a is more complex. It lies between 2.5 and 4.0. The coefficient of dielectric loss ε_r'', which is difficult to disassociate from the electrical conductivity is very dependent on the grain size distribution. It varies between $\varepsilon_r'/5$ for fine and $\varepsilon_r'/20$ for coarse soils.

2.4. Thermal properties of soils

The principal mechanism responsible for the transfer of heat in a soil is conduction (see chapter 10). Heat is therefore transmitted without displacement of material principally in the solid matrix and the interstitial water. The gaseous phase plays a negligible role. The heat flux is proportional to the negative temperature gradient:

$$\phi = -k\nabla\theta \tag{2.13}$$

where ϕ is the flux, θ the temperature and k the conductivity. It expresses the ability of the body to transmit heat. But the transfer also depends on the ability of the soil to store heat by raising or lowering its temperature. This depends on its volumic heat (heat capacity at constant volume) C_V:

$$\delta Q = C_V \, \delta\theta \tag{2.14}$$

where δQ is the variation in the quantity of stored heat. From these two properties we can define the diffusivity Γ and the thermal inertia P:

$$\Gamma = k/C_V, \qquad P = (kC_V)^{\frac{1}{2}} \tag{2.15}$$

The volumic heat of a soil is the sum of the volumic heats of the liquid and solid phases:

$$C_V = nsC_w + (V_s/V)C_s \tag{2.16}$$

where C_w, the volumic heat of the water is 4.187×10^6 SIu and where C_s, the volumic heat of the solid fraction is approximately 2.08×10^6 SIu. Therefore we note that the contribution of a given amount of water is about twice as great as that of a solid grain of the same size. The contribution of the gas phase is negligible, which corresponds to the fact that a soil with a high air content has a very weak specific heat.

Thermal conductivity varies in a very complex way relative to the different macroscopic parameters used to describe the soil. Since the end of the 1930s it has been the subject of a larger number of studies which aim at showing its relationship with humidity or with the morphology of soil. The measurement technique which has proven to be the most suitable is that which does not require a high thermal gradient. This technique consists in producing a thermal shock whereby a heated needle delivers a quantity of heat as a Heaviside function (2). The variation of temperature as a function of time quickly becomes $k \log(t)$ for a point near the needle. If one wishes to determine both properties at once, it is preferable to have two temperature measurements at different distances from the needle heat source. These points may be placed in the interior (10) or at the surface (9) of the milieu whose properties are to be measured. The determination of the two properties is obtained from a nomogram using the differences in temperature at two or more instants after the beginning of heating.

The experimental results obtained show that the two principal factors

Fig. 2.8 Relationship between volumic heat C_V, conductivity k, and diffusivity Γ, as a function of the kind of soil. From the Proceedings of IGARSS Symposium Strasbourg 1984, ESA SP215 (published by ESA scientific and technical publications branch, August 1984).

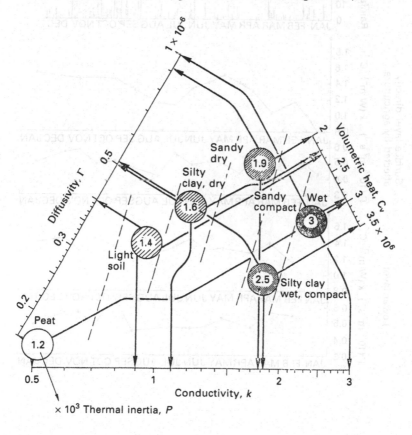

Fig. 2.9 Agricultural treatment, monthly and daily rainfall, conductivity and diffusivity for 1974 Favray, St. Martin sur Nohain, Nièvre. From ref. 10, chapter 10.

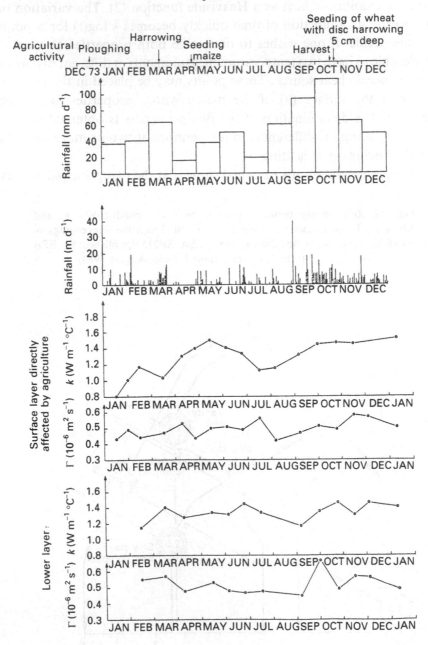

which determine thermal conductivity are grain size distribution and soil porosity, humidity playing a minor role. For archaeological structures and their environment, these results are shown in figure 2.8. This demonstrates that the two extremes are a humid sandy compact soil and a peat. The surface layer whose porosity varies as a function of agricultural activity and rainfall has properties which vary during the course of the year. This is shown in figure 2.9 for the arable and the undisturbed lower layer, in a field at St. Martin sur Nohain, Nièvre. For the latter, there are no net variations during the course of the year which are correlated with rainfall or agricultural activity. This is also true for the diffusivity of the surface layer. The conductivity of the surface follows agricultural practice and the evolution of humidity quite well.

Notes

(1) Davis, J. L., Chudobiak, W. J. 1975, *In-situ meter for measuring relative permittivity of soils*, Geological Survey of Canada Paper 75–1A, 75–9.
(2) De Vries, D. A. 1952, A non-stationary method for determining thermal conductivity of soil *in-situ*, *Soil Science*, **73**, 83–9.
(3) Hillel, D. 1971, *Soil and Water*, Academic Press, New York, NY.
(4) Hoekstra, P., Delaney, A. 1974, Dielectric properties of soils at UHF and microwave frequencies, *Journal of Geophysical Research*, **79**, 1699–1708.
(5) Rhoades, J. D., Roats, P. A. C., Prather, R. S. 1976, Effects of liquid-phase electrical conductivity, water content, and surface conductivity on bulk soil electrical conductivity, *Soil Science Society of America Journal*, **40**, 651–65.
(6) Rualt, P., Tabbagh, A. 1980, Etude expérimentale de la permittivité diélectrique des sols dans la gamme de fréquence 100 MHz–1 GHz en vue d'une application à la télédetection de l'humidité des sols, *Annales de Géophysique*, **36**, 477–90.
(7) Strunk-Lichtenberg, G. 1965, Bodenkundliche Untersuchungen an Archäologischen Objekten, die durch Luftbild-Aufnahmen entdeckt wurden, *Archaeo-Physica*, **1**, 175–202.
(8) Tabbagh, A., Jolivet, A. 1974, Procédé de mesure *in-situ* des propriétés thermiques des sols, *Science du Sol*, **4**, 268–79.
(9) Tabbagh, A. 1976, Les propriétés thermiques des sols, *Archaeo-Physika*, **6**, 128–48.
(10) Tabbagh, A. 1985, A new apparatus for measuring thermal properties of soils on rock in-situ, *IEEE Transactions on Geoscience and Remote Sensing*, **6**, 896–900.
(11) Taylor, S. A., Ashcroft, G. L. 1972, *Physical Edalphology: The Physics of Irrigated and Non-irrigated Soils*, W.H. Freeman and Company, San Francisco, Ca.

3

Aerial photography

3.1. Introduction

Aerial photography for the detection of monuments which are not visible from the ground is the oldest remote sensing technique used in archaeology. The method has been with us since shortly after the end of the First World War, and it has been used systematically in Britain from the late 1920s and in Germany, France, Belgium, and the Netherlands since the late 1950s. Most of the books on the subject deal with archaeology as seen from the air, some treat the history of the subject in considerable detail, others are essentially works on historical geography from the evidence of aerial photographs (3,8,18,52,73). There is little published in book form about the physical basis of the method. It is intended here to fill this gap. Aerial archaeology, as opposed to other methods of archaeological prospecting, can make use of data acquired for other purposes. However, the probability of spotting archaeological sites in existing vertical aerial photographs depends very much on whether or not the imagery was acquired at a propitious time of day and year or not. When aerial photographs are made specifically for archaeology, the probability of detection can be raised by many orders of magnitude, and it is this type of photography which has proven to be so highly successful in northern Europe.

The interpretation of extant vertical aerial photographs is made into something of a black art especially by geographers and photogrammetrists. Generally, the evidence available in a skilfully made archaeological aerial photograph is usually evident at a glance for the well-informed archaeologist who knows what to expect either in his or in neighbouring areas. Prolonged examination of other types of pictures usually leads to unwarranted fantasy which often cannot be confirmed by later excavation. This is especially true for features which are small relative to the physical size of the picture. It seems that if one stares long enough at a diffuse image, one's imagination will allow one to see almost anything. In this book, archaeological prospecting is considered as an exercise in the deliberate planned measurement and evaluation of physical properties which lead to the detection and mapping of archaeological sites. Therefore subjective interpretation of existing aerial photographs is not in accordance with the set goals. In those parts of the world where it is not possible to make aerial photographs specifically for archaeological purposes, readers may utilise the copious literature available for studies in remote sensing as a guide (15).

A second unique aspect of archaeological aerial photography is that the most significant discoveries have been made by experienced archaeologists or gifted amateurs who fly themselves and photograph what they observe. When flying, one sees many markings and features whose meanings are far from obvious. A selection process is carried out by the trained observer who can keep his archive from becoming cluttered with meaningless rubbish. The user of the archive must be aware that this selection process has taken place, and that it represents the subjective judgement of the observer at the time at which the picture was taken. In addition, how the picture was taken (angle of view and orientation relative to the sun) also reflects this subjectivity, and in consequence, the appearance of the site or feature.

Even though vertical mapping cameras are precision devices and the equipment used to treat the photographs afterwards is of the highest quality, the results obtained may not be as satisfying for the archaeologist as the low level obliques made at optimum times with optimum lighting. Most of the major discoveries from the air have been made with this technique. Photogrammetrists are upset by the rough and ready approach, but since archaeological sites are not precision features in the landscape like a railway line or highway, and since not all details available through excavation are visible from the air, the lack of precision in the imagery does not constitute a serious loss of information. In chapter 5 methods are discussed which partially remedy the problem of lack of geometric accuracy. It is more important to be able to see clearly what is present than to have features, which even if they can be mapped precisely, are poorly visible.

When using existing aerial photographic cover not made expressly for archaeological purposes, the selection process is much more restricted for the observer has little or no control over the appearance of the image. There are two reasons why the selection process is more difficult in the comfort of an armchair. First, the size of the features relative to the field of vision is much smaller. This is not the case when making pictures at low altitude while looking at the ground. Secondly, the colour response of the eye is far better than that of any available film, and the million years of programming of the pattern recognition computer in the human brain cannot be duplicated by any device. Nonetheless, important discoveries have been made from time to time using vertical mapping imagery and, in areas of the world where the natural conditions permit sites to be visible at most times of the year, it would be foolish not to use the material, at least for a preliminary rough survey. The main problem in other areas is that the optimum conditions for taking pictures occur within a very narrow time window, if at all, and it is relatively unlikely that vertical mapping cover will be made at the right time.

In contrast to most of the other methods of archaeological prospecting described in this book, the oblique air image made by a skilled observer conveys an enormous amount of information acquired during the small fraction of a second when the camera's shutter was open and its quality depends largely on the skill of the photographer. The data on the film are the result of a complex interaction between vegetation and soil with light. If each plant, for example, can be considered a sensor, then the number of 'measurements' made of the subsoil in the area of the photographs is very high indeed. Here lies one of the main reasons for the widespread use of the technique. It is unquestionably the cheapest form of archaeological prospecting per unit area and per working hour. It also produces the highest rate of return on money invested, and archaeologists usually accept the results without the need for detailed explanations or complex mathematical evaluation techniques.

3.2. The appearance of archaeological sites from the air

3.2.1. The high viewpoint

In the early 1920s, O. G. S. Crawford, one of the pioneers of archaeological aerial photography was the first to point out the utility of the high viewpoint afforded by a low flying aircraft (17,19). He illustrated his case by comparison with a patterned carpet the shapes in which only become apparent when seen from normal human viewing height. Seen from the vantage point of an insect, there is only a confused blur of colour. Because the intelligent viewer has prior knowledge gained through study of other methods, he consciously anticipates seeing the kinds of things he wants to see. This raises his sensitivity to faint features considerably. Once having seen some of the things he has been looking for, he remembers the range of variations and the next time has an even more sensitive detector in his head. This training process leads to some of the better known aerial archaeologists being rather like skilled hunters or gifted artists.

The features in the landscape which are the consequence of human intervention and of interest to archaeologists generally have geometric properties which differ from those of the natural surroundings. Sometimes they may differ in colour, texture, height or relative brightness or combinations thereof. If one is far enough removed from the details (but not too far) irrelevant information is subjectively ignored, and the high viewpoint reinforces visual perception by allowing one to concentrate on matters of interest. Since what is of interest to one observer may not be noticed by another, the method is not scientific, but nonetheless is highly productive! Because of this kind of phenomenon, difficult to appreciate if one has never flown an aerial survey flight, it is evident that much more information can be processed than is the case when working with still photographs in a room.

It can be shown experimentally that a series of dots will be joined by the eye to make an apparently continuous line, if the angle subtended by the dots is less than a certain threshold. This is an irreversible process, in which there is no way for the dots to be reconstructed from the information contained in the line. Many aspects of vision have this kind of property. The high viewpoint of the aerial image utilises this, although little is known about the psychological processes involved. Raising the brightness contrast of the dots aids in their separation, while its reduction can reduce discrimination to below a critical threshold. This phenomenon helps us to see geometric features in a confused pattern of dots. The situation is still more complex when viewing in colour and three dimensionally (38). For any given problem there is an optimum altitude for photography. Flying still higher will reduce quality because atmospheric effects cause diffraction and degrade image sharpness both for a camera and of the observer's view of the ground. Up to a point, however, reducing the size of an image appears to improve its sharpness, as can be readily observed by looking through binoculars turned wrong way around at a scene. Colours are also perceived as being more intense.

3.2.2. The targets ('signals')

Archaeological sites as seen from the air can be classified into a number of abstract types which can be considered independently from their period, origin or function. Defensive installations made of ditches and walls can be considered as combinations of curved and linear features, graves and tumuli as localised circular features or points, the interior features of prehistoric settlements as blobs, points, or linear boundaries and house outlines etc. With few exceptions, most of the sites sought from the air can be seen as combinations of simple geometric elements or blobs of irregular shape.

The larger the object, the greater the chance of seeing it, but perspective distortion can cause serious errors in appreciating the shapes of things in the background. In practice, it is nearly impossible to tell subjectively if a feature is rectangular or square. It is even more difficult to estimate shape correctly if irregular features are involved. Therefore, as discussed in the following chapter, correction of perspective is imperative if the oblique aerial image of a site is to be used for mapping or excavation planning. A further complication lies in the fact that the features as seen reflect a very complex interaction between the subsurface differences and the surface soil or the plants growing on it. The buried features may prove to have shapes which differ considerably in detail upon excavation, from those observed and mapped from a photograph.

3.2.3. 'Noise'

The archaeological features in the photographs are the desired information, the other features in the landscape which surround them are usually not of direct interest, since most are usually modern. However, this latter information actually dominates the picture content. If the area is agricultural, the first thing which strikes the eye are the modern field boundaries. If the terrain is reasonably level, this will usually be a rectangular pattern, except where watercourses or roads intervene. It is quite easy to separate archaeological features from this type of pattern when they differ from it in orientation. When this difference in orientation does not exist, then the presumed archaeology may be suspect, and modern agriculture may have indeed produced anomalies which can be confused with real sites (61). When the unwanted data is due to geological processes, then there are usually good criteria for separation. Geological features are only rarely of highly geometric form, they extend over wide areas and in many civilised parts of the world have been accurately mapped. One type of geological phenomenon may be difficult to separate however. This is cracking of the immediate subsurface due to intense cold in periglacial conditions in the late Quaternary period. Quite regular features can be produced by such effects and may be readily confused with those of archaeological origin.

Modern industrial society has contributed far more to changing the landscape than all other periods combined, at least in highly developed parts of the world. Although traces of the past vanish with astonishing rapidity, they sometimes reappear as a kind of modern archaeology when conditions are favourable. Thus, roads whose paths have been straightened, abandoned railway lines, wartime trenching, bomb craters, pipelines, abandoned playing fields and all of the other possible debris of modern life lies in the ground as well. It will surely be of interest to archaeologists in the distant future, but it is usually not the concern of archaeologists today. Again, one aspect helps in distinguishing this kind of phenomenon from that of the past: orientation. Property boundaries do not usually change rapidly and abandoned modern features are frequently parallel or at right angles to visible current boundaries. Furthermore, older maps of an area can usually help in cases of doubt. With most features visible from the air or in existing aerial photographs a good rule is: when in doubt, it's probably modern.

3.2.4. Aims in archaeological aerial photography

Crawford was trained as an historical geographer. His autobiography *Said and Done* (21) and his major work, *Archaeology in the Field* (20) demonstrate his strong geographical bent. St. Joseph, Bradford, and many others had strong geographical leanings. Some of the major published aerial surveys such as Agache's *La Somme Pre-Romaine et Romaine* (3) demonstrate this interest. Since the beginning in the 1920s, this historical geographical approach has dominated archaeological aerial photography in northern Europe. More recently, a tendency has arisen to see the method as an important adjunct for monument protection, or for statistical evaluation in quantitative historical geography. This usage will undoubtedly predominate in the future.

It must be stressed that the appearance or non-appearance of an archaeological site from the air is dictated by superficial geology, local climate and modern agricultural practice. Hence there are few areas where all these factors are so favourable that the majority of sites actually present in the ground become visible at one time or another from the air. In other areas, the density of sites may be equally high or higher, but little or nothing is seen. Thus aerial archaeology cannot be used as a tool for historical geography over wide areas. In areas of modern construction, almost no prospecting technique can give an adequate picture of the subsurface, yet these areas are usually of continuous historical and archaeological significance.

Distribution maps of sites found by aerial exploration are merely maps of areas suited to archaeological aerial photography, not true distributions of sites as was noted a long time ago (60). Aerial photography can only answer questions positively, that is, if something is there to be seen, it is there, but absence is no proof that something is not there. Ground and aerial surveys in one area which has been worked for nearly three decades show less than a 25% overlap, with very poor correlation in those areas where ground survey has been quite productive. This statement is, of course, a function of the number of years during which both methods have been applied, and it is possible that the correlation will rise slightly with time. No data are available on the correlation between ground survey and chance discovery through modern trenching or digging. If this were to be had, then there might be some chance of making predictions of the total number of sites in a given area. As things stand, the concept of a total site inventory is an illusion. The total picture of a large region which the historical geographer needs is not attainable. It probably can only be achieved in very small especially favourable regions. Nonetheless, the quantity of newly discovered sites is such that many monument protection services are unable to carry out the number of rescue excavations which results.

The pessimistic view of the preceding paragraph must be tempered somewhat. Observations over a long period in favourable areas may tend to saturation, that is, repeated flights and field surveys fail to bring in significant amounts of new data. This information can be used to estimate site density roughly in less favoured areas. The estimate will be pessimistic, because even in the favourable regions only a limited percentage of the total number of sites really present has been discovered. Such estimates can be completely overturned by a year with unusual drought. For example, in 1976 in northern Europe sites were discovered in areas hitherto thought unproductive, and the total quantity was much greater in favourable areas than ever before.

3.3. Crawford's site classification

3.3.1. Shadow sites

These are archaeological sites in which the remains are partly above the surface of the ground. They are visible from the air in the early morning or late afternoon sun through the shadows cast by the above-surface irregularities. Vegetation will hold and stabilise the soil over sites which have decayed. If there is no later human intervention, especially ploughing, collapsed structures can survive for long periods. In desert or near-desert areas with very low population density such as the American Southwest, many sites of this type survive. The technique of aerial archaeology for this unusual class of remains is quite different from that required in most other parts of the world (13). In temperate climates, height differences are small and the surface irregularity broad, and they cannot usually be seen well from normal viewing distances on the ground surface. A typical example of such a site is shown in figure 3.1. The picture was taken at 19:00 on 17 July, 1962. It shows a well-known iron age enclosure in north Germany. The bank and ditch are clearly visible in the setting sun. Ploughing has destroyed some of the site, but most of it can still be seen.

Optimum season and time
The contrast which delineates features in an image of this type is a function of the height of the remains, the time, date, and latitude of the photo, the angle of view and the intrinsic surface colour. Contrast is best when the picture is taken opposite to the direction of the sun's illumination, but some shadow areas may be so underexposed that details are missing. It is obvious that the appearance of such a site will change throughout the year, and rephotographing in different seasons is necessary if many complex details are to be recorded. In the spring and summer, contrast is usually best when the sun is low on the horizon in the hour or so before sunset. Depending on local climatic conditions, haze may be a problem in the early morning. In the winter with high pressure weather, fog may be a serious problem in low lying areas. The best light in winter is usually available just after the passage of a cold front giving good visibility and cloudless skies for a few hours at most.

A theory for minimum detectability given an assumed size of site and height of remains could probably be constructed for any given latitude and longitude relative to the date. For this theory, the size and contrast of a shadow produced by an undulation must be computed and compared with the geometric and grayscale resolution capabilities of the detection system. The calculation would be similar to those which can be used for the production of a shaded relief image from a digital terrain model. In the first instance, calculation might be carried out assuming a vertical

Fig. 3.1 An iron age enclosure near Sievern, Kreis Wesermünde in north Germany made visible by the setting sun. Air Photo Archive Rheinisches Landesmuseum Bonn WHVA 71.

view. The slope angle S of the topography is the change in elevation z and:

$$\tan(S) = \partial z / \partial n \qquad (3.1)$$

where n is a distance normal to the contours. With a grid of heights having spacing Δh, and with axis u positive eastward and v positive northward, slopes and exposures (slope direction) are found by second order finite differences, as suggested by Dozier and Strahler (24):

$$\left. \begin{aligned} \frac{\partial z}{\partial u} &= \frac{1}{2\Delta h}(z_{i+1,j} - z_{i-1,j}) \\[2mm] \frac{\partial z}{\partial v} &= \frac{1}{2\Delta h}(z_{i,j+1} - z_{i,j-1}) \end{aligned} \right\} \qquad (3.2)$$

which lead to:

$$\tan(S) = \left[\left(\frac{\partial z}{\partial u}\right)^2 + \left(\frac{\partial z}{\partial v}\right)^2 \right]^{\frac{1}{2}} \qquad (3.3)$$

for the slope, and for the exposure:

$$\tan(A) = -\frac{\partial z / \partial u}{\partial z / \partial v} \qquad (3.4)$$

For the minimum detectability problem in archaeological applications heights are very small, so that there are no hidden surface problems in which the shadows from one part of the site overlap another. The local solar zenith angle θ (the angle of the sun from the vertical) and the azimuth ϕ are found using the latitude ψ and longitude η, the declination of the sun δ and the longitude of the solar subpoint ω to give:

$$\cos(\theta_{sun}) = \sin(\delta)\sin(\psi) + \cos(\delta)\cos(\psi)\cos(\omega - \eta) \qquad (3.5)$$

and:

$$\tan(\phi_{sun}) = \frac{\cos(\delta)\sin(\omega - \eta)}{\cos(\delta)\sin(\psi)\cos(\omega - \eta) - \sin(\delta)\cos(\psi)} \qquad (3.6)$$

The declination of the sun δ, which ranges over $\pm 23.45°$ is obtained from ephemeris tables depending on the date. The solar longitude ω can be calculated from the local time, $\pm 15°$ per hour from local noon.

When these have been determined, then the illumination angles θ_0 and ϕ_0 of the surface with slope S and exposure A are:

$$\left. \begin{aligned} \cos(\theta_0) &= \cos(\theta_{sun})\cos(S) + \sin(\theta_{sun})\sin(S)\cos(\phi_{sun} - A) \\[2mm] \tan(\phi_0) &= \frac{\sin(\theta_{sun})\sin(\phi_{sun} - A)}{\sin(\theta_{sun})\cos(S)\cos(\phi_{sun} - A) - \cos(\theta_{sun})\sin(S)} \end{aligned} \right\} \qquad (3.7)$$

A map of the cosines of the illumination angles gives the required shaded model. Visual judgement can then be used on the displayed values to see if detectability for features of a given height and orientation at a particular latitude and longitude at a favourable time of the year has been achieved, or if the enhancement methods of chapter 4 can be applied to obtain a possible improvement. Practically, however, additional factors such as the specular and Lambertian reflection properties of the surface cover must be considered, making the problem rather formidable. This is a research problem for those who have to deal frequently with shadow sites in a particular area, and leads to the prediction of optimum time and date of photography.

Stereoscopy
Even with low obliques, height differences can be seen stereoscopically when two pictures are made at roughly the same angle in level straight flight. With vertical photogrammetric images, height differences can be measured accurately. Old-fashioned stereo cameras used two lenses and synchronised shutters to obtain two pictures simultaneously. To obtain a similar effect with a single camera, the photographer must counter his natural inclination to centre the site in the viewfinder in each photo. The two images must point along parallel rather than converging lines as far as possible. This can be achieved by concentrating on some point-like object on the distant horizon held at the top centre of the viewfinder image for both pictures. For viewing the resulting pictures, back-lighted transparencies are better than paper prints, since they record a greater range of grays and permit the user to see details in the shadows better. A zoom microscope or simple stereo viewer with modest magnification can be used.

Trees
Sites under trees pose a special problem. If the trees are deciduous, it is best to make the pictures in winter. A small amount of snow on the ground actually helps to enhance outlines. With conifers, only a very rough irregularity betrays the presence of a site, and it is usually not possible to infer much about the plan from the image unless the trees are quite small. Perhaps it may be possible to deal with this type of site using synthetic aperture radar which penetrates the tree cover and can map the air–ground interface if the frequencies used are low enough. A practical alternative is to map using classical surveying techniques on the ground after having found the site from the air.

3.3.2. Soil sites

The 'complete' destruction of a site is almost impossible, for some traces of past construction always survive in the ground. At the very least the natural order of the soil is disturbed, even if no materials which are not normally present have been incorporated. The process of further soil formation reaches stability after a time and the disturbance of the natural order may then be permanent. Pedology distinguishes between the mineral material on which the soil develops, and if the materials are not in place, such as in river valleys, the mode of transport. The effects of weather, plants, and animals produce the horizontal banding or horizons whose description takes up much of the soil scientist's time. The soil types assigned by them are essentially a shorthand description of the processes of transformation which the basic minerals have undergone. Although this descriptive terminology is of importance for agricultural applications, it is not very significant for archaeological aerial photography. Here, the mechanical structure of the soil itself and its effects on the surface colour and water retention are important. Some pedologists call this 'soil kind' as distinguished from 'soil type'. The differences which an archaeologist sees in section during an excavation are usually hidden from the air. But differences in water retention and materials transported from lower layers are visible on a bare surface, and these give rise to the soil site.

Soil colour
Perceived soil colour is a function of the spectral reflectance properties of the material. Typically, soils show spectral reflectances which increase towards the red end of the spectrum as shown in figure 3.2 which shows reflectance properties beyond that which can be recorded even on infra-red film. The reflectance is a function mainly of the moisture, organic matter, iron oxide contents, and texture as well as of the main soil materials themselves. A major compilation for many different soils is given by Stoner, Baumgardner, Biehl and Robinson (65). It is well known to every gardener that increasing organic matter darkens soil, and it is not surprising that it reddens with increasing amounts of iron oxide, usually present as one of the hydrated forms of haematite. Smith (64) states that reflectance increases for smaller particle size, but that surface roughness may predominate. He gives an extensive summary of mathematical modelling theory for bare soils and those covered with vegetation in the visible and thermal parts of the spectrum. Unfortunately

there are no observations available for archaeological soil sites discovered from the air to test the applicability of one or other of the theoretical models to practical cases. A possible use for such a study would be the prediction of photographic archaeological anomaly contrasts in soils of different kinds under different conditions. This is a different problem from that which is of interest in remote sensing applications in agriculture, such as discussed by Myers (49), where classification of soils is the aim. Such studies, however, provide the only quantitative information available.

Classification of soils by the use of spectral reflectance curves is possible, as Condit has shown (16). This study included the effects of humidity. By making measurements at five wavelengths ranging from the deep blue to the photographic infra-red, quite reasonable agreement with normal classification techniques was obtained. The best differentiation is obtainable with dry rather than wet soils. It would be instructive to use a portable spectrophotometer to monitor the change in reflectance with time on archaeological features and their surroundings in several widely differing soil contexts. Bare soils usually show a

Fig. 3.2 Reflectance curves for three dark red surface soils. From ref. 65.

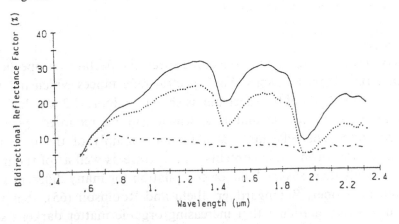

Key to Soils Data

Soil	Curve	% Organic Matter	% Fe_2O_3
Dill (Oklahoma, USA)	——	0.6	0.87
Arroyo (Spain)	••••	1.28	2.00
Londrina (Brazil)	—·—·—	2.28	25.6

spectrum with three peaks in the spectral range which can be recorded on film plus the near infra-red. One of the absorption bands, between 1.44 and 1.94 μm, is due to water. Another, at 0.9 μm, is due to iron (ferric) oxide, and another at 1.0 μm, to ferrous iron compounds. Between 2.08 and 2.32 μm is a band with high correlation for soil moisture. These properties have not been exploited in archaeological aerial photography, perhaps because economically priced high resolution sensors are not readily available for the most interesting wavelengths.

Jones and Evans (36), using the comparative Munsell notation generally accepted by soil scientists (48) state that colour changes in soils may be enhanced in black and white aerial photographs. They note that colour value (relative lightness) is more important than chroma (position in the spectrum) for subsequent exposed densities with this type of film. However, colour differences are also visible, as for example with red-sensitive films which show redder colours as lighter tones than less red colours of the same value and chroma. The contrasts are very much a function of soil moisture. For example, it is asserted that a shallow soil of unstated type on chalk has a greater contrast with its surroundings with moisture differences of 5–6% as opposed to that visible at 1–2 or 19–22%. The most reflective soils produce more marks, but differences in ploughing depth which affect the exposure of subsoil intervene strongly. Colour is also affected by grain size distribution, and a large number of stones on the surface (25% or more) produces soil marks. Surprisingly enough, surface roughness does not affect appearance very much, with soil marks showing equally well on freshly ploughed and on rolled fields. The direction of ploughing is more significant, especially when using vertical photographs. Jones and Evans say that tonal contrast with east–west ploughing is less, since light is reflected away from the camera. Freshly ploughed surfaces obscure soil marks, however.

Soil mobility
Soil is, by its granular nature, highly mobile. Wind is capable of transporting large quantities of material over considerable distances. A still more powerful agency for displacement of material is water. In addition, when in a cold, nearly frozen state, finer soils may flow like jelly down very slight slopes (solifluction). All of these transport phenomena can take place in times which are comparable with the age of a site. Ploughing, or the effects of the removal of surface vegetation

can cause transport to take place much more rapidly. Such transport mechanisms can bury a feature once on the surface, but in favourable circumstances given a slight slope, they can reveal it by removal of overburden (73).

Transport materials

Materials foreign to the site such as brick or stone may also be incorporated in sites. These can be thought of as having overly large grain size relative to the surrounding medium. Wooden remains usually degenerate completely when the soil is dry enough, and only a trace of colouring due to binding of certain minerals in complex chemical structures remains along with the disturbed soil of the holes and ditches in which posts and sleepers were placed. Even when wall materials have been removed for secondary use, fragments usually remain which alter the properties of the soil locally. Foundation packings are usually not valuable and are not removed. These can sometimes be detected by electrical methods, but if they are close enough to the surface, pieces may be brought up by ploughing and, if different enough in colour from the surrounding soil, they can be seen from the air, as in figure 3.3 by Agache (2).

Weather and agriculture

The appearance of a bare soil from the air is strongly affected by weather and by the state of agricultural practice at the moment of observation. Chemical treatment, such as application of fertiliser has little effect, but ploughing, harrowing, and rolling strongly modify surface compactness, moisture distribution and hence colour. Marling with chalk as practiced in some areas does indeed change the colour of the surface. Fertiliser and marl used to be applied by hand, and this left arc-like lines on the field. These may sometimes be seen in old aerial photographs. Modern machine application of materials is much more uniform and this is readily distinguishable from an archaeological feature.

Usually structures visible as soil colour differences can be seen for only a very short time until rain, wind, and agricultural treatment obliterate them. Much depends on the colour and water retention contrast between lower lying materials brought to the surface and their surroundings. Best visibility is usually obtained with rendzina soils which are dark brown or reddish in colour, developed on chalks or limestones which are white or pale yellow. Such soils are usually quite shallow, so that the possibility of material being brought to the surface is quite high

(74). If the site is on a slight slope leading to the washing away of surface material, quite spectacular pictures may be made, such as figure 3.4 taken on the chalks in the northern Somme valley. Another favourable location for soil sites is on clay covered peat deposits. These provide a contrast between the dark brown of the lower layer and the light gray of the surface as shown in figure 3.5 taken in Cambridgeshire. Soil sites on brownearths are much rarer for the contrast between upper and lower layers are not nearly so well marked. Figure 3.6 shows a rectangular enclosure in the Cologne Basin. In this site, a thin cover of loess lies on gravel and lighter yellow loess, the uppermost loess having been transformed into a brownearth. The direction of ploughing relative to the orientation of the feature plays a considerable role in its visibility on this type of soil. The best result occurs when the agricultural treatment is oblique relative to the feature so that the two are not confused. The water content of the surface plays an important role in the visibility of such a site. If there is too little rainfall, all dries out to a uniform brown-yellow. If there is too much, all is a uniform medium brown. The only control which the archaeological aerial photographer has over this lies in his choice of flight dates relative to recent weather in the late spring.

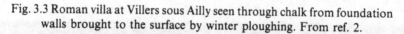

Fig. 3.3 Roman villa at Villers sous Ailly seen through chalk from foundation walls brought to the surface by winter ploughing. From ref. 2.

Fig. 3.4 (*a*) Roman villa at Grivenes before heavy rain, 11.3.64. (*b*) The same site after heavy rain, 23.3.64. From ref. 3.

(*a*)

(*b*)

Sites on sandy soils may frequently be visible throughout much of the time when there is no vegetation and little rain, since the light coloured materials are quite visible on a surface which contains large amounts of dark humus materials. In figure 3.7 the light coloured cores of a barrow group appear as round blobs, while the ditches, filled partly with surface

Fig. 3.5 Soil marks in clay, Rodden, 3 miles SE of Littleport, Cambs. Cambridge University Collection WL2 135/613849.

material and partly with the dark material of a subsurface podsol, appear
as dark rings.

Even after many years of repeated ploughing, although the site may
lose some of its crispness of outline, it is often still visible if the amount
of mixed material brought to the surface is large enough and the colour
contrast sufficient. Horizontal transport is often surprisingly minimal
and outlines reasonably crisp (74).

Optimal conditions and technique
Most soils sites are quite evanescent and can only be photographed
within a very small time window. Usually, the early spring, just after

Fig. 3.6 Rectangular enclosure in the Cologne Basin loess on gravel. Air
Photo Archive Rheinisches Landesmuseum Bonn EW7.

spring ploughing, planting, or rolling is the appropriate period. During the late autumn, stubble in the fields and the lack of rainfall produce less favourable conditions. The contrast, if visible at all, does not depend significantly on the angle of illumination, so that any time of day will do. Diffuse light is usually better than bright sunlight, and this kind of image can readily be made under light cloud cover. Since colour differences are important, a high colour contrast film may be more useful than black and white. Soil features are also frequently seen in vertical mapping cover because most cartographic agencies like to have pictures in which field and road boundaries are not obscured by the leaves of deciduous trees. However, most of the features observed are not of archaeological significance.

Fig. 3.7 Ploughed-out barrows in the lower Rhine Valley sands. Air Photo Archive Rheinisches Landesmuseum Bonn EW26.

Mapping of soil sites

The transformation of photographs of soil sites into a map is much easier than with shadow sites, for the features are perfectly flat. Outlines are sometimes rather blurred, and it is difficult to decide on the true shape if hand methods are used. The features seen in the photograph usually only vaguely reflect the actual details of the subsurface structures, since the outcast from a ditch may be seen but not the ditch itself. Soil marks due to modern activities are very numerous, and may readily be mistaken by novices for older ones, especially in vertical cover. The experienced observer is able to avoid this pitfall through his prior knowledge and experience.

3.3.3. Moisture markings

Moisture markings are really not an independent class, for the moisture on the surface is inextricably linked with soil colour and hence soil and moisture features are often present simultaneously. In this discussion, it will be assumed that there is no subsurface material of a different colour present on the surface, so that differences in appearance are attributable only to moisture differences.

The mechanical properties of soils have a considerable effect on their behaviour under conditions of varying humidity. Soil water is retained on the surface of the finest particles and this is most important for the appearance of sites on bare soils as well as for its effects on overlying crops. The time required for water to move in such a complex structure is a function of the grain size distribution with depth and the slope of the land. It is to some extent also dependent on the amount of organic material close to the surface. Direct determination of soil humidity *in situ* by neutron backscattering is possible, as well as by analysis of carefully taken samples. This type of experiment only gives the current state of moisture, but not its variability with time. Greater amounts of water usually cause the soil material at the surface to appear darker. After a uniform wetting by light rain with subsequent drying, the soil surrounding a buried feature filled with fine material will tend to dry more rapidly and features can be seen from the air as shown in figure 3.8. Further drying will cause the markings to disappear quite rapidly. The moisture mark can also disappear with further rainfall, as is usually the case in the spring in temperate climates. The site in the figure was visible for about four days. The soil was a fine grained loess on a tertiary outcrop, rather well drained, but with good water retention in the uppermost layers. On sandy soils, a moisture mark usually disappears in a matter of hours. Colour contrasts are in brown and dark yellows as with soil marks. The best moisture markings are obtained on features dug into very porous substructures such as gravels or chalks, filled with fine clay material which retains moisture very well. An example is shown in figure 3.9. Inversion of the marking with the buried feature showing

Fig. 3.8 Ring ditch visible in slowly drying field in the Rhineland. Air Photo **Archive Rheinisches Landesmuseum Bonn EV86.**

lighter than the surroundings is possible if the site is visible as a soil site as well as a moisture mark.

Optimal conditions and technique

Moisture marks can be photographed under almost any conditions, since the apparent contrast does not depend on the angle of illumination or the direction in which the picture is taken. As with soil marks, moisture marks frequently appear in vertical mapping images, where they are usually geological in origin. It is rare to see the markings in the autumn after crops have been harvested or in freshly ploughed fields. Colour film with good red response will usually give better results than black and white film. However, if the colour contrast is sufficient to be seen by eye, it will usually appear in a photograph, regardless of film type. It is probably better to take the pictures in the early morning if strong sunlight develops during the day, since the surface will dry out.

3.3.4. *Frost and snow marks*

Braasch (8), who took the picture shown in figure 3.10, has called such marks 'the poor man's thermal scanner'. When the air temperatures in

Fig. 3.9 Neolithic enclosure at L'Etoile, Somme seen as moisture marks on chalk soil. From ref. 2.

the maritime climate of northwest Europe in the winter are just around freezing point, the thermal gradients discussed in chapter 10 may become visible as differential thawing and freezing of hoar frost or a light covering of snow. Great success has been had with this type of marking in Bavaria. It is possible to fly there during the appropriate seasons and still have sufficient visibility for adequate photography. In the Rhineland, however, when conditions are right for this type of marking to appear, the ground fog and haze are usually so bad that flying is not possible.

Optimal conditions and technique
The marks persist for only a few hours after sunrise and are probably the most evanescent of all the markings discussed here. Because of the relatively poor lighting conditions, high speed films, and relatively slow shutter speeds must be used. Some reduction in image quality compared with more favourable seasons is therefore to be expected. Timing considerations are perhaps even more critical than for moisture marks, with visible contrast persisting for very short intervals indeed.

Fig. 3.10 Snow markings near Schwaßmünchen in Bavaria showing a Roman road and fields. From ref. 8.

3.3.5. The crop site

Response of growing plants to buried structures

By far the most sensitive and productive method for detecting shallow
buried archaeological structures through aerial photography is based
on the response of growing plants to soil and humidity. The roots of
many plants penetrate to depths roughly equal to their height above
ground or more and draw nourishment from depths which include the
upper parts of many archaeological features. The difference in soil and
moisture properties between the site and its surroundings can affect plant
height and colour during the growing season. This phenomenon is not
limited to temperate climates with good rainfall, but most experience
has been acquired in such areas. Most data published in the remote
sensing literature have been obtained in climates where soil moisture is
a limiting factor.

The mechanisms responsible for the production of crop sites were first
discussed in considerable detail by Riley (54,55) and much later by
Scollar (60). Although the appearance of markings in crops which are
indicative of underground structures was a well-known phenomenon to
Crawford, Keiller and other pioneers of aerial photography, it was only
the later research of Strunk-Lichtenberg (66) and Jones and Evans
(26,36,37) which provided a quantitative explanation. Detection of crop
sites has been responsible for the discovery of more buried archaeological
sites than all other prospecting methods combined. Of all the passive
methods, it is by far the most complex to analyse, being a consequence
of the interaction of growing vegetation, soil structure, and climate.
Perhaps because of this complexity, there are less quantitative experimental data available than for any physical method, most evidence being
based on qualitative observation.

There are two possible effects of a buried structure on the overlying
crop as shown in figure 3.11. Growth is either retarded or advanced.
These are idealised features shown at the peak of the growing season.
It is assumed that a dry spell has caused depletion of soil moisture
reserves, and that the plants are limited in growth capacity by lack of
humidity. Marks which appear due to excess humidity which causes
retardation of growth have also been reported in the literature, but most
sites appear when soil moisture for plant nourishment is severely limited.
This can happen more rapidly on well-drained slopes so that it is not
possible to predict appearances from rainfall statistics alone. Since plants
regulate their temperature by evapotranspiration, their water require-

ments rise when it is warm and the available soil moisture will be correspondingly depleted. Wind also increases loss of moisture and is thus also a factor of somewhat lesser importance. Therefore temperature, rainfall and wind must all be taken into account. Drainage, being a highly local phenomenon, is usually ignored. When root penetration of which 90% normally lies in the first 30 cm of soil is hindered by solid material or an impenetrable soil horizon, crops grow less well.

Fig. 3.11. The effect of a buried structure on an overlying crop, showing retardation or normal growth: (a) the spring; (b) the appearance of ears; (c) and (d) fixing and (e) ripening. From ref. 61.

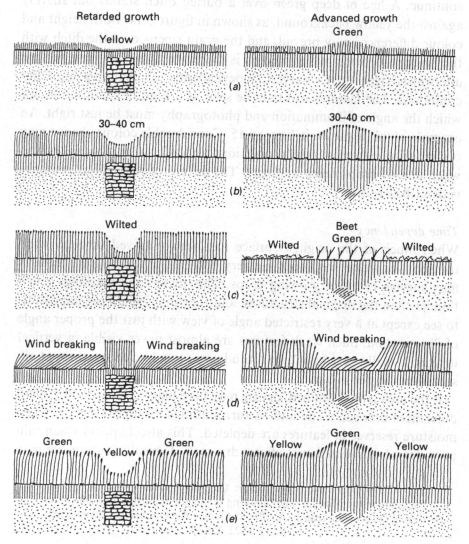

Markings over buried ditches and pits are much more common than those over walls. A typical example is shown in figure 3.12. This is because walled structures are restricted to only a few more recent archaeological periods, and furthermore, walling must survive quite close to the surface in order for it to affect growth to an appreciable extent. An example is shown in figure 3.13. Figures 3.12 and 3.13 are examples of growth limited by the availability of water relative to plant needs. A small colour difference may also accompany the effect, when less chlorophyll is produced. When a grain crop ripens, the visible contrast on a site rises dramatically provided that the moisture stress conditions continue. A line of deep green over a buried ditch stands out sharply against the yellow background, as shown in figure 3.14. Both height and colour differences may prevail, and the grain ripens over the ditch with the height difference maintained. It is therefore 'fixed' and can be seen afterwards right up to harvest, independently of changing humidity conditions. The features can then be seen as a kind of shadow site, for which the angle of illumination and photography must be just right. An example of this is shown in figure 3.15. Considerable colour change may occur during extreme drought conditions, and, of course, it always occurs when annual crops reach maturity. The nature of this colour response is described below.

Time dependency
When much of the original surface has been removed through wind transport or slope flow, only the very bottoms of the original features may survive to affect growing crops. This makes the markings much thinner than the original structures, and they may be almost impossible to see except at a very restricted angle of view with just the proper angle of illumination. Such weak features are almost never visible in vertical images. Usually very low flight altitudes are required to see them at all, sometimes under 100 m.

If the area is well drained, drying may take place quickly, and growth differences between a site and its surroundings disappear rapidly when moisture reserves in features are depleted. This also happens when rain provides adequate moisture after a dry spell. Typically, a crop site may

Fig. 3.12 A simple crop mark above the buried ditches of an Iron Age enclosure in the Moselle valley showing (*a*) height and (*b*) colour difference during the growing season. Air Photo Archive Rheinisches Landesmuseum Bonn BT22, DK43.

(a)

(b)

be visible for at most a week or so, but in exceptionally favourable circumstances, it may be photographed for as much as a month in a given year. If fixation occurs due to the death of plants over buried walls or retained height differences over a buried ditch after ripening, then visibility is maintained right up to harvesting. This can add several weeks to the time during which a site can be seen and photographed. The appearance of a crop site changes considerably with time, and it is advisable to repeat flights at intervals of not longer than a week to follow developments. When compared to all the other types of sites except the shadow site, crop sites can be photographed over longer periods and, if the soil disturbance is deep enough and the extent of the site

Fig. 3.13 The buried foundation walling of a Roman villa has retarded crop growth during a dry spell. Air Photo Archive Rheinisches Landesmuseum Bonn HJ14.

considerable, there is some chance that they will appear in vertical imagery made in early summer. The duration of appearance, the rapidity of onset and decay of contrast depend in a complex way on the interaction between plants, soil, and weather.

In many regions of northern Europe, fields are not large enough to

Fig. 3.14 Ditches of a Belgic settlement near Thorpe Achurch, Northants, in ripening grain. Cambridge University Collection ZA95 134/026824.

allow a site to be seen completely under one crop. Therefore, once a fragment of a site has been found, it is worthwhile to revisit the area in subsequent years under suitable conditions to try to acquire pictures of other portions of the site which may have been under unfavourable crops. A very few sites present such a large disturbance that they are visible almost every year, but this is much the exception.

Soil dependency

As discussed in detail in chapter 2, the water balance in an archaeological feature differs from its surroundings primarily due to the differences in soil grain size distribution and the pore volume. Small chemical differences, especially in the calcium and phosphorus compounds which are significant for plant growth, have not been demonstrated experimentally to differ in an archaeological feature compared with its surroundings. Slight differences in acidity which are attributable to the

Fig. 3.15 (*a*) Barrow ditch photographed on the 15 June 1961, showing height difference in growing grain. Air Photo Archive Rheinisches Landesmuseum Bonn AE44. (*b*) The same site three years later in August, showing height difference in ripe grain. Air Photo Archive Rheinisches Landesmuseum Bonn GS90.

(*a*)

large number of fine grains in a site feature which are capable of binding hydrogen ions have been measured (66). Increased availability of nitrogen compounds in features is proposed by Jones and Evans (36), but no experimental data have been furnished.

The increased presence of humus in a feature is another possible cause of the slight pH variation. It seems unreasonable to assume that additional mineral nourishment is available in an old buried feature, because any such non-renewable resources would have been exhausted long ago. A possible exception might be walling, where a large amount of stone and mortar might provide some additional nourishment. However, aerial photographs of crop sites over walling usually betray evidence of insufficiency rather than additional support for growth.

Fig. 3.15 (cont).

(b)

Modern agriculture and large amounts of artificial fertiliser will certainly erase any remaining variation of this kind in the upper 50 cm of soil. Hence only differences in water balance can account for the appearance of colour and height variations in growing crops over archaeological features. These differences can be obscured by irrigation or by the heavy application of nitrogen fertiliser which cause the crops to grow uniformly almost independently of soil moisture conditions.

3.3.6. Soil water balance

Water in the soil must obey the law of conservation of matter. The water content of a volume of soil can only increase with an addition from an external source due to infiltration or capillary effects, and it can diminish only by drainage or by evaporation from the surface and transpiration from plants. This balance is linked to the energy balance in a given volume. Any change requires the addition or loss of energy. The phenomenon is strongly affected by growing plants as shown in figure 3.16. Following Hillel (34) the water balance change is:

$$W_{in} - W_{out} = \Delta W \tag{3.8}$$

The quantity of water added comes as rain or snow and in some areas of the world in the form of irrigation, where W_m is the amount added by precipitation and W_i that due to irrigation, then:

$$W_{in} = W_m + W_i \tag{3.9}$$

Water is removed by runoff, drainage, evaporation, and transpiration so that:

$$W_{out} = N + F + (E + T) \tag{3.10}$$

where N is runoff which may be either positive or negative depending on whether a field gains water from or loses water to adjacent fields, F is drainage out of the root zone of the growing plants to lower layers, E is direct evaporation from the bare soil surface, and T is transpiration from vegetation. The total water balance is therefore:

$$\Delta W = W_m + W_i - N - F - (E + T) \tag{3.11}$$

The energy balance at the soil surface, where that arriving is considered positive and that leaving negative, must be zero and described by:

$$R_n + LE_t + G + H = 0 \tag{3.12}$$

where R_n is the net radiation, L is the latent heat of vaporisation of water (2.43×10^3 J cm^{-2}), E_t is the evapotranspiration, G is the energy which heats the soil and H is the energy which heats the air (the sensible heat). This equation is discussed in greater detail in chapter 10. Of interest here is the term connected with evapotranspiration. Bowen (7) shows that the ratio of the energy of sensible heat H and that of evapotranspiration is nearly constant. The Bowen ratio $\beta\,(=H/LE_t)$ is:

$$\beta = \gamma \frac{(T_2 - T_1)}{(e_2 - e_1)} \tag{3.13}$$

where T_n, $n = 1, 2$ is the air temperature at two vertical measurement points, e_n, $n = 1, 2$, is the vapour pressure at the two points, and $\gamma = c_p P_a / L\varepsilon$ is the psychrometric constant, with c_p the specific heat

Fig. 3.16 Effect of growing plants on the soil water balance. From ref. 37.

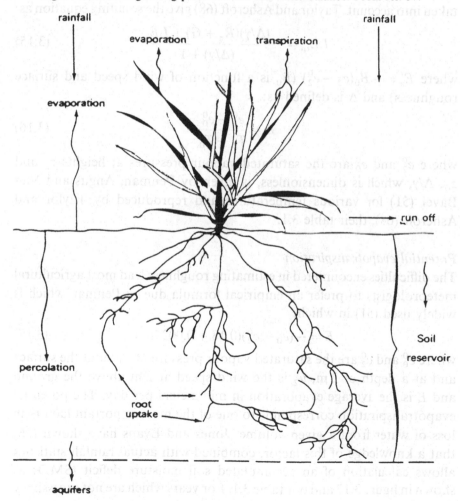

capacity of air at constant pressure, P_a the atmospheric pressure, and ε the ratio of the molecular weight of water to air. Solving the energy balance equation for H, dividing by LE_t and inserting β for H/LE_t gives the actual transpiration:

$$E_t = -\frac{1}{L}\left(\frac{R_n + G}{1 + \beta}\right) \tag{3.14}$$

This assumes a homogeneous surface completely covered by a short crop and measurements made at two points a short distance apart vertically. Differences in the height and roughness of a crop create different local wind speeds. Measurement at two points is inconvenient because these must be carried out with very high accuracy to obtain a gradient correctly. Measurement at one point is possible if aerodynamics are taken into account. Taylor and Ashcroft (68) give the resulting equation as:

$$LE_{tp} = -\frac{(\Delta/\gamma)(R_n + G) + LE_a}{(\Delta/\gamma) + 1} \tag{3.15}$$

where $E_a = -B_0(e_2 - e_2^0)$ (B_0 is a function of wind speed and surface roughness) and Δ is defined as:

$$\Delta = \frac{e_2^0 - e_0^0}{T_2 - T_0} \tag{3.16}$$

where e_2^0 and e_0^0 are the saturated vapour pressures at heights z_2 and z_0; Δ/γ, which is dimensionless, is given by Penman, Angus and Van Bavel (51) for various temperatures and reproduced by Taylor and Ashcroft (68), their table 3.1.

Potential evapotranspiration
The difficulties encountered in estimating roughness lead most agricultural meteorologists to prefer an empirical formula due to Penman which is widely used (51) in which:

$$E = -(e_0^0 - e_2^0)(0.18 + 0.0035u_2) \tag{3.17}$$

where e_0^0 and e_2^0 are the saturated vapour pressure of water at the surface and at a depth of 2 m, u_2 is the wind speed at 2 m above the ground and E is the average evaporation in millimeters per day. The potential evapotranspiration corresponds to one of the most important factors in loss of water from a given volume. Jones and Evans have shown (36) that a knowledge of this factor, combined with actual rainfall statistics allows calculation of an accumulated soil moisture deficit (SMD) as shown in figure 3.17 and our table 3.1. For years which are not excessively dry, Jones and Evans state that the use of tabulated values of potential evapotranspiration is adequate, (32) but that the daily values either calculated or obtained from a meteorological service are more accurate.

Actual SMD

The SMD occurs when potential transpiration exceeds available water.

Fig. 3.17 Accumulated SMD (PT = potential transpiration). From ref. 37.

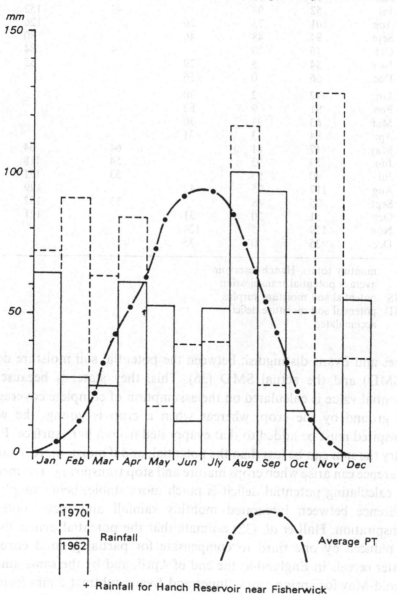

Table 3.1. *Water balance (mm) for Fisherwick*

Month	R	PT	PSMS	PSMD	Acc. PSMD
1962 Jan	64	2	62		
Feb	27	9	18		
Mar	24	33		9	
Apr	61	53	7		
May	53	81		28	30
Jun	12	93		81	111
Jul	52	93		41	152
Aug	101	75	26		126
Sept	94	48	46		80
Oct	16	20		4	84
Nov	34	5	29		55
Dec	56	0	56		
1970 Jan	72	2	70		
Feb	91	9	82		
Mar	63	33	30		
Apr	84	53	31		
May	17	81		64	64
Jun	39	93		54	118
Jul	40	93		53	171
Aug	117	75	42		129
Sept	25	48		23	152
Oct	51	20	31		121
Nov	129	5	124		
Dec	35	0	35		

R: monthly totals, Hanch reservoir
PT: average potential transpiration
PSMS: potential soil moisture surplus
PSMD: potential soil moisture deficit
Acc: accumulated

Jones and Evans distinguish between the potential soil moisture deficit (PSMD) and the actual SMD (36). This, they state, is because the potential value is calculated on the assumption of complete coverage of the ground by the crop, whereas when a crop is young, the water transpired must be added to that evaporated from a bare surface. If this is dry the sum can be less than the potential, and if moist, more. Another difference can arise when crops mature and stop transpiring. The method for calculating potential deficit is much more stable, being simply the difference between integrated monthly rainfall and mean potential transpiration. Hall *et al.* (32) estimate that the potential deficit should be reduced by one third to compensate for partial ground cover by winter cereals in England at the end of April, and by the same amount in mid-May for spring crops. Jones and Evans subtract 5 mm from the observed deficit and accumulate this for the period during which crops do not cover the ground completely. They call this the crop adjusted potential soil moisture deficit (CPSMD).

When the accumulated SMD during the growing season of the plants exceeds 40 mm, then the plants begin to suffer from moisture stress. Jones and Evans state that crop marks appear when the PSMD ranges between 50 and 65 mm. Retarded growth is responsible for the appearance of archaeological sites in crops because colour and height changes result and the leaf area per unit surface area, the leaf area index, is reduced.

Available water
Available water is affected by the granulometric and pore distribution of the soil, and samples show a distinct variation in buried archaeological ditches compared with their surroundings. This explains why, despite a soil moisture stress of over 50 mm, plants can continue for a while to grow normally over a buried ditch, maintaining normal chlorophyll content, while failing to do so where the water reserves have been exhausted. Jones and Evans state that the increased depth of soil alone in an archaeological feature cut into a subsoil of gravel is sufficient to constitute an important water reserve (36). Available water A_v can be calculated in a given case using a method given by Hodgson (35).

The water available to plants in a field is not that given by the percentage of moisture in the upper layers. In order to extract this moisture, plants must exert suction, and therefore the concept of soil water potential or the amount of work needed to move a quantity of soil water compared to that needed to move free water the same distance is introduced in quantitative crop physiology. Milthorpe and Moorby (45) define the total water potential ψ as:

$$\psi = \pi + \tau + P \qquad (3.18)$$

where π is the osmotic potential from the solute particles (independent of kind for dilute solutions), τ is the matric potential, and P is the pressure potential. P is zero in unsaturated soils, and is positive below a water table in saturated soils, π is always negative, and τ is caused by the absorptive forces of the soil matrix, especially that due to hydrogen bonding on the clay and humus colloid particles. When there is enough water, surface tension is the preponderant force, and this is inversely proportional to pore diameter.

The osmotic potential π is:

$$\pi \approx -RTC_s \qquad (3.19)$$

where C_s is the molar concentration, R is the gas constant $(8.315\,\mathrm{J\,K^{-1}\,mol})$ and T is the absolute temperature. In equation (3.18) τ is given by $\tau = -4\sigma/D$ dyne cm$^{-2} \sim -2 \times 10^2/D$ J kg^{-1} where σ is the surface tension of water, and D the pore diameter in microns.

The curve of figure 3.18 shows the relationship between matric potential and moisture content. The water held when all pores smaller than 30 μm are filled is called the field capacity giving a σ of about -10 J kg^{-1}. This is the point where drainage has stopped. When the matric potential reaches about -1500 J kg^{-1}, most plants wilt. Between these ranges water is available for growth. Obviously because of the relationship with pore size, the characteristics of a given soil vary with its composition as shown in figure 3.19.

Prediction of crop site appearance
The data required to compute the SMD can sometimes be obtained from agricultural meteorological services in most developed countries. In areas where irrigation is common, the value can be obtained on request by telephone. This may be used for a rough prediction as to when archaeological aerial photography is likely to be productive. Since the model neglects drainage, artificially drained or sandy soils or sites

Fig. 3.18 The relationship between matric potential and moisture content. From ref. 45.

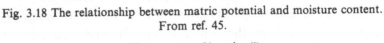

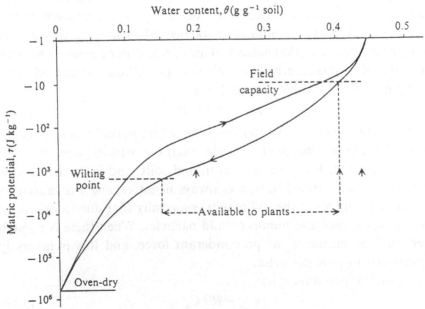

located on slopes are likely to produce crop sites earlier than those in less favoured locations. The potential evapotranspiration is dependent on temperature, which also affects the growing season, and the probability of detection also depends on this factor. Although local differences in precipitation may vary as much as three to one in Britain and Germany, Jones and Evans state that temperature is the most severe limit on crop growth in the British Isles (36). They note that rainfall maps do not show the water available to plants as defined by the difference between field capacity and wilting point. They suggest using the available water method of Hodgson and balance this against the CPSMD for a given soil depth in order to estimate when water will be exhausted as the CPSMD approaches A_v. The correlation between observed site features and this value is given for one case by them and reproduced here as table 3.2. Behaviour over many years has been monitored with respect to PSMD at one site by Evans and Jones (26). Their results are reproduced here in table 3.3, with the course of the PSMD in a given year shown as figure 3.20.

Reactions of plants to other factors
Different plants react differently when the wilting point has been reached.

Fig. 3.19 The characteristics of a given soil varying with composition. From ref. 45.

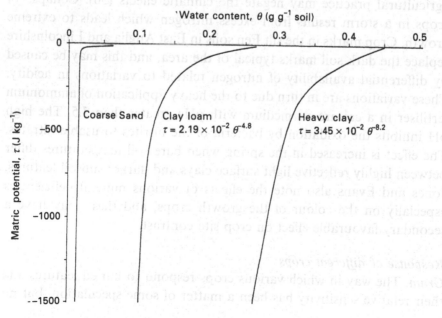

Water content, θ (g g^{-1} soil)

Coarse Sand

Clay loam
$\tau = 2.19 \times 10^{-2} \, \theta^{-4.8}$

Heavy clay
$\tau = 3.45 \times 10^{-2} \, \theta^{-8.2}$

Matric potential, τ (J kg^{-1})

Table 3.2. *Observations of the enclosure ditch marks in relation to available water* (A_v) *and crop adjusted potential soil moisture deficit* (*CPSMD*)

Date	Clarity of enclosure ditch crop marks	(A_v = CPSMD)		
		Natural soil beside ditch	Ditch infill	
			1 m	1.65 m
31 May	Not visible	26	86	146
17 June	Faint	−7	53	113
26 June	Very clear	−36	24	84
28 June[a]	Very clear (as dark mark)	−42	18	78
30 June	Very clear	−48	12	72

[a] Aerial observations, the remainder being made on the ground

Some stop growing and transpire at a lower rate, while others actually wilt. The reaction also depends on the stage of growth at the time of the deficit. Cereals need much more water before the heads of grain appear than afterward and they may reach maturity before the SMD limits growth. Deep roots which penetrate into a ground water table or an archaeological feature may provide the additional nourishment required to tide them over the difficult period.

As Jones and Evans also show, the leaf area index is strongly affected by application of nitrogen fertiliser, figure 3.21, so that intensive agricultural practice may negate the climatic effects (36). Collapse of crops in a storm results from excess nitrogen which leads to extreme growth. Crop marks in the silt Fen soils in East Anglia and Lincolnshire replace the dark soil marks typical of the area, and this may be caused by differential availability of nitrogen related to variations in acidity. These variations are in turn due to the heavy application of ammonium fertiliser in a calcareous medium with pHs greater than 7.5. The high pH inhibits the oxidation by bacteria of the nitrites to usable nitrates. The effect is increased in the spring when bare soil temperatures differ between highly reflective light surface clays and darker subsoil features. Jones and Evans also note the effects of various mineral deficiencies especially on the colour of the growth crops, and these may have a secondary favourable effect on crop site contrast.

Response of different crops
Grain. The way in which various crops respond to buried features and their relative sensitivity has been a matter of some speculation, but no

Table 3.3. *Roman fort, Glenlochar: date of photography; PSMD at that time; and clarity of marks*

Date of photography	PSMD (mm)	Visibility[a] of pattern in grass (g), cereal (c) or roots (r)			
		F1	F2	F3	F4
12.7.49	148	V. clear (g)	V. clear (r)	Mod. clear (g)	Clear (g)
5.7.51	115	Bare soil – light toned ramparts	Clear (c)	Not visible (g)	V. faint (g)
6.7.52	64	Clear (c)	V. faint – roads only (cut g)	Not visible (g)	Not visible (g)
25.7.53	24	V. faint (cut g)	Not visible (g)	V. faint (c)	Not visible (g)
15.7.55	31	V. faint – not visible (g)	V. faint (g)	Mod. clear (c)	Mod. clear (c)
30.5.57	27	Bare soil – light toned ramparts	V. faint (c)	Not photographed	V. faint – roads only (c)
1.8.57	65	Faint – ditches only (r)	V. faint (c)	Not visible (g)	Faint and mod. clear (c)
17.7.58	4	Clear (undersown clover?)	V. faint – ditches only (cut g)	Not visible (cut g)	Not visible (cut g)
25.7.62	91	Not visible (g)	Not visible (g)	Clear (c)	Not visible (g)
26.7.64	44	Not visible (g)	Not visible (g)	V. clear	Not visible (g)
29.7.66	72	(a) Bare soil – light toned ramparts (b) Not visible (g)	Not visible (g)	Not visible (g)	Mod. clear – roads only (r)
28.7.67	15	Faint – roads only (g)	V. faint – ditches only (g)	Not visible (g)	V. faint – roads only (g)
26.7.68	32	Not visible (g)	Not visible (g)	Not visible (g)	Not visible (g)
30.7.69	47	Not visible (g)	Not visible (g)	Not visible (cut g)	Not visible (g)
30.7.71	104	Not visible (g)	Not visible (g)	Not visible (g)	V. faint – roads only (g)
27.7.72	18	Faint (g)	Faint (g)	Not visible (g)	Not visible (g)

[a] V., very; Mod., moderately. F1–F4 are different points at the site.

systematic experiments have been carried out. It is difficult to distinguish different grain crops from their appearance in photographs made prior to the emergence of ears. Even then differences in black and white photographs are slight. Colour imagery is better. Some information can be gathered by examining pictures made during the ripening phase,

Fig. 3.20 The variation of PSMD over a given year (*a*)–(*g*) show different years, DP = date of photography. From ref. 26.

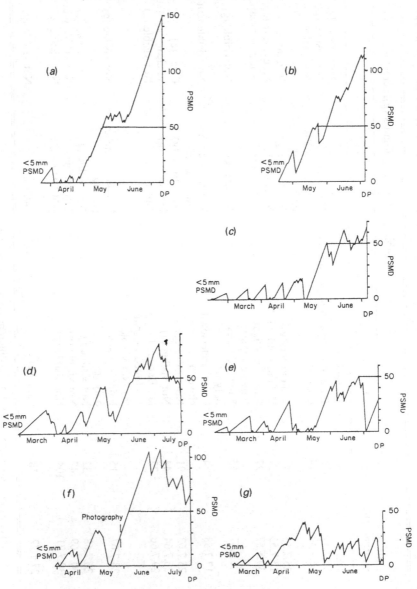

because the approximate ripening dates for each type of crop are fairly well known. Barley is usually first, then oats, followed by wheat and finally rye. The total reflectances of various types of grain crops are shown in figure 3.22 taken from data by Schmidt-Kraepelin and Schneider (59).

In these curves, the relative intensity of total reflected light which will affect a black and white film is shown from April to September. Different sorts of winter and spring grain, barley, wheat, oats and rye as well as potato, rapeseed, and beet were studied. For winter crops, planting was at the end of September. The bare fields reflect a great deal of light. In winter, the grain has grown a bit but reflection is still high. It falls abruptly in April when the plants revive and start producing the first significant amounts of chlorophyll. It is in this phase that one sees yellowing over buried walling due to late revival of the plants. Barley ripens first and this is shown by the rise in reflection for that crop during June. Green–yellow contrasts reach a peak toward the end of June in the test area. Wheat ripens next and contrast peaks in the second week of July. Rye reaches its peak during the third week of July.

Fig. 3.21 Variation in the leaf area index with the application of nitrogen fertiliser. From ref. 36.

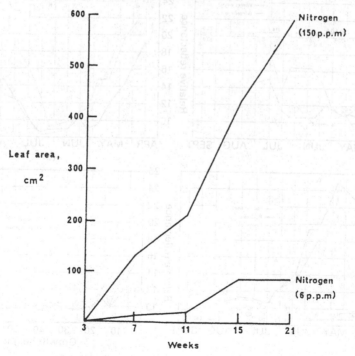

Curves for spring planted crops are similar, with some global displacement along the time axis. In the figure one sees the high reflectance of the bare fields in the middle of April. It falls uniformly for all three crops from the beginning of May to early June and the peak contrast for early yellowing over walling is probably in early May. June is likely to present a uniform dark green but height differences during growth may be preserved. The last two weeks of July produce the peak ripening contrast for barley, followed by the other two crops a week or two later. Jones and Evans note that barley has a larger leaf area index than wheat, as shown in figure 3.23 and that therefore the contrast between plants under stress and those with adequate water is greater than in wheat grown under the same conditions (36). It has not been possible to find similar data for oats. Its high sensitivity to buried features

Fig. 3.22 Total reflectivity of various grain crops in northern Europe: (*a*) winter barley, wheat, rye; (*b*) summer wheat, oats, barley; (*c*) fodder beet, potatoes, rapeseed; (*d*) natural grass, cultivated grass. (S = planting, G = first appearance of green colour, A = development of ears, B = bloom, R = ripening, E = harvest). From ref. 59.

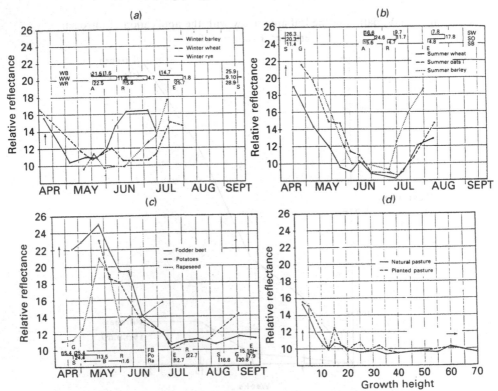

makes it probable that it also has a higher leaf area index than wheat.

Other data which were not available at the time of writing are the complete spectral responses for the four common grain types in spring and winter varieties in northern and western Europe over the whole of the growing season.

Root crops. Fodder and root crops like beet or potatoes follow quite different courses. The fields become progressively darker from mid-May onwards after the bare fields have been covered with leaves. Peak green is reached during the first week of July, and there is little change in reflectance until harvest. Potatoes usually yellow first, but the rise in reflectance is not as dramatic as with the grain crops. Sensitivity to moisture is much less, and it is only in times of great drought that archaeological features can be seen. Jones and Evans note that since the leaves of potatoes die at the time when the available water in shallow soils is exhausted after mid-July, contrast is reduced. Crop marks in beet require a very high SMD because of the water reserves in the plants themselves and the shading effect of their broad leaves. Sites have been found in the Rhineland and in Britain in late September and early October in unusually dry autumn weather.

Grass. Grass goes through intensive colour changes. It lies dormant in the winter, turning quite yellow. It revives in the spring, and walling very near the surface may retard new growth if there is little rainfall. Afterward, it is unusual to see any kind of crop site in grass. If allowed to grow to make hay, it is usually harvested in late May or early June in northern Europe. During the growth phase between 5 and 25 cm walls may be visible, but ditches rarely produce a response from grass

Fig. 3.23 The leaf area indices for various crops. From ref. 36.

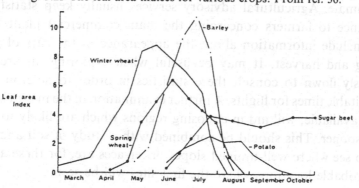

growing in permanent pasture. In a drought, grass may produce another response in the late summer from almost any type of feature, but this is a rare occurrence. Jones and Evans note that sites in grass are much less common than those in cereals and attribute this to the fact that the preponderance of grassland in Britain is planted in areas where the SMD seldom exceeds the critical 50 mm. They note that marks are produced in grass when the SMD is about 10 or 20 mm greater than for cereals in the same area.

Agricultural practice and spurious markings
The effect of agricultural treatment after growth has started may be considerable. Many artificial marks are produced by spraying for weed and insect control, and the novice may mistake some of these for archaeological features. Ploughing or spraying in a rectangular spiral pattern is also frequent, and this produces markings. Drains are often visible in low lying areas, and these in fact may have a favourable effect on the probability of a crop site. They are not readily confused with archaeological features either. Deforestation leaves considerable traces in the soil when trees are mechanically uprooted. Many pit-like structures which are visible from the air can be attributed to this cause. Irrigation with rotating or oscillating sprayers produces very characteristic and unmistakable markings which may obscure parts of archaeological sites in the fields so treated (chapter 4, figure 4.19).

Choosing optimal conditions and regions
Plots of SMD for a whole country like that published by Jones and Evans and reproduced here in figure 3.24 are not available for other parts of the world, but it seems likely that the general pattern in temperate regions will be similar, with differing dates and intensity, depending on local climate. Agricultural advisory services usually keep statistics of importance to farmers concerning the main commercial plant types. These include information about the appearance of the ears of grain, ripening and harvest. It may be useful when exploring an area not previously flown to consult these statistics in order to determine the most suitable times for flights. A further examination of the microclimatic data, if available, will aid in choosing regions which are likely to show SMDs sooner. This should be combined with a study of soil and relief maps to see where well-drained slopes or terraces are, for these are the most probable locations where the first sites will be seen.

The happy fact that not all crops ripen at once makes it possible to distribute the flying work load over most of the growth season and somewhat thereafter. It is also advisable to visit each promising region at least four times during the growth season to catch each of the grain crops at their peak. Unfortunately, due to the decline in the horse population, oats, which were mentioned by the earliest aerial archaeologists as being the most favourable crop have became rare in most areas. Barley grown for beer production has also declined somewhat.

Fig. 3.24 Variation of SMD over Britain in a given year. From ref. 26.

Most distressing, in northern Europe at least, has been the widespread introduction of maize which is able to ripen during the short summers. This crop shows little or no response to buried structures. The Common Market agricultural policy with regard to imported feed materials such as soya and maize which has allowed these to be sold at low prices has kept the maize acreage at a modest level except at times of very high dollar exchange rates. It seems unlikely that this will continue in the long run. Very deep rooted crops such as lucerne (alfalfa) have occasionally shown crop sites, but the acreage planted is small, and it is not known how sensitive this plant really is. Secondary crops of fodder grasses, clover, or turnips can be surprisingly effective, because the late summer is usually quite dry. Weed growth after harvesting of the main crop may also be productive in some areas.

3.3.7. *Kinds of soil and their effects*

In an early qualitative study (60) the dependence of crop sites on the kinds of soil in a given region was noted. Jones and Evans describe in detail the differences attributable to soils as a function of available water (36). They note that the main effect of different kinds of soil lies in the manner in which water is released. This is confirmed by the curves in figure 3.19. There is not much difference in the available water in the upper 50 cm in all soils in a given microclimatic region. In sandy soils, plants exhaust available water easily and abruptly, leading to sudden and severe deficits with a consequent rapid appearance of crop sites. On sites containing more clay, plants can extract water with a gradual reduction of growth rate, and crop marks develop more slowly. In all cases leaves are affected first, then stems and lastly roots. The variation in leaf area index is the cause of most crop marks due to moisture stress, prior to colour changes connected with permanent wilting. Although the critical 50 mm SMD is significant for the onset of crop marks, observation of a considerable number of sites by Jones and Evans show that natural crop sites of pedological origin are correlated with values almost twice as high or more. Most were on relatively shallow soils, less than 60 cm deep. At 36 archaeological sites where crop marks have been photographed, all the sites were on gravels of river or glacial origins, with soil depth less than 76 cm. From this, Jones and Evans contend that shallow soil depth is more significant than particle size distribution.

At all but a very few localities, crop marks did not appear until the

SMD was 40 mm or more. Marks were seen every year when the SMD was between 50 and 65 mm. The total range of SMDs was 40–295 mm, but two thirds of all pictures were taken when the SMD was between 50 and 150 mm. Only 20% were recorded at the higher 150–200 mm, and were faint at still higher SMDs. This, Jones and Evans attribute to the fact that the whole crop was under moisture stress, even over the archaeological features. They note that the patterns showing the most detail were photographed at 150 mm. This explains why only the early phases of a very dry year like 1976 were highly productive of crop sites over almost all of northern Europe.

3.3.8. Spectral properties of vegetation

The spectral reflectance properties of crops at various wavelengths is of importance for the choice of photographic materials to obtain maximum contrast in the resulting images. For agricultural crops, the most exhaustive treatment is due to Gausman and coworkers (28,29). This data covers many plants which are not grown in northern Europe, and some crops which are of interest in this area are not represented. Reflection is specular at strongly absorbing wavelengths in the visible region, Lambertian at wavelengths greater than 750 mm, and non-Lambertian in the far infra-red. The mechanism in the visible region is primarily due to chlorophyll absorption, with a rise at 550 nm in the green. In the near infra-red, from 750 to 1350 nm, reflectances are affected by multiple scattering in the leaf structure. A number of studies, typified by that of Ripple who gives an extensive bibliography (56), demonstrate convincingly the correlation between leaf water content and radiation in this range. The phenomenon has been exploited by Rigaud and Hersé (53), using black and white infra-red film which is sensitive only to the wavelengths between 730 and 930 nm to obtain images of buried stone structures under plant cover which had been subject to a prolonged drought followed by abundant rainfall in mid-November.

The wavelengths beyond 930 nm have not been exploited in archae-ological aerial photography because, as noted elsewhere, suitable scanning equipment is not readily available and films cannot be used. Most of the data in the remote sensing literature concerns these regions. Plant spectral behaviour in the visible is typified by the curves of figure 3.25. The main sources of colour are the chlorophylls which absorb about 70–90% of 450 nm blue and 690 nm red light. They mask xanthophylls, carotenes and anthocyanin. Absorption is smallest at 550 nm, with a reflection peak of 20% or less resulting. When chlorophyll production declines either upon maturity or because of moisture stress, the other dye substances become visible, as shown in figure 3.26. The phenomenon is more pronounced in the near infra-red for some genera than for others. The rise from 40 to 60% reflectance shown in figure 3.27 is not very different in relative terms to the decline in visible green from 15 to 10% or a bit less. According to Myers (49) Russian studies have shown that the red and blue parts of the spectra have similar values for 80 different plant species studied. Reflectance measurements, he notes, have been used to follow changes in leaf chlorophyll content and that carotenes and xanthophylls were used to estimate dry biomass. Reflectance is also a function of the angle of incidence of light, with a complex interplay of spectral and Lambertian properties. This accounts for the frequent observation of an optimum viewpoint for the recording of a crop site relative to the angle of illumination.

Damage to plants from insects and fungus infection has been monitored for many years by the use of photography in the near infra-red using false colour film. Comparisons between near infra-red and visible red have been used in evaluating crop biomass from satellite data, a matter of considerable economic and political significance.

Polarisation measurements reported by Egan (25) show considerable

Fig. 3.25 Plant spectral behaviour in the visible and near infra-red regions. From ref. 49.

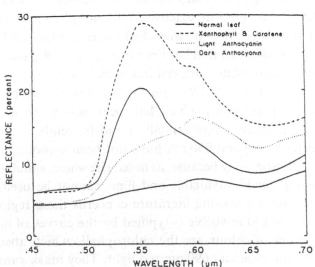

differences for a number of samples of interest. Polarisation is also linked to moisture content at shorter wavelengths. This phenomenon has not been exploited consciously in archaeological aerial photography, but the use of polarising filters with colour film to reduce haze is almost universal, and if properly directed, may actually enhance crop site contrast.

3.4. Cameras and films

3.4.1. Historical note

The pioneering work of Major Allen in southern Britain was carried out with a homemade camera using a 4 × 5 in cut film (4). Crawford and Keiller used similar equipment (18). The use of large format (4 × 5 or 5 × 5 in) black and white films continued after the Second World War, encouraged by the ready availability of cheap war surplus hand-held cameras like the American K20 and K24 and the British F24. Only

Fig. 3.26 Decline of chlorophyll production at maturity or under moisture stress and its effect on the spectral behaviour in the visible and near infra-red regions. From ref. 49.

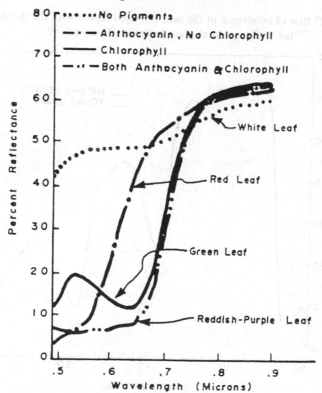

Williamson in Britain with the F117, Fairchild in America with the F505, and Linhof (Aerotechnika) in Germany produced hand-held, large format, aerial cameras afterward, and these at prices out of reach for all but a very few archaeological institutions.

In part, the use of war surplus material made a virtue of necessity, for the early 6 × 6 cm and 35 mm format cameras had viewfinders which were highly inconvenient for use in the air and the quality of film left much to be desired. Probably, the introduction of good quality, single lens, reflex 35 mm cameras in the 1960s did much to encourage workers with modest budgets to adopt this format. Not surprisingly, the high resolution and contrast of the early Kodachrome film led to its adoption by most who favoured this technique for want of better equipment. The perfection of 6 × 6 cm format cameras some of which could be fitted with 70 mm film cassettes and motor drives in the late 1960s and early 1970s, as well as limited production of pure 70 mm cameras, produced higher image quality, although the technology of 35 mm improved considerably.

Fig. 3.27 Rise of reflectance in the near infra-red region with the decline in the visible green at plant maturity. From ref. 49.

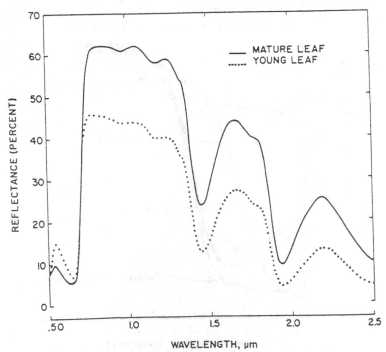

Although the vast bulk of aerial archaeological imagery has been produced with small formats, the growing necessity for accurate site mapping has led to the acquisition of used professional photogrammetric equipment of large 23 × 23 cm size in several instances. This type of camera requires a larger aircraft than that usually used, one especially equipped for the purpose. Such a course of action is only possible for organisations with good financial backing and full time personnel. The approach has the advantage that normal photogrammetric methods can be used afterwards. It has the disadvantage that oblique pictures must be made anyway to obtain the required high contrast, and frequent time consuming changes in flight altitude may be required, wasting precious moments if crop or soil conditions are in a critical phase. An interesting compromise which has seen considerable interest in the non-mapping remote sensing community, is vertical mounting of one or more electrically driven small format cameras between the wing strut and door of a high winged monoplane (57). Images thus acquired are usually vertical enough for conventional photogrammetry, and calibrated cameras can be used. Costs are within the reach of almost any group.

3.4.2. Lenses and filters

Modern lens design is a complex computational compromise. The designer has a large number of variables at his disposal. The number of glass elements making up the lens, their spacings, front and back curvatures and refractive indices at all spectral wavelengths of interest must be chosen to give optimum results commensurate with the constraints of the planned cost of production and testing. Lenses made for non-photogrammetric cameras must be capable of cheap mass production, but nonetheless give subjectively satisfactory results in terms of apparent image sharpness at the picture taking distances which are normally of interest to amateur photographers. When such lenses are used for archaeological aerial photography, the conditions which prevail are not those for which the lens has been designed and the performance may be less than that desired. This is especially true if quantitative evaluation of the images is subsequently required. Lenses for modern professional aerial cameras are optimised for quite different conditions. Focus is optimised for planned flying altitudes, and the variation of air density with height taken into account. Old aerial cameras, especially those used in the pioneer days of archaeological aerial photography, had lenses which were little different from those available for terrestrial photography, and correction for various aberrations and distortion was poor. This can be readily seen if one examines an original negative of the time.

Nonetheless, if high quality, modern lenses for small format cameras are carefully chosen, aerial performance can be adequate for reasonably accurate mapping of archaeological sites. Image sharpness with hand-held cameras depends not only on lens performance, but also on shutter type and speed relative to the flight altitude and choice of film. A modern lens with a focal length between one and two times the size of the image diagonal, when operated at two or three F stops larger than maximum opening, will not usually be the major factor in limiting overall sharpness. A detailed discussion of the various factors involved is given in chapter 4.

It is not possible to discuss the theory of optics and its relation to photographic lens construction here. A good survey can be found in Baker (6) and in Gruner (31). These authors also give extensive bibliographies for further reading and a history of the development of the subject. A survey of many lens types and some details of practical lens construction can be found in Dejager (23) accompanied by a useful discussion of the spectral properties of filters. Practical properties of commercially available filters are given in ref. 39. The choice of useful filters depends on the type of film used, the spectral signature of the features which are to be photographed relative to their surroundings and to some extent on the time and season of picture taking. Some useful suggestions are made by Slater *et al.* (62).

Resolution
The two aspects of lens performance which are of principal concern to users of archaeological aerial photography are resolution or sharpness and the geometric distortion of the image relative to an ideal projection. Modern analysis of sharpness uses Fourier transform theory as discussed in chapter 4. A good non-mathematical survey of the methods applied to lenses is given by Brock (10,11). Under the highly simplifying assumption that a lens is radially symmetrical, the modulation transfer function (MTF) in cycles per millimetre describes the sharpness with a two-dimensional graph. For a typical small format camera lens, this function is shown in figure 3.28. It can readily be seen that for the lens in this example, optimum sharpness is achieved at about three F stops greater than that of maximum opening. Performance at the centre of the image is much better than away from it as shown in figure 3.29. At

only 15° the sharpness is less than half of that at the image centre. This
drop in sharpness away from the centre is less for long focal length
lenses, as shown in figure 3.30. However, such lenses are less useful for
oblique aerial photography destined for later photogrammetric evaluation,
since the images show a much smaller area of terrain for a given flight
altitude. The consequence is that scarce geometric control points, as
discussed in chapter 5, are rarer still. Although the archaeological features
are more visible because they fill the image frame, they are more difficult
to transform into maps afterwards.

Geometric distortion

When an archaeological aerial photograph is subjectively examined, the
observer is usually unaware of any geometric distortion in the picture.
However, lenses do not produce an ideal perspective transformation at
the film plane of a camera. Simple Seidel lens theory predicts two types
of distortion which are of third order which lead to outward and inward
displacement of an image point as shown in figure 3.31, called pin-
cushion and barrel distortion. These higher order effects result from an
expansion of the sine in the refraction law:

$$\sin(i) = i - \frac{i^3}{3!} + \frac{i^5}{5!} - \frac{i^7}{7!} + \cdots \tag{3.20}$$

Fig. 3.28 The MTF of a small format camera lens as a function of lens
opening (55 mm f1.2 lens 24 × 36 mm format). From ref. 11.

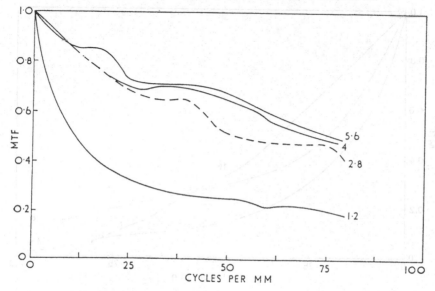

Aerial photography

where *i* is the angle of incidence. The first order terms are used in Newtonian optics, and the second term in the expansion gives the Seidel analysis of the aberrations of a lens system. The higher fifth and seventh order terms are used in the Schwarzchild theory. Pin-cushion distortion

Fig. 3.29 MTFs on and off the lens axis (50 mm f2 lens, 24 × 36 mm format; T = tangential, R = radial). From ref. 11.

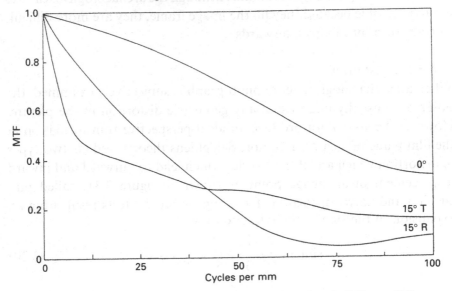

Fig. 3.30 Off-axis sharpness as a function of focal length (280 mm f4.8, 24 × 36 mm format). From ref. 11.

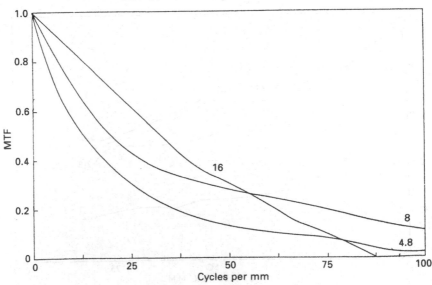

is a positive function of the third power of the field angle from the optical axis of the system, barrel distortion a negative third power function. A good lens design attempts to balance the inward and outward displacements so that there is zero displacement at one or more points across the image surface. Other commercial considerations in lens design for non-photogrammetric purposes usually attach less importance to these conditions, and this kind of distortion which is radially symmetric is of considerable magnitude in most small format cameras. A typical distortion curve for such a camera is shown in figure 3.32 taken from Hatzopoulos (33). It can be seen that the distortion at 10 mm from the image centre reaches almost 60 μm for this typical lens focused on a test target at a distance of 1 m. Distortion at large distances is similar. The image scale in a vertical image is the focal length divided by the flight altitude, so that for the nominal 50 mm lens shown at a typical flight altitude of 500 m the scale at the centre of the image plane is 1:10000. Therefore 60 μm represents a distance error on the ground of 0·6 m, a figure which is probably acceptable for most archaeological photogrammetric purposes provided that the data from the corners of the image where distortion rises to much higher values are not used.

A secondary type of distortion is due to the lens not being perfectly centred and perhaps somewhat tilted relative to the image plane. This type of distortion can be very serious in cameras which do not have rigid bodies, or in older single lens reflex cameras where the lens was deliberately displaced in order to allow room for mirror swing. This distortion is tangential, and in combination with radial distortion, produces the irregular image shape shown in figure 3.33.

Fig. 3.31 (*a*) Pin-cushion and (*b*) barrel distortion of an image. From ref. 6.

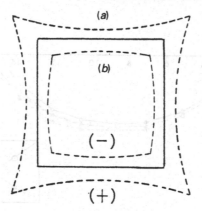

Calibration

It is possible to compensate for considerable image distortion by calibrating the camera. Many modern 35 mm single lens reflex cameras and most 6 × 6 cm cameras can be calibrated to give astonishingly good photogrammetric results, so that the effects of radial distortion on the image geometry are reduced to values insignificant for most archaeological purposes. Older cameras, especially the large format Second World War surplus models which were the mainstay of aerial archaeology in Britain and Germany up to the end of the 1960s, had very considerable radial and tangential distortion at the image plane. Most of these historical relics were unfortunately not preserved by their owners, so that *ex post facto* calibration is no longer possible. But if they are available, then calibration to determine distortion, focal length and principal point should be carried out to enable older films made with them to be used with higher accuracy.

Several calibration methods are used in conventional photogrammetry. Camera manufacturers and standards laboratories use precision goniometers or arrays of collimators to measure lens distortion. A survey of these techniques is given by Livingston *et al.* (42).

For camera systems with lens, body, and shutter assembled into a

Fig. 3.32 Radially symmetric distortion in a small format camera lens. From ref. 33.

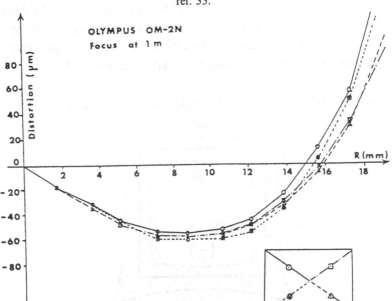

unit, a field of objects which are in known positions is photographed and the image(s) measured. For close range imagery, beads strung on wires attached to plumb bobs in oil baths are used as target points whose positions are accurately measured with a precision theodolite (72). These targets give good results on small and medium format cameras at short distances. For aerial survey cameras, highly visible targets can be laid out on the ground at accurately measured points. Some photogrammetric research institutions have constructed such test sites and it may be possible to obtain the measurement data from them. Alternatively, the stars can be photographed on a clear night. With correction for atmospheric refraction this gives quite good calibration over the entire

Fig. 3.33 Combined tangential and radial image distortion. From ref. 6.

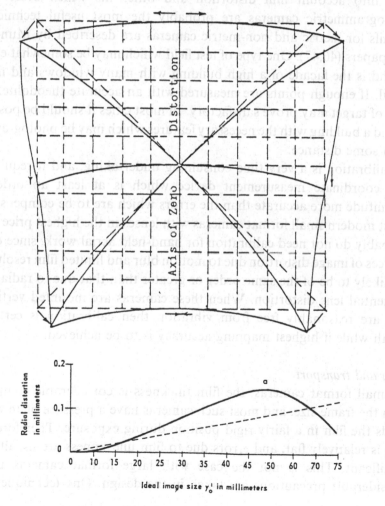

Radial distortion in millimeters

Ideal image size y_o' in millimeters

image. The calculations required are described by Case (14). If accurate equipment is available for the measurement of the star images on the films, this is probably one of the simplest methods which can be used for the cameras typical of archaeological aerial photography. Another possibility is self-calibration using a large number of images containing many points with known coordinates. This approach, simultaneous multiframe analytical calibration (12), it not suitable for hand-held oblique imagery because the number and positional quality of the known image points is unlikely to be good enough. It is definitely usable for star images, especially if the camera is rotated through a number of known principal positions to obtain the multiple images. If good target fields like stars are available, then finite element methods which also take into account film distortion and other ills which beset non-photogrammetric cameras are probably the most useful techniques. Details for metric and non-metric cameras are described by Munjy in two papers (46,47). One type of test field which may be somewhat easier to find is the facade of a high building with many windows and much detail. If enough points are measured with an accurate theodolite, this type of target may prove satisfactory. In most cities it should be possible to find a building with the necessary features which may be photographed from some distance.

Calibration is a very time consuming undertaking, and it requires a film coordinate measurement device which is at least an order of magnitude more accurate than the errors which are to be compensated. Most modern small format cameras with lenses in the highest price class probably do not need calibration for hand-held aerial work, since other sources of image distortion due to motion blur and limited film resolution are likely to be of the same order or greater than that due to radial and tangential lens distortion. When these cameras are mounted vertically and are reasonably free from vibration then calibration is certainly worth while if highest mapping accuracy is to be achieved.

Film and transport
In small format cameras, the film thickness is considerable compared with the frame size, and most such cameras have a pressure plate which holds the film in a fairly rigid position during exposure. Therefore, the film is relatively flat, and errors due to film unevenness are usually not significant. This is not the case with large format cameras unless considerable precautions are taken in the design. One technique is to

use an optically flat glass plate against which the film is pressed by a rigid back. Reseau marks at accurately known positions inscribed on the glass allow compensation during photogrammetric evaluation for thermal changes in the film after development. A similar mark for the optical axis (principal point) of the system is also helpful. When such a glass plate is used, the lens design must take its distance, thickness and refractive index into account or optimum image sharpness is not obtained. This was not the case for several of the popular war surplus cameras which had such plates. Large format vertical mapping cameras have elaborate vacuum backs which hold the film by suction to an accurately machined surface. They also produce reseau markings by projection of light through small holes in the pressure plateau on to the film.

Film which is transported across a glass plate system is often scratched by the inevitable adhering dirt particles, and the static electricity produced by the friction across the plate sometimes produces lightning-like streaks on the images. But many small format cameras also produce scratches due to film transport, usually because their owners neglect the cleaning of the frame and back pressure plate which is necessary from time to time.

Shutters

The most appropriate shutter for aerial photography is the between-the-lens central type. This shutter exposes the entire image at once. The Williamson F117 which was designed in the 1960s for hand use had a rotating central shutter of very high efficiency. The blade shutters used in some 6 × 6 cm cameras do not have as high a light efficiency, but are quite satisfactory.

Most small format 35 mm cameras have focal plane shutters and this is especially true of those in the higher price range for which really good lenses are available. Unless the shutter speed is quite high, the time taken to traverse the film will not be small relative to the displacement of the image on the film plane due to movement of the aircraft, and a type of distortion results which is quite difficult to eliminate by calculation unless a very large number of known image and ground points are visible in the frame. Some of the Second World War surplus cameras, notably the British F24 and the American K24 also had slow focal plane shutters, as did the Fairchild F505 produced for the Vietnam War by the US. Images produced with these cameras may show larger residual errors when used with the photogrammetric mapping techniques described in chapter 5, especially when the pictures were made at slow shutter speeds under poor light conditions. It is usually worthwhile, as described in chapter 4, to use a high speed film having a somewhat lower resolution in order to be able to use the highest possible shutter speed for a given lens opening and light level.

3.5. Photographic film as a detector and information carrier
3.5.1. Introduction

Photographic films are familiar, stable and readily available commodities. Even when using them for scientific purposes, few users know or are aware of the immense scientific and engineering effort which has gone into their development in over 100 years of progress. In this book, the discussion of the physics of each prospecting technique also includes a discussion of the properties of the detectors used for the physical phenomenon measured. The physics of photographic materials, their engineering, calibration, and detailed properties are subjects with such a vast literature that a detailed treatment is not possible here.

Hence to keep this chapter within reasonable proportions compared with the others, the discussion will be highly compressed. Readers who wish to know more must be prepared for a considerable reading programme. The best introductions, with a thorough survey of the literature, are the many editions of C. E. K. Mees and T. H. James monumental *The Theory of the Photographic Process*, which cover all aspects of the subject except the actual engineering techniques involved in film production, and include exhaustive bibliographies (44). Revisions at intervals of roughly a decade have appeared since the first edition in 1942. The second edition contains a very complete review of the history of photographic science and many details of practical importance whereas the third and fourth editions are more oriented toward the fundamental physics and chemistry of the matter.

Survey volumes which are directed toward the practical needs of the user rather than the producer of photographic materials are the *SPSE Handbook of Photographic Science and Engineering* edited by W. Thomas (70) and *Neblette's Handbook of Photography and Reprography Materials, Processes and Systems* edited by J. M. Sturge (67), a perennially revised cookbook.

Modern methods of evaluation of films and optical systems as detectors in aerial photography are given in two standard works by G. C. Brock, *The Physical Aspects of Aerial Photography* (10) and *Image Evaluation for Aerial Photography* (11) and in J. C. Dainty and R. Shaw's *Image Science* (22). A good survey of photographic materials for photogrammetric applications is given in chapter 6 of the fourth edition of the *Manual of Photogrammetry* edited by R. G. McKinney (43) and in chapter 6 of the second edition of the *Manual of Remote Sensing* (62). Both of these also include copious bibliographies with references up to the early 1980s. A useful short guide describing current materials available from one manufacturer is the *Kodak Data for Aerial Photography* (40) which appears in a new edition at irregular intervals when products become available. The materials described there and in supplemental reference booklets published under such various titles as *Kodak Aerial Films and Photographic Plates* or *Physical Properties of Kodak Aerial Films* or *Physical and Chemical Behavior of Kodak Aerial Films* (41) can be considered to be the current state of the art, and more importantly still, most of them can be readily obtained.

Although the fundamental physics and chemistry of film processes are well understood and adequately published, constant improvement in film engineering results in change which makes any discussion of a specific material outdated before a book is printed. The materials can, however, be classified into a number of types which have remained available on the market for several decades. Their general properties and applications will be treated in the following sections.

3.5.2. Physics, chemistry, and engineering of films

Mechanical construction of film

A photographic film in its simplest form, as shown in figure 3.34 consists of a transparent plastic supporting medium (the film base), a gelatin–silver halide emulsion coating which will produce the image when exposed to light and chemically developed afterward, and in many instances a backing coating which helps to prevent curling by compensating for the different thermal and humidity properties of the emulsion and base. The backing coating is sometimes tinted to provide light absorption (antihalo backing), so that the reflection of light from the lower base surface back into the emulsion is reduced. The dyes in this coating are usually soluble in the processing chemicals and hence they are not visible in the final image. In amateur black and white films, the base material is often permanently dyed to reduce reflection from the rear film surface. In professional aerial films, an additional hard coating is usually placed in front of the emulsion to prevent scratches during film transport in the camera and during processing. Colour films have a large number of coatings. For example, in one particular commercial product 18 are used. Details of the engineering behind such complex structures are not available in the open literature.

Silver halide crystals

The peculiar properties of the silver halides (chloride, bromide and iodide) when exposed to light were discovered in the first half of the nineteenth century. They have remained the basis for conventional photography ever since. The physics of the phenomena were first explained after the development of quantum mechanics and advances in knowledge of the solid state. In contrast with the other types of detectors used in archaeological prospecting, photography is a hybrid mixture of physics, chemistry, and engineering, with a little bit of cookery as well. Silver halides used in photographic emulsions are solid solutions of two or three of the halides which form microcrystals of cubic form, figure 3.35. These are suspended in a polymer gelatin matrix. The microcrystals are precipitated in the liquid state. Details of the manufacture of modern emulsions are closely guarded trade secrets, but generally the crystals are precipitated by mixing solutions of an alkali or ammonium halide with silver nitrate in a gelatin solution which creates an emulsion. The properties of the crystals and hence the film depend critically on

Fig. 3.34 A simplified cross-section through a photographic film. From ref. 40.

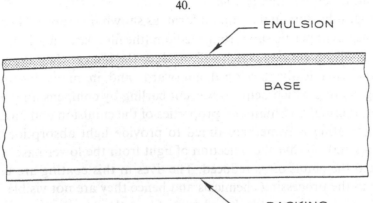

the amounts of the different halides, the speed of mixing, the temperature, the properties of the gelatin used and the treatment which follows the mixing phase. Considerable detail can be found in the second chapter of Mees and James (44). The smallest grains are of the order of 5 nm, the largest 50 μm, although 5 μm is the maximum size usually found in commercial films.

Silver halide crystals have high ionic conductivity and the micro-crystal gelatin polymer has quite remarkable surface properties as well. Dislocations and grain boundaries are present over considerable distances in the crystals, and these allow electron trapping with the formation of metallic silver in deformation regions. The absorption of light energy leads to the production of free electrons and holes and to excited electronic states in the crystal. The quantum mechanical explanation for the processes and their significance are described in chapter 1 of Mees and James. Optical absorption and production of excited states is primarily in the blue and ultra-violet wavelength regions for mixtures of silver halides, but the spectral response of photographic emulsions is extended to longer wavelength regions by the addition of special dyes.

Fig. 3.35 Silver halide crystal mixtures of cubic form. From ref. 44.

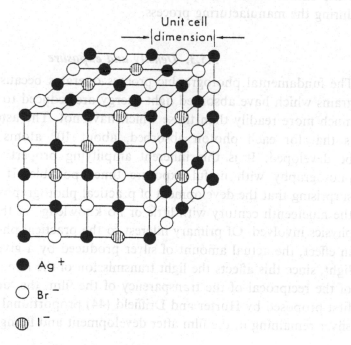

Unit cell
|← dimension →|

● Ag$^+$

○ Br$^-$

◍ I$^-$

Silver halides grains

During manufacture, recrystallisation and recombination permit the production of clumps of crystals of a desired particle size distribution. The particle size–frequency distribution in the emulsion has a great effect on the light sensitivity (the speed) of the film and on its resolution. The grain size–frequency distribution, usually measured by scanning electron microscope techniques, follows a log–normal distribution:

$$\left. \begin{array}{ll} \phi(x) = 0 & x \leqslant \alpha \\[2mm] \phi(x) = \dfrac{1}{(x-\alpha)\sigma(2\pi)^{\frac{1}{2}}} \exp\left\{ -\dfrac{1}{2}\left[\dfrac{\log_e(x-\alpha)-\mu}{\sigma} \right]^2 \right\} & x > \alpha \end{array} \right\} \quad (3.21)$$

where x is the particle size, μ is the average particle size, σ is the standard deviation, and α is a constant. The grain size–frequency distribution for emulsions is shown in figure 3.36. For emulsions which adhere to the log–normal distribution rule, the logarithm of the cumulative grain size–frequency and the logarithm of the grain size are related linearly. Practical emulsions do not obey this rule exactly, having distributions which are made up of a number of straight line regions as shown in figure 3.37. These are from various components of the mixtures which are due to the different rate constants, recrystallisation, twinning and other effects which take place during the growth of the microcrystals during the manufacturing process.

3.5.3. *Density and exposure*

The fundamental photographic process operates because silver halide grains which have absorbed light energy are reduced to metallic silver much more readily than those which have not. The astonishing thing is that for each photon absorbed, about 10^8 atoms of silver can be developed. It is this inherent amplifying property which makes photography with useful exposure times possible. It is even more surprising that the development of practical photography took place in the nineteenth century with little or no knowledge of the fundamental physics involved. Of primary interest to the practical photographer, is, in effect, the actual amount of silver produced by a given exposure to light, since this affects the light transmission of the film. The logarithm of the reciprocal of the transparency of the film, the 'density' D is, as first proposed by Hurter and Driffield (44) proportional to the mass of silver remaining in the film after development and fixing, although this is to some extent a function of the grain size distribution. Since most commercial emulsions have a considerable spread in grain size distribution, the proportionality with respect to silver mass is not strictly constant. Since all films have a finite emulsion thickness, absorption and scattering of light by the upper layers produces a more complicated situation.

Fig. 3.36 The grain size–frequency distribution for a number of simple film emulsions showing log–normal distribution. From ref. 44.

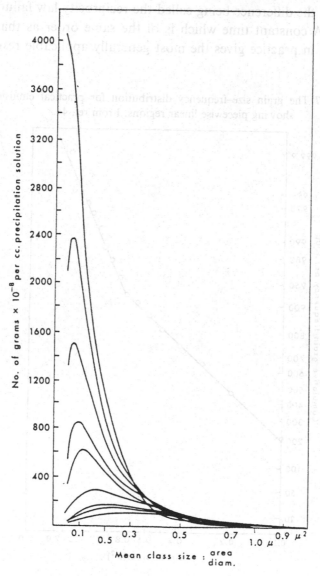

   The amount of light received is usually measured in metre candle seconds, although it might equally well be expressed in the number of quanta. This quantity is the 'exposure' E. By exposing sections of a strip of film for different times and plotting D against E, the classical Hurter and Driffield (H + D) curve, figure 3.38 is produced. It is more convenient to obtain this curve by exposing the film for a constant time through a tapered glass wedge which contains an absorbing dye. This modulates the intensity rather than the time of the exposing light. The curves obtained in this way do not agree with those made by varying the time of exposure, the difference being called the reciprocity law failure of the given film. A constant time which is of the same order as that which will be used in practice gives the most generally applicable result. The

Fig. 3.37 The grain size–frequency distribution for practical emulsions showing piecewise linear regions. From ref. 44.

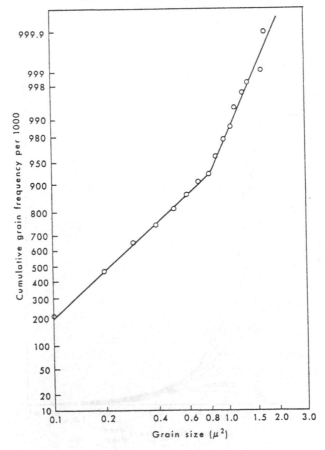

straight line portion of the log–log H + D curve is usually the region of correct exposure E:

$$D = \gamma(\log_{10} E - \log_{10} i) \tag{3.22}$$

in which i is the exposure when the projection of the straight line part of the curve reaches zero ('inertia'). The slope of the curve, γ, is the standard measure of contrast. The range of light values which can be recorded on the straight line part of the curve is the 'latitude' of the film. For materials used in low level oblique aerial photography, this must usually be quite large, while films for high altitude photography are usually chosen with higher contrast to overcome the effects of haze and to have less latitude. In general, the higher contrast films have a smaller spread in grain size distribution and smaller grains.

Spectral response

Silver halide emulsions inherently respond to lighter or shorter wavelengths only. The limit is 420 nm for silver chloride, 470 nm for silver bromide and 510 nm for a mixture of silver bromide and iodide. This response is shown in figure 3.39. Addition of dyes to the emulsion extends the sensitivity as shown in the figure. This is called spectral sensitisation, a process which was discovered by chance in 1873 by Vogel. Progressive improvements through research into sensitising dyes have led to an

Fig. 3.38 The Hurter and Driffield curve. From ref. 44.

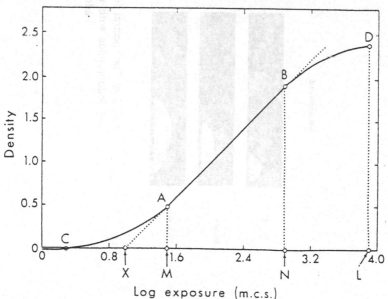

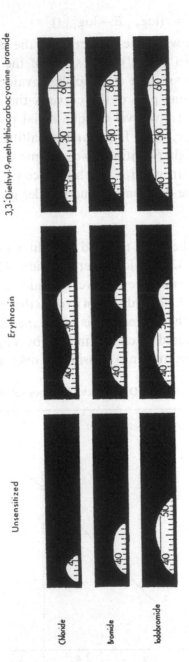

Fig. 3.39 The spectral response of a mixture of silver chloride, bromide, and iodobromide with the extension of spectral sensitivity through the addition of dyes. From ref. 44.

extension of spectral response well into the near infra-red region. The sensitisation by dyes is attributable to the absorption of energy or electrons by dye molecules on the surface of the silver grains and the transfer of this energy or electrons from the excited dye molecule to the silver halide, an extremely complex and only partially understood process. For a detailed discussion see Mees and James, chapters 11 and 12 (44). Grossly put, the dye absorbs a light quantum, undergoes a transition from its ground state to an excited singlet state which is at an energy level comparable with the conduction band of silver bromide. The excited electron passes from the dye to the crystal where it is trapped and acts just as if the crystal had absorbed a light quantum directly.

The spectral sensitivity of a typical medium speed aerial black and white film is shown in figure 3.40. It is reasonably uniform over the visible spectrum from 400 to 700 nm. Usually such a film will be used with a yellow filter which cuts off all wavelengths below 500 nm. Black and white infra-red sensitive materials have a response as shown in figure 3.41 which extends out to about 925 nm. This film is usually used with a deep red filter which blocks all light below 690 nm, so that the response is essentially restricted to the near infra-red only.

Development

The excited states of the silver grains are only the preliminary to the

Fig. 3.40 The spectral sensitivity of a typical medium speed aerial black and white film. From ref. 40.

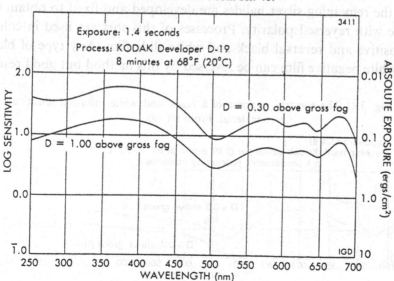

formation of the photographic image. The image itself is produced by bathing the emulsion in a solution which is a reducing agent for silver halide to metallic silver. The excited grains affect the rate of the reaction by decreasing the height of the energy barrier for the reduction process, and the difference in developed silver in exposed and unexposed areas depends on the difference in reduction rate. This phenomenon can be explained as a result of the probability of the transfer of an electron from a developing agent molecule to a silver ion in the crystal. The process is autocatalytic, that is, the transfer gives rise to a further exchange of a second electron adjacent to the excited site and so on. The result is that the entire grain is reduced to metallic silver within a short time.

Reversal

With normal negative emulsions, exposure to light produces a greater density of silver and a corresponding negative image. However, by developing a film but not fixing it, a positive image can be created. Fixing is the removal of silver halides, which have not absorbed light quanta, by means of an ammonium or sodium thiosulphate bath. In reversal processing the film is given a second exposure to light after the first development has reduced the primarily exposed grains to metallic silver. This silver partially prevents light from reaching the silver halides which were not initially exposed. Usually a bleaching agent is used after the second exposure to remove the metallic silver produced by the first, then the remaining silver halides are developed and fixed to obtain the image with reversed polarity. Processes of this sort are used in colour diapositive and reversal black and white slide film. Any type of black and white negative film can be reversed by this method but good results

Fig. 3.41 The spectral response of a black and white infra-red sensitive material. From ref. 40.

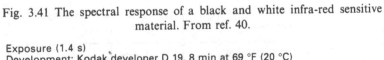

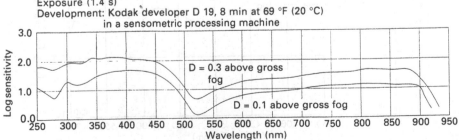

are usually obtained only with films which were especially designed for the purpose.

Tone reproduction

The reproduction of a scene by means of a photographic film can be described as a function of two spatial variables, with the value of the function related to the amount of light reflected toward the camera at a given point. With a black and white film which registers light over a wide spectral range, this is the sum of all of the spectral components. In a colour film which, as described below, uses layers sensitive to red, green, and blue light, each layer records light over a narrower band. Ideally, the response at each point follows the H + D curve of figure 3.38. The response of a grain of a given size follows a normal probability distribution relative to the number of quanta of light received. Large grains are more likely to respond to fewer photons because of their greater cross-section, larger number of dislocations and the higher probability of interception of a photon, whereas the smaller grains respond at higher light levels. With a well-constructed grain size distribution, the manufacturer can make a film which has a nearly linear response over a very wide range of light levels. However, the price paid is the presence of large grains which affect the image structure.

In the linear part of the curve, the density of the developed image is proportional to the logarithm of the exposure, and $\Delta D/D \log_{10} E$ is the average gradient G, for the interval $\Delta \log_{10} E$. The gradient of the curve is not constant outside the linear region. The tangent of the angle of the linear part of the curve to the abscissa was christened γ by Hurter and Driffield, and in this region it is a constant. When it is unity, there is a one to one reproduction of exposure and density. The slope of the curve is strongly affected by the time, temperature and the concentration of the developing solutions as shown in figure 3.42. The relationship is roughly exponential with $\gamma = \gamma_\infty(1 - \exp(-kt))$ where t is the time of development, and differs from developer to developer, as shown in figure 3.43, and k depends on the material, the developer and the temperature (44).

For any given material, there is a maximum γ which can be reached in prolonged development and which is again characteristic of the grain size distribution present. Rendition of dark areas of the scene is usually best judged subjectively when the darkest portion of interest lies at the point on the curve where the slope is about 0.3 of the slope in the linear

Fig. 3.42 The slope of the H+D curve as affected by temperature and the time of development. From ref. 44.

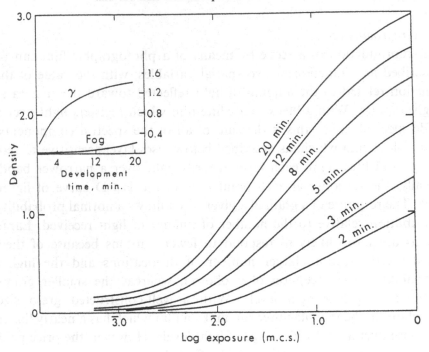

Fig. 3.43 The slope of the H+D curve as affected by different developers. From ref. 44.

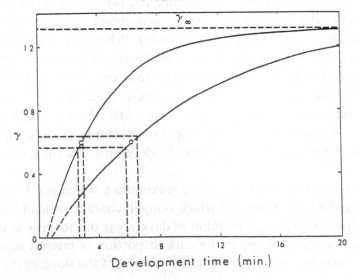

range and this criterion is used as definition of the 'speed' of a film as the reciprocal of the exposure required to obtain 30% of the average gradient.

It is customary to look at photographs on paper, not on film. The characteristics of photographic paper emulsions are quite different from those of films. The maximum whiteness which can be obtained depends on the reflective qualities of the paper and its treatment, and the maximum blackness on the highest value of density which can be reached. This also depends on the surface, with diffuse reflection density having values of 1.3–1.7 for most materials. A glossy surface paper gives a higher value of maximum black than a matt surface one. The available density values are much less than a film can reproduce. Hence, to produce a pleasing print, it is customary to develop the film to a γ less than unity so that the limitations on the density range of the paper are compensated by an equally limited density range in the film.

The subjective impression of brightness in an image at a given point depends on the brightness of the surrounding local area. Thus, an interpretation of the quality of reproduction based on density measurements alone will not be perceived as correct. The conditions of observation also affect matters strongly. There is an enormous difference in the observed tone reproduction of paper prints and that of diapositives viewed in a darkened room. Furthermore, the light levels in an outdoor scene are about a 100 times that of the indoor conditions under which the image is viewed. Under these conditions, the light reflected from the white parts of the image is about the same order of magnitude as that received by the negative in the darkest portions of the scene! And yet the observer sees white, not black because the eye compensates for low light levels over an enormous range. Matters are still more complex when colour or moving images are involved. In many instances of interest in this book, the image is viewed not on paper or film, but on a cathode ray tube which has its own set of characteristics as an active light source which are quite different from that of film. The theory of tone reproduction in films under many different possible conditions is covered in detail in chapter 22 of Mees and James. For subjective properties of cathode ray tube display see ref. 27.

The objective reproduction of luminance is much easier to describe mathematically. Luminances in scenes can be measured with a telescopic photometer constructed so that the amount of stray light received is small. In a typical outdoor scene, the distribution of luminances can be reasonably well fitted with a log–normal curve as shown in figure 3.44. The illuminance in a small area of film resulting from such luminances has two components, one from the part of the scene being imaged, I_0, the other from stray light reflected and scattered from parts of the camera and the lens surfaces, I_f, so that $I = I_0 + I_f$. For parts of the scene near the lens axis and at a great distance, the relation between luminance L and illuminance is given by:

$$I_0 = LT/(4f^2) \qquad (3.23)$$

where f is the F stop of the lens used and T is its transmittance which is always less than 1. Off-axis, light is to some extent cut off by the lens barrel (vignetting factor H) and the relationship between the focal length of the lens F and the image distance V enters into the relationship along with the off-axis angle θ, so that the expression (3.23) becomes:

$$I_0 = LT(1/4f^2)(F/V)^2 H \cos^4(\theta) \qquad (3.24)$$

The stray light inside the camera is nearly uniform and adds to this relationship. Typical values are $T = 0.92$, $(F/V)^2 = 0.97$, $H = 1.0$ and

Fig. 3.44 The distribution of luminances in an average outdoor scene. From ref. 44.

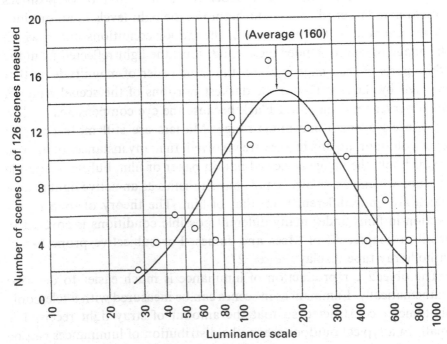

$\theta = 12°$. The density response of the film is logarithmic, so that for these parameters:

$$\log_{10} I = \log_{10}(2.2L/f^2 + I_f) \tag{3.25}$$

whereas ideal reproduction should be a straight line given by:

$$\log_{10} I = \log_{10}(2.2L/f^2) \tag{3.26}$$

The difference is plotted in figure 3.45. It can be seen that stray light effectively reduces the slope of the transfer characteristic curve and hence the contrast in the low luminance part of the curve. This means that the contrast in shadow regions of the scene is reduced. One of the methods described in chapter 4 is designed to compensate for this problem. Similar curves exist for reproduction of images on paper, and the overall effect of the transfer characteristics of each of the components of the imaging

Fig. 3.45 The difference between ideal reproduction of luminances in the scene and actual reproduction with a film. From ref. 44.

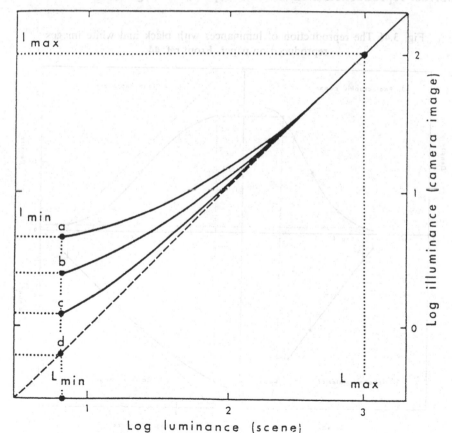

process on paper is shown in figure 3.46. The amount of stray light
which is strong enough to affect the film in the camera itself depends
on the level of luminance in the scene, so that for outdoor scenes the
stray light may be two to three times greater than for interior scenes
with small areas of high luminance. For old cameras which did not have
lenses with anti-flare coatings the effects of stray light are more significant
than for modern equipment.

For films exposed within their linear range and for prints made within
the reproduction range of the paper, the image contrast can be varied
by the choice of contrast grade of the latter as shown in figure 3.47. The
experienced darkroom technician will choose a proper paper contrast
grade instinctively upon examination of the projected negative on the
enlarger easel. Variation in exposure time will produce variation in print
quality, with acceptable quality being obtained over a fairly narrow
range for each contrast grade. For colour materials the problem of
correct reproduction is far more complex, although some of the basic

Fig. 3.46 The reproduction of luminances with black and white images
reproduced on paper. From ref. 44.

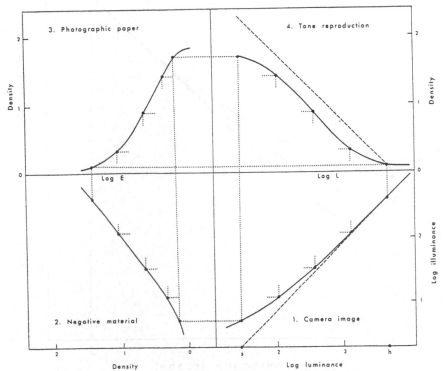

concepts for black and white materials also apply. For aerial images made under hazy conditions, the scattering of light reduces the contrast of shadow detail just like stray light within the camera. The effect is shown in figure 3.48. Filters which remove the more highly scattered blue and ultra-violet components therefore increase the contrast in the final result which can also be improved by use of higher contrast printing papers. The higher contrast inherently available in positive transparencies also helps.

3.5.4. Colour

Colour imagery has been extensively used in archaeological aerial photography since the early 1960s. One major reason for its introduction was not connected with the advantages of colour rendition. Considerations of convenience when flying alone in a small aircraft led to widespread use of 35 mm cameras. Normal black and white films available at the time for these cameras were of lower inherent contrast than those used

Fig. 3.47 Variation of image contrast with choice of printing paper. From ref. 44.

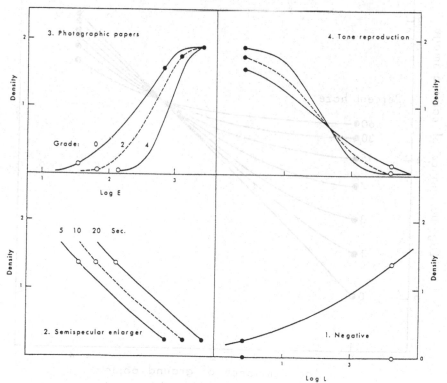

at larger formats and did not have particularly high resolution. But Kodachrome diapositive film had high contrast, especially in the green–yellow region, and was almost grainless when compared with black and white counterparts. Furthermore, machine processing by Kodak is highly uniform and requires no effort on the part of the user. Colour films from other manufacturers did not have these properties, but were sometimes used anyway. The necessary compromise which the decision to employ colour slide material entailed was sweetened by the pleasant realistic appearance of the results. Black and white publication of the spectacular results in the Somme valley by Agache (2) was satisfactory, although the pictures have less nuanced grayscales than images made with large format black and white film would have had. As a consequence, use of

Fig. 3.48 The reduction of contrast in shadows due to scattered light in an aerial image made under hazy conditions. From ref. 44.

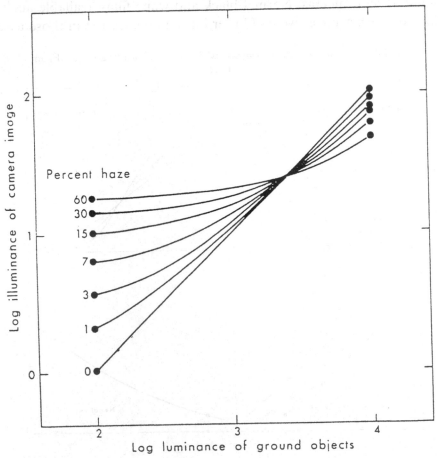

colour slide material became widely prevalent among solo fliers and others and has continued to the time of writing. As a result, there are probably more colour slides of archaeological sites on Kodachrome film than on all other taken together. The importance of this volume of archaeological information is tremendous. However, small format colour slides are more difficult to use as a quantitative tool in the way one can use films made specifically for aerial photography and photogrammetry.

Colour films are multilayered marvels of modern chemical engineering. All use a subtractive process, whereby light passes through a set of layers of dyed material, each of which absorbs a portion of the spectrum. An idealised structure is shown in figure 3.49. A blue sensitive layer comes first, then a green one, and finally a red one. The order is dictated by three considerations. The topmost layer is not dye sensitised, so that it responds only to blue light. But all dye sensitised silver halides are sensitive to blue light so that a yellow filter layer (minus blue), which is removed in the development process, is interposed between the blue and the lower red–green sensitive layers. Secondly, the focal length of a lens tends to be slightly longer towards the longer wavelengths, so that the image is sharper if the red layer is last. In addition, the silver halide top layers absorb and scatter some of the light as it passes on to the lower layers, and this scattering is least for the longer wavelengths. The layers produce yellow, magenta, and cyan dyes when developed, and these absorb blue, green, and red light. The combination of absorptions can reproduce a wide range of colours even though the absorption curves of the dyes overlap considerably. The retina of higher primates contains cone cells which contain protein pigments of three different spectral sensitivities. These sensitivities also overlap widely as shown in figure 3.50 (50), but the nervous system is finely tuned to detect the ratios of the three components rather than their absolute values so that we are able to differentiate a very large numbers of colours, especially when

Fig. 3.49 A simplified diagram of the construction of a colour film. From ref. 6.

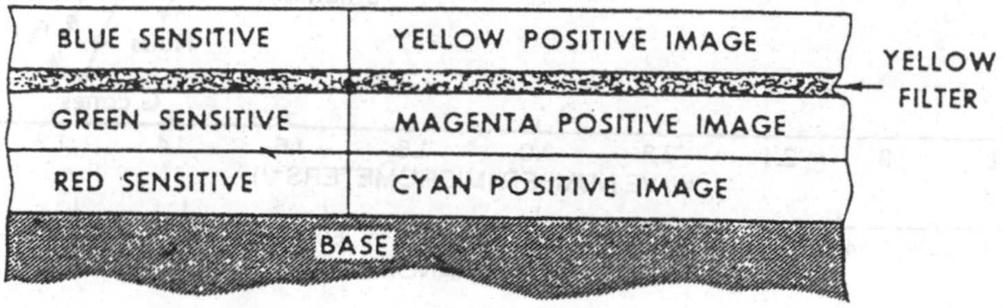

BLUE SENSITIVE	YELLOW POSITIVE IMAGE	
		YELLOW FILTER
GREEN SENSITIVE	MAGENTA POSITIVE IMAGE	
RED SENSITIVE	CYAN POSITIVE IMAGE	
BASE		

presented as comparisons. Typical spectral sensitivities of the three layers for an aerial colour reversal film are shown in figure 3.51(*a*). It can readily be seen that the sensitivities of the three layers overlap considerably, as to the transmission characteristics of the dyes produced upon development, as seen in figure 3.51(*b*). The eye is remarkably tolerant of these overlaps and quite realistic colour will be perceived. Because of the spectral overlap of the sensitised silver halides and the resultant dyes, colour film material cannot be used for quantitative colorimetric purposes. It is therefore not feasible, for example, to attempt to obtain data about crops whose spectral signatures are known through measurements of the colour densities on an aerial film even though the dyes track well, figure 3.51(*c*). Only a scanning spectrophotometer with narrow band filtering can be used for this purpose.

Colour positive transparencies are produced by arranging for the amount of dye to be less when more light has been received by a given layer, by means of a reversal process using a second exposure of the image after development in a black and white developer. Then the remaining silver halides are developed in one or more colour developers which produce the desired dye in each layer in proportion to the amount of residual silver. The silver is removed by a bleach after the dyes have

Fig. 3.50 The overlapping of the spectral sensitivities of the cone cells in the retina of a higher primate. From ref. 50.

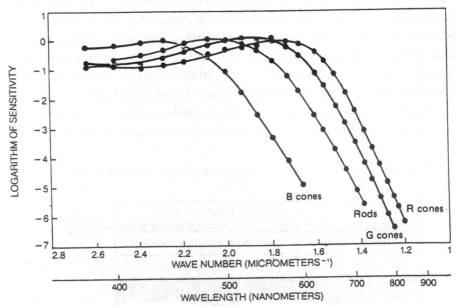

been formed. In most commercial colour films part of the dye molecule is incorporated directly in the emulsion, in a carrier material or with attached ballasting groups to prevent diffusion in wet gelatin. These are usually dispersed mechanically in the presence of a homogeniser. Such films which are the most frequently encountered types in the commercial market usually have lower resolution than does Kodachrome, where the specific dye couplers are in separate colour developers for each layer, but their processing is much simpler and can be carried out in modestly equipped laboratories. The dye concentration which can be reached is limited to the original amount incorporated per unit volume, whereas

Fig. 3.51 (*a*) The spectral sensitivities of the three layers of an aerial colour reversal film. (*b*) The transmission characteristics of the dyes produced upon development. (*c*) Dye response tracking. From ref. 40.

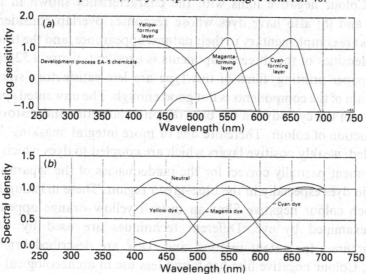

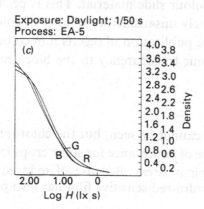

in the Kodachrome process dye deposits of high concentrations are possible. Most modern materials use dyes produced by oxidative condensation of colour developing agents such as one of the *p*-phenylenediamines. The amount of metallic silver regulates the process and determines the quantity of dye produced. The chemistry and physics of the dyes are discussed in the literature but details of modern commercially available materials are not usually revealed by their manufacturers. Details of the organic chemistry of patented processes which are in common use are also given by Thirtle (69).

Colour negatives are produced without reversal exposure. A bleaching bath removes the metallic silver after the dyes have been formed. The response of the three components to coloured light of different intensities tracks well, so that colour rendition is maintained in dark and light areas. Colour negative films with the characteristics shown in figure 3.52(*a*) and (*b*) also have dyes whose responses overlap considerably. Colours are complementary to their natural appearance, and the tracking of the densities of the three components is excellent, figure 3.52(*c*).

For paper printing, films of this type are less satisfactory since the separation of the components is not good enough. The unwanted spectral response of the cyan dye in the blue region desaturates and distorts the reproduction of colour. Therefore one or more integral 'masking' layers are added: weakly positive layers which are coupled to dyes which upon development partially correct for the inadequacies of the separation of the main dyes especially in the blue–green region. These masking layers give such colour negative films an overall yellow–orange appearance when examined by eye. Different techniques are used by various manufacturers for patent reasons, and these are described briefly by Thirtle. Colour negative film has found less use in archaeological aerial photography than colour slide material. This is perhaps because of the high cost and relatively unsatisfactory quality of colour paper prints in the past. In addition, publication of results if in colour, usually requires submission of a colour transparency to the block maker rather than a colour negative.

False colour
Near infra-red light cannot be seen, but the chlorophyll response in the near infra-red may be of importance for some crops in revealing moisture or other forms of plant stress. Because of reduced scattering at long wavelengths in air, infra-red sensitive film has also proven useful in low

altitude photography under hazy early morning or winter conditions. One way to render the infra-red visible is to use a false colour method. In this approach, a three layer film is used with sensitivities extending out to 900 nm, as shown in figure 3.53(*a*). The yellow forming layer has a spectral response well into the visible light red, a magenta forming one peaks in the visible deep red, and a cyan layer extends well out into the near infra-red. This last layer also has a strong response in the green, blue and ultra-violet regions, and this is usually suppressed by photographing through a yellow filter which removes all light with wavelengths shorter than 500 nm. Upon development, contrasts between the visible

Fig. 3.52 (*a*) The spectral sensitivities of the three layers of an aerial colour negative film. (*b*) The transmission characteristics of the dyes produced upon development. (*c*) Dye response tracking. From ref. 40.

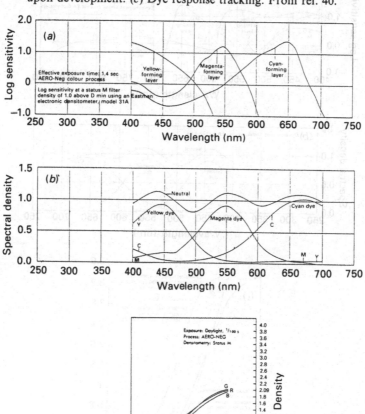

green–red and the infra-red are made visible as differences in blue-greens and reds, the latter modulated by the cyan dye of the infra-red sensitive layer, as shown in figure 3.53(*b*). The colour dyes do not need to track at all, as shown in figure 3.53(*c*).

3.5.5. *Image structure*

Individual photographic grains, when developed and viewed under an electron microscope, grow whiskers of metallic silver which clump

Fig. 3.53 (*a*) The spectral sensitivities of the three layers of an aerial false colour infra-red film. (*b*) The transmission characteristics of the dyes produced upon development. (*c*) Dye response tracking. From ref. 40.

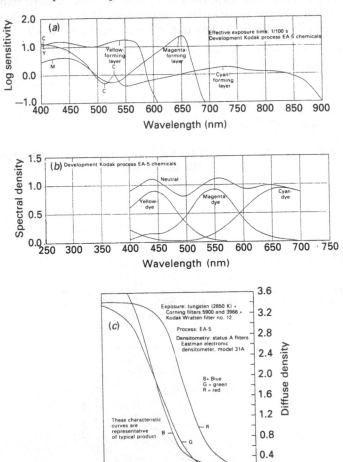

together in a furry ball around each grain. In a cross-section through the emulsion, increasing exposure is seen as an increasing depth of irregular silver shapes, looking somewhat like a cross-section of soil. The vertical density distribution and the size distribution of the grains determine the 'granularity' of the image which becomes visible at considerable degrees of enlargement. This granularity can be considered to be 'noise' as defined in chapter 4, and it sets a limit to the resolution of fine structure in the image. Measurement with a microdensitometer shows that a Gaussian distribution of densities over the film surface is typical. The successive density deviations are uncorrelated when the scanning aperture of the densitometer is large compared with the size of the grains. The visual sensation of graininess correlates reasonably well with this rule, as formulated by Van Kreveld when an appropriate aperture size is used.

Since form features in an archaeological aerial photograph are seen either as fine lines or points, the degree to which a film is able to resolve them is of interest. Other factors, as described in chapter 4, such as the movement of the aircraft with resultant blurring, and the limitations of lens resolution are sometimes more significant in determining the ultimate visibility of fine structures. But especially when light levels are low and fast films with correspondingly larger grains are needed, film resolution is important too.

Structures in an image can be analysed in terms of spatial frequencies as described in chapter 4. The distribution of photographic densities as a function of distance is then expressed as a set of orthogonal functions, usually sines and cosines. If the mean value of the density is given by b_0 and the amplitude of sinusoidal variation about this by b_1, then the modulation M of the distribution at a given spatial frequency is given by:

$$M = \frac{B_{max} - B_{min}}{B_{max} + B_{min}} = \frac{(b_0 + b_1) - (b_0 - b_1)}{(b_0 + b_1) + (b_0 - b_1)} = \frac{b_1}{b_0} \tag{3.27}$$

for a sinusoidal test object giving a luminance distribution $G(x)$ as a function of distance x of $G(x) = b_0 + b_1 \cos(2\pi v x)$, where v is the spatial frequency (chapter 4).

The modulation transfer function (MTF) for a typical aerial black and white film is shown in figure 3.54. The ordinate represents the spatial frequency in cycles per millimetre, and the abscissa the response. Fine grained higher resolution films have MTFs which maintain a higher value for higher spatial frequencies, as shown in figure 3.55. Colour films have different MTFs for the blue, green, and red sensitive layers, with the response dropping for the lower layers since light must pass through the grains of the upper emulsion before it reaches them. In addition, the red sensitive layer is inherently less sensitive, so that a compensating larger grain size distribution may be used. These differences are shown for a typical colour film in figure 3.56, which also shows that the modulation transfer characteristic of a black and white film depends to a slight extent on the colour of the exposing light. Since the response of film is non-linear, resolving power also depends on the degree of

Fig. 3.54 The MTF of an aerial black and white film. From ref. 40.

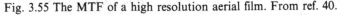

Fig. 3.55 The MTF of a high resolution aerial film. From ref. 40.

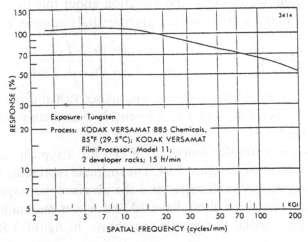

exposure to some extent. Optimum density of 0.85 above minimum density is mentioned in the literature (44), but the aerial photographer is usually not in a position to control this. The contrast ratio in the modulation of the light from the scene affects the measurement of the MTF for similar reasons. For aerial films it is customary to use a low contrast usually 1.6:1, rather than the much higher contrasts used in measuring other types of films.

The modulation transfer characteristic, being mainly determined by the three-dimensional grain size distribution of the developed silver image, is little affected by development techniques, although some developers which reduce grain size chemically may actually lower the high spatial frequency response. It is better practice to use different films and standard development rather than to influence the modulation transfer characteristic via development, since this is not as reproducible.

Dyes which absorb scattered light are sometimes added to emulsions to raise the high frequency end of the modulation transfer curve. Sensitivities are correspondingly reduced due to light loss. The dye image in a colour film depends on the developed silver grains, and the structure is similarly random. But since the shapes of the dye structures are not the same as those of the silver grains from which they are derived, there are some differences. The green sensitive part of the image to which the eye is most responsive is the source of most of the graininess seen in colour images and dyes are sometimes included in the uppermost blue

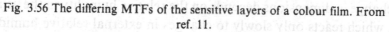

Fig. 3.56 The differing MTFs of the sensitive layers of a colour film. From ref. 11.

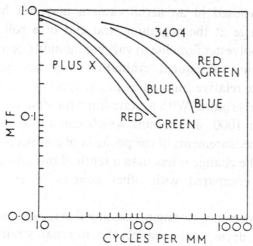

sensitive layer to reduce light scattering before it reaches the green sensitive layer below. These are removed during processing. Most colour films have lower high frequency response than black and white films because of their multilayer structure and the incorporation of ballast to keep the colour couplers in place.

3.5.6. Dimensional stability of negatives and prints

When accurate plans are to be made from aerial archaeological images, then the dimensional stability of the film material used is significant. The base material used in amateur films has been almost exclusively cellulose acetate while that used in professional aerial films is polyethylene terephthalate (polyester) since the mid-1950s. Temperature changes affect the base which expands and contracts in a reversible fashion. But these changes are usually of minor importance compared with changes in relative humidity. This question is discussed in some detail by McKinney (43) and for non-photogrammetric materials by Adelstein (1). Details of typical currently available commercial aerial films can be found in ref. 41.

The change in linear dimension with a change in relative humidity for a typical film thickness is shown in figure 3.57. It can readily be seen that the acetate films have a much larger coefficient of humidity expansion than polyester ones. Most of the change takes place in the gelatin layer which then applies stress to the base. If the base is thicker, then resistance to change is greater as shown in the curves of figure 3.58. The change in dimensions depends on the relative humidity in the gelatin itself, which reacts only slowly to changes in external relative humidity. Hence, a film exposed in an aerial camera may not have the same dimensional change at the beginning and end of a roll. Dimensional changes even in polyester film due to varying humidity conditions during picture taking and subsequent evaluation may be significant, since temperatures and relative humidities in the aircraft are quite different from those on the ground. With acetate films the differences can readily attain 5 parts in 1000, an amount which cannot be neglected when making precise measurements of the position of control points on a film. With polyester, the change is less than a tenth of this amount, a quantity which is small compared with other sources of error in position determination.

Changes in dimensions during processing due to wetting of the gelatin and subsequent drying can change the internal strains in the film compared with that which existed after manufacture, and even with polyester films, a variation of about 3 parts in 10000 is observed. More significant changes have been observed with acetate based films, and these also suffer from ageing variation due to evaporation of residual solvents. Up to 0.5% shrinkage after storage has been reported (71). Photographic papers exhibit large dimensional changes after processing, orders of magnitude more than film, with shrinkage after wetting and drying. Resin coated papers are much better in this respect, but are still too unstable for any accurate photogrammetric work.

   A far more distressing change in the dimensions of acetate materials is due to the fact that the films are stretched in length during manufacture so that the humidity, thermal and solvent-loss effects are different for the width and length directions. For papers this effect is still greater, with factors of 4 reported for the humidity coefficient in the width direction compared with the length, and factors of 2 in length during processing shrinkage. All of this indicates the perils which may be encountered when using paper prints for measurement purposes when making large scale plans of sites. When new photographs are made, it

Fig. 3.57 The change in linear dimension against relative humidity for a typical film. From ref. 70.

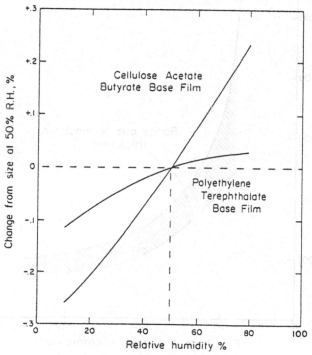

is best to use properly designed aerial films in the larger formats which permit use of greater base thickness. Such films are sufficiently stable that the effects mentioned can be ignored in all but architectural aerial photogrammetry. The cameras used for these films will have rigid bodies, plane film supports and sometimes reseau plates in the film plane, and are readily calibrated so that all sources of distortion can be greatly reduced.

Colour films reportedly suffer more from dimensional changes than do black and white ones because of the greater thickness of the multilayer gelatin emulsions. When existing photographs must be geometrically corrected and transformed into maps, the effects of the base material and the emulsion thickness must be considered qualitatively in evaluating probable accuracy. If, for example, only 35 mm colour slides are available, then probably the only way to minimise these effects is to use as large a number of control points and as many images as possible with a geometric correction technique which gives quantitative information

Fig. 3.58 The effect of film base thickness on dimensional stability. From ref. 43.

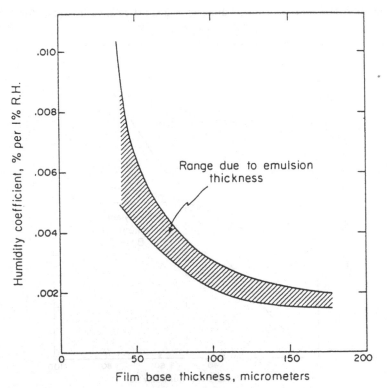

about the residual errors at all known points. This will significantly increase the time required for manual intervention during such correction and the result will almost always be less accurate than if proper stable materials had been used in the first place. It is much easier to prevent the problems discussed at the time the picture is taken by the choice of proper materials and cameras than to deal with them afterwards by calculation. But most frequently one must work with old pictures from the archives, and probable dimensional changes must be taken into account as far as possible.

3.5.7. *Archive quality*

Nearly all aerial photographs of archaeological crop, soil, and moisture marks are unique records. The original negative or colour positive films therefore constitute documents which must be preserved for an unlimited period, as far as possible. Silver halide emulsions are potentially capable of quite long stability if they are stored at relatively low temperatures of less than 21°C. Moderate humidity of between 40 and 50% is essential with still lower humidity better. High humidity will encourage growth of fungus on gelatin, resulting in serious damage. Even at moderately high humidities, small red or yellow spots may appear after a long time, due to minute deposits of coloured colloidal silver. Potassium iodide in the fixing bath or a gold protective treatment is usually recommended. The ANSI standard for archive film PH1.28-1969 provides useful guidelines for films on acetate bases, and for best permanence, the standards of PH1.41-1971 for archive records on polyester bases should be followed when permanent storage is essential.

The major cause of problems with black and white film is the presence of hydrogen sulphide and sulphur dioxide in the atmosphere which ultimately cause loss of a silver image through tarnishing. If the films are inadequately washed after processing or if the wash water is too cold, residues from the fixing baths, as shown in figure 3.59, may remain in the emulsion to release damaging sulphur in later years. Treatment with an oxidising solution which transforms the thiosulphates in fixing baths to sulphates is useful, especially if augmented with potassium bromide to promote oxidation (71). A protective treatment which coats the silver grains with gold is probably the best method for achieving the greatest permanence, but is seldom employed. Black and white negatives on polyester base films have much greater permanence than paper prints because, as shown in figure 3.60, it is not possible to remove the thiosulphate from the fixing bath completely. It is also quite important that the fixing baths be reasonably fresh, and that no excessive quantities of silver thiosulphate compounds be present, or stains will appear after long periods of time. Two fixing baths used sequentially and then testing for residual silver is recommended (40,70). Quantitative tests for residual thiosulphates are also described in ref. 40 and in ANSI standard PH4.8-1971.

The stability of the dyes in colour images is far less than that of the metallic silver in black and white films, but no reliable data are available. For colour images of the highest importance which are to have a lifetime greater than four years, it is recommended (40) to make colour separation master negatives and treat these as archive black and white films. Colour transparencies which are of archival importance should never be projected, and storage conditions with respect to humidity and temperature

Fig. 3.59 Washing time and temperature and the retention of fixing salts in film. From ref. 44.

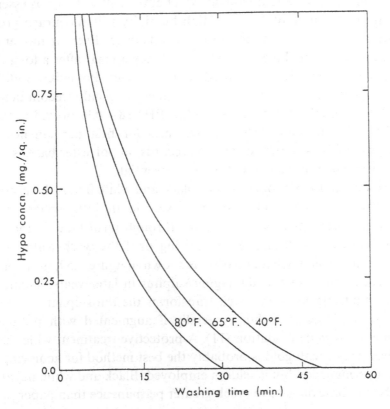

must be carefully maintained. The colour balance will shift with time because the dyes in the various layers react differently, and there is no way to restore this balance accurately, even with the numerical methods described in chapter 4. Colour prints on paper fade rapidly and should never be used for any application requiring long term stability. Kodachrome transparencies may possibly have greater lifetimes than other films processed by local laboratories because of the careful uniform processing techniques used by Kodak. Detailed quantitative data on the keeping qualities of colour films is not available.

Most archaeological aerial photographers neglect all of these matters, and the majority leave processing to a commercial laboratory, the quality of whose work is not controllable. As a result, many older pictures have defects which provide a rich field for the methods described in chapter 4, many of which would be unnecessary if suitable precautions had been taken in choice of film and cameras.

Fig. 3.60 Washing time and temperature and the retention of fixing salts in paper. From ref. 44.

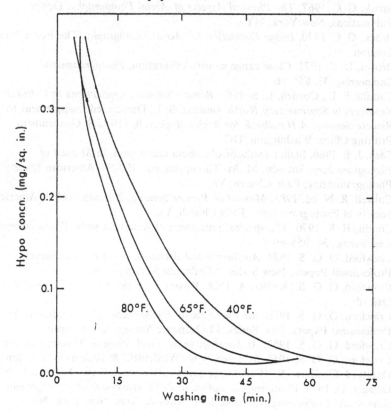

Notes

(1) Adelstein, P. Z. 1973, Physical properties of photographic materials, *SPSE Handbook of Photographic Science and Engineering*, 487–91, John Wiley & Sons, New York, NY.

(2) Agache, R. 1970, Détection aérienne de vestiges protohistoriques Gallo-Romains et Médiévaux dans le bassin de la Somme et ses abords, *Bulletin de la Société de Préhistoire du Nord*, 7, 230 Musée d'Amiens, Amiens.

(3) Agache, R. 1978, *La Somme Pré-Romaine et Romaine*, Société des Antiquaires de Picardie, Amiens.

(4) Allen, G. W. G. 1984, Discovery from the Air, *Aerial Archaeology*, 10, 37–92.

(5) Avery, T. L., Lyons, T. R. 1981, *Aerial and Terrestrial Photography for Archaeologists*, supplement to *Remote Sensing: A Handbook for Archaeologists and Cultural Resource Managers*, 7, 48 U.S. Government Printing Office, Washington, DC.

(6) Baker, J. G. *et al.* 1980, Elements of photogrammetric optics, in *Manual of Photogrammetry*, 4th edn, C. C. Slama, ed., 103–185, American Society of Photogrammetry, Falls Church, Va.

(7) Bowen, I. S. 1962, The ratio of heat losses by conduction and by evaporation from any water surface, *Physical Review*, 27, 779–87.

(8) Braasch, O. 1983, *Luftbildarchäologie in Süddeutschland*, Gesellschaft für Vor- und Frühgeschichte in Wüttemberg, Aalen.

(9) Bradford, J. 1957, *Ancient Landscapes*, Bell & Sons, London.

(10) Brock, G. C. 1967, *The Physical Aspects of Aerial Photography*, Dover Publications, New York, NY.

(11) Brock, G. C. 1970, *Image Evaluation for Aerial Photography*, The Focal Press, London.

(12) Brown, D. C. 1971, Close range camera calibration, *Photogrammetric Engineering*, 37, 855–66.

(13) Camilli, E. L., Cordell, L. S. 1983, *Remote Sensing: Applications to Cultural Resources in Southwestern North America*, B. L. Daniels, ed., supplement to *Remote Sensing: A Handbook for Archaeologists*, 8, 110 U.S. Government Printing Office, Washington, DC.

(14) Case, J. B. 1966, Stellar methods of camera calibration, in *Manual of Photogrammetry*, 3rd edn, M. M. Thompson, ed., 180–93, American Society of Photogrammetry, Falls Church, Va.

(15) Colwell, R. N. ed. 1983, *Manual of Remote Sensing*, 2nd edn, Vol. 1, American Society of Photogrammetry, Falls Church, Va.

(16) Condit, H. R. 1970, The spectral reflectance of American soils, *Photogrammetric Engineering*, 36, 955–66.

(17) Crawford, O. G. S. 1924, *Air Survey and Archaeology*, Ordnance Survey Professional Papers, New Series, 7 Ordnance Survey, London.

(18) Crawford, O. G. S., Keiller, A. 1928, *Wessex from the Air*, The Clarendon Press, Oxford.

(19) Crawford, O. G. S. 1929. *Air Photography for Arhcaeologists*, Ordnance Survey Professional Papers, New Series, 12 Ordnance Survey, Southampton.

(20) Crawford, O. G. S. 1953, *Archaeology in the Field*, Phoenix House, London.

(21) Crawford, O. G. S. 1955, *Said and Done*, Weidenfeld & Nicholson, London.

(22) Dainty, J. C., Shaw, R. 1974, *Image Science*, Academic Press, New York, NY.

(23) Dejager, D. 1973, Photographic optics, in *SPSE Handbook of Photographic Science and Engineering*, 159–256, John Wiley & Sons, New York, NY.

(24) Dozier, J., Strahler, A. H. 1983, Ground investigations in support of remote sensing, in *Manual of Remote Sensing*, 2nd edn, 962–6, American Society of Photogrammetry, Falls Church, Va.

(25) Egan, W. G. 1970, Optical stokes parameters for farm crop identification, *Remote Sensing of the Environment*, **1**, 165–80.

(26) Evans, R., Jones, R. J. A. 1977, Crop marks and soils at two archaeological sites in Britain, *Journal of Archaeological Science*, **4**, 63–76.

(27) Farrell, R. J., Booth, J. M. 1975, *Design Handbook for Imagery Interpretation Equipment*, Boeing Aerospace Company, Seattle, Wa.

(28) Gaussman, H. W., Allen, W. A., Wiegand, C. L., Escobar, D. E. 1971, Leaf light reflectance, transmittance, absorptance, and optical and geometrical parameters for eleven plant genera with different leaf mesophyll arrangements, in *Proceedings of the 7th Symposium Remote Sensing of the Environment*, University of Michigan, Ann Arbor, Mi.

(29) Gaussman, H. W., Escobar, D. E., Everitt, J. H., Richardson, A. J. 1978, The leaf mesophylls of twenty crops, their light spectra, and optical and geometrical parameters, *SWC Research Report*, No. 423, **88** Rio Grande Soil and Water Research Center, Weslaco, Tx.

(30) Goddard, M. C. 1973, Color photography, in *SPSE Handbook of Photographic Science and Engineering*, 444–8, John Wiley & Sons, New York, NY.

(31) Gruner, H. 1966, Elements of photogrammetric optics, in *Manual of Photogrammetry*, 3rd edn, M. M. Thompson, ed., 67–132, American Society of Photogrammetry, Falls Church, Va.

(32) Hall, D. G. M., Reve, M. J., Thomasson, A. J., Wright, V. F. 1977, *Water Retention, Porosity and Density of Field Soils*, Soil Survey Technical Monograph, **9** Soil Survey of England and Wales, Harpenden.

(33) Hatzopoulos, J. N. 1985, An analytical system for close-range photogrammetry, *Photogrammetric Engineering & Remote Sensing*, **51**, 1583–8.

(34) Hillel, D. 1971, *Soil and Water, Physical Principles and Process*, Academic Press, New York, NY.

(35) Hodgson, J. M. ed. 1974, *Soil Survey Field Handbook*, Soil Survey Technical Monograph, **5** Soil Survey of England and Wales, Harpenden.

(36) Jones, R. J. A., Evans, R. 1975, Soil and crop marks in the recognition of archaeological sites by air photography, *Aerial Reconnaissance for Archaeology*, Research Report, **12**, 1–11 Council for British Archaeology, London.

(37) Jones, R. J. A. 1979, Crop marks induced by soil moisture stress at an Iron Age site in midland England, U.K., *Archaeo-Physika*, **10**, 656–68, Rheinland Verlag, Köln.

(38) Julesz, B. 1971, *Foundations of Cyclopean Perception*, University of Chicago Press, Chicago, Il.

(39) Kodak, Eastman, *Kodak Filters for Scientific and Technical Photography*, Eastman Kodak Co., Rochester, NY.

(40) Kodak, Eastman, 1982, *Kodak Data for Aerial Photography*, 5th edn, Kodak Publication, M-29, Eastman Kodak Co., Rochester, NY.

(41) Kodak, Eastman, 1982, *Physical Properties of Kodak Aerial Films, Physical and Chemical Behavior of Kodak Aerial Films*, Kodak Publication M-62, M-63, Eastman Kodak Co., Rochester, NY.

(42) Livingston, R. G. *et al.* 1980, Aerial cameras, in *Manual of Photogrammetry*, 4th edn, C. C. Slama, ed., 247–73, American Society of Photogrammetry, Falls Church, Va.

(43) McKinney, R. G. *et al.* 1980, Photographic materials and processing, in *Manual of Photogrammetry*, 4th edn, C. C. Slama, ed., 305–66, American Society of Photogrammetry, Falls Church, Va.

(44) Mees, C. E. K., James, T. H. 1966, *The Theory of the Photographic Process*, 3rd edn, The Macmillan Company, New York, NY.

(45) Milthorpe, F. L., Moorby, J. 1974, *An Introduction to Crop Physiology*, Cambridge University Press, Cambridge.

(46) Munjy, R. A. H. 1986, Calibrating non-metric cameras using the finite-element method, *Photogrammetric Engineering & Remote Sensing*, **52**, 1201–5.

(47) Munjy, R. A. H. 1986, Self-calibration using the finite element approach, *Photogrammetric Engineering & Remote Sensing*, **52**, 411–18.

(48) Munsell Color Company 1954, *Munsell Soil Color Charts*, Munsell Color Company, Baltimore, Md.

(49) Myers, V. I. 1983, Remote sensing applications in agriculture, in *Manual of Remote Sensing*, 2nd edn, R. N. Colwell, ed., 2111–28, American Society of Photogrammetry, Falls Church, Va.

(50) Nunn, B. J., Schnapf, J. L., Baylor, D. A. 1984, Spectral sensitivity of single cones in the retina of Macaca Fascicularis, *Nature*, **309**, 264–6.

(51) Penman, H. L., Angus, D. E., Van Bavel, C. H. M. 1967, Microclimatic factors affecting evaporation and transpiration, *Irrigation of Agricultural Lands*, R. M. Hagan ed., **11**, 483–505.

(52) Piccarreta, F. 1987, *Manuale di Fotogravia Aerea: Uso Archeologico*, 'L'Erma' di Bretschneider, Roma.

(53) Rigaud, P., Hersé, M. 1976, Télédétection photographique infrarouge à partir de ballons captifs, *Comptes Rendus Académie des Sciences*, **303**, 1703–8.

(54) Riley, D. N. 1944, The technique of air archaeology, *The Archaeological Journal*, **101**, 1–16.

(55) Riley, D. N. 1987, *Air Photography and Archaeology*, Duckworth, London.

(56) Ripple, W. J. 1986, Spectral reflectance relationships to leaf water stress, *Photogrammetric Engineering & Remote Sensing*, **52**, 1669–76.

(57) Roberts, A., Griswold, L. 1986, Practical photogrammetry from 35-mm aerial photography, *Photogrammetric Engineering & Remote Sensing*, **52**, 501–8.

(58) Rose, C. W. 1966, *Agricultural Physics*, Pergamon Press, Oxford.

(59) Schmidt-Kraepelin, E., Schneider, S. 1966, Landeskundliche Luftbildauswertung im mitteleuropäischen Raum: Luftbildinterpretation in der Agrarlandschaft, *Schriftenfolge des Instituts für Landeskunde*, **7**, 154 Bundesanstalt für Landeskunde und Raumforschung, Bad Godesberg.

(60) Scollar, I. 1964, Physical conditions tending to produce crop sites in the Rhineland, *Colloque International d'Archéologie Aérienne, Paris 1963*, 39–48, SEVPEN Ecole Pratique des Hautes Etudes, Paris.

(61) Scollar, I. 1965, *Archäologie aus der Luft*, 55, Rheinland Verlag, Köln.

(62) Slater, P. N., Doyle, F. J., Norman, L. F., Welch, R. 1983, Photographic systems for remote sensing, in *Manual of Remote Sensing*, 2nd edn, R. N. Colwell, ed., 269–72, American Society of Photogrammetry, Falls Church, Va.

(63) Smith, L. P. 1967, *Potential Transpiration, Technical Bulletin*, Ministry of Agriculture, Fisheries and Food, **16**, Her Majesty's Stationery Office, London.

(64) Smith, J. A. 1983, Matter-energy interaction in the optical region, in *Manual of Remote Sensing*, edn, R. N. Colwell, ed., 98–101, American Society of Photogrammetry, Falls Church, Va.

(65) Stoner, E. R., Baumgardner, M. F., Biehl, L. L., Robinson, B. F. 1980, *Atlas of Soil Reflectance Properties*, Research Bulletin, Purdue Agricultural Experiment Station, **962** Purdue University, West Lafayette, Ind.

(66) Strunk-Lichtenberg, G. 1965, Bodenkundliche Untersuchungen an Archäologischen Objekten, die durch Luftbild-Aufnahmen entdeckt wurden, *Archaeo-Physika*, **1**, 175–202.

(67) Sturge, J. M. ed. 1977, *Neblette's Handbook of Photography & Reprography Materials, Processes & Systems*, 7th edn, Van Nostrand Reinhold, New York, NY.

(68) Taylor, S. A., Ashcroft, G. L. 1972, *Physical Edaphology: The Physics of Irrigated and Non-irrigated Soils*, W.H. Freeman & Company, San Francisco, Ca.

(69) Thirtle, J. R. 1973, Chemistry of color processes, in *SPSE Handbook of Photographic Science and Engineering*, 549–68, John Wiley & Sons, New York, NY.

(70) Thomas, W. 1973, *SPSE Handbook of Photographic Science and Engineering*, John Wiley & Sons, New York, NY.

(71) Umberger, J. Q. 1973, Photographic chemistry, stability of the silver image, *SPSE Handbook of Photographic Science and Engineering*, 531–4, John Wiley & Sons, New York, NY.

(72) Veress, S. A., Tiwari, R. S. 1976, Fixed frame multiple camera system for close range photogrammetry, *Photogrammetric Engineering & Remote Sensing*, **42**, 1195–210.

(73) Wilson, D. R. 1982, *Air Photo Interpretation for Archaeologists*, Batsford, London.

(74) Wilson, D. R. 1989, The formation and appearance of archaeological soil marks, *Into the Sun, Essays in Air Photography in Archaeology in Honour of Derrick Riley*, D. Kennedy ed., John R. Collis Publications, Sheffield

4

Archaeological image enhancement

4.1. Introduction

4.1.1. Images as information records

Since the invention of photography, motion pictures and television we have come to take pictures for granted. Our lives are flooded with images created by mechanical, chemical, or electronic means, and we have grown accustomed to thinking about them as a reflection of reality. Images serve in archaeology as records of non-repeatable phenomena such as photographs of excavations or as pictures made of buried sites from the air under unusual climatic conditions. But they may also be used as a technique for displaying quantitative data such as may be obtained through geophysical prospecting. Images contain not only the qualitative information which we readily appreciate when looking at them, they also contain data which can be extracted or incorporated by measurement and calculation. The quantitative treatment of visual or visualisable data is the subject of the modern discipline, image processing.

When we wish to work quantitatively with a picture, we must step back a bit from that which seems to be obvious and penetrate more deeply into its physical make-up much in the way that Renaissance artists analysed perspective or pointillist painters in the nineteenth century thought about colour. The psychological and neurophysiological counterpart to this kind of analytical image processing is the low level vision of the specialised literature (29). The human role in this kind of analysis remains important because only the mind can grasp the content of the image as a whole and provide *a priori* external information for understanding it. Attempts to produce similar performance using machine methods, with pattern recognition or image understanding have not been successful in archaeology. The analytical or numerical approach describes succinctly what is happening in the microcosmos of the picture rather than its global significance. A numerical method can analyse, modify, or displace brightness, colour, and contrast in an image, but at the moment only the human mind can decide that the object pictured is of archaeological significance and use that information in context.

An image can be thought of as containing two types of information: the first is contained within its physical properties which can be measured, the second is suggestive to the observer of information external to the image itself. This enables him to make decisions about the data conveyed. In this chapter we are only concerned with the first type of information, independent of the context in which the picture has been made. Context dependent information extracted with external *a priori* knowledge is a research theme and proven methods are not available for archaeological data. Most of the techniques to be described here are quite general and apply to images made using various methods including those which do not use film as a recording medium, such as magnetic and thermal prospecting data from archaeological sites, or do not use visible light such as x-ray images of archaeological objects. But the vast majority of archaeological images which are in need of computer treatment have been recorded on film and these will be of primary concern in this chapter.

Many of the methods to be described can be used without full knowledge of the underlying mathematics, and many of them are available as ready-made programs for computers with facilities for image processing. But understanding the mathematical basis of these techniques provides much deeper insight, and therefore a brief outline will be given. Much more detail can be found in one of the many excellent surveys of the subject (1,10,11,14,18,38,41). In order to explain what the computer can do with a picture, some fundamentals of conventional photography beyond that discussed in chapter 3 will be reviewed. Then computational methods which duplicate or surpass classical photographic techniques will be discussed. Examples will be taken from archaeological aerial photography because more is known about treating this kind of image and there are probably more of them in need of treatment, but the techniques apply to almost any kind of image regardless of source.

4.1.2. Some photographic preliminaries

In archaeological aerial photography light from the sun falls on the landscape and is reflected by the vegetation and ground surface to the observer in an aeroplane. Sunlight contains all colours, but some of these are absorbed by the plants or by the ground surface and others are returned by reflection. How plants and different types of soils return or absorb coloured light is described in chapter 3. Greater detail is given in ref. 50. Chlorophyll absorption is strongest at *ca* 400 nm wavelength (blue), and hence this wavelength is a good indicator of how much chlorophyll is present. Because of this, non-red sensitive orthochromatic films gave good results in early archaeological aerial photography. A wavelength of 520 nm (green) correlates well with chlorophyll concentration while 550 nm (yellow–orange) is the so-called chlorophyll hinge point which is insensitive to the concentration of this substance. Chlorophyll also has a strong reflective component in the near infra-red which may sometimes be of interest for certain conditions as described in ref. 16 and in chapter 3.

Black and white films which respond to light of all colours are usually used with a yellow or orange filter to remove bluish haze, or with a cyan/blue–green filter to enhance green–yellow vegetation contrast on crop sites. Colour transparencies are usually made through an ultra-violet filter to prevent the backscatter of short wavelength radiation giving a bluish cast to the images unless the film itself provides this filtering in one of its layers. A highly red sensitive film (not infra-red) will give lower contrast when used with a deep yellow or orange filter than does a normal panchromatic film without a filter or with a light yellow filter, for it does not respond to the wavelengths of interest shorter than 520 nm.

A simplified mathematical model of the photographic process
The sensitive surface of the negative at the image plane is activated in proportion to the amount of light received. Exposing the film can be modelled following Pratt (38) by:

$$X(C) = k_x \int C(\lambda)L(\lambda)\,d\lambda \tag{4.1}$$

where $X(C)$ is the integrated exposure, $C(\lambda)$ and $L(\lambda)$ are the spectral response of light and film, k_x is a function of the time for which the shutter is open and the size of the lens opening. The film is developed chemically, and a concentration of microscopic silver grains is left after processing which appears black. The transmittance of the film is related to the density of the metallic silver by an exponential law:

$$\tau(\lambda) = \exp[-d_e D(\lambda)] \tag{4.2}$$

where $D(\lambda)$ is the density as a function of the wavelength of the exposing light and d_e is proportional to the exposure. $D(\lambda)$ is usually nearly constant for visible light when using panchromatic film. In the $\log_{10}$ units customarily used in photography, density is a function of the logarithm of transmittance ($d_x \propto$ inverse exposure):

$$d_x D(\lambda) = -\log_{10}[\tau(\lambda)] \tag{4.3}$$

In enlarging, the negative is placed in a carrier and, with a light source behind it, an image is projected through a lens onto a sheet of paper with a photosensitive layer. The paper reproduces the light and dark areas of the original scene in white and black tones, opposite to those of the negative. The reflectivity r_0 of the print as a function of incident light wavelength is proportional to silver density with:

$$d_x D(\lambda) = \log_{10}[r_0(\lambda)] \tag{4.4}$$

For the central portion of the approximately linear exposure curve:

$$d_x = \gamma\{\log_{10}[X(C)] - K_F\} \tag{4.5}$$

where as described in chapter 3, γ is the slope of the linear region and K_F is the point where the line crosses the log exposure axis. In the linear part of the curve, the transmittance is:

$$\tau(\lambda) = K_\tau(\lambda)[X(C)]^{-\gamma D(\lambda)} \quad \text{and} \quad K_\tau(\lambda) \propto 10^{\gamma K_F D(\lambda)} \tag{4.6}$$

The sensitometric curve
As explained in chapter 3, the slope of a sensitometric curve is determined by the type of film and by the development time at a constant temperature. The subjective impression of picture contrast is related to the slope of the sensitometric curve, whereas the impression of brightness is related to the position on the curve of a given part of a scene. Steeper slopes give the impression of higher contrast.

The quantitative relationship between film blackening and light falling on a photographic film may be recorded as a graph, the H+D curve described in chapter 3, measured with a densitometer. Precision densitometers operate by comparing film transmittance with that of a dyed glass wedge, moving the latter until a visual balance is obtained either with a dual field optical system which combines the light paths of both

film and wedge or with a dual photocell arrangement. Simple densitometers rely on the calibrated response of a photocell or photodiode
only. Usually a number of test films are exposed for the same length of
time and developed for different lengths of time at various constant
temperatures. Transmittance is displayed as a curve of density versus
exposure light intensity. Since the range of light values which a film can
reproduce is usually quite large, a dual logarithmic scale is used. A
density of 1.0 represents a ten-fold increase over the film base density.
The eye also perceives brightness almost logarithmically over a very
wide range so that equal density steps are seen as nearly equal brightness
steps. For a typical black and white film there is a nearly linear relationship between density and exposure to light over a fairly considerable
range, but light values occur in most scenes which are either too dark
or too bright to be recorded in the linear region.

With photographic paper prints, incident light passes through the
sensitised surface, is reflected by the white paper behind it and passes
once again through the sensitive layer. Because of this double passage
and the loss of light on reflectance at the paper surface, the range of
reproducible gray values and the linearity of response is not as large in
a negative or in a film transparency where light must pass only once
through the material. For best quality prints of normal terrestrial scenes,
it is customary to develop the negatives to a slope of less than 1:1 to
allow for a wide exposure range. Correspondingly, higher contrast is
used in the enlarging paper to create an impression of the original
brightness values. For pictures made at high altitudes, the scattering
effect of the dust in the intervening atmosphere reduces contrast. To
compensate for this, professional aerial films usually have a higher
contrast than do amateur films, but if desired they can be developed for
shorter times with diluted developer, lower temperature, or less active
developers to obtain lower contrast. Usually the less active developer
gives the best results when taking pictures at low altitudes where
contrasts are about the same as in ground photography.

The higher the contrast, the smaller the range of light values which
can be accommodated on the straight line region. A high degree of
blackening due to overexposure on the upper part of the curve cannot
be reproduced well by the limited exposure range of typical enlarging
papers. Underexposure results in loss of information on the lower part
of the curve. Hence, the higher the contrast, the more critical the exposure
time and lens opening becomes because the available straight line portion
of the curve is reduced in range.

A colour film model

Colour films are more complicated. At least three sensitive layers which respond to light of different colours are involved, each of which has its own sensitometric curve. Modern colour films have many more layers whose combined effect gives a performance approaching that of an ideal three layer film. The exposures of the red, green, and blue sensitive layers are:

$$X_R(C) = d_R \int C(\lambda) L_R(\lambda)\, d\lambda, \qquad X_G(C) = d_G \int C(\lambda) L_G(\lambda)\, d\lambda,$$

$$X_B(C) = d_B \int C(\lambda) L_B(\lambda)\, d\lambda \qquad (4.7)$$

where the ds are constants of proportionality such that exposure is equal for a reference white light source. The transmittance is the product of the main dyes used, usually cyan, magenta, and yellow:

$$\tau_T(\lambda) = \tau_{TC}(\lambda)\tau_{TM}(\lambda)\tau_{TY}(\lambda) \qquad (4.8)$$

The relative density of each dye is:

$$\tau_{TC(\lambda)} = 10^{[-cD_{NC(\lambda)}]}, \qquad \tau_{TM(\lambda)} = 10^{[-mD_{NM(\lambda)}]}, \qquad \tau_{TY(\lambda)} = 10^{[-yD_{NY(\lambda)}]} \qquad (4.9)$$

The non-linear relationship between the layers leads to a shift in colour balance in very dark or light parts of a picture (see chapter 3). Colour negative films show much more uniform properties at low and high light levels and more nearly resemble black and white films. In the middle of each curve the density versus exposure can be approximated by:

$$c = K_{FC} + \gamma_C \log_{10}(X_R), \qquad m = K_{FM} + \gamma_M \log_{10}(X_G),$$

$$y = K_{FY} + \gamma_Y \log_{10}(X_B) \qquad (4.10)$$

as in (4.5). The complete colour film model is shown in figure 4.1.

Where commercially processed colour films are used, one has a choice of film type only. The variation in colour films depends not only on type, but also on light and subject colour quality which are not controllable.

In summary, recording of data on film through photography is a highly non-linear complex process using techniques which have evolved since the early part of the nineteenth century, not all of which are fully explained in the limited literature available. Information which is not recorded in the linear portion of the H + D curve can only be imperfectly recovered, if at all. Many phenomena are not under the direct control of the photographer, and, unfortunately, few users pay attention to the possibilities which do exist through choice of film type, exposure, and subsequent laboratory treatment.

4.2. The image as numerical data

4.2.1. Getting a picture into the computer

Scanners

A densitometer which is automated to measure the transmission of a film or the reflectance of a paper print over its entire surface is required for input to a computer. In one traditional type of apparatus, the drum scanner, figure 4.2 and figure 4.3, the film is fastened over a rectangular opening in a cylindrical metal drum which is rotated by a motor coupled to a shaft encoder which gives precise data on angular position. A fine beam of light is focused on the back of the film through a microscope lens. It passes through the plastic base material and the developed silver layer and forms a tiny tight spot on the sensitive surface, figure 4.4. A highly sensitive photomultiplier or semiconductor detector 'sees' this spot through a second microscope lens via an aperture plate and the photoelectric current is measured and converted to digital form. The size of the aperture plate determines the size of the area measured. The current is a direct measure of film density. By rotating the drum, a line of density measurements can be made along the path which the light spot traverses. When the line is finished, the photocell and light source are moved a small distance equal to the spot size by a lead screw coupled to the driving motor and the process is repeated. In this way the film surface is scanned completely. The mechanical arrangement is very similar to that of precision lathe.

In a typical drum scanner, the area seen by the photocell is adjustable

Fig. 4.1 A colour film model. From ref. 38.

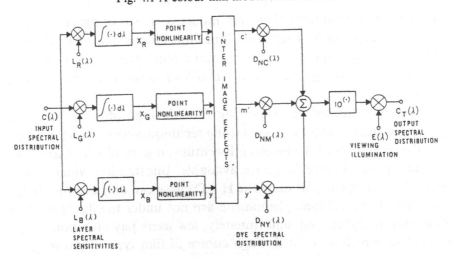

in steps between 12.5 and 100 μm by means of the pair of rotating aperture plates mounted in the light path. For such a small rapidly moving spot, the electronics must be able to measure the photocurrent accurately within a few microseconds before the drum rotates an appreciable distance. If the density reading time is short relative to the rotational speed of the drum and the size of aperture, the photocell appears to see a nearly stationary point. The value of the photocell

Fig. 4.2 A drum film scanner.

Fig. 4.3 Functional schematic of a drum film scanner. From ref. 33.

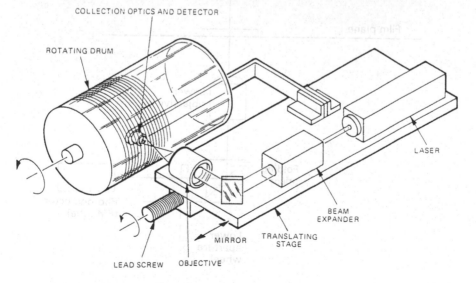

current which corresponds to a particular value of density in the image is transformed to a digital value which can be readily stored. An opening on the drum opposite to that over which the film is fastened is used to provide a zero density reference pulse. This is used to compensate for zero drift in the logarithmic photocurrent amplifier, the light source, and the photocell. For very exact density measurement, a neutral filter of accurately known density can be placed over another opening and the pulse obtained from it used to calibrate the high end of the density range. Many more refinements in drum scanner construction are known to the few manufacturers of these devices which, being made to order, are extremely expensive. There is little information about the details in the published literature.

Fig. 4.4 Schematic view of the optical system of a drum or flatbed scanner using collimated light.

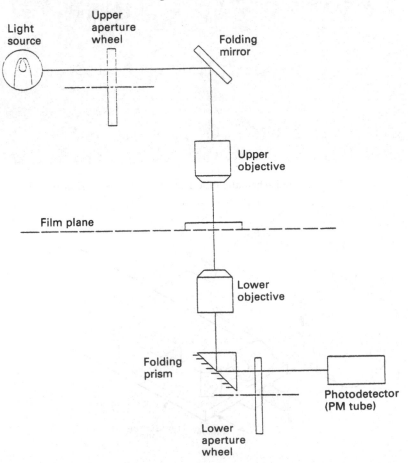

The process of digitising a single reading must be completed before the drum moves to the next spot position. Because of the inertia of the drum, movement is continuous even when stepper motors are used instead of continuous drives with shaft encoders. Digitised film densities are the gray values of individual picture elements (pixels). Many other devices for digitising pictures have been invented during the last few decades. These vary in complexity, accuracy, and of course, price. Most modern scanners are based on semiconductor devices which scan one axis electronically. These scanners do not have a collimated light beam passing vertically through the image and hence are not as accurate in density measurement because of the Callier effect (9). They are much faster and cheaper than a drum scanner, and are quite adequate when very high density measurement accuracy is not required. The Callier Q factor is a measure of the degree of light scattering in a heterogeneous medium. It is defined as:

$$Q = D_s/D_d$$

where D_s is the specular density and D_d the diffuse density of the film. Since almost all media have some degree of turbidity, Q is always greater than 1. For dye mixtures in gelatin, used as neutral density filters, Q is very nearly 1, and a photographic image at average density has a Q of about 1.4 so that considerable error in density measurement is introduced in darker parts of an image if non-collimated or off-axis light, which is more highly scattered, is used.

Gray level resolution and film 'noise'
Experiments have shown that the unaided eye can readily distinguish about 60 discrete gray levels in a typical black and white picture of high contrast. Digitising must be carried out to at least 30 levels if contouring effects at the margins of level transitions in large uniform gray areas are not to be seen. Usually, digitising is done to 256 gray levels corresponding to photographic densities of 100 or 1000 to 1. The latter allows a wider margin for very dark and light parts of the picture but with coarser steps since a number between 0 and 255, which can be stored in one byte is produced regardless of the density range in most machines. No normal film can record 255 distinct gray levels. Most aerial films do not have a usable density range much above 100:1 or so. The maximum number of measurable steps is influenced by the 'grain' of the emulsion produced by clumps of developed silver (2). These random grains scatter the light of the scanner and statistical uncertainty about a given gray value exists

at a given resolution. Two factors are of importance: (1) the light intensity reaching the photocell during the measurement period which is photon limited for small apertures; (2) the graininess of the film. The result is that a recorded value of gray level varies slightly about a mean value. The variation is a complicated function of grain size distribution, mean density, scanner aperture, scan light intensity, and the statistical properties of the detector employed. A well-designed scanner should not be photon limited. A simple approximation lumps all of these in a Gaussian distribution for film grain 'noise' fluctuation as:

$$p[D(x, y)] = [2\pi\sigma_D^2(x, y)]^{\frac{1}{2}} \exp\left\{-\frac{[D(x, y) - u_D(x, y)]^2}{2\sigma_D^2(x, y)}\right\} \quad (4.11)$$

where the exposed film density $D(x, y)$ varies about a mean density $u_D(x, y)$, with a variance $\sigma_D^2(x, y)$. The average density is taken by the scanner window around (x, y) with a standard deviation:

$$\sigma_D(x, y) = \alpha[u_D(x, y)]^\beta \quad (4.12)$$

where β is a constant usually between $\frac{1}{2}$ and $\frac{1}{3}$, and α is empirically set to:

$$\alpha = 0.66[a/A]^{\frac{1}{2}} \quad (4.13)$$

where a is the average grain size, and A is the window area. The Gaussian density as given by equation (4.11) can be decomposed into a mean density and a normally distributed random component $N(x, y)$:

$$D(x, y) = u_D(x, y) + \alpha[u_D(x, y)]^\beta N(x, y) \quad (4.14)$$

Thus film grain noise depends on exposure and is additive in density, so that the process is non-linear.

The film transmittance can be stated as:

$$\tau(x, y) = \tau_0(x, y) \times 10^{[\sigma_D(x,y)N(x,y)D(\lambda)]} \quad (4.15)$$

the product of the noise free transmittance and a noise factor. Film grain noise is thus multiplicative in the intensity domain, and this is of importance when designing methods to improve image quality as will be seen below. The model for image recording and scanning is shown schematically in figure 4.5.

A typical measurement in the middle gray range of the drum scanner described above has a standard deviation of 7 units in the 0–255 range with high speed aerial film so that about 36 gray levels are actually distinguishable at this single pixel level to one standard deviation in a single reading. A check with a gelatin reference film of density 1.0 which has no grain shows a standard deviation of 1 in 255. This shows that the gray level resolution is limited by film grain noise and not by the photon noise in the optical system. The eye responds to average properties of local area, so that gray levels can be distinguished to better than one standard deviation, and the total number of perceived grays appears higher than that measured. A film with lower speed and finer grain will give better results than the worst case chosen for the experiment.

Digitising quality should always be limited by film properties rather than by the performance of the scanner. The scanned pixels represent the photographic densities of the film over a range which is 2.0 or 3.0 D. Most information in aerial films suitable for low level oblique photography is contained in the density range between 0.5 and 1.5 D. Terrestrial images usually contain most of their information between 0.2 and 0.9 D. The 1:100 range of the scanner is therefore used to obtain maximum differentiation of gray values. Other types of scanners do not have such a great dynamic range or precision in gray level measurement. However, given the large number of pixels, the eye tolerates quite severe departures from the optimum and little difference can be seen in final results if the quantization is somewhat coarser.

Geometric resolution requirements for aerial images
For a large size (>130 mm) aerial image on professional black and white film, 50 μm is a useful scanning spot size producing about 2400 readings per line on a 5 in aerial film. With 70 mm film, 25 μm can be used to produce a similar number of readings. The larger format images are square, so that about 2400 scan lines are needed. This gives almost six

Fig. 4.5 A model for image scanning and recording. From ref. 38.

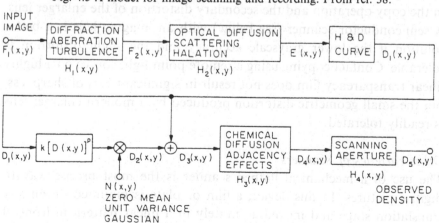

million readings of image density over the film surface. The scanner
described above can deal with an area up to 230×230 mm, so that
several images may be scanned at the same time if printed together on
a contact copy transparency or one vertical 230 mm image can be dealt
with. At 50 μm resolution, the scan takes about 20 min, with the drum
rotating at 4 revolutions per second. At 25 μm, 80 min are required.
Non-mechanical scanners are usually much faster. If smaller size films
are used, smaller spot sizes are needed, but larger batches of pictures
can be printed on a suitable transfer film and scanned simultaneously.
The user must pick out the individual images from the large scan file
afterward and transfer them to separate files for further processing. If
all pictures can be given the same treatment, then the batch can be
processed at once, treating the large scan as a single image. This technique
saves considerable time when dealing with production quantities but
costs a larger amount of disk storage.

Resolution to about 1 part in 2000 has proven to be enough for most
practical applications in image enhancement and is adequate for the
simple cartographic needs of archaeology. Resolution as low as 1 part
in 500 is adequate for enhancement of significant parts of an image,
allowing much smaller computers and cheaper scanning devices to be
used if the whole image need not be treated. For publication, pictures
should be digitised to at least 1200 lines when reproduced at normal
page sizes. At this resolution the line structure cannot be seen at normal
viewing distances and contrasts. When small images (35 mm) are
involved, it may be difficult to achieve high resolution with mechanical
scanning of the original film. In this case one solution is to enlarge
optically to a transfer film, accepting the slight information loss inherent
in the copy operation and the secondary distortion of the enlarger lens.
A semiconductor scanner (see below) with an imaging lens is a better
solution if the lower grayscale accuracy and the slight distortion is
tolerable. Contact copying using a remote point light source on a highly
linear transparency film does not result in significant loss of sharpness,
but the small geometric distortion produced by a modern enlarger lens
is readily tolerated.

Other varieties of scanners
The use of a mechanical flatbed scanner is the most precise way to
digitise pictures. In this device, a film or plate is mounted on an x–y
translation stage and moved accurately in a linear pattern in front of

the scanning optics, figure 4.6. Flatbed scanners are used for precision cartography and in astronomical image processing since they can readily be made accurate to a fraction of a wavelength of light. The price paid is slow scanning speed, for rapid translation, acceleration, and deacceleration of the scanning table are not easily achieved together with fractional micrometre accuracy.

The drum scanner is considerably faster but has lower geometric accuracy. This is because it relies on the stiffness and dimensional stability of the film around the drum opening to maintain a true cylindrical shape and stay in focus. With a flatbed, the film is pressed against an optically flat glass plate, an inherently stable arrangement. Both devices are the products of precision optical engineering, are produced by hand in small quantities, and are consequently too expensive for most archaeological establishments. The drum scanner is, in fact, adequate for precision photogrammetric work, as Trinder (53) has shown. A typical high quality photogrammetric image has a point spread function (defined below) of about 25 μm at two standard deviations. The smallest aperture of a typical drum scanner is 25 μm, comparable to the spread function. The overall geometric accuracy has been experimentally shown to be about

Fig. 4.6 A high precision dual axis flatbed scanner. From ref. 33.

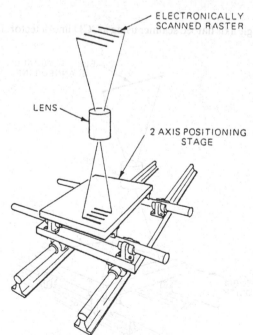

ELECTRONICALLY
SCANNED RASTER

LENS

2 AXIS POSITIONING
STAGE

0.18 pixels, which is more than good enough for any archaeological image. In fact, practice has shown that 50 μm is almost always adequate for larger aerial photos.

Another type of scanner contains a semiconductor image line scanner (CCD array) combined with a mechanical mirror or lens system which moves the array across the image, figure 4.7. This scanner takes up to several seconds to scan 2500 lines. Horizontal resolution up to 4096 points is available at higher cost. Such scanners have the advantage of closely controlled geometry and fairly good resolution of gray level information, although each cell in the array requires program correction for the best results. Production quantities are fairly high and costs are reasonable compared with high resolution drum or CRT (cathode ray tube) scanners. Modern versions based on the mass produced mechanics used in photocopying machines are reasonably priced and have good geometric accuracy. Unfortunately, the cheapest ones are produced for scanning paper black and white reflection paper copy only with consequent further reduction of grayscale reproduction range and geometric accuracy.

The cheapest scanner is a television camera. A transparency is placed in front of a uniform light source or paper copy on a uniformly

Fig. 4.7 A single axis flatbed scanner using a CCD line detector. From ref. 33.

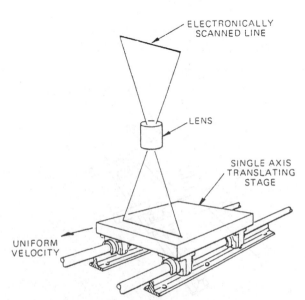

illuminated easel. The signals from the camera can be digitised and coupled to a computer using a cheap commercial interface (frame grabber). Television cameras are economical devices for scanning a picture because they are made in large quantities, and the electronics for coupling them to the computer may actually be more costly than the camera itself. The measurement of photographic density via a television camera is inherently less accurate than that obtained in the rotating drum densitometer, and the geometric accuracy for a vidicon is poor. About 30–40 gray levels can be recorded with about 20 of them in the range of interest, and geometric accuracy is seldom better than 1 part in 200. By digitising a film placed in front of a plate with accurately placed calibration marks (a reseau plate) and using these with an interpolation procedure, the errors in the camera geometry can be corrected to about 1 part in 1000 (12). By averaging the pixel values at each point in an image, the noise at each point decreases with the square root of the number of averaged frames, so that the quantising error per pixel is reduced. High resolution slow scan television cameras have also been made giving 1000–2000 line resolution at an intermediate price. These have given way to semiconductor devices which offer similar precision at much lower cost. The results obtained with a television camera when dealing with critical pictures which have a very large brightness range may be disappointing and the device is not to be recommended if more money is available.

A scanner which can give good geometric and grayscale accuracy contains a precision television display tube (CRT) which displays a constant raster, the light from which is focused on the image by a mirror or a lens, and passes onto a photocell for digitising, figure 4.8. CRT scanners are easily as expensive or more so than mechanical scanners, but are more rapid and have no moving parts. They are especially useful for scanning colour slides since the lens allows reduction of the scanned area with resulting high resolution directly at the original. Colour separation is easy if the CRT has a white phosphor and a colour filter wheel is used. Gray levels are recordable to about 1 part in 50–60 or more, but tend to be inaccurate in dark areas since the light does not pass vertically through the film in a collimated beam. Scan geometry can usually be controlled to about one part in several thousand but high accuracy comes at a very high price, higher, in fact, than for a mechanical scanner. Such scanners are still widely used in commercial colour plate making systems for the printing industry.

   The use of a mechanical scan to obtain a sweep of the line array over the image can be replaced by a square semiconductor array with high resolution and no defective cells. These are currently produced in small quantities and are not available at reasonable prices (51). Television cameras for semiprofessional use with square array semiconductor targets are available and provide a useful alternative to the old-fashioned vidicon television camera systems in terms of geometry and grayscale accuracy. However, it is very difficult to fabricate a square array with more than about 500 elements on a side with no defects. When used in a television camera, a few defective or low sensitivity elements will go mostly unnoticed, but in a scanner this is very objectionable. The cameras used in the NASA Space Telescope have four arrays with 800×800 elements without defects, and the prices are appropriately astronomical.

Scanning colour films
Colour films now widely used in archaeological photography present a special problem for scanning. The colour information is contained in the densities of the yellow, magenta, and cyan dyes of the picture. Narrow band filters in the scanner are needed to separate the three colour density components. Some scanners use a rotating filter wheel and pass over the image three times, once for each colour component. This has the

Fig. 4.8 A CRT film scanner or recorder. From ref. 33.

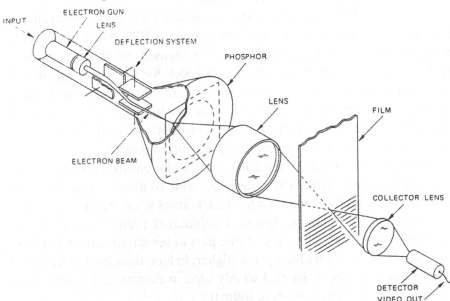

advantage that the detector is identical for all components. The disadvantage lies in obtaining accurate repositioning of the scanning element for each pass so that the pixels for the three components coincide exactly. Higher price scanners use a beam splitter and three matched detectors or three switched light sources and a beam combiner.

The amount of data in a scanned colour image is three times that in a black and white picture. The data are not all equally significant, since the eye is much more sensitive to small changes in green than in red and blue. The colour separation filter centre wavelength and bandwidth are different for the overlapping dye response of each film, and must be optimised if the best results are to be obtained. Performance is usually optimised for the magenta dye which controls the green. The eye can see small differences in colour balance best in this range. Tunable filters using linear interference filters or tunable laser light sources are used in the highest priced instruments.

In addition to the methods described, there are a number of other devices which have been produced to digitise images and transfer the pixels to a computer which have been discussed in the specialised literature. Undoubtedly with progress in electronics, more will be invented in the future and some are already produced at costs which are within the reach of a modest archaeological establishment.

4.2.2. The digitised picture

If we look at the highly enlarged surface of a scanned image as reproduced on film, we can see the actual pixels as small squares with slightly different levels of gray, figure 4.9. Each pixel represents the numerical value of the average transmission of a scanned transparency at a given point. Usually, the images are stored on disk in the computer before further processing. Disks may hold numerical data from dozens of images, depending on the scanning resolution and picture size. All or part of the picture must be held in the computer's main memory for immediate processing. The type of operations which are to be performed on the picture have a considerable effect on where data must be stored, for this determines the rapidity with which they must be accessed. There is a hierarchy of storage arranged in order of decreasing access times. Magnetic tape or optical disk is slowest, magnetic disks more rapid, and main memory the most rapid. In some very small machines, it used to be useful to use the picture itself in the scanner as its own data storage, for access to any part of the image was faster that way than if recorded on tape. This approach had the disadvantage that repeatability of access in terms of positional and grayscale accuracy was not good unless a very high quality scanner was available. With cheap mass storage devices readily available for even the smallest machines, this is no longer necessary. In the distant past, some limited kinds of electronic treatment were carried out directly on signals coming from a scanning device without any digitising at all. These methods, which are still occasionally encountered, have very restricted application because better techniques require exact numerical values for gray levels or access to pixels in widely separated parts of the image at the same time. Such so-called 'analog' machines are very fast, if not very flexible. They were widely used in automatic enlargers and printers, especially for colour work but they are now museum pieces.

Fig. 4.9 Surface of a scanned image reproduced on film showing individual pixels. The areas are perceived by the eye as being uniformly gray at four different degrees of resolution.

4.2.3. Picture output: displays, film writers

Displays

After digitising, a picture can be thought of as a rectangular array of numbers or as three such arrays if the picture is in colour. The side lengths of the array are equal to the number of lines scanned and the number of pixels in each line, with gray values usually limited to a range of 0–255. Numerical processing techniques operate on this array of numbers to produce a new array which must be made visible for the user. Techniques which do not modify the geometry of the image produce one output pixel for every input pixel. Methods which modify geometry may produce fewer or greater numbers of pixels depending on whether the image is enlarged or reduced. Intermediate results must be displayed so that decisions can be made about further steps to be taken. After all operations have been completed, a new negative or transparency is often required for reproduction by normal photographic means.

It is customary to control numerical treatment by displaying intermediate results on a display device which allows one to see the pixel values converted back to black and white or colour pictures. The picture is usually displayed line by line with quality much superior to that of normal television, but less than possible with film output. Such raster display devices are connected directly to the processing computer, and many modern versions have sufficient internal computational power to allow the user to try simple numerical methods directly on the screen almost instantaneously. Some personal computers have sufficiently high resolution for this kind of use if fitted with a high resolution graphics adapter card. The display system has its own memory and the image is usually transferred from disk via computer memory to display memory. In some computers, display memory is an extension of the main computer memory itself and can be addressed directly for all operations. The memory is dual ported, with one port addressed by the computation circuitry and the other by the video circuitry which converts the pixels to voltages which can drive a CRT. For very quick results on film, the video signals from the raster display memory may be applied to a CRT and an instant film camera. Although some very cheap display systems are available with resolutions under 250 pixels per line, visual results first appear to be satisfying when the resolution approaches 500 pixels. More than 1000 pixels per line usually cannot be seen at normal viewing distances, so that resolution need seldom be higher than this for most applications. Where many very fine lines are present, even higher resolution may be of interest.

User interaction

The display system is the control centre through which user sees and guides all operations. It has a keyboard for entering data and choosing functions and usually has a 'trackball', 'joystick', graphic 'tablet', or 'mouse' which permits the user to point to objects of interest on the screen using a cursor which appears superimposed on the picture. It is moved about the screen via one of these devices at will. The cursor is typically used to give the position of fixed points in an image for later use in geometric correction programs, or to pick out areas of interest for special processing. The function keys of a personal computer can also be used for this purpose.

Picture output (hardcopy) devices

Permanent copies of results are usually needed at a much higher resolution than the display system provides. Photographing the display system is a poor substitute. The most expensive approach is to use a rotating drum device, mechanically similar to that used in scanning. For the highest quality in making colour printing plates a flatbed film recorder is used. An unexposed film is placed on the drum or stage and a very fast electronic source exposes it with brief flashes of light point for point as the drum turns. The light source and the position of the drum are computer controlled with the intensity or number of the flashes made proportional to the numerical value of the result pixels. In some high priced film recorders for the printing industry, the film to be exposed is held on the inside of a semicircular drum by vacuum. The computer controlled light source is reflected by a moving rotating mirror or prism on a lead screw and exposes the film surface. Light intensity is modulated by pulsing or with a Kerr cell, figure 4.10. Results comparable to a flatbed film recorder are achieved at a somewhat lower cost for the mechanics.

Another method for exposing an output picture on film uses a precision CRT as a light source, like that in a CRT scanner. In fact a CRT scanner itself can be used, if a light-tight film transport casette is available. Such devices are almost as expensive as a mechanical one if accuracy is high, but are faster in operation though less precise in the control of exposure and geometry. Rapid output devices attached to the display are similar in principle, but lower in resolution. Colour is usually obtained with a white light source and exposure through filters.

The most elaborate drum film writers have three light sources passing simultaneously through filters for this purpose or tuned dye lasers may be used for highest writing speed. Three runs are made with a single light source using a different filter each time in lower cost machines.

Computer controlled colour photocopying machines are an interesting future alternative for production of fast colour hardcopy, but the saturation and reproduction ranges of the dyes used are not as good as those obtainable in colour transparencies or in paper prints made from colour negatives.

Accuracy of digitising

There are three fundamental measurements which can be made on a film with a scanner. The measurement of density is one, the two coordinates on the film at which the measurement is made are the others. One great advantage of computer methods as opposed to conventional photographic or photogrammetric techniques is that the measurement need be made only once at each point and the result stored on disk. Afterwards, each measurement of density and position can be recalled exactly from storage as required without any mechanical error either in

Fig. 4.10 A curved film plane recorder. From ref. 33.

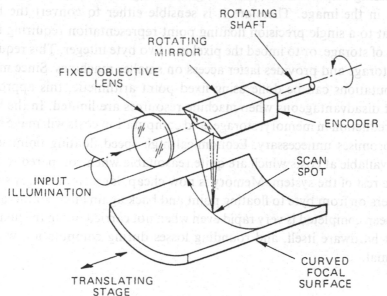

position or gray value. In other words, repeatability is perfect. This is not the case with the traditional mechanical devices used in photogrammetry, although these have attained very high accuracy through decades of refinement. It is certainly not the case with purely photographic methods of treatment of gray values, where chemical methods are involved. The numerical approach can duplicate or surpass all of the classical techniques in accuracy and repeatability, and for a few important operations, in speed. The computer deals with the image as a large array of numbers whose values and positions in the array are known at all times, and where any given value can be accessed with times of the order of a fraction of a millionth of a second. Operations are usually carried out pixel by pixel, in contrast with optical methods, where the whole or a large part of an image is dealt with at once. Purely optical methods are usually much faster than numerical techniques, but are not as readily controlled and often require high precision equipment and skilled personnel which may be much more expensive than a computer–scanner–display–recorder system. A few multiprocessor systems which can work on many pixels at the same time exist, but these are not likely to be available for archaeological use.

Dynamic range of computation and rounding errors
Pictures scanned to byte (0–255) precision cannot be put through many processing steps without producing rounding errors which appear as noise in the image. Therefore it is sensible either to convert the byte format to a single precision floating point representation requiring four bytes of storage, or to imbed the pixel in a two byte integer. This requires less storage and provides faster access on smaller machines. Since many computations can be done using fixed point arithmetic, this approach is not disadvantageous when machine resources are limited. In the long run, reduction in memory, storage, and computation costs will make such compromises unnecessary. Economical high speed floating point units are available at costs which are quite reasonable when compared to that of the rest of the system. Memory is now cheap, large, and fast. A single conversion from byte to floating point and back again after computation has been completed is very rapid even when not carried out in the floating point hardware itself, and rounding losses during computation will be minimal.

4.2.4. Software

For all computer operations, clearly defined algorithms are required so that programs can be written. Generalised mathematical techniques must usually be rearranged for step by step calculation if they are to be carried out efficiently. The algorithm is programmed to operate on the data of the image and usually produces a new image. Data may have to be restructured for efficiency in either speed of processing or optimum use of machine storage facilities, and the type of data structure used depends very much on the available equipment, the computer's operating system and the algorithms needed. Optimal data structures are highly dependent on machine architecture and resources, as well as on the choice of algorithm to achieve a given aim.

In a practical image processing installation, many useful methods may have already been programmed and can be chosen by the user. He must have some idea, however, about the capabilities of the many known techniques, and which to choose for his problem. When using an existing installation or commercial software package, the user will have to read the documentation thoroughly before beginning unless he has someone to guide him through the operation of a few programs which may be of interest to him. The image processing literature is vast, and it is only possible to discuss a few of the most frequently needed techniques here. Several tens of thousands of articles have appeared during the last two decades (39,40), of which only a small number give algorithms which are used repeatedly.

There are many systems now in existence or on the market which vary greatly in available equipment and software. Unfortunately standardisation of software and usage does not exist. If an installation for archaeological purposes is planned, then skilled personnel may be required to set up and operate the system, and one should be prepared for a very considerable financial outlay over a long period of time unless a 'turnkey' system is acquired. There are several image processing software packages commercially available including a few for personal computers, and the user should look into the possibility of implementing one or more of these in a new system before writing any programs (24). Some complete image processing systems are delivered directly with the software accompanying the hardware and this is the method of choice for the financially potent unskilled buyer.

4.3. Elementary statistical properties of images

Pixels are usually represented by a single byte, a positive number between 0 and 255. The number may describe either greater brightness or greater darkness, a choice made by the system designer. In a typical system which is film oriented, higher valued numbers mean greater density in the scanned picture. Systems working with images obtained from satellite data usually use the opposite convention. The polarity of the image can be readily inverted to display a positive, since it is difficult for the eye to interpret detail in a negative image. Image polarity inversion is best carried out in the display system itself rather than on the data. Most displays permit this to be done with negligible calculation time. If inversion of the data itself is required, this can be readily programmed as a table look-up operation in the scanner control program itself if desired.

The values of the pixels of a picture are the basic units for all further operations. The simplest operation is that of counting the number of times pixels with a given gray value occur in a picture or a part of it. These statistics can be computed for the whole picture or in a local area of interest. For example, the average gray value in a given area of a picture is obtained by adding up all the gray values and dividing by the number of pixels counted. Let $f(x, y)$ be the brightness at point (x, y) in an image. Examine a small square region around this point of size $(2k + 1) \times (2k + 1)$. It has an average gray value:

$$m(x, y) = \frac{1}{(2k + 1)^2} \sum_{u=-k}^{k} \sum_{v=-k}^{k} f(x - u, y - v) \qquad (4.16)$$

This average is perceived as the brightness of the area. If one plots the number of times each gray value occurs against the gray value one obtains a histogram, figure 4.11. This histogram of a small uniform area contains only one peak whose centre value corresponds to the brightness of that area. The standard deviation of a small area is:

$$\sigma(x, y) = \left\{ \frac{1}{(2k + 1)^2} \sum_{u=-k}^{k} \sum_{v=-k}^{k} [f(x - u, y - v) - m(x, y)]^2 \right\}^{\frac{1}{2}} \qquad (4.17)$$

The standard deviation of a small area corresponds to perceived local contrast. A picture with widely separated peaks and a large global standard deviation is perceived as having high global contrast. The separation of two peaks in a pixel histogram is a measure of the difference in brightness in two areas, figure 4.12. A Student t test is convenient for describing the separation of the peaks in an area (25).

In samples taken from a Gaussian distribution, the ratio t of the sample mean and the distribution mean μ to the estimated standard error is:

$$t = \frac{(\bar{x} - \mu)[n(n-1)]^{\frac{1}{2}}}{[\sum (x - \bar{x})^2]^{\frac{1}{2}}} \qquad (4.18)$$

where n is the number of samples. The distribution of t and tests for significance are given in the statistical literature (25) and can be calculated to see if differences between two peaks are significant or not.

In a histogram obtained over a large area, the spread of the multipeaked histogram reflects the apparent overall contrast. A fast algorithm for computation of the mean and standard deviation is thus required. Simple statistics can be calculated with algorithms which avoid redundant computation and can later be most useful in modifying an image (23,31,48,55).

Fig. 4.11 The single peak histogram of pixel values of a small area of uniform grayness.

NUMBER OF PIXELS	10007		
M E A N	113 86	M E D I A N	114 00
VARIANCE	32 43	25% QUARTILE	110 00
STD DEVIATION	5 66	75% QUARTILE	118 00
SKEWNESS	4 23	ABS DEVIATION	4 42
KURTOSIS	2 55	ENTROPY	4 53

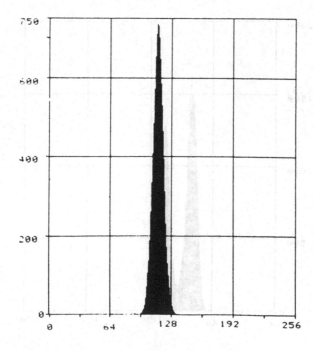

Another important statistic is the central value in a histogram, the median. Let P_{xy} be the sample cumulative distribution function of $\{f(y-u, x-v)|u = -k, \ldots, 0, \ldots, k; v = -k, \ldots, 0, \ldots, k\}$. Then the local median is $M(x, y) = \mu$ such that $P_{xy}(\mu) = \frac{1}{2}$. Half the gray levels are greater, half less than this value. For symmetrical histograms, the median is the same as the average, but with a multipeaked or skewed histogram, they may differ considerably. The median is obtained easily by counting the number of values starting from the lowest, until half the maximum, known from the size of the measured area, is reached. There is an elegant algorithm for computing the median in a small square area moved across the image by updating the previous median which makes operations using medians almost as fast as those using averages (19). A still faster method for obtaining medians and quantiles (p. 153), even for data not limited to byte pixel ranges, the *P*2 algorithm of Jain and Chlamtac (23),

Fig. 4.12 A double peaked histogram measured in an area overlapping two adjacent areas of uniform grayness.

```
NUMBER OF PIXELS      40000
M E A N               98 50      M E D I A N           94 00
VARIANCE             227 38      25% QUARTILE          85 00
STD  DEVIATION        15 48      25% QUARTILE         113 00
SKEWNESS               0 14      ABS  DEVIATION        13 31
KURTOSIS               1 44      ENTROPY                5 51
```

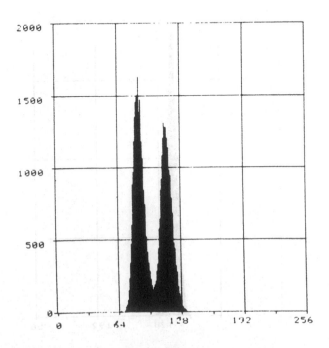

can be used with profit when a wide range of data requiring floating point calculation is needed. It may also be faster than the Huang, Yang, Tang algorithm (19) for byte data if carefully programmed.

An order statistics measure for dispersion analogous to standard deviation but similar in computation to the median is the difference between the first and third quartiles of the histogram $Q(x, y) = v - \lambda$, where $P_{xy}(v) = \frac{3}{4}$ and $P_{xy}(\lambda) = \frac{1}{4}$. The first quartile is the gray value which is greater or equal to a quarter of the total number of gray values in the histogram. The third quartile is at the three-quarter point. This measure is the interquartile difference. Together with the median, it may be made the basis of a number of useful image processing operations (48). A simple extension is a dispersion measure based on any desired quantile. The 2.5 and 97.5% quantiles $P_{xy}(v) = 0.975$ and $P_{xy}(\lambda) = 0.025$ have proven useful for display of archaeological magnetic prospecting data (49); see also chapter 8. They represent something close to the conventional statistical choice of two standard deviations for values lying within the compact zone of a unimodal distribution. Values outside this range are considered to be outliers. Order statistics based measures are useful for processing with small moving windows which do not contain enough pixels for more conventional measures of dispersion like the variance to be stable.

4.4. Mathematics of image analysis

4.4.1. Representation by means of orthogonal functions

An image forms a data set with two dimensions and one or more amplitudes: x, y coordinates and the gray value or three colour components. For digitised images all values are usually positive integers, there being no negative light. Scan coordinates are usually given in standard matrix representation where rows and columns are numbered from the upper left corner. But more generally, the values can be assumed to be simply real signed numbers. A system of orthogonal functions to represent this data set will be defined. In general, these functions are complex. The grey values can be approximated by a linear combination of these functions. In one dimension functions ϕ_i defined in the range $a \leq x \leq b$ are said to be orthogonal if they obey the condition:

$$\int_b^a \phi_i(x)\phi_j^*(x)\,\mathrm{d}x = K_i\delta_{ij} \tag{4.19}$$

where * implies the complex conjugate and $\delta_{ij} = 0$ for $i <> j$ and $\delta_{ij} = 1$ for $i = j$. They are orthonormal if $K_i = 1$. A system of orthonormal functions can be defined by:

$$\hat{\phi}_i(x) = \phi_i(x)/K_i^{\frac{1}{2}} \qquad (4.20)$$

There is no loss of generality if it is assumed that the (ϕ_i) are normalised and defined in the interval $[0, 1]$.

Consider for a function f a linear combination of orthonormal functions:

$$\hat{f}(x) = \sum_{i=1}^{N} c_i \phi_i(x) \qquad (4.21)$$

where:

$$c_i = \int_0^1 f(x)\phi_i^*(x)\,dx \qquad (4.22)$$

The (ϕ_i) are complete in the sense defined by Courant and Hilbert (8), if any piecewise continuous function $f(x)$ can be represented by a linear combination such that the mean square error:

$$I_N = \int_0^1 \left| f(x) - \sum_{i=1}^{N} c_i \phi_i(x) \right|^2 dx < \varepsilon \qquad (4.23)$$

is less than a given arbitrarily small positive number ε for sufficiently large N. With real images we have to deal with discrete data sets having a finite number of terms. This means in one dimension that $f(x)$ is known (measured) at discrete points x_i, $i = 1, 2, \ldots, N$.

If:

$$(1/N) \sum_{j=1}^{N} \phi_i(x_j)\phi_k^*(x_j) = \delta_{ik} \qquad (4.24)$$

then equations (4.21) and (4.22) become:

$$c_i = (1/N) \sum_{j=1}^{N} f(x_j)\phi_i^*(x_j) \qquad (4.25)$$

$$f(x_j) = \sum_{i=1}^{N} c_i \phi_i(x_j) \qquad (4.26)$$

These equations define a discrete transformation and its inverse using the functions. The transforms are performed as matrix multiplications. Define the matrix Φ by:

$$\Phi_{ij} = \phi_i^*(x_j) \qquad (4.27)$$

then one obtains with $f = {}^t(f(x_1), \ldots, f(x_N))$:

$$C = \left(\frac{1}{N}\right)\Phi f, \qquad f = \Phi^+ C, \qquad \Phi\Phi^+ = \text{Identity matrix} \qquad (4.28)$$

where $^+$ signifies the Hermitian conjugate, whereby the opposite sign for the imaginary part of a complex number is taken.

A number of general theorems hold:

(1) If the mean square error is $I_N(\gamma)$ for piecewise continuous f:

$$I_N(\gamma) = \int_0^1 \left| f(x) - \sum_{i=1}^N \gamma_i \phi_i(x) \right|^2 dx \qquad (4.29)$$

then $I_N(\gamma)$ is a minimum for $\gamma_i = c_i$ where c_i is defined by equation (4.22).

The mean square error depends very much on the kinds of functions used for the approximation of the given data. The more closely the shapes of the functions approach the geometry of the data, the smaller the error will be. The most commonly used functions used in image processing are sines and cosines because for these functions a number of other useful properties are also satisfied. They do not approach the shape of the digitised pixels in an image when digitisation is coarse, but if the scan is fine enough, then sines and cosines are quite satisfactory. For widely spaced data such as that obtained in magnetic prospecting, this may not be the case.

(2) Parseval's formula for discrete data sets:

$$\sum_{i=1}^N |c_i|^2 = (1/N) \sum_{i=1}^N |f(x_i)|^2 \qquad (4.30)$$

(3) Convergence for piecewise continuous f:

$$\sum_{i=1}^N |c_i|^2 \leq \|f\|^2 \qquad (4.31)$$

where:

$$\|f\| = \int_0^1 |f(x)|^2 dx \qquad (4.32)$$

The series only converges for functions which are square integrable. For other functions $\|f\|$ has no meaning.

In Fourier analysis sines and cosines are used, and the following extremely useful additional properties hold for function f and transform F:

(4) The shift theorem:
   If:
$$f(x) \leftrightarrow F(k) \tag{4.33}$$
   then:
$$f(x - x_0) \leftrightarrow F(k) \exp(ikx_0) \tag{4.34}$$

(5) The convolution theorem:
   If:
$$f(x) \leftrightarrow F(k), \qquad g(x) \leftrightarrow G(k) \tag{4.35}$$
   then:
$$\int_{-\infty}^{\infty} f(y)g(x - y) \, dy \leftrightarrow F(k)G(k) \tag{4.36}$$

The shift and convolution theorems are only satisfied for very few functions in addition to sines and cosines.

4.4.2. The point spread function

If a class of operations applied to the data obeys:
$$O[af + bg] = aO[f] + bO[g] \tag{4.37}$$
for all images f and g, and for all constants a and b, the result of applying O to f is said to be linear. Many operations are possible where this relationship does not hold and these are said to be non-linear. A convenient operation designed to investigate linear operations on pictures is the point source function. The light from a far-off star, if the earth's atmosphere did not intervene, might be thought of as such a point source of light. Applying such a light source to an optical system causes it to be modified because of the properties of the system, and the output is the point spread function. A point source can be thought of as the limit of a set of images which are spatially restricted. A simple set is:
$$\text{rect}(x, y) = \begin{cases} 1 & |x| \leq \tfrac{1}{2} \quad \text{and} \quad |y| \leq \tfrac{1}{2} \\ 0 & |x| > \tfrac{1}{2} \quad \text{or} \quad |y| > \tfrac{1}{2} \end{cases} \tag{4.38}$$

Let:
$$\delta_n(x, y) = n^2 \, \text{rect}(nx, ny), \qquad n = 1, 2, \ldots \tag{4.39}$$
then δ_n has the value n^2 inside a small square of side length (nx, ny) and zero elsewhere. From this follows:
$$\int_{-\infty}^{\infty}\!\!\!\int \delta_n(x, y) \, dx \, dy = 1 \tag{4.40}$$

When n is allowed to go to infinity, the limit is a Dirac delta function. Here $\delta(x, y)$ is infinite at $(0, 0)$ and zero elsewhere. Consider:

$$\int\int_{-\infty}^{\infty} g(x, y)\delta_n(x, y)\, dx\, dy \tag{4.41}$$

which is the average of $g(x, y)$ over the $(1/n) \times (1/n)$ square at the origin. Therefore, at the origin itself:

$$\int\int_{-\infty}^{\infty} g(x, y)\, \delta(x, y)\, dx\, dy = g(0, 0) \tag{4.42}$$

in the limiting case. If δ is shifted by a small amount (a, b) so that $\delta(x - a, y - b)$ is used instead of $\delta(x, y)$, the values of $g(x, y)$ at (a, b) is:

$$\int\int_{-\infty}^{\infty} g(x, y)\, \delta(x - a, y - b)\, dx\, dy = g(a, b) \tag{4.43}$$

Also it can be shown that:

$$\int\int_{-\infty}^{\infty} \exp[-i2\pi(ux + vy)]\, du\, dv = \delta(x, y) \tag{4.44}$$

which relates the properties of δ to the properties of sines and cosines, since the real and imaginary parts are:

$$\exp(-ix) = \cos(x) - i\sin(x) \tag{4.45}$$

The maxima and minima of the sine and cosine functions define a set of lines at right angles to each other in a plane. These go through a fixed number of transitions along the axis as shown in figure 4.13 which can be expressed in cycles per unit distance. These are the spatial frequencies along the axes. The real and imaginary parts taken together produce a two-dimensional pattern with frequencies u and v. The whole pattern:

$$F(u, v) = \exp[-i2\pi(ux + vy)] = \cos[2\pi(ux + vy)] - i\sin[2\pi(ux + vy)] \tag{4.46}$$

has a spatial period $(u^2 + v^2)^{-\frac{1}{2}}$.

Thus any function can be composed of a linear combination of patterns with this structure. The amount of each pattern taken can be expressed as a set of weights $F(u, v)$ which is called the frequency spectrum of x and y. Any image cannot only be thought of in terms of a set of orthogonal functions, it can also be thought of as a linear sum of a set of point sources, so that:

$$f(x, y) = \int\limits_{-\infty}^{\infty}\int f(\alpha, \beta)\, \delta(\alpha - x, \beta - y)\, d\alpha\, d\beta \qquad (4.47)$$

If the operation O is shift invariant then:

$$O[f(x, y)] = O\left[\int\limits_{-\infty}^{\infty}\int f(\alpha, \beta)\, \delta(\alpha - x, \beta - y)\, d\alpha\, d\beta\right]$$

$$= \int\limits_{-\infty}^{\infty}\int f(\alpha, \beta) O[\delta(\alpha - x, \beta - y)\, d\alpha\, d\beta] \qquad (4.48)$$

which is the same a saying that the sum of the responses to each source is the same as the response to the sum of the sources. Thus where the

Fig. 4.13 The spatial frequencies along the x and y axes in the plane. From ref. 41.

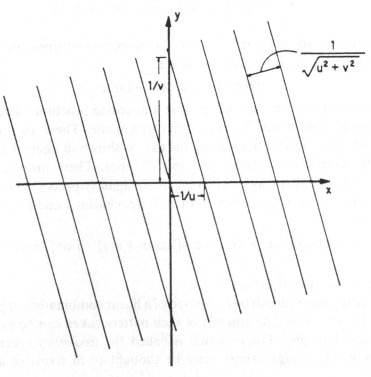

response $h(x - \alpha, y - \beta)$ to a point source δ located at (α, β) is:

$$g(x, y) = \int\int_{-\infty}^{\infty} f(\alpha, \beta)h(\alpha - x, \beta - y) \, d\alpha \, d\beta \qquad (4.49)$$

the integral obeys the convolution property above with $O[f]$ given by g.
If $f(x, y)$ is a circular symmetric function with $r = (x^2 + y^2)^{\frac{1}{2}}$:

$$f(x, y) = f(r, 0), \text{ briefly } f(r) \qquad (4.50)$$

then the frequency spectrum is also circularly symmetric and is equal to:

$$F(u, v) = F(\rho) = 2\pi \int_0^{\infty} rf(r)J_0(2\pi r\rho) \, dr \qquad (4.51)$$

where $\rho = (u^2 + v^2)^{\frac{1}{2}}$ and $J_0(x)$ is a zero order Bessel function of the first kind:

$$J_0(x) = (1/2\pi) \int_0^{2\pi} \exp[-ix \cos(\theta - \phi)] \, d\theta \qquad (4.52)$$

with $\theta = \arctan(y/x)$, and $\phi = \arctan(v/u)$.
The inverse transformation is given in this case as:

$$f(r) = 2\pi \int_0^{\infty} \rho F(\rho)J_0(2\pi r\rho) \, d\rho \qquad (4.53)$$

and the transform pair is called a Hankel transform of zero order whose importance in a totally different context is demonstrated in chapters 6 and 9. Because of the circularly symmetric properties of many optical systems, this transformation is of importance in their analysis.

If two functions $f(x, y)$, $g(x, y)$ are simultaneously present, one can define their cross correlation as the expected (mean) value of their product:

$$C_{fg}(x', y') = \int\int_{-\infty}^{\infty} f^*(x - x', y - y')g(x, y) \, dx \, dy \qquad (4.54)$$

For real functions, convolution differs from cross correlation in that a rotation through $180°$ is carried out before the product is taken. In Fourier transformation notation, cross correlation can be expressed as:

$$F\{C_{fg}\} = F^*\{f\}F\{g\} = F^*(u, v)G(u, v) \qquad (4.55)$$

whereas convolution is:

$$F\{f_1(x, y)\}F\{f_2(x, y)\} = F_1(u, v)F_2(u, v) \qquad (4.56)$$

Let the response of an imaging system be given by its point spread function $h(x, y)$ and the input or scene be $f(x, y)$. Then the output of the imaging system is:

$$g(x, y) = f(x, y)*h(x, y) \qquad (4.57)$$

where * stands for convolution. If the image is sampled at a sampling interval in (x, y), then the sampled output image is:

$$g_s(x, y) = g(x, y)\, \text{comb}(x, y) \qquad (4.58)$$

where:

$$\text{comb}(x, y) = \sum_{m=-\infty}^{\infty} \sum_{n=-\infty}^{\infty} \delta(x - m, y - n) \qquad (4.59)$$

and δ is the Dirac delta. The sampling comb function is shown in figure 4.14.

If a function is cross correlated with itself, this is called autocorrelation. It is defined as:

$$R_f(x', y') = \int\!\!\!\int_{-\infty}^{\infty} f^*(x - x', y - y')f(x, y)\, \mathrm{d}x\, \mathrm{d}y$$

$$= \int\!\!\!\int_{-\infty}^{\infty} f(x + x', y + y')f^*(x, y)\, \mathrm{d}x\, \mathrm{d}y \qquad (4.60)$$

Fig. 4.14 Sampling of an image. From ref. 41.

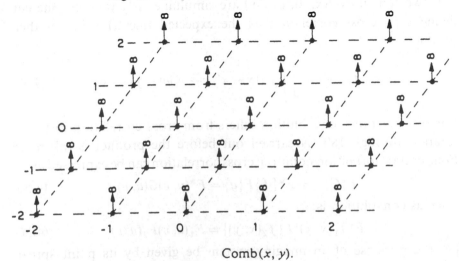

Comb(x, y).

and this can be expressed in Fourier transform notation as:

$$F\{R_f\} = |F(u, v)|^2 \qquad (4.61)$$

This is the power spectrum of the function. Thus the Fourier transform of the autocorrelation function is the power spectrum, a historically important result, since the power spectrum in classical signal analysis is readily measured.

It was stated above that for discrete data the transform operations can be expressed in matrix form. Let $[P]$, $[Q]$ be two real, square, orthogonal matrices. When the matrices are symmetric (where the matrix and its transpose are identical), then the inverses and the matrices are identical. For a sampled picture represented by matrix $[f]$ a transform is given as:

$$[F] = [P][f][Q] \qquad (4.62)$$

For such matrices, a discrete version of the Fourier transform can be defined as:

$$F(u, v) = \sum_{m=0}^{M-1} \sum_{n=0}^{N-1} P(u, m)f(m, n)Q(n, v) \qquad (4.63)$$

It has an inverse:

$$[f] = [P]^{-1}[F][Q]^{-1} \qquad (4.64)$$

and the convolution of two square matrices M and N can be obtained by simply multiplying their discrete Fourier transforms and taking the inverse. This is of considerable advantage when both matrices contain large numbers of elements, because there are very fast library algorithms and special computer hardware available for obtaining the discrete Fourier transform of a matrix in a very short time (1,18,20,41).

4.4.3. Filtering

For any set of orthogonal functions, an operation is possible which linearly modifies the spectrum. In the case of a discrete set of functions this means the selection of a set of weights to be applied to the transform F of (4.62). The spectrum is modified by multiplying each term by a weight. Undesired spectral components may be considered 'noise' by analogy with the acoustic case, and desired ones 'signals'. If the spectral components are sufficiently separated in spatial frequency, then filtering may allow separation of desired from undesired components, with resultant improvement in the ratio of signal to noise. The operation may be written as a product in the transform domain:

$$G(u, v) = A(u, v)F(u, v) \qquad (4.65)$$

Where $A(u, v)$ is called the filter transfer function. In the case of images it is two-dimensional. If it is constant over the entire image the operation is called space invariant. To obtain the filtered data, the inverse transform of the modified spectrum must be taken. Convolution can be carried out by multiplication in the Fourier domain of the transform of the data and the transform of a second matrix, the convolution kernel. This is exactly equivalent to filtering, whereby the weights of the filter function are the Fourier transform of the kernel. When the kernel is large and when it is possible to store all data in fast, directly addressable storage, it is faster to filter in the transform domain than with a convolution. When the kernel is small or when only a small part of the image can be stored in main memory, then convolution is faster than transform filtering. The break-even point between the two methods also depends on how the input/output system for mass storage is organised. Some techniques to be described below make convolution methods more attractive for small machines which are not equipped with special hardware.

4.5. Enhancement of archaeological images

4.5.1. Correction of errors of exposure and development

The typical archaeological aerial photographer pays little attention to technicalities, alas! Photographic exposure is frequently left to the automatic exposure meter in the camera. In older pictures, exposure was a matter of judgement or based on a hand-held exposure meter. The automatic exposure meter, designed for terrestrial use, usually gives a incorrect measure of the amount of light available when used for aerial photography, since it responds to light scattered by dust in the atmosphere and to some sky light as well. Experienced photographers correct for these effects by setting the film speed indicator to a slightly different value from that which would be used on the ground. With colour transparency film, the tendency is sometimes to underexpose deliberately in order to obtain greater colour saturation in green and brown tones. When processing black and white films, darkroom technicians sometimes do not adhere to constant development times and temperatures, although modern automatic machines are not subject to this problem. Often, however, even in the machines, the developer chemicals are not always fresh, and this is a great problem with some colour film processing plants where films which arrive at the beginning of the working week have better colour balance than those received at the end when chemicals may no longer be fresh or replenishment has been improperly carried out.

Films are frequently under or overdeveloped, after having been under or overexposed, and the colour balance of transparencies is then no longer correct. These defects mean that a significant part of the information for the scene has been recorded on the non-linear bottom of the exposure curve, or perhaps highlights are pushed into the non-linear upper parts of the curve. With under or overdevelopment, the dynamic range is too low to reproduce the light values which exposed the films. Underexposed or underdeveloped pictures are characterised by an apparent lack of contrast. Since extremely dark parts of a scene did not reflect enough light to affect the silver halide grains, there is an irrecoverable information loss in these areas. Overexposed parts of a picture have an abnormally high concentration of silver grains after development. These tend to clump together to give the impression of graininess when printed. In colour transparencies, the situation is somewhat different. Overexposed parts of the picture contain little or no information, being lighter than the rest, and underexposed parts contain incorrect colour balance since the non-linear part of the film curve is not well matched for the three colour components. Graininess is less of a problem with these films unless high speed colour films are used. With special infra-red sensitive films which use shifted (false) colours to reproduce part of the visible and near infra-red, exposure errors affect colour balance too, but the observer often fails to notice this because he has no intuitive appreciation of the colour which appears after development.

In an underexposed or underdeveloped black and white image, any mechanical defects on the surface of the film such as scratches, dirt or lint particles become highly visible when these pictures are printed with high contrast to compensate for a low contrast negative. Defects in colour transparencies are especially visible in clear highly exposed areas. Colour negatives behave similarly to black and white films in this respect.

The degree of development of a film also depends on the mechanical movement of the film or the developer during development. If movement is insufficient, development over the surface is not uniform. This effect is often seen in older sheet films developed by hand. It is also occasionally seen in large format aerial films developed in reel-to-reel development tanks. It is unusual in 35 mm films developed in a tank, and almost never occurs in machine processed colour films, unless the machine was incorrectly serviced or adjusted.

All of the problems described can be grouped together as exposure and development errors. It would be nice if they never happened, but they do, and usually in combination. In the case of colour films, the possibilities for problems are multiplied by the fact that they have many sensitive layers which respond to the red, green, and blue components of the available light, and to ultra-violet light as well. In addition, in most colour films, especially colour negatives, there are other layers which are intended to compensate for limitations in the transmission and colour separation characteristics of the dyes left on the film support after development. The other layers (masks) appear to be orange in a colour negative. Not infrequently, commercial processing laboratories fail to adjust or clean their machines properly, so that shifts in colour balance and density occur, or the surfaces are scratched by dirt in the machines. Exposure errors made by the photographer also produce unwanted changes in colour balance, which can be partly compensated for by filtering when using colour negatives, but which cannot be corrected in a transparency. Computer controlled enlargers use several image processing algorithms for the automatic correction of colour balance, but these are almost never disclosed by the manufacturers of such machines. Correction is usually based on global colour statistics, and it may be unsatisfactory in archaeologically critical parts of the image.

Global modification of picture histograms
Since the pixel average represents the brightness of the picture and the standard deviation the contrast, it is relatively easy to understand what happens when brightness and contrast are globally modified for the entire picture. In conventional photography this is as if another grade of enlarging paper and different printing times were used. A global brightness change implies shifting the whole histogram to higher or lower values. It is accomplished numerically by adding or subtracting a constant number to all the pixels in the picture. Multiplying each gray value in a pixel by a constant will not only shift the histogram but spread it out, raise the contrast and shift the brightness, figure 4.15(a)–(c).

If the multiplier is less than 1, the contrast is diminished. A modification of contrast by multiplication must be followed by subtraction or addition to keep average brightness constant, figure 4.15(*d*) and (*e*). Such a combined operation is reversible by division and addition or subtraction. The result is perceived as having changed the exposure and contrast as with changing the grade and exposure of an enlarging paper, but the precision and repeatability of the operation is perfect, and it can be carried out again and again from picture to picture if required.

Histogram modification can be carried out experimentally in the hardware of most modern computer displays until the user is satisfied, then applied to the stored data. If multiplication by a constant greater than 1 or addition is used, a result may exceed 255, conversely with subtraction fall below 0, and information is lost unless special precautions are taken to prevent this from happening by storing the data temporarily at higher precision. If the operation is carried out only in the display, no original data are lost. In most displays, the pixel values are kept in a memory. These values are used as addresses for a table (video look-up table). The table values determine the visible gray on the CRT. For each pixel value in display memory there is a corresponding table value. Performing all operations only on the table is very rapid because only very few entries are involved. The display memory contents are not altered and no data can be lost.

Usually there is no simple modification of the whole histogram which will produce a satisfactory contrast and brightness change. In some cases, the archaeological information is confined to a small part of the picture, and it is possible to carry out the transformation on this part only by giving the program information about the limits of an area on which a transformation is to be carried out. This is best done using the display system's cursor controller and delineating a significant area. An appropriate program must connect the points with a closed polygon, and then the histogram transformation is carried out within these limits only, figure 4.16(*a*) and (*b*). Details in an adjacent area can be transformed in the same way with different parameters, and finally the whole area of interest retransformed so that it appears to have uniform brightness and contrast. This is a rather tedious process, but it has the advantage of being computationally quite simple and hence within the capabilities of the smallest machines.

Numerically what has been done is similar to the 'dodging' which a skilled darkroom worker carries out when he varies the amount of light

Archaeological image enhancement

Fig. 4.15 (*a*) Positive image prior to global histogram modification. Air
Photo Archive, Rheinisches Landesmuseum, BU14 (neolithic house). (*b*)
Positive image with a constant subtracted from the histogram. (*c*) Positive
image after subtraction of constant and multiplication by 2. (*d*) Histogram
of the source negative. (*e*) Histogram of the negative with constant subtracted
from negative. (*f*) Histogram of the negative after subtraction of constant
and multiplication by 2.

(*a*) (*b*) (*c*)

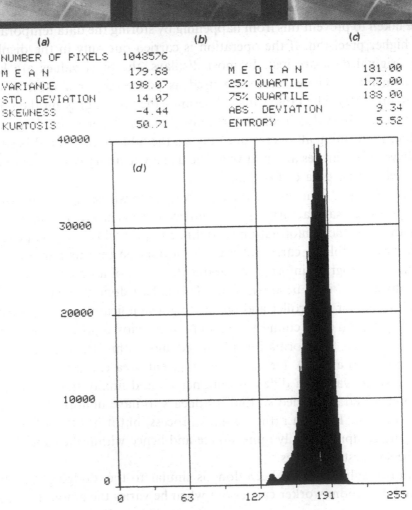

NUMBER OF PIXELS	1048576		
M E A N	179.68	M E D I A N	181.00
VARIANCE	198.07	25% QUARTILE	173.00
STD. DEVIATION	14.07	75% QUARTILE	188.00
SKEWNESS	-4.44	ABS. DEVIATION	9.34
KURTOSIS	50.71	ENTROPY	5.52

(*d*)

reaching the paper under the enlarger in different parts of his picture using his hands or small opaque objects on sticks or wires. He may also increase contrast locally by agitating the developer more in areas of greater required contrast, or use a variable contrast enlarging paper and vary the filter times with exposure in different areas. Such techniques are difficult to reproduce and require a great deal of manual skill. Using an interactive computer system the process requires no manual dexterity and is much simpler to control.

It is usually not possible to find a linear transformation which produces a satisfactory appearance for the whole picture if its histogram is multi-peaked. In this case it may be more satisfactory to transform the histogram into one with a standard single peaked form with a Gaussian

Fig. 4.15 *(cont.)*

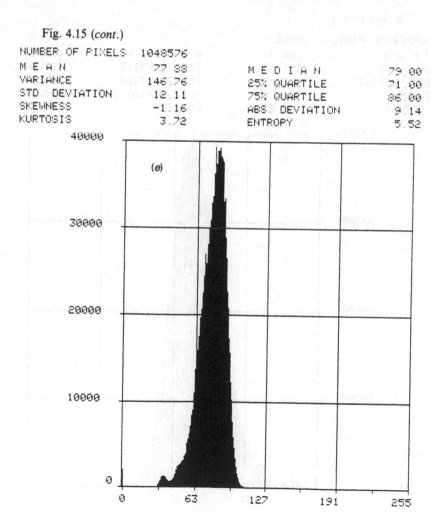

distribution:

$$\phi(x) = \frac{1}{\sigma(2\pi)^{\frac{1}{2}}} \exp\left[-\frac{1}{2}\left(\frac{x-\mu}{\sigma}\right)^2 \right] \tag{4.66}$$

or some other suitable shape found by experiment using the display transformation table. In this case, the histogram transformation is read back into the computer after a desirable solution has been found and used to compute new values for an output picture. This is a general histogram transformation by table look-up. Linear transformation can also be done in this way by suitable loading of the tables: the new values are simple multiples of the original, plus a constant. In most small computers it is faster to use a table look-up method even for linear

Fig. 4.15 (*cont.*)

NUMBER OF PIXELS	1048576		
M E A N	104.94	M E D I A N	107.00
VARIANCE	551.43	25% QUARTILE	91.00
STD. DEVIATION	23.48	75% QUARTILE	121.00
SKEWNESS	-0.81	ABS. DEVIATION	18.11
KURTOSIS	1.26	ENTROPY	5.52

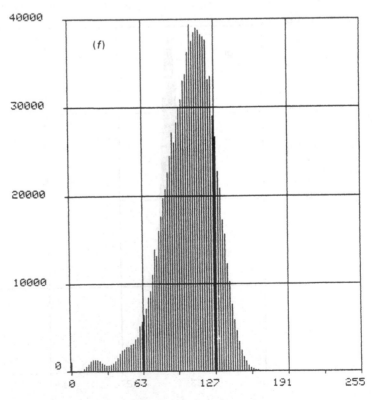

Fig. 4.16 Area of significance selectively modified by histogram transformation. Air Photo Archive, Rheinisches Landesmuseum, FI63.

(a)

(b)

transformation since multiplication of the original data pixels is slow in comparison. Variations on this method which appear in the specialised literature (21,22) call for the production of an output histogram which is maximally flat but usually the appearance of the result of this operation is not very pleasing.

Our experiments have shown that the eye appreciates images whose statistical distributions approach that of a normal distribution spread out over the reproduction range of the output device with two standard deviations over the total range of grays. The new distribution will have a different mean and standard deviation and these are usually best chosen by trial and error.

One method, suggested by Gonzalez (11), constructs the transformation in two steps. If p_x is the original probability density function of the image, and p_y the desired one, then first transform the original distribution to a uniform one with the transformation:

$$s = T(x) = \int_0^x p_x(t)\, dt \qquad (4.67)$$

where t is a variable of integration. This is simply the cumulative distribution function. The desired distribution could also be transformed to a uniform distribution by a similar transformation:

$$v = U(y) = \int_0^y p_y(t)\, dt \qquad (4.68)$$

First, the image is transformed using s, then the inverse of the specified distribution is applied to get $y = U^{-1}(s)$. The operation is combined by taking the transformation:

$$y = U^{-1}[T(x)] \qquad (4.69)$$

but since the inverse of U is not usually known, a discrete table is used.

There are a number of ways to construct the discrete table of the inverse function but one of the simplest is to construct a table, $\text{Tab}(x)$, usually with 256 elements for byte pixels which is filled with the values of a cumulative distribution. When p_y is Gaussian:

$$\text{Tab}(x) = \text{Tab}(x-1) + \exp\left[\frac{1}{2}\left(\frac{x-\mu}{\sigma}\right)^2\right] \qquad (4.70)$$

This table is then normalised by dividing each element by the value of the last entry, $\text{Tab}(255)$. The inverse table is constructed by binary search, using each value in the picture histogram in turn and searching for the nearest argument in the normalised function table which would have produced it. These values are then applied to the picture data.

Almost any mathematical function can be used to modify a histogram. One simple function might assign all values in the table above a given level to black and all those below it to white. The resulting transformed picture would be in pure black and white with no intervening gray values. This process is thresholding. It occasionally is useful as a simple technique for extracting traces of features in areas of nearly constant background gray level. It will not work if the background level varies. If this is the case, it is desirable to use an adaptive threshold (17) which is based on the local mean and standard deviation of features and filtered background. Most frequently, both distributions are locally Gaussian, and optimum separation is obtained using a threshold T for $\sigma_1 \neq \sigma_2$:

$$T = \frac{\mu_1 \sigma_2^2 - \mu_2 \sigma_1^2}{\sigma_2^2 - \sigma_1^2} + \sigma_1 \sigma_2 \left[\frac{(\mu_1 - \mu_2)^2}{(\sigma_2^2 - \sigma_1^2)^2} + \frac{2 \ln(\sigma_2/\sigma_1)}{\sigma_2^2 - \sigma_1^2} \right]^{\frac{1}{2}} \qquad (4.71)$$

A library of frequently used transformations can be kept as a small data file and these may be loaded at choice from a displayed menu into the display look-up table in a single operation taking no apparent time at all. New useful tables constructed during the course of experimentation can be added to the file and recalled by number or name at will.

It is useful to have some simple aids which allow the user to see what he ought to do when carrying out histogram transformation. Although it is possible to experiment with various parameters and look at the results quickly, numerical aids shorten the time required for experimentation. The most useful allows rapid display of the histogram in an area of variable size along with some useful statistical values, another shows the densities present along an arbitrary line or curve drawn through features of interest, figure 4.17, and a third shows the variation of average, standard deviation and median with areas of varying sizes in different parts of a picture, figure 4.18(a) and (b). For the techniques described in the next section, it is most important to know how variations in brightness and contrast relate to the size of a local area in different parts of a picture.

All the histogram operations discussed are used only on pixels in a single input image and produce one output image. It is also useful to have a program which can produce any mathematical combination of pixels from more than one image to generate a single output picture. In this case, typically:

$$O(x, y)_{\text{mix}} = F_1(x, y)_{\text{R}} + F_2(x, y)_{\text{G}} + F_3(x, y)_{\text{B}} \qquad (4.72)$$

where $F_{1,...,n}$ are arbitrary histogram transformations of n component pictures, e.g. red, green, blue, etc. A typical applicaton for this kind of program is modifying the balance of a colour picture. The red, green, and blue components will have been scanned separately. Each can be treated individually to modify contrast and brightness, and the result combined using variable amounts of each component to produce a new colour picture with enhanced contrast in one or more colour ranges. For example, increasing the ratio of green to the sum of green plus some red (green–yellow) will increase the colour contrast of green features on a background often encountered in archaeological aerial photography so that:

$$O(x, y) = \frac{\alpha I(x, y)_G}{I(x, y)_R + I(x, y)_G} \tag{4.73}$$

where α is a constant which controls green–yellow contrast.

Appropriate software tools must be available for accurately measuring the colour components of features and backgrounds before this kind of operation can be used intelligently. A display system capable of showing the images in full natural colour at high resolution and a scanner capable of accurate colour separation of the original are needed. Since the dyes

Fig. 4.17 The densities present along an arbitrary curve through features of interest.

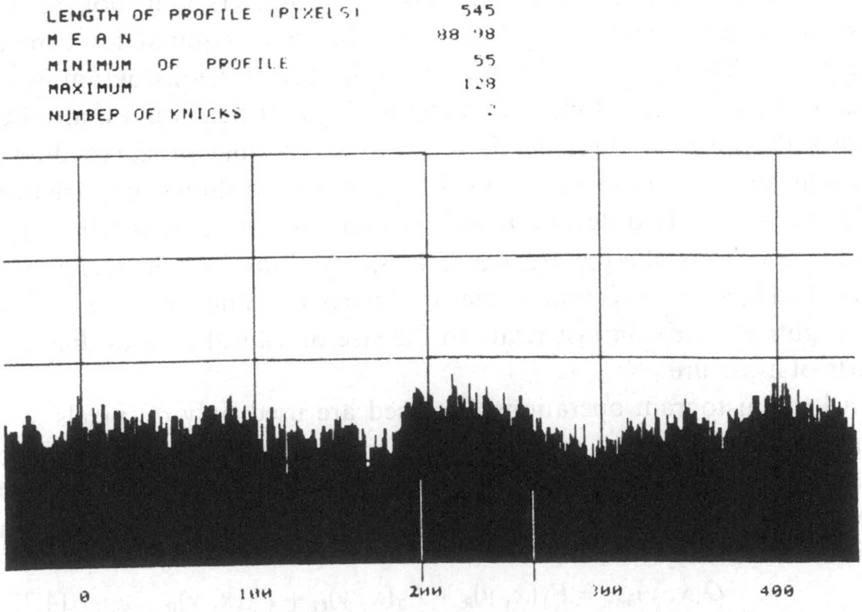

```
LENGTH OF PROFILE  (PIXELS)        545
M E A N                           88 98
MINIMUM   OF   PROFILE             55
MAXIMUM                           128
NUMBER OF KNICKS                    2
```

used in colour films have overlapping absorption wavelengths, the colour sensitive layers respond to light at wavelengths other than that for which they are intended. The use of narrow band optical filters in the scanner with wavelengths chosen to optimise the separation of the components is necessary. The bandwidths and centre wavelengths of the filters depend

Fig. 4.18 The variation of average, standard deviation and median with the size of the area of interest: (a) on a high contrast boundary; (b) in a uniform area.

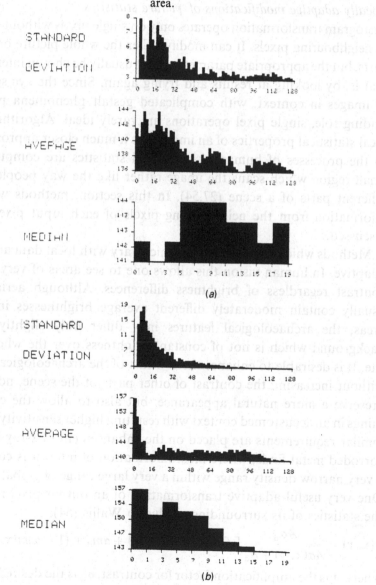

on the type of colour film to be scanned. The most expensive scanners have automatic control of the filter parameters via software for each film type. Even after scanning with such a device there is some residual response overlap which can be corrected algebraically by appropriate combinations of the three channels. Commercial colour scanners for the printing industry incorporates these features in their software. Details are usually not published.

Locally adaptive modifications of picture statistics

Histogram transformation operates only on single pixels without reference to neighbouring pixels. It can modify either the whole picture or selected parts, but the appropriate parameters must usually be chosen interactively, that is, by looking at results and trying again. Since the eye sees parts of images in context, with complicated gestalt phenomena playing a leading role, single pixel operations are rarely ideal. Algorithms using local statistical properties of an image give a much closer approximation to the processes of human vision. Local statistics are computed in a small region which scans the image rather like the way people look at different parts of a scene (27,54). In this section, methods which use information from the neighbouring pixels of each input pixel will be discussed.

Methods which use parameters which vary with local data are termed adaptive. In human vision this allows one to see areas of very different contrast regardless of brightness differences. Although aerial scenes usually contain moderately different average brightnesses in various areas, the archaeological features may differ only slightly from a background which is not of constant brightness over the whole of the site. It is desirable to enhance the contrast of the archaeological features without increasing the contrast of other parts of the scene, not only to preserve a more natural appearance, but also to allow the eye to see things in an accustomed context with resulting higher sensitivity to detail. Similar requirements are placed on the enhancement of x-ray images of corroded metal finds, where most information of interest is confined to a very narrow density range within a very large range of global densities. One very useful adaptive transformation of an output pixel relative to the statistics of its surroundings is due to Wallis (54):

$$g(x, y) = \frac{A\sigma_d}{A\sigma(x, y) + \sigma_d} [f(x, y) - m(x, y)] + \alpha m_d + (1 - \alpha)m(x, y) \quad (4.74)$$

where A is the amplification factor for contrast, σ_d is the desired standard deviation, $\sigma(x, y)$ is the standard deviation in the current window, $f(x, y)$ is the pixel centre of the current window, $m(x, y)$ is the mean of the current window, α is the factor controlling portion of the observed mean, and m_d is the desired mean. This reduces or increases contrast and brightness locally to reduce or increase the total range. Dark parts of the picture are pushed toward the middle range of grays, as are light parts. Contrast is increased in very dark or very light areas, and in areas where contrast is inherently low, and left unchanged where inherently high. The moving window must have at least 400 pixels for the local statistics to be stable enough for good results. The operation is repeated for every pixel in the image. It is quite similar to automatic 'dodging' of a picture when using variable contrast enlarging paper, but does not depend on human skill. An improvement made to the published algorithm removes redundant calculations and makes the operation much faster (46). The amelioration of the overall appearance of the archaeology which this technique provides is considerable, figure 4.19(a) and (b). The changes in light and dark areas can be combined with table look-up techniques to modify the effects given by the mathematical formulation of the algorithm thus achieving subtle control with compensation possible for almost any imaginable film or display response curves. For example, the constants in the above equation may be replaced with table look-up values, using non-linear table entries.

It is possible to compute the local brightness (average) and contrast (standard deviation) in a large area (window) scanned through a picture with little effort (31,46). Two lines of the picture corresponding to the top and bottom of the window are kept in core, and one line is used for an output buffer. Two long integer accumulator arrays are used to store the current sums of the pixels over the vertical length of the window for all window positions in the line. As the window is moved, it is merely necessary to add the two new values at the upper and lower right corners to the accumulator and subtract the upper and lower left values to update the average, and to square the values to update the variance. When the line is completed and the window moved down to the next line, the window must be initialised and the process repeated. Four additions/subtractions plus the initialising overhead are required to update the average, four squaring operations, one square root and a division are needed for the standard deviation. Floating point accumulators may be substituted for the long integer arrays if the machine used has more rapid conversion from byte to that format than to long integer.

Fig. 4.19 (*a*) An image prior to treatment with the Wallis algorithm. Air Photo Archive, Rheinisches Landesmuseum, HF36. (*b*) After treatment with the Wallis algorithm.

(*a*)

(*b*)

An adaptive filter with a similar design, very useful for noise reduction in the middle gray range of an image is given by Hou (17) extending the work of Lee (27). Let $g(x, y) = f(x, y) + n(x, y)$ where $f(x, y)$ is an ideal image, $n(x, y)$ is additive noise, and $g(x, y)$ is the resulting image obtained by combining them. It is assumed that $n(x, y)$ is random with variance σ^2 and zero mean. Let the local spatial averages approximate ensemble averages, f_1 the low pass filtered image, and define means and variances as:

$$\overline{n(x, y)} = 0, \quad \overline{n(x, y)n(l, m)} = \sigma_n^2 \delta_{x,l} \delta_{y,m}, \quad \overline{f(x, y)} = f_l(x, y) \Bigg\} \quad (4.75)$$
$$\sigma_f^2 = \overline{[f(x, y) - f_l(x, y)]^2} = \sigma_g^2 - \sigma_n^2$$

Then the minimum least square error estimated image is given by:

$$\hat{f}(x, y) = rf_l(x, y) + (1 - r)g(x, y) \quad (4.76)$$

where:

$$r = \sigma_n^2/(\sigma_f^2 + \sigma_n^2) \quad (4.77)$$

If the image is noise free then $r = 0$ and:

$$\hat{f}(x, y) = g(x, y) = f(x, y) \quad (4.78)$$

If the signal is absent then $r = 1$ and:

$$\hat{f}(x, y) = f_l(x, y) \quad (4.79)$$

A bandpass (combined high and low pass) filtered image is constructed to extract edges and suppress high spatial frequency noise and this is used as the first estimate of the signal. The difference between this and the noisy original image is taken and the estimated signal and noisy variances calculated from this and the edge image. This gives the needed data for r. The procedure is iterated until the noise variance is below a desired small number.

4.5.2. *Problems with image sharpness*

Motion blur

The image projected on the film plane moves to some extent during the exposure interval in an aircraft, and to a lesser extent in ground photography made without a tripod. With a hand-held camera, the photographer unconsciously tends to compensate somewhat for movement by keeping the camera pointed at some central point of interest, but in so doing adds additional random movement (hand jitter). Even if the shutter speed is very high, a small amount of residual movement remains whose direction and amount vary from picture to picture. In low level aerial imagery, relative ground movement is high, and with low shutter speeds, especially in winter pictures made at low light levels, blur is noticeable even at modest degrees of enlargement. With hand-held oblique pictures, the blurring is most noticeable in the image foreground where relative movement is greatest. Modern professional aerial mapping cameras move the film to compensate for aircraft motion, but such equipment is seldom at the disposal of archaeologists. With hand-held near verticals made in a tight turn, the actual camera motion is rotary, and blur may be greater at the edges than in the centre of the picture.

Movement of the camera or movement of the scene can be analysed in the same way. Analysis in the literature (14) has been given for the moving scene case.

The image at a point actually recorded over a time t is $g(v_i, t)$, such that the recorded image during an exposure time T is:

$$g(v_i) = \int_0^T g(v_i, t) \, dt \qquad (4.80)$$

It can be shown that at a fixed time t the instantaneous reflectivity $f(v_o, t)$ of object $v_o = (x_o, y_o, z_o)$ equals $g(v_i, t)$. Therefore the time integral can be written as:

$$g(v_i) = \int_0^T f(v_o, t) \, dt \qquad (4.81)$$

Thus, if the object point moves, then a number of object points are imaged onto one image point during the exposure period. If something is known about the motion of the camera, then the time integral of the moving object can be converted into a positional integral over an equivalent stationary object. Following Hall (14) let the instantaneous object positions be described in parametric form:

$$x_o(t) = p_1(v_i, t), \qquad y_o(t) = p_2(v_i, t), \qquad z_o(t) = p_3(v_i, t) \qquad (4.82)$$

When each p has an inverse q, then:

$$t = q_1(x_o, v_i) = q_2(y_o, v_i) = q_3(z_o, v_i) \qquad (4.83)$$

The object motion path is assumed to be a curve r, where:

$$r = [x_o(t), y_o(t), z_o(t)] \qquad (4.84)$$

and from elementary calculus, the length of a path element ds is:

$$ds = [(dx_o/dt)^2 + (dy_o/dt)^2 + (dz_o/dt)^2]^{\frac{1}{2}} \, dt \qquad (4.85)$$

and the time integral (equation (4.81)) is changed to a line integral:

$$g(v_i) = \frac{\int_{S_0}^{S_T} f(v_o) \, ds}{[(dx_o/dt)^2 + (dy_o/dt)^2 + (dz_o/dt)^2]^{\frac{1}{2}}} \qquad (4.86)$$

where the limits of integration S_0 and S_T are:

$$\left.\begin{array}{ll} S_0 = [x_0^2 + y_0^2 + z_0^2] & t = 0 \\ S_T = [x_0^2 + y_0^2 + z_0^2] & t = T \end{array}\right\} \quad (4.87)$$

And to remove the blur from an image degraded by movement, equation (4.86) must be solved for $f(v_o)$. But to do this one must know the positions S and the velocities of movement V. This is extremely difficult for the general case. For the limited but practical case where the motion is linear across the film plane then with f the instantaneous image:

$$g(x, y) = \int_0^T f[x - \alpha(t), y - \beta(t)]\, dt \qquad (4.88)$$

where $\alpha(t)$ describe and $\beta(t)$ the motion parametrically. In the Fourier transformation domain a filter can be constructed. Taking the Fourier transform of equation (4.88) gives:

$$G(u, v) = \int\limits_{-\infty}^{\infty}\!\!\int g(x, y) \exp[-i2\pi(ux + vy)]\, dx\, dy \qquad (4.89)$$

Using a transformation for $\alpha(t) = V_x$ and $\beta(t) = V_y$, then $\xi = x - \alpha(t)$; $x = \xi + \alpha(t)$; $\eta = y - \beta(t)$; $y = \eta + \beta(t)$ and:

$$G(u, v) = \int_0^T \int_{-\infty}^{\infty} f(\xi, \eta) \exp[-i2\pi(u\xi + v)]\, d\xi\, d\eta$$
$$\times \exp\{-i2\pi[u\alpha(t) + v\beta(t)]\}\, dt \qquad (4.90)$$

then:

$$G(u, v) = F(u, v) \int_0^T \exp\{-i2\pi[u\alpha(t) + v\beta(t)]\}\, dt \qquad (4.91)$$

With $H(u, v)$ equal to integral part of this expression, then $G(u, v) = F(u, v)H(u, v)$ which is a linear filtering operation. When the motion is horizontal, $\alpha(t) = Vt$ and $\beta(t) = 0$, and:

$$H(u, v) = \int_0^T \exp(-i2\pi Vut)\, dt = \sin(\pi VTu)/\pi Vu = T\, \mathrm{sinc}(\pi VTu) \quad (4.92)$$

which is nothing other than an optimum high pass filter using the velocity and time parameters in its construction. In practice, however, since the velocities and times are seldom known it is best to construct a set of filters with assumed velocities and exposure times and try them in a small area of the image containing a known sharp point until one is found which yields the best result. This can then be applied to the whole image. The calculations when a sinc function is used are formidable for a small machine, and simpler methods with a displaced centre point as discussed below may produce adequate results. Alternatively, the separable spline approximations to a sinc function described in chapter 5 can be used with considerable reduction in computation time and little difference in appearance.

Lens blur and film resolution limitations
All aerial pictures are made at distances which require the camera lens to be focused at infinity. Professional aerial lenses such as used in large format hand-held or vertical cameras cannot be focused, and thus cannot be accidentally set to something other than infinity. With amateur cameras, this may happen unintentionally. In ground imagery, the experienced photographer will focus on a point of maximum interest or use the depth of field indicator on his lens to get good average sharpness. If light levels are low, the experienced photographer will use the highest possible shutter speed to reduce motion blur, thus requiring a large lens opening. Even the best of photographic lenses are not sharp at their largest opening, especially at the corners of the image. Usually, best lens sharpness is obtained between two and three F stops higher than the largest opening, but light conditions on cloudy or winter days do not always allow the use of such small apertures with high shutter speeds. In such cases it is better to use a high speed film, and to accept the slightly greater graininess and lower resolution in the image which results. Although modern high quality lenses impose fewer restrictions than those used in the past, defects due to lens blur become apparent at high enlargement convolved with motion blur. Older archive pictures are frequently unsharp. The lenses were of inferior quality, and blur is apparent even at modest degrees of enlargement. Fortunately, these cameras also had very large film formats which means that considerable enlargement is seldom needed. Lens blur has radial and tangential components, as well as non-circularly symmetric contributions of lesser importance.

The image intensity i_i at point (x_i, y_i) and corresponding object intensity i_o at point (x_o, y_o) are related by a superposition equation:

$$i_i(x_i, y_i) = \int\int_{-\infty}^{\infty} h(x_i, y_i; x_o, y_o) i_o(x_o, y_o) \, dx_o \, dx_o \qquad (4.93)$$

If the simplifying assumption is made that the function does not vary in space over the image plane, then in the Fourier domain, the image and object intensity fields are:

$$I_o(\omega_x, \omega_y) = H(\omega_x, \omega_y)I_i(\omega_x, \omega_y) \tag{4.94}$$

where $I_o(\omega_x, \omega_y)$ is the Fourier transform of the light intensity at the object plane, and $I_i(\omega_x, \omega_y)$ at the image plane.

An optical transfer function $H(\omega_x, \omega_y)$ can be defined as:

$$H(\omega_x, \omega_y) = \frac{\int\limits_{-\infty}^{\infty}\!\!\int h(x, y) \exp[-i(\omega_x x + \omega_y y)]\, dx\, dy}{\int\limits_{-\infty}^{\infty}\!\!\int h(x, y)\, dx\, dy} \tag{4.95}$$

The absolute value of this function is the MTF as defined in chapter 3. For a misfocused ideal lens, the function falls off rapidly, as shown in figure 4.20. In real lens systems, the point spread function varies over the image plane and is non-separable. In Fourier space with the optical transfer function defined as in equation (4.95), the function for an aperture a is:

$$H(\omega_x, \omega_y) = J_1(a\rho)/a\rho, \qquad \rho = (\omega_x^2 + \omega_y^2)^{\frac{1}{2}} \tag{4.96}$$

where J_1 is a Bessel function of the first order.

Fig. 4.20 The MTF of a misfocused lens. The numbers on the curves show the degree of misfocus. From ref. 38.

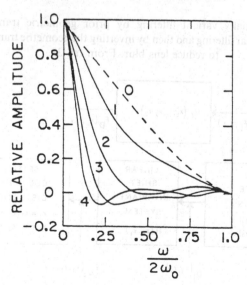

This is an extremely complex case. Andrews and Hunt treat these matters in considerable detail (1). Two approaches to solving the problem of restoring images degraded by lens distortion or misfocus have been cited in the literature. In the first, the image is broken up into smaller blocks, each of which is subject to linear high pass filtering with parameters a function of block position. This method is not too difficult to implement roughly as a modification of crispening (see below) using a different box size in different parts of the image, together with different relative positions of the central point. For more elaborate filters it would be quite difficult to carry out in a small machine. As Pratt (38) points out, there is no way of knowing just how large or small to make the image blocks, but a rule of thumb tells us that they ought to be of a convenient size for input/output operations using direct access file

More elegant methods have been suggested by Sawchuk (43,44,45) who recommends carrying out a geometric transformation on the picture to produce a spatial coordinate distortion, then filtering linearly and finally inverting the distortion. This technique is possible only if the coordinate transformation operator has a unique inverse, such as is the case for some kinds of lens distortion. The scheme is shown in figure 4.21. Restoring sharpness by calculation is equivalent to trying to equalise the fall-off of the composite modulation transfer function as far as this is possible given the limits of our knowledge of the characteristics of film, lens and movement of the camera.

Fig. 4.21 Spatially variant filtering by prior geometric transformation followed by linear filtering and then by inverting the geometric transformation to reduce lens blur. From ref. 38.

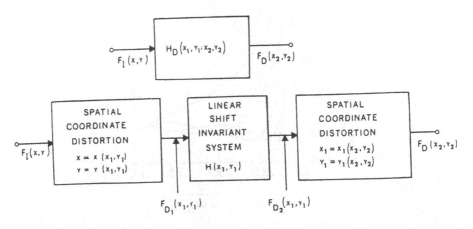

Film resolution is usually inversely correlated with film speed. Most film manufacturers can supply curves which show the resolution given by different emulsions as illustrated in chapter 3. Resolution is expressed in terms of the MTF which shows the relationship between the film's ability to record closely spaced lines as a function of the contrast between the lines. Lenses also have MTFs which can be measured in a well-equipped optical laboratory. Total resolution in an image is given by the convolution of the MTFs of the film, the lens, and motion blur, if any. The combined MTF falls off much more rapidly than that of the film alone (36), being the product of the ordinates of the two curves in figure 4.22, for example.

Simplified numerical reduction of blur
The blur due to motion and lens imperfection are mixed inextricably in most images. In addition, the film scanner has a fixed aperture which usually does not resolve the film grain, and this is effectively averaged. If a fine enough aperture is used, the scanner does not contribute to blur. With modern cameras motion blur predominates in low level oblique aerial images, lens blur and film grain blur with scanner averaging are more noticeable in pictures taken with older cameras. These imperfections limit resolution of the finest details, especially in the background of oblique images and near the corners of older pictures. Because there is usually no way of separating the different types of blurring they can only be treated heuristically, as if all from one cause.

In crispening (38) the local average in a small window is partially subtracted from each pixel followed by normalisations of brightness by multiplication:

$$g(x, y) = w[f(x, y) - \alpha m(x, y)] \tag{4.97}$$

The normalisation constant depends on window size and the fraction of the local average subtracted. It is designed to keep overall brightness and contrast constant. The size of the local window used to compute the moving average must be chosen as a function of the amount of blur. If one thinks of blur as spreading a point source of data in the scene into a comet-like feature with a central area and a tail, then the window has to be about as big as the feature. One might imagine that a local average can be computed exactly over the irregularly shaped blurred point. Subtracting part of this from the point will tend to remove some of the blur. In practice the shape of the blur varies from one part of the picture to another and the eye cannot see much difference between a rough average computation over a square window and the exact computation using the true shape when the blur is compared with image size. The window size is best chosen experimentally for a small part of the image containing known point-like or line-like objects while watching the results on the display. The mixture of subtracted average window and original pixel value controls the degree of crispening and this must

Fig. 4:22 Combined MTF (*a*) lens (55 mm micro-Nikkor) and (*b*) film (Kodak SO-115). From ref. 36.

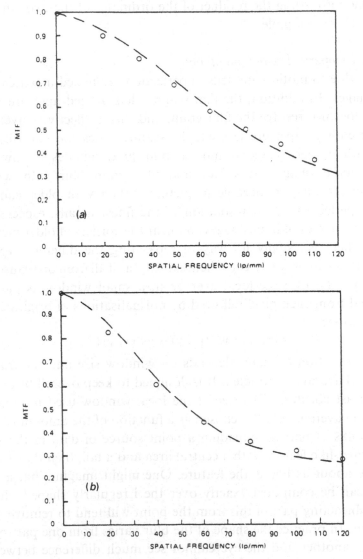

also be selected interactively. The process cannot be carried very far or light and dark banding appears around the edges of objects with high contrast. However, quite satisfactory results can usually be obtained with a little patience. Since the average in a square window can be computed very efficiently, this is one of the most rapid digital treatments at our disposal. If, instead of using the average, the median value in the window is used for the process, banding is much less for a given degree of crispening (48) but calculation times are somewhat longer.

It is assumed that the average or median is partly subtracted from each pixel in the centre of the moving window. This will reduce symmetrical blur only. If the window is displaced laterally and vertically relative to the output pixel, then asymmetrical blur such as that produced by motion in a single direction can be reduced. In an oblique image, the relative motion on the film is not uniform from foreground to background, and hence this will only work within a limited area. Usually the subtracted window is square in shape if only lens blur is to be treated, but it may be useful to make it rectangular with the long axis in the direction of motion and the centre displaced if motion blur is believed to be the culprit. In practice the apparent improvement in sharpness is impressive, but the effect is more psychological than real as comparison of the greatly magnified portion of a sharpened and unsharpened image reveals, figure 4.23(*a*) and (*b*).

Processing with more complicated filter functions on high resolution images can be very time consuming on a small machine. One very useful technique for reducing computation time by using approximations to such functions has been published by a number of authors (35,42,52). The original image is low pass filtered with a simple function and then sampled down to an image of half the original number of pixels on a side. This, in turn, is filtered again and resampled. This rapid process is continued as far as desired. The result is a hierarchical pyramid of multiresolution images, as shown in figure 4.24. Filtering operations can now be carried out on one or more of the hierarchical images using simple local operators. The hierarchical images are then recombined to obtain the original resolution. For rather large kernels and bandpass filtering, O'Gorman and Sanderson (35) show that the number of computations required is very much less than fast Fourier transform techniques performing the same operation. Since the simple filter kernels are separable in the x and y directions, even a further improvement is possible. This method requires no special hardware. It is conservative with regard to the number of computations and has modest input/output overhead.

Deblurring may be improved if instead of using a square or rectangular window, an empirical inverse convolution approximating the shape of

Fig. 4.23 (*a*) Magnified part of an unsharp image before crispening. (*b*) After crispening.

(*a*)

(*b*)

the blur of a point source is used. Roughly, blur can be considered to be locally Gaussian in shape, with perhaps the two-dimensional form of the bell-shaped curve being an ellipse oriented along a radius from the centre of the image if coming mainly from the lens. Burt (6,7) gives a particularly elegant method for convolving with Gaussian kernels in the space domain based on the multiresolution method which is much faster than Fourier transform filtering. If the separable filters are of different sizes, then an ellipsoidal convolution is approximated. Rotating the image prior to filtering may be necessary if the blur is not parallel to one of the scan axes which will seriously compromise the speed advantage of the method.

A 5 × 5 generating kernel is used. This is iteratively applied to produce an image pyramid which begins with the original image. Where $g_0(x, y) = I(x, y)$ is the original image and the lowest level of the pyramid,

Fig. 4.24 Multiresolution pyramid representation of an image with a sampling rate = 2. From ref. 35.

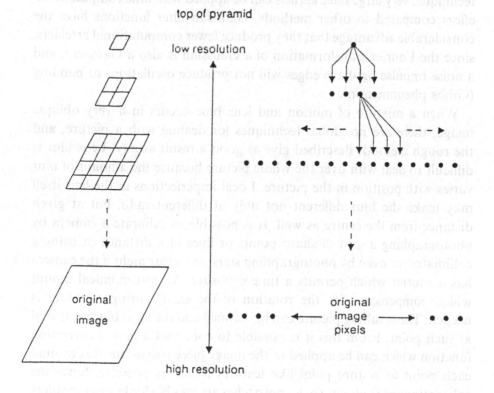

subsequent levels are given by:

$$\hat{g}_l(x, y) = \sum_{m=-2}^{2} \sum_{n=-2}^{2} w(m, n)\hat{g}_{l-1}(2y + m, 2x + n) \qquad (4.98)$$

The weights $w(m, n)$ are separable. The separate weights $\hat{w}(m)$ are symmetric. The transformation can be applied sequentially in x and y directions with three weights (7). If $\hat{w}(0) = a$, $\hat{w}(1) = \hat{w}(-1) = b$, $\hat{w}(2) = \hat{w}(-2) = c$, then the constraints are satisfied when $b = \frac{1}{4}$, $c = \frac{1}{4} - a/2$, and a is a free parameter. A very close approximation of a Gaussian is achieved when $a = 0.4$ to give $w(0, 0) = 0.4$, $w(1, -1) = 0.25$ and $w(2, -2) = 0.05$ as can be seen from figure 4.25. This kind of operation can be rapidly carried out on a small machine even if the number of points in the image is very large. The technique is very similar to unsharp masking which has been used for many years in blockmaking to improve the sharpness of images and to allow fine lines to be reproduced in the pictures with a wide variety of line widths (28). With the pyramid technique, very large filter kernels can be applied with little computational effort compared to other methods. Gaussian filter functions have the considerable advantage that they produce fewer computational artefacts, since the Fourier transformation of a Gaussian is also a Gaussian, and a noise impulse or sharp edges will not produce oscillations or banding (Gibbs phenomenon).

When a mixture of motion and lens blur occurs in a very oblique image, there are no global techniques for dealing with a picture, and the rough methods described give as good a result as any. Lens blur is difficult to deal with over the whole picture because the amount of blur varies with position in the picture. Local imperfections in the lens itself may make the blur different not only at different radii, but at given distance from the centre as well. It is possible to calibrate a camera by photographing a grid of sharp points or lines at a distance or using a collimator, or even by photographing stars on a clear night if the camera has a shutter which permits a time exposure. An astronomical mount which compensates for the rotation of the earth during exposure is needed. The resulting picture can be scanned and the local blur measured at each point. From this it is possible to construct a global correction function which can be applied to the image piecewise in the area around each point to restore point-like features as far as possible. Since the calculations are tedious, such approaches are feasible for large computers only. Unfortunately, many of the cameras which produced problem images are no longer available for calibration and simpler trial and error methods on the material in the archives must suffice.

Losses of resolution due to film grain as well as problems imposed by atmospheric turbulence, dust, or haze are not as difficult to deal with, and they have received attention in the technical literature devoted to satellite and astronomical photography. The most useful approach is

Fig. 4.25 An approximation to a Gaussian convolution kernel for fast filtering in the space domain. From ref. 7.

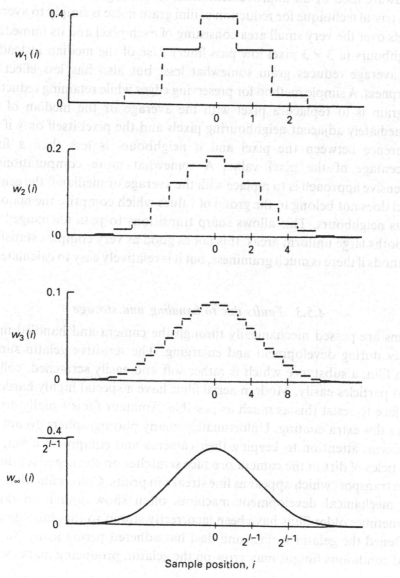

Sample position, i

equivalent to a three-dimensional statistical analysis of the data, with estimates being made of the most likely gray value at each point for all points in a perturbed image, given some knowledge of the physics of image formation (5). Results have only been published for relatively low resolution images even when large computers are used since the number of equations to be solved is huge. If satellite or very high level aerial photography is important for archaeology in the region of interest, then these methods will have to be considered and either special high speed hardware used or an improvement made in the algorithms.

A trivial technique for reduction of film grain noise is simply to average pixels over the very small area consisting of each pixel and its immediate neighbours (a 3×3 pixel low pass filter). Use of the median instead of the average reduces grain somewhat less, but also has less effect on sharpness. A simple method for preserving edges while retaining reduction of grain is to replace a pixel with the average or the median of the immediately adjacent neighbouring pixels and the pixel itself only if the difference between the pixel and it neighbours is less than a fixed percentage of the pixel value. A somewhat more computationally expensive approach is to replace with the average or median if the central pixel does not belong to the group of values which comprise the majority of its neighbours. This allows sharp transitions to pass unchanged but smooths large uniform areas. It is not as good as very complex statistical methods if there is much graininess, but it is relatively easy to calculate.

4.5.3. *Faults due to handling and storage*

Films are passed mechanically through the camera and handled many times during development and enlarging. The sensitive gelatin surface of a film, a substance which is rather soft and easily scratched, collects dust particles easily. Modern aerial films have a special highly hardened surface to resist this as much as possible. Amateur films usually do not have this extra coating. Unfortunately, many photographers do not pay sufficient attention to keeping their cameras and equipment clean, and particles of dirt in the camera produce scratches on the negatives during film transport which appear as fine streaks in prints. Colour films handled by mechanical development machines often show similar markings. Sometimes older films have been incorrectly stored so that humidity has softened the gelatin surface and dust has adhered permanently. In very bad conditions fungus may grow on the gelatin, producing marks which cannot be removed. This problem is often encountered with older colour transparencies which were mounted as slides before they thoroughly dried. Colour slide originals may have been passed through projectors and suffered from the strong light and heat which causes fading of dyes. Black and white negatives which were not sufficiently washed after processing retain traces of chemicals which with time cause the image to fade or turn brown. All of these unfortunate phenomena are observed in film archives.

Removal of scratches, dust and dirt
In conventional 'retouching' techniques for dealing with scratches, dust, and dirt, a skilled technician uses an airbush with a colouring agent on the final print or, if very skilled, he may work on the original negative with ink and a knife. This is a very dangerous operation demanding great dexterity, and it cannot be applied to other than very large negatives. Original negative retouching ought never to be used on archive quality aerial or excavation photographs, for it may result in destruction of an irreplaceable document. Automatic 'retouching' by calculating the local statistics of a moving window over the digital image is effective. If the histogram of a small area around a dust particle or a scratch is examined, in addition to the usual peak corresponding to the average gray level in the window, there is a second and sometimes a third small peak well removed from the first containing fewer pixels. This is due to exceptionally light or dark pixels which are produced by the dirt. The dirt peak can usually be identified by eye without too much difficulty. Removing it mathematically is not complicated. A central pixel in a moving window is compared with the local standard deviation and replaced by the average of the window if two standard deviations from the average are exceeded. This is successful only if the dirt pixels constitute a very small percentage of the pixels which form the average. It is usually true if the window is large compared with the size of the scratches or dirt. If the window is big enough, the average and standard deviation will be stable even in the presence of dirt pixels. This method reflects normal statistical practice for removal of 'outliers', for only 5% of pixels will lie outside the two standard deviation band by chance alone. The calculations are quickly done and give good elimination of dirt and scratches in areas of uniform gray.

If there are any objects in the picture with high contrast sharp edges, the method above will tend to remove those edges. In many archaeological aerial photographs this is not serious but occasionally one wishes to retain such sharp points for further use in dealing with the geometry of the image. If this is so, then an analogue of the above method using medians and quantile differences is preferable. In this approach, the offending pixels are replaced by the median value of a window if they lie outside the 2.5% and 97.5% quantiles. An example is shown in figure 4.26(a) and (b) with dirt removed in figure 4.26(c). Other quantile values can be chosen interactively if these do not produce satisfactory results. Medians and quantiles are not affected very much by sharp transitions at high contrast objects even when relatively small windows are used. For data limited to values between 0 and 255, the quantile difference is quite stable with only 25 values (5 × 5 window). The standard deviation needs at least 400 values for less than 2% error (21 × 21 window).

The price which has to be paid for quantile methods lies in increased computing time. It is considerably quicker to compute averages and standard deviations than it is to compute medians and quantile differences

Fig. 4.26 (a) An image made from a dirty negative. Air Photo Archive, Rheinisches Landesmuseum, EL. (b) An enlarged view of the dirt (inset). (c) After enhancement with the Wallis algorithm, crispening and, dirt removal by quantile difference filtering (insets).

(a)

Fig. 4.26 – *cont.*

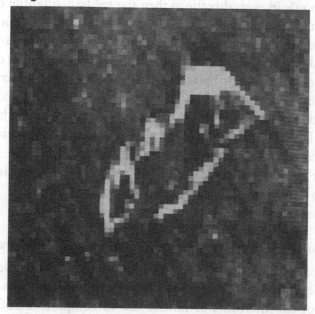

(b)

(c)

in square windows. If a large cross-shaped window is used (with only one line of pixels in each direction) then the median–quantile difference computation is nearly as fast as the average–standard deviation method, but more lines of picture have to be simultaneously present in fast storage – a disadvantage only in some very small computers. A number of picture lines equal to the number of pixels in the vertical arm of the cross must be stored. The cross-shaped window technique (34) gives almost as good a result as the square window on uniform areas and somewhat better results near large high contrast objects. The degree of dirt reduction can readily be controlled by varying a single multiplier for the standard deviation or by varying the quantiles. In general, an interactive trial is called for to determine the multiplier or quantile choice.

4.5.4. *Converting grayscale pictures to drawings* (*edge extraction*)

Although a photograph conveys a vast amount of data in the millions of picture elements which it contains, only a very small portion is usually of archaeological interest. For further use in comparisons of plans or for putting together the jigsaw puzzle of a site repeatedly photographed under different light and crop conditions in different years, it is necessary to convert the images to a black and white drawing. When this is done by hand, it requires tedious tracing of all features. There is a considerable element of subjectivity which then enters into the result because different observers will see details differently and will be more or less skilled in tracing. Usually an idealisation of the wiggly lines produced by hand is wanted even though the original image may only have contained rather fuzzy features which were difficult to follow.

It is quite simple to convert a black and white image to a drawing by calculation. What is required is a technique for removing the gray levels and replacing them with black or white pixels at features where the gray level changes abruptly. Essentially the problem often reduces to the construction of a high pass filter with thresholding for lines and edges. There is probably no topic which has received more attention than this in the image processing literature, and one is faced with a difficult choice as to which method to use (10,11,14,26,30,35,37,38,41).

There are simple techniques for extracting edges and lines using moving averages or medians in windows of various sizes. These are not as effective as some more sophisticated methods described below but they are very fast. The simplest of all is to subtract the moving average of neighbouring pixels completely (in crispening only part is subtracted) from each pixel over a sufficiently large window, check the result to see if the remainder is above a certain level and if it is, replace it with a black pixel, if not, with a white one. The result, shown in figure 4.27(*a*) and (*b*), is not bad at all considering the simplicity of the technique. Archaeological features in oblique aerial photographs unfortunately do not tend to have clear edges as do field boundaries and other modern structures, and hence the method does not tend to extract the data of interest unless feature contrast is high. Furthermore, the size of the window has an effect on the thickness of the resulting lines. Dirt and scratches will also be preferentially extracted along with desired features. Therefore, before applying any edge extractor, the image must be pretreated, usually with a dirt/scratch reducing algorithm. It is useful to sharpen the image before using the dirt remover. Then it may be desirable to normalise contrast and brightness so that all further steps can work on pixels with the best possible contrast for the features of interest. Feature extraction is used at the end of the treatment sequence when the archaeology has been rendered as visible as possible in an environment which is as uniform as possible.

An improvement on the simple average subtraction detector is obtained by first smoothing the pretreated picture with averaging or a median filter over a very small window. This removes the very small local randomness in pixel values in lines and edges of interest and makes the result less ragged in appearance. Further treatment with rather more complicated algorithms which look for adjacent pixels to replace ragged edges with smooth lines may be of interest.

One of the most useful edge extractors is based on a non-linear weighted average of the eight nearest neighbours of a central pixel due to Sobel and described by Duda and Hart (10). The x and y directions are treated separately with weights:

$$X = \frac{1}{4}\begin{bmatrix} -1 & 0 & 1 \\ -2 & 0 & 2 \\ -1 & 0 & 1 \end{bmatrix},$$

(4.99)

$$Y = \frac{1}{4}\begin{bmatrix} 1 & 2 & 1 \\ 0 & 0 & 0 \\ -1 & -2 & -1 \end{bmatrix},$$

Fig. 4.27 (*a*) Original image prior to edge extracton. Air Photo Archive, Rheinisches Landesmuseum, HH31. (*b*) Edge extraction by box filtering and thresholding.

(*a*)

(*b*)

and the output edge image is taken as:

$$g(x, y) = (X^2 + Y^2)^{\frac{1}{2}} \qquad (4.100)$$

and thresholded. This edge extractor is not very noise sensitive and takes the direction of the edge into account, but it is computationally more intensive than the use of averaging methods because of the square root operation at the end. Fast Sobel algorithms which eliminate redundant operations and which are carried out with only six additions and subtractions have been published (15,26). These techniques are also usefully applied to multiresolution images as described above. Algorithms using order statistics and statistical tests based upon them are less sensitive to noise in the image (3,4). In one of these methods, the difference of medians rather than the difference of averages is used, and for images with spot-like defects, such detectors are superior. Furthermore there is less sensitivity to the orientation of edges and edge boundaries are thinner, since the median changes value more abruptly on passing over an edge than does the average. This means that the edge location is more accurately determined. Computation times are considerably longer, of course, than with simpler methods.

4.5.5. *Enhancement methods for colour pictures*

Some of the forgoing discussion of techniques for picture enhancement applies to colour imagery. There has been widespread use of false colour techniques, and most modern display systems allow the showing of any black and white image in false colour without any difficulty. For this purpose, the individual pixel values are used as entries in a table of colour values, so that a particular colour is assigned to a particular black and white value. Although widely used in satellite imagery, the method when applied to archaeological aerial photographs has been disappointing. Grayscale enhancement algorithms which work on local statistics seem to give better results. False colour displays seem to give more appealing results with data obtained from non-photographic sources such as that acquired by magnetic prospecting. Perhaps this is because the observer has acquired a definite idea of what to expect in normal imagery, and this expectation is disturbed by false colours. A common choice for false colouring uses a pseudospectral order from deep blue to deep red, with one end or the other assigned to a zero value pixel. By crowding or spreading the pseudospectrum, small differences in pixel values are rendered more visible. Some display systems have facilities for interactive adjustments of their video look-up tables which make this kind of false colouring easy.

Application of the treatment techniques for black and white to colour pictures has lagged, mainly due to the lack of suitable equipment for scanning and film reproduction with high colour quality. This situation is changing rapidly. Reasonably priced, high resolution, colour display boards and monitors are already available for personal computers, and good quality colour scanners and film recorders have come down drastically in price with the introduction of desktop publishing. One important development has been the introduction of hue–saturation– intensity (HSI) coding of the colour image, instead of the red, green, blue (RGB) hardware related scheme (56) typical of the past. In RGB, the colour images are separated into their red, green, and blue components as described above, and these components are separately manipulated. It has been observed that the techniques of local adaptive brightness and contrast modification can be applied to the RGB components of a colour image separately to produce a considerable improvement in the visibility of a feature. But it is more useful to combine treated components in various ways to enhance colour contrast (57). These methods must take the cross correlation of the colour components into account, a subject which requires considerable investigation.

The much simpler HSI approach is closely related to the non-linear properties of human colour vision. It is intuitively difficult to make decisions about processing parameters in RGB, because the human eye does not respond linearly to these components, and the display system has a square law response of brightness for each of the colour components. Hue is an attribute describing a pure colour such as red, yellow, etc. It is equivalent to the rotation of a vector centred on the pure white illuminant of the CIE (Commission International de l'Eclairage) chro- maticity diagram. By convention the zero angle of the vector is taken between pure white and maximum red. Saturation reflects the greater or lesser amount of pure white combined with a given hue. The intensity is independent of colour. These attributes are perceived independently, and statistical properties such as a hue histogram for a picture are meaningful. The uniform colour space of the CIE 1976 chromaticity diagram is obtained from a mixture of the RGB primary colours with the non-linearity of the eye taken into account. The RGB data from a scanner can be transformed into the HSI space of the CIE 1976 norm by:

$$H = \frac{90 - \arctan(F/3) + \begin{cases} 0, & G > B \\ 180, & G < B \end{cases}}{360}, \qquad G \neq B$$

$$S = (1 - \min(R, G, B))/I$$

$$I = (R + G + B)/3$$

(4.101)

where $F = [2(R) - G - B]/(G - B)$. Intensity is simply the relative proportion of equally weighted RGB components. Hue is an angle between 0 and 360°. Saturation varies from 0 to 1 along the hue vector. This transformation is best carried out in specialised hardware, but it can also be done in software if need be. Operations to improve sharpness, extract edges or correct contrast are performed on the intensity component. Colour enhancement is done on saturation and colour correction on hue, using histogram transformation or linear transformation. Methods to reduce dirt apply to colour imagery as well. Operations such as median filtering are carried out on the intensity component only.

If a colour scanner is not available, a hybrid photographic and digital method can be used. Colour separation negatives are made for the red, green, and blue components using narrow band colour filters and a low speed panchromatic black and white transfer film normally supplied to blockmakers. Almost any blockmaker can execute the order. After processing, black and white colour separation positives are written to a black and white film writer and combined photographically to make a single colour negative for the production of colour prints. The method suffers from the information loss in the many transfer steps and difficulty in registering the images mechanically. It is time consuming, so that only a few images have been processed experimentally. Experience has shown that using a cyan filter to reduce the red component of a colour slide gives some contrast improvement in archaeological aerial photographs containing crop sites with modest differences in green–yellow. Highest quality colour film output unfortunately requires an extremely expensive film writer.

4.6. Some practical considerations in picture processing
4.6.1. Computational feasibility of enhancement methods
Input/output and mass storage problems
When pictures are scanned at resolutions high enough to make scan lines invisible, that is, between 1500 and 4000 lines, files with several

million data points are obtained. These must be stored so that rapid retrieval of the data is possible for quick calculation. For example, if calculations are carried out with floating point arithmetic, intermediate files with 16 megabytes result from 2000 pixels per line square images. Most smaller machines have physical disk transfer times of the order of 1.5–3.0 megabytes per second, but the operating system and the high level language interface to it usually reduces this by half an order of magnitude. Therefore the simple transfer of 16 megabytes to main storage and back to disk again may take many seconds even on a single user system. If there are other simultaneous disk operations going on concurrently in a multitasking or virtual memory system, then the transfers for large files may take much longer than many simple picture processing operations. If three colour files are involved, the input/output times may represent a very large fraction of treatment times.

After calculation, a result image must be produced which is usually as large as the original and needs as much storage. And if intermediate results are to be preserved in case a step has to be repeated with other parameters, even more must be stored. Fortunately, mass storage peripheral devices for smaller computers are cheap enough for this to be feasible. Usually two disk drives are needed so that data can be loaded from one, computed and passed to the other, if possible with overlapping of the reading, computing, and writing when the operating system and the hardware support this. Operations may be further speeded up if the operating system allows reading or writing blocks of lines of data from the disks in a way which minimizes the system overhead required to find them. For most machines the data files should be contiguous, that is, all the blocks needed should lie in adjacent locations on the physical storage media so that mechanical movement of the recording heads is minimal. Machines with virtual storage and scatter–gather storage control remove this problem from the user's control and attempts to write blocks directly to the disks may actually slow things down. If possible, direct access to a given line or block of image data for reads and writes should be requested by the program asynchronously in advance of need. Other useful techniques are doubly buffered input/output, and circular buffering where applicable to the given algorithm. Since most algorithms require several sequential passes through the data, disk caching, whereby most recently used blocks are retained in main memory, is not very useful.

These kinds of operations may require additions to the operating system and perhaps even the hardware of some machines. A useful hardware feature is always a very large physical memory with a linearly addressable address space so that the entire input picture is in fast storage. When the standard input, output, and storage arrangements available in most general purpose machines are used, these may make some kinds of treatment agonisingly slow. But modifying hardware or the operating system is not recommended if one wants to transport painstakingly developed software to another machine at some time in the future. In the long run, it is probably cheaper to buy a faster machine and use the operating system as the manufacturer intends, than to invest in the labour of extensive modifications which will be useless in the next machine generation.

Arithmetic operations
Almost all of the techniques discussed in this chapter can be carried out with a relatively small number of arithmetic operations. A typical simple statistical algorithm will require a few dozen machine operations per image point, but even this means a very large number considering the amount of data. Personal computers cannot usually cope with the computational and input/output load imposed by high resolution pictures, although elementary operations can be carried out well if skilfully programmed (32).

Many pictures require a dozen passes or more through the data, so that the situation with regard to input/output is not favourable. For modest machines, a practical upper limit to an algorithm is about 50 operations per point on a 3000 line image with a dozen data passes. If lower resolution is used, then more arithmetic operations per point are feasible. It often is useful to select only those parts of a picture which are of archaeological interest to work on in order to keep the calculation times within reasonable limits. Most software packages have features for selecting a working area within a picture, and some kinds of software allow this to have an arbitrary shape.

A hardware coprocessor chip or high speed arithmetic board is almost essential. An array processor is of little advantage for the algorithms described here, but it is useful if operations are carried out using Fourier transform techniques. Array processors are available as plug-in components at quite reasonable cost and have floating point operation times which are usually more than an order of magnitude better than on the system on which they are installed. A large locally addressable memory and control processor which operates independently of the main computer under separate program control characterises some of them, and much of the software needed for frequently used algorithms is supplied by the manufacturer. Array processors are usually optimised for vector operations and other highly structured algorithms. Maximum throughput depends on how efficiently the data can be gotten into and out of the processor. The most efficient arrangement makes the array processor memory part of the main machine memory. Less efficient designs use transfer data via a parallel interface whose transfer times impose a maximum limit on a possible gain in speed.

4.6.2. Order of operations on pictures

Picture processing requires many separate steps with examination of intermediate results, rather than one run through a monster program which does everything. The choice of steps and their order must be under the control of a skilled observer who sees the intermediate results and can backtrack and select a different course of treatment if the result is unsatisfactory. The original scan is stored as a file, and all intermediate results produce intermediate files which should not be overwritten as processing proceeds. This method, although very wasteful of disk storage, allows one to backtrack and try a different kind of treatment if an error of judgement in the choice of an algorithm or a parameter has been made. Automating a processing sequence is not only difficult, but, in fact, dangerous, since a long sequence applied without examining intermediate results usually produces an unsatisfactory end product. Therefore, the processing of pictures is not conveniently done on large remote computers with a long wait between the intermediate steps. The volume of input/output to be passed back and forth is usually such that the user will quickly lose patience unless he has the large machine almost to himself.

Experience has shown that some orders of operation produce better results than others. It is usually best to carry out any sharpening operations first. This way one is using all the sharpness in the original data before its possible loss through subsequent operations. The sharpened image usually shows dust and dirt more than the original, so that following the sharpening, the dirt removing algorithms should be used. Then locally adaptive compensation for brightness and contrast may follow. These should always be used on clean images. It is possible that a feature extraction phase may be useful prior to geometric transformation and combination of images. Enhancement and geometric operations can be run simultaneously in large machines along with scanning and film output operations. Keeping track of what is going on is feasible. For such purposes special monitor programs, tailored to the hardware and software of the machine, as well as to the needs and level of understanding of the user, must be available. Such monitors must be flexible enough that changes in machines, user needs, skill, and available software can be readily made without writing a program each time. A monitor command file for each user which can be changed at will by the system manager is one technique which enables this. It allows selection of operations by menu, and simultaneous processes are run without the user having to keep track of things himself. A useful addition is the command stack utility available for many machines which permits easy backtracking and repetition of previously given commands using the function keys.

4.6.3. Storage reduction

When the treatment on a small window is satisfactory, then it can be tried on the full resolution image, perhaps running simultaneously while experimenting on another picture. One sets up a kind of picture pipeline whereby a set of images is scanned, a previously scanned image group is trial processed on small windows, a still earlier group processed at full resolution, and the ultimate output images transferred to an output device. The degree of simultaneity which is sensible depends very much on the hardware, the operating system of a given installation, and the skill of the user in keeping track of what is going on. The software for such a complex system may become elaborate, and skilled programming personnel are needed for development and maintenance. Personal computers whose operating systems do not permit the running of tasks simultaneously are unsuitable for the pipeline method.

4.6.4. History of treatment

It it wise to connect the history of the treatment of a picture or sequence of pictures with the files which contain the data. A trivial technique to use in a small machine with modest numbers of pictures is to choose the file names in such a way as to reflect the treatment history. Large numbers of pictures require a formal record keeping system and perhaps a database to keep track of things. Much work still remains to be done to find an optimum organisation based on menu selection, simultaneous operations of various processes synchronisation of tasks, backtracking, and perhaps even an expert system to advise unskilled users on paths to take for best results.

Notes

(1) Andrews, H. C., Hunt, B. R. 1977, *Digital Image Restoration*, Prentice-Hall, Englewood Cliffs, NJ.
(2) Billingsley, F. C. 1979, Noise consideration in digital image processing hardware, in *Picture Processing and Digital Filtering*, T. S. Huang, ed., 249–82, Springer Verlag, Berlin.
(3) Bovik, A. C., Munson, D. C. 1986, Edge detection using median comparison, *Computer Vision, Graphics, and Image Processing*, **33**, 377–89.
(4) Bovik, A. C., Huang, T. S., Munson, D. C. 1987, The effect of median filtering on edge estimation and detection, *IEEE Transactions Pattern Analysis and Machine Intelligence*, **9**, 181–94.
(5) Burch, S. F., Gull, S. F., Skilling, J. 1983, Image restoration by a powerful maximum entropy method, *Computer Graphics and Image Processing*, **23**, 113–28.
(6) Burt, P. J. 1981, Fast filter transforms for image processing, *Computer Graphics and Image Processing*, **16**, 20–51.
(7) Burt, P. J. 1983, Fast algorithms for estimating local image properties. *Computer Vision, Graphics, and Image Processing*, **21**, 368–82.
(8) Courant, R., Hilbert, D. 1953, *Methods of Mathematical Physics*, Interscience, New York, NY.
(9) De Palma, J. J., Schiano, M. L. 1973, Filters, in *SPSE Handbook of Photographic Science and Engineering*, 257–330, John Wiley & Sons, New York, NY.
(10) Duda, R. O., Hart, P. E. 1973, *Pattern Classification and Scene Analysis*, John Wiley & Sons, New York, NY.
(11) Gonzalez, R. C., Wintz, P. 1977, *Digital Image Processing*, Addison-Wesley, Reading, Ma.
(12) Green, W. B. 1983, *Digital Image Processing, A Systems Approach*, Van Nostrand Reinhold, New York, NY.
(13) Gubbins, D., Scollar, I., Wisskirchen, P. 1971, Two dimensional digital filtering with Haar and Walsh transforms, *Annales de Géophysique*, **27**, 85–104.
(14) Hall, E. L. 1979, *Computer Image Processing and Recognition*, Academic Press, New York, NY.
(15) Hashimoto, M., Sklansky, J. 1987, Multiple-order derivatives for detecting local image characteristics, *Computer Vision, Graphics, and Image Processing*, **39**, 28–55.
(16) Hibbs, A. R., Wilson, W. S. 1983, Satellites map the oceans, *IEEE Spectrum*, **20**, 10, 46–54.
(17) Hou, H. S. 1983, *Digital Document Processing*, John Wiley & Sons, New York, NY.
(18) Huang, T. S. 1979, ed. *Picture Processing and Digital Filtering*, 2nd edn, Springer Verlag, Berlin.
(19) Huang, T. S., Yang, G. J., Tang, G. Y. 1979, A fast two dimensional median filtering algorithm, *IEEE Transactions Acoustics, Speech, Signal Processing*, **27**, 13–18.

(20) Huang, T. S. ed. 1981, *Two Dimensional Digital Signal Processing*, Vol. 1: *Linear Filters*, Vol. 2: *Transforms and Median Filters*, Springer Verlag, Berlin.

(21) Hummel, R. 1975, Histogram modification techniques, *Computer Graphics and Image Processing*, **4**, 209–24.

(22) Hummel, R. 1977, Image enhancement by histogram transformation, *Computer Graphics and Image Processing*, **6**, 184–95.

(23) Jain, R., Chlamtac, I. 1985, The P2 alogrithm for dynamic calculation of quantiles and histograms without storing observations, *Communications of the Association for Computing Machinery*, **28**, 1076–85.

(24) Joint System Development Corp. 1985, *SPIDER: Subroutine package for image data enhancement and recognition*, JSDC 1-chome, Toranomon, Minato-ku, Mitsui, Tokyo.

(25) Kendall, M. G., Stuart, A. 1961, *The Advanced Theory of Statistics*, Vol. 1, 374–7, Griffin, London.

(26) Lee, C. C. 1983, Elimination of redundant operations for a fast Sobel operation, *IEEE Transactions Systems Man Cybernetics*, **13**, 242–5.

(27) Lee, J. S. 1981, Refined filtering of image noise using local statistics, *Computer Graphics and Image Processing*, **15**, 380–9.

(28) Levi, L. 1974, Unsharp masking and related image enhancement techniques, *Computer Graphics and Image Processing*, **3**, 163–77.

(29) Levine, M. D. 1985, *Vision in Man and Machine*, McGraw-Hill, New York, NY.

(30) Marr, D., Hildreth, E. 1980, Theory of edge detection, *Proceedings Royal Society of London, Ser. B*, **207**, 187–217.

(31) McDonnell, M. J. 1981, Box filtering techniques, *Computer Graphics and Image Processing*, **17**, 65–70.

(32) Meyers, H. J., Bernstein, R. 1985, Image processing on the IBM personal computer, *Proceedings of the IEEE*, **73**, 1064–70.

(33) Montuori, J. S. 1978, Image scanner technology, *Perkin-Elmer Technical News*, **7**, 2, 7–19.

(34) Narendra, P. M. 1981, A separable median filter for image noise smoothing, *IEEE Transactions Pattern Analysis and Machine Intelligence*, **3**, 20–9.

(35) O'Gorman, L., Sanderson, A. C. 1987, A comparison of methods and computation for multi-resolution low- and band-pass transforms for image processing. *Computer Vision, Graphics, and Image Processing*, **37**, 386–401.

(36) Ostrem, J. S., Falconer, D. G. 1981, A differential operator technique for restoring degraded signals and images, *IEEE Transactions Pattern Analysis and Machine Intelligence*, **3**, 278–84.

(37) Peli, T., Malah, D. 1982, A study of edge detection algorithms, *Computer Graphics and Image Processing*, **20**, 1–21.

(38) Pratt, W. K. 1978, *Digital Image Processing*, Wiley-Interscience, New York, NY.

(39) Rosenfeld, A. 1973, Progress in picture processing, 1969–1971, *Computing Surveys*, **5**, 81–108.

(40) Rosenfeld, A. 1972–, Progress in picture processing, yearly bibliography, *Computer Graphics and Image Processing*

(41) Rosenfeld, A., Kak, A. C. 1982, *Digital Image Processing*, John Wiley & Sons, New York, NY.

(42) Rosenfeld, A., ed. 1984, *Multiresolution Image Processing and Analysis*, Springer Verlag, Berlin.

(43) Sawchuk, A. A. 1972, Space variant image motion degradation and restoration, *Proceedings of the IEEE*, **60**, 854–61.

(44) Sawchuk, A. A. 1973, Space variant system analysis of image motion, *Journal of the Optical Society of America*, **63**, 1053–63.

(45) Sawchuk, A. A. 1974, Space variant image restoration by coordinate transformation, *Journal of the Optical Society of America*, **64**, 138–44.

(46) Scollar, I. 1978, Progress in aerial photography in Germany and computer methods, *Aerial Archaeology*, **2**, 8–18.

(47) Scollar, I., Weidner, B. 1979, Computer production of orthophotos from single oblique images or from rotating mirror scanners, *Aerial Archaeology*, **3**, 17–28.

(48) Scollar, I., Weidner, B., Huang, T. S. 1984, Image enhancement using the median and the interquartile distances, *Computer Vision, Graphics, and Image Processing*, **25**, 1–18.

(49) Scollar, I., Weidner, B., Segeth, K. 1986, Display of archaeological magnetic data, *Geophysics*, **51**, 623–33.

(50) Smith, J. A. 1983, Matter-energy interaction in the optical region, in *Manual of Remote Sensing*, 2nd edn., R. N. Colwell, ed., 98–101, American Society of Photogrammetry, Falls Church, Va.

(51) Smith, R. W., Tatarewicz, J. N. 1985, Replacing a technology: the large space telescope and CCDs, *Proceedings of the IEEE*, **73**, 1221–35.

(52) Tanimoto, S., Pavlidis, T. 1975, A hierarchical data structure for picture processing, *Computer Graphics and Image Processing*, **4**, 104–19.

(53) Trinder, J. C. 1987, Measurements on digitized hardcopy images, *Photogrammetric Engineering and Remote Sensing*, **53**, 315–21.

(54) Wallis, R. 1977, An approach to the space variant restoration and enhancement of images, in *Proceedings, Symposium on Current Mathematical Problems in Image Science*, Monterey, CA, Nov. 1976, reprint, *Image Science Mathematics*, C. O. Wilde, E. Barett, eds., Western Periodicals, North Hollywood, Ca.

(55) West, D. H. D. 1979, Updating mean and variance estimates, an improved method, *Communications of the Association for Computing Machinery*, **22**, 532–4.

(56) Wilson, A. 1988, Einführung in die HSI-Farbbildverarbeitung, *Design & Elektronik*, **8**, 114–18, Markt & Technik, Munich.

(57) Wojick, Z. M. 1985, A natural approach in image processing and pattern recognition, *Pattern Recognition*, **18**, 299–326.

5

Geometric transformation of archaeological aerial photographs and mapping

5.1. Introduction

The negative image is ideally a flat surface. The photographed scene is usually not flat and in oblique archaeological aerial photography the camera is usually at an unknown angle and position relative to the scene so that there is a complex relationship between the geographical location of a site on the ground and the position and shape of the site in the image. In addition to possible distortions in the recording of light intensities, a geometric transformation of the data has also taken place. The mathematical properties of the geometric transformation from a scene to a flat surface represent a classical problem of computer graphics (18). The inverse problem, obtaining a map of the surfaces from the information in the photograph, is the main subject of photogrammetry (70).

Three angles, ω, the rotation around the x axis, ϕ, the rotation around the y axis, and κ, the rotation around the z axis, plus three position coordinates X_j, Y_j, X_j denote the orientation and location from which the picture was taken. They are almost always unknown. The camera can be calibrated for the focal length of the lens and the exact position of the intersection of the optical axis of the lens with the image plane (principal point) can be measured. Specialised photogrammetric cameras sometimes have a glass plate in the film plane against which the film is pressed so that it lies flat, or a vacuum system which ensures that the film is held flat by air pressure against a metal plate. Reseau marks are sometimes scribed on the glass plate so that possible film distortion due to temperature changes between the time of exposure and processing may be compensated. The amateur cameras used almost exclusively in European aerial archaeology have none of these features.

It is possible to compute a mathematical transformation from image to scene to obtain the position and height of any point in an image when two or more images are available and the true positions and heights of at least two points, plus the height of one additional point in the image sequence is known and a sufficient number of other corresponding points (whose coordinates need not be known) are visible in both images.

Conversely, in computer graphics, the shape of an object which is to be displayed on a CRT is known. What is wanted are the gray values at any point on the screen, given an arbitrary position for the object as if seen at the centre of the screen. This problem is directly solvable since the transformation required can be computed easily from the positional information. In the case of an archaeological aerial photograph, one has only one or more images and a map. When there are four or more identifiable points in the map which are also visible in a single image, then the transformation parameters can be computed without knowledge of the camera constants if it is assumed that the terrain is flat. If six points are identifiable both in the picture and on the map, this is also adequate for a mildly undulating surface with knowledge of the heights of the points. If more than one image is available, then it is possible to carry out the calculations with as little as five known points even for hilly terrain.

5.2. The image as a projection

5.2.1. Perspective

Perspective allows us to see the relationships between features in the film and in the scene. If we have two images taken at some distance apart from each other, then we can see the depth relationships directly with a stereoscope. In highly populated parts of the world, monuments rarely survive above ground and there are no height differences in the visible remains to be taken directly into account. Usually we need a precise plan of the archaeological features visible in the image. But a plan is a subjective interpretation of the data in the picture, and the picture itself transformed to one of uniform scale and vertical orientation may be desirable. The result of such a geometric transformation of the picture or pictures to a pseudovertical picture at a definite scale, is an orthophoto. Orthophotos are usually made by optomechanical methods in photogrammetry from precision vertical mapping pictures. Making orthophotos from archaeological oblique pictures is a much more difficult problem.

5.2.2. Maps

A map is a projection to a plane of the surface of the earth. When the area under consideration is small, then the curvature of the earth is negligible, and geographic features project with parallel lines to the map. When the map covers a large area, then the earth's curvature must be taken into account, but this problem does not usually occur in images of archaeological interest. Mapped features are idealised abstractions of the true vertical representation as seen in an orthophoto, with some details emphasised and others suppressed, depending on the map makers' view of their significance. A vertical aerial photograph, typically made with a wide angle lens does not have the data projected vertically to a plane like an orthophoto or a map even if it may superficially resemble one. Only the region immediately under the camera is seen as a true vertical image. Away from the image centre, the features are distorted where ground relief is present. In addition the aeroplane carrying the camera rolls and pitches slightly and even the centre of the picture does not constitute a true vertical view of the terrain. For the hand-held oblique images which are the mainstay of archaeological aerial photography, there is an even more complex relationship between the shapes of features and their true plans, for no part of the image is vertical.

5.2.3. Transformation of sites from photographs to maps or orthophotos

Fortunately archaeological sites are rarely built on the sides of cliffs, and even for hill forts the occupied area is relatively flat compared with the surroundings. In pictures of the fields of the river valleys in northern Europe where the vast majority of aerial discoveries have been made, the sites and the terrain may be hilly but usually not mountainous. For such pictures there are relatively simple techniques for transposing points in the image to points in the maps. To make orthophotos from these, the problem is a bit more tedious. With sites located in rugged terrain, where visible points of known position are found at significantly different heights from the site features, the problem requires considerably more mathematical treatment. In this case the generation of an orthophoto is a major problem especially when details must be taken from a number of pictures made at different times and with varied equipment. It is possible to carry out this task entirely by calculation without specialised machinery, and for very oblique images, this method is preferable.

If one has a set of pictures of an archaeological site taken at different times these usually show different details which should be enhanced, analysed, and recorded at known scales and orientation in one resultant enhanced orthophoto with a map superimposed on it. Even if a large scale map is not available, it may be possible to obtain conventional orthophotos of the area upon which the archaeological oblique images may be superimposed or if all else fails, vertical aerial cover may be used which itself can be converted into an orthophoto to which archaeological detail from another set of pictures may be added.

To do all this it is necessary to measure the film coordinates of many points which are identical in different pictures. These points need not all be known from maps or ground survey, it is enough to know that they are the same points in the landscape. If there are enough of them, then there is only one possible set of relative orientations of the photograph sequence which could have produced the points at the locations seen in the pictures.

Field boundaries which are visible in aerial photographs of any kind do not necessarily reflect ownership boundaries. However, if archaeological monuments visible in aerial photographs are to be protected, it is usually necessary to inform local authorities and owners, using geometrically corrected data as evidence. Therefore, if the results of geometric correction are to carry legal weight in monument protection applications, maps of adequate scale, which show ownership boundaries, into which the archaeological data can be incorporated must be available and if not, they must be made.

5.3. Mathematics of picture geometry

5.3.1. Scanned image rectification

Mathematically, a picture is a function of two variables. Its domain is a bounded region of the plane. The values of the two independent variables are coordinates describing a location inside the domain and the function value gives the gray level of the picture at that location. The geometric transformation problem is, generally stated: given an input picture $f(x, y)$, produce an output picture $g(x, y)$ where geometric properties have been altered in some desired fashion.

The scanned lines of the input image constitute one coordinate system and those of a scanned map another. The desired orthophoto can be thought of as the result of a transformation which relates the two coordinate systems. The orthophoto looks like a skewed quadrilateral imposed on the map if the terrain is flat, figure 5.1. If the terrain is not flat then the quadrilateral has curved sides and is distorted like a rumpled carpet, figure 5.2. In effect, the rasters of the map and picture are rotated, skewed, changed in scale, and distorted in shape relative to each other.

A transformed input point will only rarely coincide exactly with an output raster point. Therefore the gray value of an output pixel must be estimated from points in the input image which are in the neighbourhood of the location of the output point referred to the input coordinate system. Using a direct transformation from image to output map coordinates would produce some areas which contain no pixels and others where pixels are doubled. Therefore for complete rectification the indirect transformation from map to image is carried out.

It is usually not necessary to calculate the exact transformed position in the input image for every output point. Instead, the exact coordinates of the intersections of a coarse grid, typically with 32 or 64 pixel spacing in the output image are calculated for the input image, and these points used as a basis for interpolating linearly within the grid meshes to obtain the other positions.

Fig. 5.1 Pixels of rectified photo overlaid on the map (flat terrain).

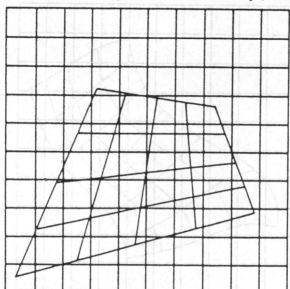

Interpolation of gray values in images

At each position in the output image the local gray value must be
estimated, either by taking the nearest corresponding input image point,
by linear interpolation from the four nearest input points, or by more
complex interpolation from a larger number of surrounding input points
(6). A sampled one-dimensional signal $g_s(x)$ is:

$$g_s(x) = g(x)\,\mathrm{comb}(x) \tag{5.1}$$

where:

$$\mathrm{comb}(x) = \sum_{n=-\infty}^{\infty} \delta(x-n)$$

δ is the Dirac delta and n is a variable of summation, and a signal
reconstructed from the sampled signal $g_r(x)$ is the convolution of an
interpolation function $r(x)$ with the sample:

$$g_r(x) = g_s(x)\mathbin{*}r(x) \tag{5.2}$$

where the interpolation function is the response to a sampled input and
$*$ denotes convolution. The aim is to use an $r(x)$ which gives an output
signal which is not degraded relative to the input signal. Equation (5.2)
can also be written as:

$$g_r(x) = \sum_{n=-\infty}^{\infty} g(n)r(x-n) \tag{5.3}$$

Fig. 5.2 Pixels of rectified photo overlaid on the map (uneven terrain).

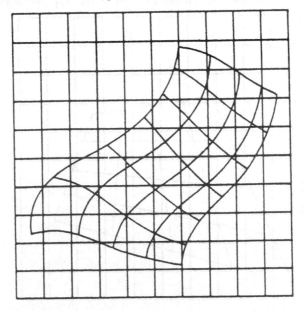

This means that for a given point x, the degree to which nearby signal samples contribute to the output signal is a function of the spread of the interpolation function. An ideal reconstruction of the input preserves all spatial frequencies in the data up to a given limit. It can readily be shown that in the space domain, this is equivalent to a convolution with $r(x)$:

$$r(x) = \text{sinc}(x) = \sin(\pi x)/\pi x \qquad (5.4)$$

The sinc function is the ideal interpolator. It is not limited in spatial extent and if used in two dimensions, excessively large numbers of convolution coefficients would be necessary to describe it well. This would lead to a very large number of multiplications when computing. If resolution in the input scanned image is less than 1000 lines then the use of nearest neighbour interpolation produces staircase-like effects in high contrast lines of the input image. Separable bilinear interpolation from four nearest input neighbours to a desired output point produces some blurring. In this case the interpolation weight w is inversely proportional to the distance x or y between the point to be interpolated and its nearest neighbour pairs which are separated by a distance X or Y in each axis:

$$w = \begin{cases} 1 - |x|/X, & |x| < X \\ 0 & |x| \geq X \end{cases} \qquad (5.5)$$

It can be shown that this triangular function in space has a response in the frequency domain of a squared sinc function:

$$f(x) = x \, \text{sinc}^2(\omega x/2\pi) \qquad (5.6)$$

There is therefore a considerable loss of high spatial frequencies compared with the ideal sinc interpolator which has a rectangular transfer function in the frequency domain. However, for pictures with a very much larger number of lines than the eye can resolve at normal viewing distances, the loss of resolution will not be noticed. It is therefore possible to trade off the much faster computation time of the separable bilinear interpolator against the larger number of pixels which must be treated. If the picture must undergo many enhancement processes prior to geometric transformation, the gain in time here may well be lost elsewhere.

It has been shown by Simon (69) that for typical high resolution pictures, a near-sinc weighted average of 16 near neighbour input pixels is good enough for interpolating the value of an output pixel for most purposes. The weights used must vary, depending on the location of the transformed output point coordinates relative to the input raster. The total amount of computation needed to obtain an orthophoto with about the same size as the input picture is very considerable indeed, about 50 floating point arithmetic operations for each of 4–9 million output points. Bernstein (6) demonstrated that a family of piecewise cubic polynomials given by:

$$r(x) = \begin{cases} (\alpha + 2)|x|^3 - (\alpha + 3)|x|^2 + 1, & |x| < 1 \\ \alpha|x|^3 - 5\alpha|x|^2 + 8\alpha|x| - 4\alpha, & 1 \le |x| < 2 \\ 0, & \text{otherwise} \end{cases} \qquad (5.7)$$

can be applied to approximate the sinc function and that this produces smooth results. The family of interpolators is shown in figure 5.3. Park and Schowengerdt (50) show that the optimal value for α is -0.5. The value given by Bernstein is -1. No difference between images treated

Fig. 5.3 (a) Sinc function (part) and cubic spline approximation. (b) Five members of the parametric cubic convolution interpolator family for various values of α, giving response r as a function of distance x. From ref. 50.

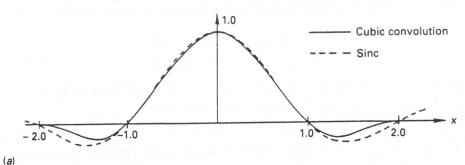

(a)

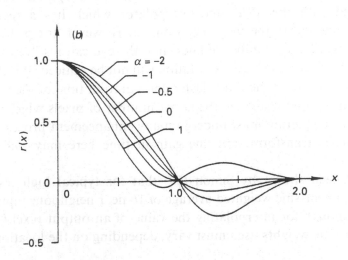

with both values can be seen when the resolution is greater than 1500 lines or so. These interpolators are separable in x and y, and the coefficients may be precomputed and stored in a table. Experiments have shown that a table with 200 is adequate. Nonetheless, a very considerable amount of computation per output image point is required.

Transformation mathematics
Geometric transformations are coordinate transformations given by two functions of x and y:

$$x = X(u, v), \qquad y = Y(u, v)$$

and the inverses

$$U = u(X, Y), \qquad V = v(X, Y)$$

(5.8)

The simplest possible transformation is a shift, obtained by adding a constant, a and b, to each coordinate:

$$x'_p = x_p + a, \qquad y'_p = y_p + b$$ (5.9)

Another very simple transformation is obtained by multiplying by a constant r in x and s in y, which results in a scale change:

$$x'_p = rx_p, \qquad y'_p = sy_p$$ (5.10)

A rotation through an angle θ is given by:

$$x'_p = x_p \cos(\theta) + y_p \sin(\theta), \qquad y'_p = -x_p \sin(\theta) + y_p \cos(\theta)$$ (5.11)

A scale change followed by rotation and shift is an affine transformation.

5.3.2. *A mathematical model of projection in a camera*

In a camera, the light rays which are reflected from a scene pass through the centre of the lens to the film. In the ideal case, the rays are straight lines and are not distorted by the lens and atmospheric refraction, the film is a perfect plane, and a perfect point in the scene results in a perfect point on the film. In the real world all this is not true, but here these distortions will be neglected. Under projection, the only invariants are the cross ratio and the coincidence of points and straight lines. Areas, angles, distances, parallelism, and proportion are not invariant. When the film and object planes are parallel then only the scale changes.

For figure 5.4, the object point, film point, and centre of projection lie on a straight line so that:

$$\alpha_p \mathbf{v} = \mathbf{w}$$

where:

$$\mathbf{v} = \mathbf{Op} \qquad \text{and} \qquad \mathbf{w} = \mathbf{OP}$$ (5.12)

The picture coordinate system is right handed and Cartesian, with the origin at O, with the x and y axes parallel to the film plane and with the principal point at $(0, 0, -f)$, where f is the focal length of the lens. In the following development, picture coordinates are given with small letters and ground coordinates with capitals. From equation (5.12) it follows that:

$$\alpha_p \begin{pmatrix} x_b \\ y_b \\ -f \end{pmatrix} = \begin{pmatrix} X_b \\ Y_b \\ Z_b \end{pmatrix}$$

with: (5.13)

$$p = \begin{pmatrix} x_b \\ y_b \\ -f \end{pmatrix} \qquad P = \begin{pmatrix} X_b \\ Y_b \\ Z_b \end{pmatrix}$$

relative to the picture coordinate system. In oblique pictures, the local scale α_p varies.

Fig. 5.4 A model for photographic projection: O = centre of projection, f = focal length of lens, H = picture principal point, τ = total rotation relative to the ground, p = image point, P = object point, N = nadir.

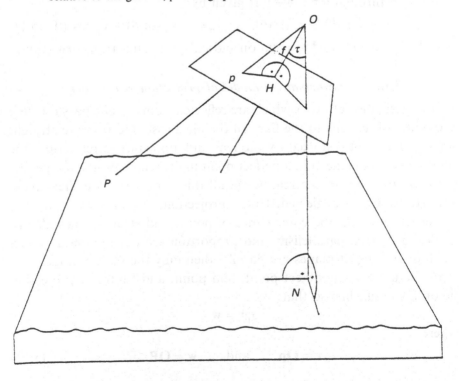

5.3.3. The collinearity equations

For pictures of archaeological interest, the curvature of the earth is insignificant and therefore:

$$\alpha_p \begin{pmatrix} x_b \\ y_b \\ -f \end{pmatrix} = \eta R \left[\begin{pmatrix} X \\ Y \\ Z \end{pmatrix} - \begin{pmatrix} X_0 \\ Y_0 \\ Z_0 \end{pmatrix} \right] \qquad (5.14)$$

where (X, Y, Z) are the map coordinates of P and (X_0, Y_0, Z_0) are the map coordinates of the centre of projection. R is an orthogonal rotation matrix and η is the scale factor. In pictures taken with non-calibrated cameras the principal point is not known and may only be estimated by means of measurements from the corners or sides of the picture. Thus x_b, y_b are measured relative to a coordinate system which is displaced in the film plane. Therefore:

$$\begin{pmatrix} x \\ y \\ 0 \end{pmatrix} - \begin{pmatrix} x_p \\ y_p \\ f \end{pmatrix} = \frac{\eta}{\alpha_p} R \left[\begin{pmatrix} X \\ Y \\ Z \end{pmatrix} - \begin{pmatrix} X_0 \\ Y_0 \\ Z_0 \end{pmatrix} \right] \qquad (5.15)$$

where (x_p, y_p, f) are the coordinates of the principal point and $(x, y, 0)$ are the coordinates of the film point p. If the ground coordinates are to be derived from the film coordinates then:

$$\begin{pmatrix} X \\ Y \\ Z \end{pmatrix} - \begin{pmatrix} X_0 \\ Y_0 \\ Z_0 \end{pmatrix} = \frac{\alpha_p}{\eta} {}^tR \left[\begin{pmatrix} x \\ y \\ 0 \end{pmatrix} - \begin{pmatrix} x_p \\ y_p \\ f \end{pmatrix} \right] \qquad (5.16)$$

where ${}^tR$ is the matrix transpose.

Let n picture points (x_i, y_i) with n corresponding ground points (X_i, Y_i, Z_i) be known then:

$$\begin{pmatrix} x_i \\ y_i \\ 0 \end{pmatrix} - \begin{pmatrix} x_p \\ y_p \\ f \end{pmatrix} = \mu_i R \left[\begin{pmatrix} X_i \\ Y_i \\ Z_i \end{pmatrix} - \begin{pmatrix} X_0 \\ Y_0 \\ Z_0 \end{pmatrix} \right], \qquad i = 1, \ldots, n \qquad (5.17)$$

This is a system of three equations. To eliminate μ_i, one divides the first two equations by the third:

$$x_i - x_p + \frac{f[m_{11}(X_i - X_0) + m_{12}(Y_i - Y_0) + m_{13}(Z_i - Z_0)]}{m_{31}(X_i - X_0) + m_{32}(Y_i - Y_0) + m_{33}(Z_i - Z_0)} = 0$$

$$y_i - y_p + \frac{f[m_{21}(X_i - X_0) + m_{22}(Y_i - Y_0) + m_{23}(Z_i - Z_0)]}{m_{31}(X_i - X_0) + m_{32}(Y_i - Y_0) + m_{33}(Z_i - Z_0)} = 0 \qquad (5.18)$$

with $R = (m_{ij})_{i=1,...,3, j=1,...,3}$. These are the collinearity equations, the fundamental equations of photogrammetry. Ground points are projected on the film as:

$$T: \begin{pmatrix} X \\ Y \\ Z \end{pmatrix} \rightarrow \begin{pmatrix} -\dfrac{f[m_{11}(X_i - X_0) + m_{12}(Y_i - Y_0) + m_{13}(Z_i - Z_0)]}{m_{31}(X_i - X_0) + m_{32}(Y_i - Y_0) + m_{33}(Z_i - Z_0)} + x_p \\ -\dfrac{f[m_{21}(X_i - X_0) + m_{22}(Y_i - Y_0) + m_{23}(Z_i - Z_0)]}{m_{31}(X_i - X_0) + m_{32}(Y_i - Y_0) + m_{33}(Z_i - Z_0)} + y_p \end{pmatrix}$$

(5.19)

5.3.4. Simple methods for transposing points from a single image to a map

For flat terrain the simplest methods take advantage of the fact that when the plane of the earth is projected onto the plane of the picture, straight lines such as roads and field boundaries remain straight. If the terrain is hilly, this is not the case, as can be readily seen in a picture of a known straight road passing directly over a hill, seen at an angle. It appears to be curved, the curvature depends on the point and angle of view. The features of an archaeological site are similarly distorted. There are several approaches to the problem in flat terrain which are relatively easy to apply without computation (60).

The first uses graphical techniques to find the intersections of lines and site features, the second requires a special enlarger which has a film carrier, lens, and projection board which can be tilted independently of each other over a considerable range. Alternatively, special optical sketching devices may be used if tilt is not too severe or if the area of interest to be transferred to a map is not too large. When terrain is hilly and the camera angle is very far from the vertical, these methods will not give accurate results.

When the archaeological features of interest are quite complex and have much local detail, then point by point transfer by graphical means becomes very tedious and is rather subjective. Either an optically or a computer generated orthophoto superimposed on a map is the most elegant solution.

Graphical techniques
Graphical methods are based on the relationship between points in the image and known points on a map between which lines are drawn on

tracing paper laid over both. For these techniques at least four points must be visible in the picture whose locations are known on the ground or found on the map, and they must be well spaced out toward the corners of the picture. The method is accurate to about 1 part in 200 at best, so that lens and all other image distortion factors are usually negligible in importance if the terrain is quite flat and the known points (control points) well spaced and really point-like. Graphical methods are quite good enough for site sketching and general inventory purposes on maps with scales of 1:10000 or smaller, or for locating cuttings in a large scale excavation on a site plan at 1:5000 or so. They introduce too much error for small digs which must be precisely located so as to save labour, or for the legal protection of a site where distances from field boundaries must be known with high precision.

The simplest technique is the paper strip method shown in figure 5.5. Four points a, b, c and d must be identified in the picture as being

Fig. 5.5 The paper strip technique. From ref. 60.

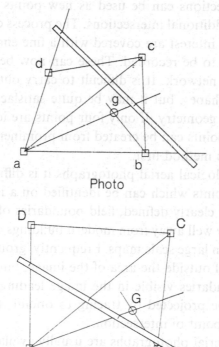

identical with A, B, C and D in the map. The lines connecting point a with all the other points are drawn, including a line to a feature point g whose position in the map is to be determined. A strip of paper with a straight edge is used. It is placed over the image such that the edge intersects the lines ad, ac, ag, and ab. Marks are made at the intersection points with a pencil. The strip is then placed on the map and adjusted such that the marks of ad, ac and ab coincide with the lines AD, AC and AB. A line can now be drawn from A intersecting the mark made where ag crossed the strip. The process is repeated from point b in the photograph, and B in the map, giving point G at the intersection of lines AG and BG. It is quite tedious to construct more than a small number of points with this method.

Another graphical technique requires five identifiable points in the photograph and map, see figure 5.6. A tracing is made of all possible lines joining the visible points both in the photograph and map. It is a property of such a Möbius network that the intersection of lines in the network correspond in the image and map when the terrain is flat. At least five well-distributed points not on a line or quadrilateral must be present. The intersections can be used as new points to draw further lines which create additional intersections. The process can be continued until the features of interest are covered with a fine enough network to cover all structures to be recorded. These can now be sketched in the map relative to the network. It is difficult to carry out this process for complex irregular shapes, but it may be quite satisfactory for features with simple regular geometry. If only four points are identifiable in the map, five or more points can be created from prominent image features with the paper strip method first.

In many archaeological aerial photographs it is difficult to find even four well-defined points which can be identified on a map. Field roads tend not to be very clearly defined, field boundaries often even less so, and sites are usually well away from modern buildings or features which appear accurately on large scale maps. Frequently, ground points visible on the map are well outside the area of the image, but straight lines of roads or field boundaries visible in the image leading to them can be seen. These must be projected by tracing to obtain the fictive picture coordinates of the point of intersection.

Archaeological aerial photographs are usually available as black and white prints on paper. Maps are usually purchased on the same medium. If control points are determined graphically, then it is most important to use dimensionally stable positive prints and maps. Many older enlarging papers which are not resin impregnated change dimensions on drying by as much as 1% or more, unequally in height and width. Paper maps are also unreliable for accurate coordinate measurement, especially if they are made from a transparency on a printing machine which heats the paper after exposure. Compensation for paper map error in small areas is possible by drawing a grid over the map to connect coordinate points on the boundary at regular intervals, measuring the actual distances on the map carefully and correcting by interpolation for the coordinates of measured points. It is better to use maps on stable plastic film if they can be obtained, and still better to scan these in a high resolution drum scanner which then provides a data set on which very accurate measurements can be made. If computer generated orthophotos are subsequently made using control points obtained from

Fig. 5.6 The Möbius network method. From ref. 60.

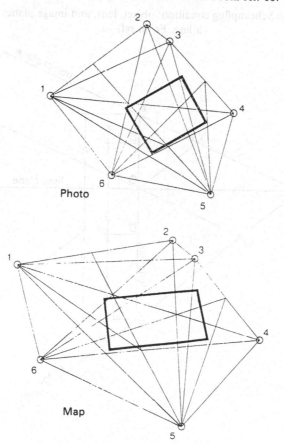

scanned maps, then they can be merged directly with these maps with small error.

5.3.5. *Optical rectification of oblique images*

If the degree of tilt from the vertical is small then it is possible to compensate for it by an appropriate countertilt in the base board of the enlarger, the film carrier, and the enlarger lens. Projections of the negative plane, the base board and the lens must meet in a line, the Scheimpflug condition, figure 5.7. This is difficult to achieve with normal enlargers. Special enlargers are made for photogrammetry which allow rapid adjustment of the necessary angles and rotations with displacements as much as 85° from the vertical. For extreme tilts, multiple projection systems have been built using reflection of the enlarger beam from several orientable mirrors. Such a multiple 'rectifier' requires considerable skill

Fig. 5.7 The Scheimpflug condition: object, lens, and image planes meet in a line. From ref. 70.

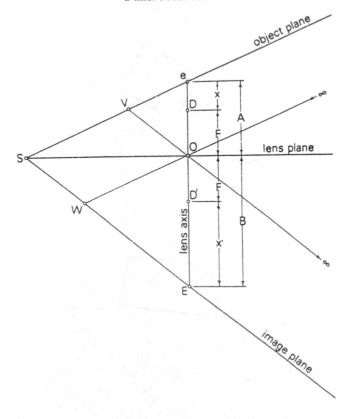

in use. They are not routinely produced and would probably not be much cheaper than a small image processing system. The enlargers which provide a mechanical method of obtaining the Scheimpflug condition have been available to photogrammetrists for many years. They are found in most photogrammetric establishments where they are used to make orthophoto maps from verticals taken even over rugged terrain by rectifying pieces of several pictures. Some even incorporate automatic dodging and exposure control to compensate for the uneven illumination of the photographic paper on the tilted enlarging easel. These devices require an accurately known principal point in the image and six points visible in both image and map.

Special orthophoto machines which run under computer control, are available. A transparency (colour or black and white) is mounted on a rotating drum and scanned with a small aperture. The light passes through a series of computer controlled prisms and zoom lenses which rotate and scale the scanner aperture and project it onto an unexposed film. In this way, the orthophoto is built up strip by strip rather quickly. The calculations for the movements of the optics are done beforehand on a large machine and are stored on magnetic tape. This, in turn, is read by a small local processor to run the motor drives for the optics. There are several advantages in using this type of device when making orthophotos from mapping verticals. A colour original is passed without colour distortion to the output film and processing is more rapid than if the original were scanned for three colour components and the orthophoto made by pure calculation. The disadvantage lies in the cost for a specialised instrument which can be used for no other purpose. The device can work with only one original at a time, whereas a fully computerised system can deal with several pictures simultaneously stored as disk files. Before using an orthophoto machine, control points in the images must be measured and a mathematical model of the terrain height at many points must be constructed.

The technique makes use of the fact that for small segments of a picture, the distortion at a local point can be corrected approximately by rotation and a change of scale only, since the skew in a small piece of a vertical image is negligible. Optical devices which are capable of dealing directly with multiple photographs taken from random points of view with non-mapping cameras typical of aerial archaeological obliques are not available.

An economical alternative to the orthophoto enlarger for occasional use is one of the many optical sketchmasters which are based on the principle of the camera lucida. The user sees the photograph and the map at the same time through an arrangement of prisms and lenses. Made primarily for sketching local detail in a vertical mapping photograph, sketching devices can be used on oblique images if the size of detail to be sketched is small compared with the whole picture. Most camera lucida arrangements cannot deal with the large amount of lateral tilt which is unfortunately present in many hand-held pictures made during a steeply banked turn. The method suffers from subjectivity of interpretation in sketching features, but if the scale of the map is small, this is not significant.

5.3.6. Computer methods for single images

Control point identification and correspondence between images and maps
Photogrammetrists who work with high quality images use precision measuring devices to survey high contrast targets laid on the ground in the area of interest as control points. This technique is almost never applicable when dealing with archaeological oblique imagery. Instead, one must make use of those features which can readily be seen like houses, electric pylons, road and field boundary intersections, etc. It is best to display images and maps whose control points are to be matched on a pair of high resolution displays to allow interactive selection of points and the positioning of an image cursor as accurately as possible by hand. Quick enlargement of the image by pixel replication is a feature provided by many modern displays and is a considerable help. Automatic control of cursor grayscale polarity which causes the cursor to have maximum contrast with its surroundings is another.

A major problem with all photogrammetric approaches arises from the need for determining the correspondence between points visible in images and maps. The accuracy of the determination of the coefficients of an image transformation depends in a highly critical way on the accuracy of the determination of the film coordinates of the control point.

When real images are examined, sharply defined high contrast points have to be used such as the corners of houses. A corner of a house seen in a second image taken at a different angle looks quite different because of differences in illumination and possible occlusion. In the map, the house has been drawn in still another fashion, namely as a plan. Therefore it is not possible to find an automatic method for determining correspondences between useful control points in extremely oblique images and maps. The possibilities for error are manifold. As shown in figure 5.8, a set of identical houses visible in the background are highly suitable for use as control points, but no automatic method can identify and choose a particular house and its map representation because they all look alike.

There is a considerable advantage in using a matching algorithm which takes over the fine control of the process once a rough match between image pairs has been found interactively. Matching algorithms have been published which use the data from a window in one image and attempt to maximise the cross-correlation function (see chapter 4) between it and the second image. This technique cannot be applied to an image and a map. For example, roads usually appear white in images and black in maps, and field boundaries are in different shades of gray.

Wiesel (78), who studied the problem of matching two images projectively distorted relative to each other, recommends displaying one window in red, the other in green on a colour display, the best match being observed when a maximum of yellow is present. In Wiesel's technique, the transformation parameters of the image to be corrected are computed in an iterative loop, and the cross-correlation function for the grayscale windows maximised. This approach only applies to two images, not to an image and a map. In the latter case, the black and white map cannot be correlated with the grayscale image. But the display technique for observing the quality of the correlation can be applied when low resolution maps and imagery are used, edges can be extracted from the picture and a binary correlation technique employed.

The simplest approach is template matching (57). The edge extracted image control point window is compared with a trial position in the map. Whenever pixels are of the same polarity in both, a counter is incremented. The window is then moved systematically and the position with maximum count is sought. It is intuitively obvious that this technique is extremely sensitive to any geometric distortion in the image relative to the map. It is also sensitive to noise in the image which will produce spurious mismatches and it is computationally intensive since for a template size of $M \times M$ in an $N \times N$ image $(N - M + 1)^2$ possible match locations must be examined and M^2 comparisons be performed at each trial location. When noise and geometric distortion are present, there may be several positions of apparent best match. If the transformation of the image is placed in an iterative loop and the edge extracted

Fig. 5.8 (*a*) Oblique airphoto of a Roman marching camp near Rheinberg in the Rhineland. Airphoto Archive, Rheinisches Landesmuseum HF28. (*b*) Map of the area of the photograph with corrected photograph superimposed. (Orthophoto made using interactively determined control point correspondence between picture and map of identical houses.)

(*a*)

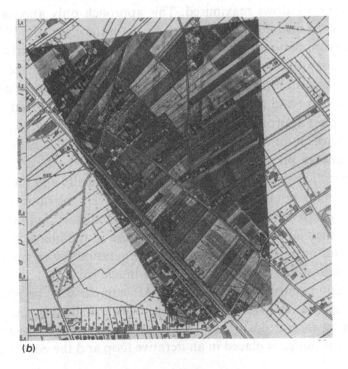

(*b*)

window transformed at each iteration, the best match computed, and the transformation recomputed, convergence to an optimum may be expected in most cases. Noise cleaning of the edge extracted image prior to this step is essential.

Several techniques have been developed to reduce the computational load. Barnea and Silverman suggest ordering the image point pairs in a non-repeating random sequence and computing a measure of mismatch which, when a given threshold is reached, causes abandonment of further computation at the position of the template (5).

Hall (26) suggests constructing a set of images which have decreasing resolution and are therefore smaller in size for both the picture and the search area. Positions of best match are then computed in the coarsest resolution image, and then in higher resolution images a search is carried out only in the neighbourhood of these matches. This reduces the number of trial positions to $K \log_2 (N - M + 1)^2$ possible search positions, where K is a constant dependent on the number of levels of resolution in the set. The images with lower resolution can be created by low pass filtering and the edge extraction required can be executed in the same operation if one of the simple bandpass edge extractors described in chapter 4 is used. A simple adaptive noise cleaning can be incorporated in the same pass. Obtaining a preliminary match interactively on a very high resolution display goes a long way to reducing the complexity of the problem.

Rosenfeld and Kak (57) as well as Hall discuss the problem of matching two images in considerable detail. The methods described in these references are appropriate for inserting the data from an oblique aerial image in an already extant orthophoto made from vertical pictures taken at roughly the same time. These ideal conditions are rarely encountered in archaeological problems since it is unlikely that vertical and oblique imagery will be simultaneously available. Pictures made at different times of the year or in different years cannot usually be matched well when using the grayscale data. If edges are extracted to avoid this problem, the methods discussed above are appropriate.

At present, there is no easy way of using the other geometric information visible in pictures, namely linear or curved structures, to help in parameter estimation. In contrast with vertical aerial cover made for mapping purposes, the typical archaeological oblique image contains very few accurately defined points which correspond to map information. Some rather unexpected results from a limited study of multiple images lead us to believe, that for oblique imagery with sparse and poorly defined control points, currently available photogrammetric methods for multiple images from the same series may actually degrade precision compared with the single image case rather than improving it. Methods for single images with a small number of control points will therefore be discussed first, and then the multiple image case will be treated.

Church's equations

One of the oldest algorithms for determining the unknown parameters of equation (5.18) is due to Church (13). It assumes that the focal length of the lens and the principal point are known exactly. Using the notation of figure 5.9, three points in an image a, b, c are measured relative to the principal point p. For these three points, map coordinates at A, B, C are known. Rays joining A, B, and C with a, b, and c pass through

Fig. 5.9 The geometry of space resection using the Church algorithm. From ref. 79.

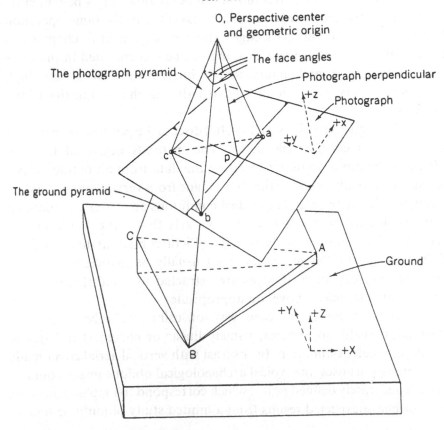

the projection centre at O as shown. Then the fundamental properties
of the figure require:

Angles:

$$AOB = aOb, \quad BOC = bOc \text{ and } AOC = aOc$$

Assigning:

$$K_1 = \cos(AOB) \qquad k_1 = \cos(aOb)$$
$$K_2 = \cos(BOC) \qquad k_2 = \cos(bOc)$$
$$K_3 = \cos(AOC) \qquad k_3 = \cos(aOc)$$

then:

$$K_1 = k_1 \qquad K_2 = k_2 \qquad K_3 = k_3 \qquad (5.20)$$

A film pyramid is formed by the projection centre and three points on
the film. A ground pyramid is formed by the projection centre and the
corresponding three points on the ground. The angles are computed at
the projective centre for the known film coordinates. The angles are also
computed for an assumed projective centre in space for the known ground
coordinates. Church then solves the resulting system of equations via
Cramer's Rule and iterates on the differences between the two sets of
angles until the film and ground pyramids are brought into register by
satisfying equation (5.20).

Four iterations are usually enough. Full details are in Church's original
publication, and in other handbooks of photogrammetry. The method
has the advantage that the calculations can be carried out by hand or
with a very simple program. It has the disadvantage that information
from only three points is used, and with extremely oblique images, four
solutions are possible as Fischler and Bolles have shown (17). Given
three corresponding points, a solution without iteration for the unknown
parameters of equation (5.18) is possible. This calculation, due to Fischler
and Bolles, will be described later.

Three control points

In conventional photogrammetry, the focal length f, and the coordinates
of the principal point x_p, y_p are called the inner orientation. They are
constants for a given camera. The camera position in space is (X_0, Y_0, Z_0).
The rotation of the camera about the x axis is ω, about the y axis is ϕ,

and about the z axis, κ. For conventional vertical photography, all external parameters have approximately known values which are controlled in flight. Therefore, the solution of the collinearity equations by Newton iteration or a method derived from it is possible.

The non-linear collinearity equations have six unknowns: three translational and three rotational parameters. This is the exterior orientation of conventional photogrammetry. For each control point, two collinearity equations are obtained. Therefore, an approximation for the transformation of the ground surface to the image plane can be solved when three control points are in the image, giving six equations in six unknowns. The control points may not lie on a line, as will be discussed later.

An iterative method for obtaining a solution of the collinearity equations due to Schmid is described by Tewinkel *et al.* (74) and was utilised in a solution of the problem for archaeological oblique images in the early 70s (60). In order to solve the equations, the height of each of the ground control points must be known. The technique used was as follows: Eight plausible starting positions and rotations around the centre of gravity of a convex figure circumscribing the ground control points were chosen, assuming that the camera was pointed downward at an angle of 45° and flight altitude was 500 m. Iteration was carried out on each of these positions in turn. The results were examined for plausibility in terms of the resultant flight altitude, and the residual error when more than three points were available.

Later, Fischler and Bolles (17) showed that four solutions to the equations are possible when only three points are available. Even when four points which do not lie in the same plane are at hand, more than one solution can be obtained. If the 'true' solution lies in the vicinity of a 'false' one, then the plausibility calculations will not help. A method which allows iterative improvement in the estimation of the camera rotations, position, and focal length will be described below. To use this technique, one must have height information for every point of interest. For this purpose, a numerical model of the ground must be available. Interpolation points for this digital terrain model can be replaced with height estimates from a contoured map of the area with some reduction in accuracy even if the ground surface has considerable height differences. The error attributable to use of contour information is usually small compared with other error sources encountered in archaeological oblique imagery.

Four control points

If one examines transformations of the type:

$$T: \begin{pmatrix} X \\ Y \\ Z \end{pmatrix} \rightarrow \left(\frac{t_{11}X + t_{12}Y + t_{13}Z + t_{14}}{t_{31}X + t_{32}Y + t_{33}Z + 1}, \frac{t_{21}X + t_{22}Y + t_{23}Z + t_{24}}{t_{31}X + t_{32}Y + t_{33}Z + 1} \right) \quad (5.21)$$

with $t_{ij} \in \mathbb{R}$, where $\mathbb{R}$ are the real numbers, one sees that transformations like (5.19) belong to this class. However (5.21) has one additional independent parameter. The number of independent parameters is reduced when the terrain is flat, that is:

$$Z = aX + bY + c \quad (5.22)$$

Then the transformation T has the following simple form:

$$T: \begin{pmatrix} X \\ Y \end{pmatrix} \rightarrow \left(\frac{s_{11}X + s_{12}Y + s_{13}}{s_{31}X + s_{32}Y + 1}, \frac{s_{21}X + s_{22}Y + s_{23}}{s_{31}X + s_{32}Y + 1} \right)$$

with: (5.23)

$$s_{i1} = \frac{t_{i1} + at_{i3}}{t_{33}c + 1}, \quad s_{i2} = \frac{t_{i2} + bt_{i3}}{t_{33}c + 1}, \quad s_{i3} = \frac{t_{i4} + ct_{i3}}{t_{33}c + 1}, \quad i = 1, \ldots, 3$$

For a ground point (X_i, Y_i, Z_i) with corresponding picture coordinates (x_i, y_i), the following equations are obtained:

$$\left. \begin{array}{l} x_i X_i s_{31} + x_i Y_i s_{32} + x_i - X_i s_{11} - Y_i s_{12} - s_{13} = 0 \\ y_i X_i s_{31} + y_i Y_i s_{32} + y_i - X_i s_{21} - Y_i s_{22} - s_{23} = 0 \end{array} \right\} \quad (5.24)$$

This was derived by Hallert (27) among others and first used by Palmer for archaeological aerial photographs (49).

If n control points are known, the following system of equations results:

$$\begin{vmatrix} X_1 & Y_1 & 1 & 0 & 0 & 0 & -x_1 X_1 & -x_1 Y_1 \\ \vdots & \vdots & \vdots & \vdots & \vdots & \vdots & \vdots & \vdots \\ X_n & Y_n & 1 & 0 & 0 & 0 & -x_n X_n & -x_n Y_n \\ 0 & 0 & 0 & X_1 & Y_1 & 1 & -y_1 X_1 & -y_1 Y_1 \\ \vdots & \vdots & \vdots & \vdots & \vdots & \vdots & \vdots & \vdots \\ 0 & 0 & 0 & X_n & Y_n & 1 & -y_n X_n & -y_n Y_n \end{vmatrix} \begin{pmatrix} s_{11} \\ s_{12} \\ s_{13} \\ s_{21} \\ s_{22} \\ s_{23} \\ s_{31} \\ s_{32} \end{pmatrix} = \begin{pmatrix} x_1 \\ \vdots \\ x_n \\ y_1 \\ \vdots \\ y_n \end{pmatrix}$$

$$(5.25)$$

At least four control points are required, and no heights need be known in order to solve this system of equations to obtain the projective transformation parameters. The control points may not lie on a line. A least squares technique may be used to minimise residual errors when $n > 4$ points are available. The control points must be well distributed in the image, and if the lens used had little geometric distortion far from the picture centre, this method can yield reasonably good results on relatively flat or gently sloping terrain. Its major virtue is that neither the camera constants nor the approximations for the camera position and rotations are required. The technique is also extremely useful for obtaining a first solution which can serve as a starting point to be refined by one of the other methods. Burnside *et al.* (8) and Palmer in Great Britain have used the method extensively with good results on maps with a scale of 1:10000. The limits of this method except in flat terrain would have been quickly apparent if the 1:2500 map had been used.

The map coordinates of the corners of a rectangular area of interest in an input image which is to be transformed to an orthophoto can be solved for analytically. By transforming (5.24) one obtains:

$$\left.\begin{array}{l}(x_i s_{31} - s_{11})X_i + (x_i s_{32} - s_{12})Y_i = s_{13} - x_i \\ (y_i s_{31} - s_{21})X_i + (y_i s_{32} - s_{22})Y_i = s_{23} - y_i\end{array}\right\} \tag{5.26}$$

Six control points

When the terrain is hilly, the transformation parameters of (5.21) can be derived in a similar fashion to (5.25) to obtain:

$$\begin{pmatrix} X_1 & Y_1 & Z_1 & 1 & 0 & 0 & 0 & 0 & -x_1 X_1 & -x_1 Y_1 & -x_1 Z_1 \\ \vdots & \vdots & \vdots & \vdots & \vdots & \vdots & \vdots & \vdots & \vdots & \vdots & \vdots \\ X_n & Y_n & Z_n & 1 & 0 & 0 & 0 & 0 & -x_n X_n & -x_n Y_n & -x_n Z_n \\ 0 & 0 & 0 & 0 & X_1 & Y_1 & Z_1 & 1 & -y_1 X_1 & -y_1 Y_1 & -y_1 Z_1 \\ \vdots & \vdots & \vdots & \vdots & \vdots & \vdots & \vdots & \vdots & \vdots & \vdots & \vdots \\ 0 & 0 & 0 & 0 & X_n & Y_n & Z_n & 1 & -y_n X_n & -y_n Y_n & -y_n Z_n \end{pmatrix} \begin{pmatrix} t_{11} \\ t_{12} \\ t_{13} \\ t_{14} \\ t_{21} \\ t_{22} \\ t_{23} \\ t_{24} \\ t_{31} \\ t_{32} \\ t_{33} \end{pmatrix} = \begin{pmatrix} x_1 \\ \vdots \\ x_n \\ y_1 \\ \vdots \\ y_n \end{pmatrix} \tag{5.27}$$

when six control points are available. The parameters t_{ij} can be obtained only when the control points do *not* lie in a plane. If the height of all other unknown points can be estimated from a digital terrain model or a contoured map, then this technique allows geometric transformation of pictures with considerable height differences. It also has the advantage that no approximations for the camera parameters and orientation are required. One method for the derivation of these equations is given by Rogers and Adams (56), another by Hall (26). If more than six control points are available, then least squares methods to minimise residual errors are useful.

To determine the area of the map in which the picture area which is to be rectified lies, one can make use of the similarity with the four point method. Compute the mean height of the control points Z_m. The six point transformation can then be approximated by:

$$T:\begin{pmatrix} X \\ Y \end{pmatrix} \rightarrow \left(\frac{t_{11}X + t_{12}Y + t_{13}Z_m + t_{14}}{t_{31}X + t_{32}Y + t_{33}Z_m + 1}, \frac{t_{21}X + t_{22}Y + t_{23}Z_m + t_{24}}{t_{31}X + t_{32}Y + t_{33}Z_m + 1} \right)$$

$$(5.28)$$

Many control points

The idea behind equation (5.21) can be given an interesting variation if the height coordinates are replaced by a simple bivariate polynomial of the form:

$$Z = \sum_{i=0}^{N} \sum_{j=0}^{i} a_{ij}X^{i-j}Y^j \qquad (5.29)$$

Collecting the depending parameters gives a transformation of the form:

$$T:\begin{pmatrix} X \\ Y \end{pmatrix} \rightarrow \begin{pmatrix} \dfrac{u_{11}X + u_{12}Y + u_{13}\left(\sum_{i=2}^{N} \sum_{j=0}^{i} a_{ij}X^{i-j}Y^j \right) + u_{14}}{u_{31}X + u_{32}Y + \left(\sum_{i=2}^{N} \sum_{j=0}^{i} a_{ij}X^{i-j}Y^j \right) + 1} \\[4ex] \dfrac{u_{21}X + u_{22}Y + u_{23}\left(\sum_{i=2}^{N} \sum_{j=0}^{i} a_{ij}X^{i-j}Y^j \right) + u_{24}}{u_{31}X + u_{32}Y + \left(\sum_{i=2}^{N} \sum_{j=0}^{i} a_{ij}X^{i-j}Y^j \right) + 1} \end{pmatrix}$$

$$(5.30)$$

This approach, due to an unpublished idea of Weidner, has the great advantage that the use of terms of higher order allows the correction of non-perspective distortion as well. Such distortion may be due to film shrinkage, imperfect lenses, etc. Iterative refinement is to be recommended for this method, that is, increase N in (5.30). However for $N > 1$ one needs M control points, where:

$$M = 4 + 3N/4 + N^2/4 \qquad (5.31)$$

whose heights need not be known. For a second order, 7, for a third order, 9 and for a fourth order 11 points are required. Unfortunately it is not usual to find so many control points in a typical archaeological image. In addition, a polynomial terrain model is unsatisfactory, especially when the control points lie outside the area to be rectified. A minor disadvantage of the method is that instead of a system of linear equations as in the four and six point methods, a non-linear system results.

To obtain a linear equation system, the approach can be generalised to:

$$T:(X, Y) \to \left(\frac{\sum_{i=0}^{N} \sum_{j=0}^{i} a_{ij}^1 X^{i-j} Y^j}{1 + \sum_{i=1}^{N} \sum_{j=0}^{i} a_{ij}^3 X^{i-j} Y^j}, \frac{\sum_{i=0}^{N} \sum_{j=0}^{i} a_{ij}^2 X^{i-j} Y^j}{1 + \sum_{i=1}^{N} \sum_{j=0}^{i} a_{ij}^3 X^{i-j} Y^j} \right) \qquad (5.32)$$

The transformation parameters are the solutions to the following overdetermined system of equations:

$$\mathbf{A} \cdot \mathbf{a} = \mathbf{v}$$

where:

$$
\mathbf{A} = \begin{pmatrix}
1 & X_1 & Y_1 & \cdots & X_1 Y_1^{N-1} & Y_1^N & 0 & 0 & 0 & \cdots & 0 \\
1 & X_2 & Y_2 & \cdots & X_2 Y_2^{N-1} & Y_2^N & 0 & 0 & 0 & \cdots & 0 \\
\vdots & \vdots & \vdots & \vdots & \vdots & \vdots & \vdots & \vdots & \vdots & & \vdots \\
1 & X_n & Y_n & \cdots & X_n Y_n^{N-1} & Y_n^N & 0 & 0 & 0 & \cdots & 0 \\
0 & 0 & 0 & \cdots & 0 & 0 & 1 & X_1 & Y_1 & \cdots & X_1 Y_1^{N-1} \\
0 & 0 & 0 & \cdots & 0 & 0 & 1 & X_2 & Y_2 & \cdots & X_2 Y_2^{N-1} \\
\vdots & \vdots & \vdots & \vdots & \vdots & \vdots & \vdots & \vdots & \vdots & & \vdots \\
0 & 0 & 0 & \cdots & 0 & 0 & 1 & X_n & Y_n & \cdots & X_n Y_n^{N-1}
\end{pmatrix}
$$

and:

$$
\mathbf{a} = {}^t(a_{00}^1, a_{10}^1, a_{11}^1, \ldots, a_{N,N-1}^1, a_{NN}^1, a_{00}^2, a_{10}^2, a_{11}^2, \ldots,
$$
$$
a_{N,N-1}^2, a_{NN}^2, a_{00}^3, a_{10}^3, a_{11}^3, \ldots, a_{N,N-1}^3, a_{NN}^3) \qquad (5.34)
$$
$$
\mathbf{v} = {}^t(x_1, x_2, \ldots, x_n, y_1, y_2, \ldots, y_n)
$$

For $N = 1$, one obtains the four point transformation. For a polynomial of degree N, $3(N + 2)(N + 1)/2 - 1$ parameters must be determined. It follows from this that nine control points are required for $N = 2$ and fifteen are needed for $N = 3$, a severe handicap.

To prevent a numerically singular system of equations from developing, the columns in the matrix can be incorporated step by step, and with each new column a test made to see if there is a linear relationship between the new column and the rest. If this is so, then the related parameters are set to zero and the column is removed from the matrix.

To determine the coordinates of the area of interest in the map, the close relationship between this technique and the four point technique can be used by simply neglecting the higher order terms.

The global polynomial method
A simple estimate of the heights of a surface can be described by a series of polynomials in X and Y:

$$f(X, Y) = \sum_{r+s \leq p} a_{rs} X^r Y^s \tag{5.35}$$

where p is the 'order' of the surface and with a number of coefficients

$$\begin{pmatrix}
0 & -x_1 X_1 & -x_1 Y_1 & \cdots & -x_1 X_1 Y_1^{N-1} & -x_1 Y_1^N \\
0 & -x_2 X_2 & -x_2 Y_2 & \cdots & -x_2 X_2 Y_2^{N-1} & -x_2 Y_2^N \\
\vdots & \vdots & \vdots & & \vdots & \vdots \\
0 & -x_n X_n & -x_n Y_n & \cdots & -x_n X_n Y_n^{N-1} & -x_n Y_n^N \\
Y_1^N & -y_1 X_1 & -y_1 X_1 & \cdots & -y_1 X_1 Y_1^{N-1} & -y_1 Y_1^N \\
Y_2^N & -y_2 X_2 & -y_2 Y_2 & \cdots & -y_2 X_2 Y_2^{N-1} & -y_2 Y_2^N \\
\vdots & \vdots & \vdots & & \vdots & \vdots \\
Y_n^N & y_n X_n & -y_n Y_n & \cdots & -y_n X_n Y_n^{N-1} & -y_n Y_n^N
\end{pmatrix} \tag{5.33}$$

given by the order of the polynomial chosen:

$$P = \frac{(p + 1)(p + 2)}{2} \tag{5.36}$$

as derived in Ripley (55).

The coefficients a_{rs} are chosen to minimise:

$$\sum_{i=1}^{n} [Z(X_i, Y_i) - f(X_i, Y_i)]^2 \tag{5.37}$$

It is also possible to imagine the terrain as a two-dimensional manifold which is homeomorphically mapped in $\mathbb{R}^2$. This is only true when every point on the ground has a corresponding picture point, which is by no means the case in reality.

If there is a photograph and a ground surface for which a homeomorphism g exists:

$$x = g_1(X, Y), \qquad y = g_2(X, Y) \tag{5.38}$$

then one can approximate g_1, g_2 with bivariate polynomials of the form:

$$x = \sum_{i=0}^{N} \sum_{j=0}^{N-i} K_{ij}^1 X^i Y^j$$

$$y = \sum_{i=0}^{N} \sum_{j=0}^{N-i} K_{ij}^2 X^i Y^j \tag{5.39}$$

The rectification parameters K_{ij}^1 are obtained by the solution of the following system of equations:

$$
\begin{pmatrix}
1 & Y_1 & Y_1^2 & \cdots & Y_1^N & X_1 & X_1 Y_1 & X_1 Y_1^2 & \cdots & X_1^{N-1} Y_1 & X_1^N \\
1 & Y_2 & Y_2^2 & \cdots & Y_2^N & X_2 & X_2 Y_2 & X_2 Y_2^2 & \cdots & X_2^{N-1} Y_2 & X_2^N \\
\vdots & \vdots & \vdots & & \vdots & \vdots & \vdots & \vdots & & \vdots & \vdots \\
1 & Y_n & Y_n^2 & \cdots & Y_n^N & X_n & X_n Y_n & X_n Y_n^2 & \cdots & X_n^{N-1} Y_n & X_n^N
\end{pmatrix}
$$

$$
\begin{pmatrix}
K_{0,0}^1 \\
K_{0,1}^1 \\
\vdots \\
K_{0,N}^1 \\
K_{1,0}^1 \\
\vdots \\
K_{1,N-1}^1 \\
\vdots \\
K_{N-1,0}^1 \\
K_{N-1,1}^1 \\
K_{N,0}^1
\end{pmatrix}
=
\begin{pmatrix}
x_1 \\
x_2 \\
\vdots \\
x_n
\end{pmatrix}
\tag{5.40}
$$

K_{ij}^2 are obtained similarly. At least $(N+2)(N+1)/2$ control points (without heights) are required to solve the equations. This method as given by Bernstein (6) is described by Hall (26) with a slightly different polynomial approximation.

The method has a number of disadvantages for archaeological aerial photographs. It is not based on a model of the real situation when using a camera. Therefore, for a small number of control points, significantly poorer results than those for the four point method are obtained. The results are especially erroneous when the control points lie outside the area to be rectified. The precision with which the distortion function is approximated is not necessarily raised by using a higher order polynomial. In interpolation theory it is well known that for an interval $[a, b]$ and any interval divisions δx_m, one can find a continuous function f such that the interpolation polynomial, $P_{\delta x_m}$, when $m->\infty$ does not converge to $f(x)$ (Faber's Theorem).

Polynomial methods of this kind are used for satellite pictures because the rectification function must take a lot of factors into account, such as atmospheric refraction, movement of the satellite during the scan of the image area, curvature of the earth, etc. which are not of interest here. In comparison with archaeological aerial photographs, satellite images have a large number of control points, but even then, the experiments of Welch *et al.* (77) have shown that residual errors actually increase when a polynomial of degree greater than two is used.

A variant of this technique due to Goshtasby (22,23) uses local two-dimensional polynomials of order three (splines), and fits them to the known data points. This technique has the advantage that errors are locally confined and do not spread over the whole image. The spline functions are usually chosen to ensure a smooth fit with continuity in first, and sometimes in second derivatives.

Despite all these disadvantages, the method may sometimes be useful for archaeological aerial photographs for improving roughly rectified pictures: First, the image is roughly rectified with one of the other techniques such as the four point method. Then one tries to find as many points as possible with approximate correspondences in the image and the map in the rectified area. Finally, using these control points (no heights are needed), one computes a secondary correction function which may be either a polynomial or a spline function to stretch the image to fit all the corresponding points to the map. Another possible application of this technique might be the rectification of old bird's eye drawings or paper plans of vanished city walls and pre-modern fortifications to bring them into correspondence with modern maps. It has also found use in the geometric correction of thermal scanning images of archaeological sites (chapter 10). These do not obey projective transformation laws and are more comparable with satellite imagery.

5.3.7. *An example of rectification in hilly terrain with different methods*

Figure 5.10 shows a Roman *éperon barré* camp with *tutuli* near Alpen, Kreis Moers, photographed in 1970, and later excavated by Bakker (4). Seven control points were available. A digital terrain model with 40 support points was constructed using the contour lines of the 1:5000

Fig. 5.10 A Roman camp near Alpen in the Rhineland, enhanced oblique image. Airphoto Archive, Rheinisches Landesmuseum HH35.

base map. Height differences of up to 20 m are in the area to be rectified. The focal length of the lens was nominally known but the camera was uncalibrated and the principal point had to be estimated from the sides of the picture frame. Figure 5.11 shows the best result. It was started with the three point method of Fischler and Bolles followed by iterative improvement. Figure 5.12 shows the result with the four point method. The considerable errors resulting from neglect of height differences and the choice of map coordinate system are immediately visible. Shifting

Fig. 5.11 The three point method of Fischler and Bolles followed by iterative improvement using the known heights from the map.

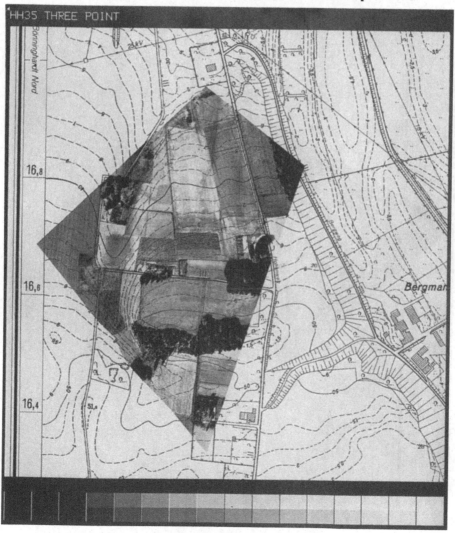

the centroid of the map coordinates to the centroid of the control points improved things considerably as shown in figure 5.13. The six point method gave the next to worst result because it was numerically unstable, but visually the result is quite pleasing, as shown in figure 5.14. Shifting the map coordinates to the middle of the control point area would make stability much better and give the method higher rank.

The non-linear multipoint method using a second degree polynomial shown in figure 5.15 gives a visually much more pleasing result than the four point method, but it is not as good as that of the three point

Fig. 5.12 The four point method. Height differences have been neglected.

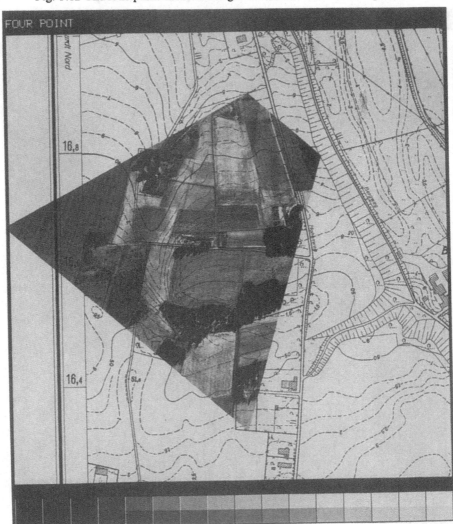

technique. This is because the method does not use a very precise terrain model. If no height information from the map had been available, this would have given the best result.

The result for the polynomial method in figure 5.16 is not at all bad in the foreground of the image, but it goes to pieces in the further distance. The polynomial approach, so favoured for satellite imagery and contained in most image processing program packages, is usually the most unfavourable for archaeological pictures.

Fig. 5.13 The four point method with the centroid of the map coordinates shifted to reduce error.

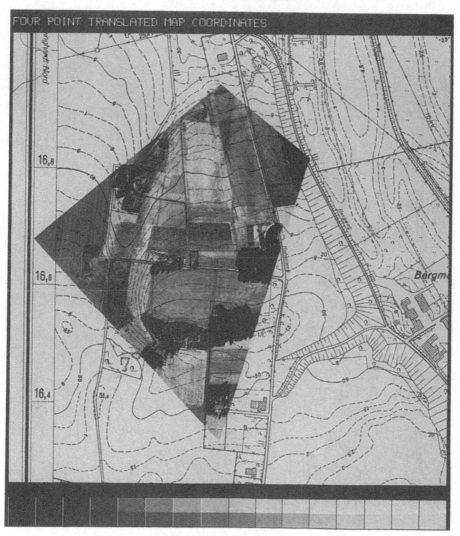

5.3.8. *Computer methods for multiple images*

An archaeological site is usually photographed from five to ten times or more on a single mission, and it may be rephotographed in other seasons. It is natural to assume that the information in additional photographs can be used to improve the accuracy of rectification of a single image.

There are two approaches for dealing with multiple images which are standard techniques in conventional photogrammetry: in one, the

Fig. 5.14 The six point method without shifted map centroid coordinates leads to a pleasing result.

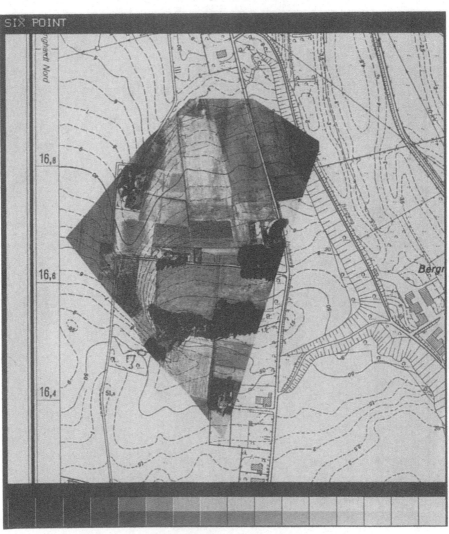

parameters of each photograph are determined separately, and using the three or six point method, the ground coordinates for corresponding points in image pairs can be calculated. It is also possible to determine the rectification parameters from two images at the same time, if one makes use of the relationships between them. In this case, the *coplanarity* equations are used, and the problem solved in two steps to determine relative and absolute orientation. Whether or not this raises the accuracy with which the parameters are obtained for a single image depends very

Fig. 5.15 Weidner's non-linear multipoint method using a second degree polynomial.

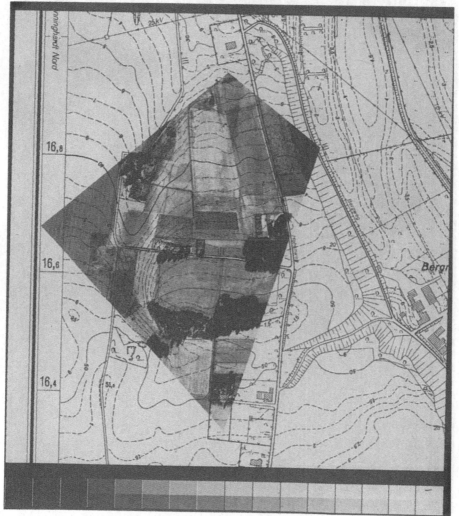

much on the distribution of errors in the control and corresponding points in the set of images. This is a theme for further research.

The coplanarity equations
If one examines two images of a scene, then an additional condition for the parameters of external orientation is obtained which constrains the vectors $\mathbf{O'A}$, $\mathbf{O''A}$ and $\mathbf{O'A''}$ of figure 5.17 to lie in a plane. This means:

$$\det(\mathbf{O'A}, \mathbf{O''A}, \mathbf{O'O''}) = 0 \tag{5.41}$$

Fig. 5.16 The polynomial method of Bernstein showing large errors in the image background.

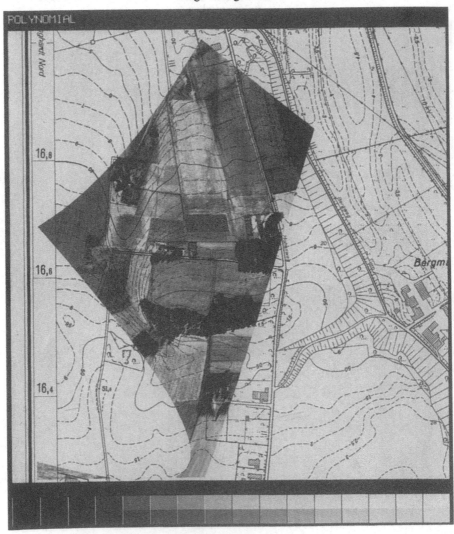

Following equation (5.17):

$$\mathbf{O'A} = \begin{pmatrix} X_a - X_0' \\ Y_a - Y_0' \\ Z_a - Z_0' \end{pmatrix} = \mu_a' R' \begin{pmatrix} x_a' - x_p' \\ y_a' - y_p' \\ -f' \end{pmatrix}$$

$$\mathbf{O''A} = \begin{pmatrix} X_a - X_0'' \\ Y_a - Y_0'' \\ Z_a - Z_0'' \end{pmatrix} = \mu_a'' R'' \begin{pmatrix} x_a'' - x_p'' \\ y_a'' - y_p'' \\ -f'' \end{pmatrix} \qquad (5.42)$$

and since $\mu_a' \neq 0$, $\mu_a'' \neq 0$ it follows from equation (5.41):

$$\det\left(R' \begin{pmatrix} x_a' - x_p' \\ y_a' - y_p' \\ -f' \end{pmatrix} \quad R'' \begin{pmatrix} x_a'' - x_p'' \\ y_a'' - y_p'' \\ -f'' \end{pmatrix} \quad \begin{pmatrix} X_0' - X_0'' \\ Y_0' - Y_0'' \\ Z_0' - Z_0'' \end{pmatrix} \right) = 0 \qquad (5.43)$$

This condition gives an additional non-linear equation with twelve unknowns if the camera constants are known. It is not possible to determine all transformation parameters for both photographs by solving the coplanarity equations for twelve or more unknowns. The last column vector is the difference between two vectors. Only the column vector differences but not the vectors themselves can be calculated. Furthermore the determinant is invariant with any rotation of the three column vectors

Fig. 5.17 The coplanarity condition for two images.

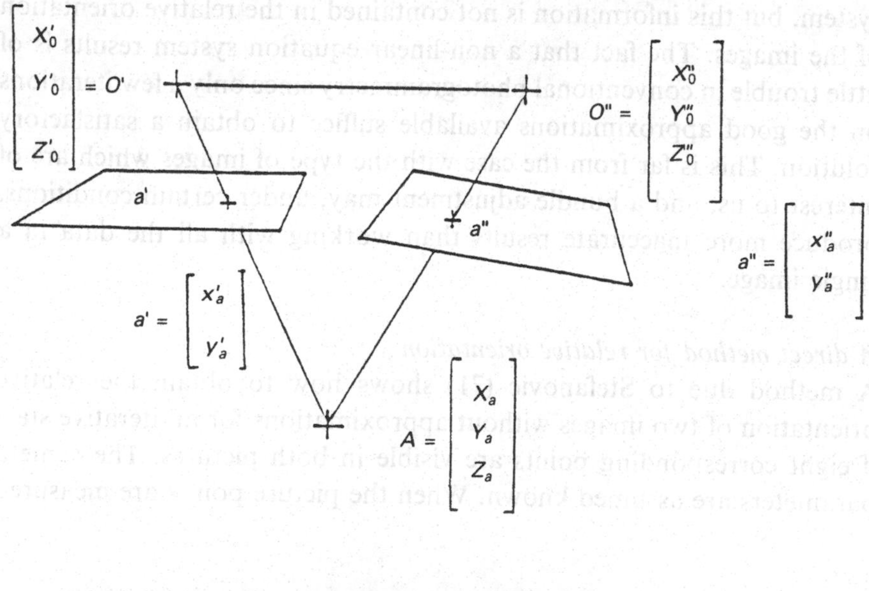

in the determinant and:

$$\det(v_1, v_2, v_3) = \lambda \det(v_1, v_2, v_3) \qquad (5.44)$$

Since these determinants are to be equal to zero, this means that the length of one of the vectors is not important. Thus it can be shown that there are seven dependent unknowns in the coplanarity equations. It is usual in photogrammetry to choose seven unknowns freely and solve for the remaining five to obtain the relative orientation. For example this reduces the equations to:

$$\det\left(R \begin{pmatrix} x'_a - x'_p \\ y'_a - y'_p \\ -f' \end{pmatrix} \begin{pmatrix} x''_a - x''_p \\ y''_a - y''_p \\ -f'' \end{pmatrix} \begin{pmatrix} a_1 \\ a_2 \\ 1 \end{pmatrix} \right) = 0 \qquad (5.45)$$

Relative and absolute orientation

The remaining seven unknowns will be calculated later via so-called absolute orientation. The relative orientation can be visualised in assuming an arbitrary position for the first image with a fixed distance to the second. Then one displaces and rotates the first picture until all corresponding rays through points visible in both meet, a procedure which photogrammetrists call a bundle adjustment, because the bundles of rays are brought into coincidence. Afterwards, this model is shifted, rotated, and scaled relative to the real situation (absolute orientation). It is thus evident why there are seven dependent parameters in equation (5.43). They are dependent on the choice of the ground coordinate system, but this information is not contained in the relative orientation of the images. The fact that a non-linear equation system results is of little trouble in conventional photogrammetry since only a few iterations on the good approximations available suffice to obtain a satisfactory solution. This is far from the case with the type of images which are of interest to us, and a bundle adjustment may, under certain conditions, produce more inaccurate results than working with all the data in a single image.

A direct method for relative orientation

A method due to Stefanovic (71) shows how to obtain the relative orientation of two images without approximations for an iterative start if eight corresponding points are visible in both pictures. The camera parameters are assumed known. When the picture points are measured relative to the known principal point, then equation (5.45) simplifies to:

$$\det\left(R\begin{pmatrix} x'_a \\ y'_a \\ -f' \end{pmatrix}\begin{pmatrix} x''_a \\ y''_a \\ -f'' \end{pmatrix}\begin{pmatrix} a_1 \\ a_2 \\ a_3 \end{pmatrix}\right) = 0 \tag{5.46}$$

whereby one of the a_i can be freely chosen.

The expression $\det(\mathbf{u}, \mathbf{v}, \mathbf{w}) = 0$ is equivalent to:

$$^t\mathbf{u}\begin{pmatrix} 0 & -v_3 & v_2 \\ v_3 & 0 & -v_1 \\ -v_2 & v_1 & 0 \end{pmatrix}\mathbf{w} = 0 \tag{5.47}$$

and therefore equation (5.46) can be rewritten as:

$$^t\mathbf{x}''RB\mathbf{x}'' = 0 \tag{5.48}$$

with:

$$\mathbf{x}' = \begin{pmatrix} x'_a \\ y'_a \\ -f' \end{pmatrix}, \qquad \mathbf{x}'' = \begin{pmatrix} x''_a \\ y''_a \\ -f'' \end{pmatrix}, \qquad B = \begin{pmatrix} 0 & -a_3 & a_2 \\ a_3 & 0 & -a_1 \\ -a_2 & a_1 & 0 \end{pmatrix}$$

B is skew symmetric, has rank 2 (unless $a_1 = a_2 = a_3 = 0$) and has the property:

$$B^3 = -(a_1^2 + a_2^2 + a_3^2)B \tag{5.49a}$$

Let U be the matrix ^tRB from (5.48), then:

$$^t\mathbf{x}'U\mathbf{x}'' = 0 \tag{5.49b}$$

or

$$\mathbf{cu} = 0 \tag{5.49c}$$

where $\mathbf{c}$ is the row vector

$$\mathbf{c} = (x''_a x'_a, \, y''_a x'_a, \, -f''x'_a, \, x''_a y'_a, \, y''_a y'_a, \, -f''y'_a, \, -x''_a f', \, -y''_a f', \, f''f')$$

and $\mathbf{u}$ the column vector

$$\mathbf{u} = {}^t(u_{11}, u_{12}, u_{13}, u_{21}, u_{22}, u_{23}, u_{31}, u_{32}, u_{33})$$

A vector $\mathbf{c}$ can be created for every point pair and these vectors collected as the matrix C. The homogeneous system of equations: $C\mathbf{u} = 0$ results.

When the two pictures were not taken from exactly the same place, the matrix $U = {}^tRB$ has rank 2 and is not null. This implies that C may not have maximal rank, that is, $\text{Rank}(C) < 9$. Generally C has rank 8, but there are some exceptions in practice, for example, when all x''_i and

y_i'' are affinely linear dependent on x_i', y_i', which means:

$$\begin{pmatrix} x_i'' \\ y_i'' \end{pmatrix} = \begin{pmatrix} a_{11} & a_{12} \\ a_{21} & a_{22} \end{pmatrix} \begin{pmatrix} x_i' \\ y_i' \end{pmatrix} + \begin{pmatrix} b_1 \\ b_2 \end{pmatrix} \tag{5.50}$$

This happens when both film planes are parallel. Then C has rank less than six, as is also the case when all the points (x_i', y_i') lie in a line. If the rank of C is too small, then the method described here cannot be used.

To solve the homogeneous equation system $Cu = 0$, one selects one of the u_{ij} and assigns it the value 1. This is equivalent to the free choice of one of the a_i in equation (5.46). Thus the system of equations:

$$C_0 u_0 = -c \tag{5.51}$$

is created, in which c is the column of C which is related to u_{ij}; C_0 is the matrix C without column c and u_0 the vector u without u_{ij}. This overdetermined equation system can be solved through least squares.

Since the relative orientation is a problem with five independent variables, it ought to be possible to find three further conditions for u which are independent of $Cu = 0$. From $U = {}^tRB$ it follows that:

$$\begin{aligned} {}^tUU^tU &= ({}^tBR)({}^tRB)({}^tBR) \\ &= {}^tBB^tBR \\ &= (-B)B(-B)R \\ &= -b^2BR \qquad \text{because of (5.49a)} \end{aligned}$$

with

$$\begin{aligned} b^2 &= (a_1^2 + a_2^2 + a_3^2) \\ &= \tfrac{1}{2}\operatorname{tr}(B^tB) \end{aligned} \tag{5.52}$$

and thus:

$$\begin{aligned} {}^tUU^tU &= b^2\,{}^tU \tag{5.53a} \\ {}^tU(U^tU - b^2I) &= 0 \tag{5.53b} \\ {}^tUD &= 0 \tag{5.53c} \end{aligned}$$

with

$$D = (U^tU - b^2I)$$

Since the trace of a matrix is invariant under orthogonal transformation, then:

$$b^2 = \tfrac{1}{2}\operatorname{tr}(U^tU) \tag{5.54}$$

Equation (5.53c) has nine non-linear constraints on u_{ij}. The column vectors of D are vectors from the kernel of ${}^tU$. Since ${}^tU$ has rank two, the rank of D is one at most. On the other hand D has at least rank one, unless $a_1 = a_2 = a_3 = 0$. Therefore it follows that three out of nine equations (5.53c) are independent.

Stefanovic (71) recommends adding the column vectors of D, whereby the following system of equations results:

$$\begin{pmatrix} u_{11} & u_{21} & u_{31} \\ u_{12} & u_{22} & u_{32} \\ u_{13} & u_{23} & u_{33} \end{pmatrix} \begin{pmatrix} \sum\limits_{k=1}^{3} \sum\limits_{j=1}^{3} (u_{1j}u_{kj} - \frac{1}{2}u_{kj}^2) \\ \sum\limits_{k=1}^{3} \sum\limits_{j=1}^{3} (u_{2j}u_{kj} - \frac{1}{2}u_{kj}^2) \\ \sum\limits_{k=1}^{3} \sum\limits_{j=1}^{3} (u_{3j}u_{kj} - \frac{1}{2}u_{kj}^2) \end{pmatrix} = \begin{pmatrix} 0 \\ 0 \\ 0 \end{pmatrix} \quad (5.55)$$

This non-linear system of equations is independent of the control point coordinates. This is a requirement which matrix U must satisfy.

Stefanovic suggests first solving $Cu = 0$, then improving the result iteratively until the conditions of equation (5.55) are satisfied. It is thought to be more sensible, however, to consider the secondary and control point conditions simultaneously with the calculation of U, that is, to minimise:

$$Q = \sum_{i=1}^{N} \left(\sum_{j=1}^{8} c_{ij} \sum_{j}(u_{11}, \ldots, u_{32}) - c_{i9} \right)^2$$

$$+ \sum_{m=1}^{3} \lambda_m \left(\sum_{n=1}^{3} u_{nm} \sum_{k=1}^{3} \sum_{j=1}^{3} (u_{nj}u_{kj} - \frac{1}{2}u_{kj}^2) \right) \quad (5.56)$$

with:

$$\sum_{j}(u_{11}, \ldots, u_{32}) = u_{(j-1)\text{div}\,3+1,(j-1)\text{mod}\,3+1}$$

where N is the number of corresponding points.

This expression assumes that u_{33} is the largest element in U. If another element is larger, then Q can be written in a very similar fashion. However, non-linear optimising methods require starting values, and calculating an approximation for U from $Cu = 0$ seems a good idea.

Stefanovic suggests another method for calculating U when approximations for the desired parameters are known. This method is iterative and needs only five corresponding points in both pictures. This is accomplished by using the technique described below for the iterative solution of generalised normal equations. Since normal equations are usually badly conditioned, this cannot be recommended.

Instead, compute the matrices R and B from U. From $T = {}^tUU$ the components of B can be computed (except for sign) as:

$$T = {}^tUU = {}^tBR^tRB = {}^tBB$$

$$\Rightarrow T = \begin{pmatrix} a_3^2 + a_2^2 & -a_1a_2 & -a_1a_3 \\ -a_1a_2 & a_3^2 + a_1^2 & -a_2a_3 \\ -a_1a_3 & -a_2a_3 & a_2^2 + a_1^2 \end{pmatrix} \tag{5.57}$$

Following Stefanovic one calculates the a_i from $T = (t_{ik})$ in a numerically convenient way. Let u_{ij} be the matrix element with the largest absolute value; u_{ki} is the largest absolute value outside of row i and column j. The third line m can be obtained through:

$$m = 6 - (i + k) \tag{5.58}$$

Then one calculates B from:

$$\left. \begin{aligned} a_m &= \pm \left(\frac{-t_{mm} + t_{ii} + t_{kk}}{2} \right)^{\frac{1}{2}} \\ a_i &= -(1/a_m)t_{mi} \\ a_k &= -(1/a_m)t_{mk} \end{aligned} \right\} \tag{5.59}$$

For the computation of R, $U = {}^tRB$ is transformed to $B = RU$. Then one can determine the rotation matrix R using the method of Thompson and Schut as described below (75,59). The method calculates a rotation matrix R from vectors v_i, V_i for which:

$$V_i = Rv_i, \qquad i = 1, \ldots, n, \tag{5.60}$$

is valid.

There must be at least two linearly independent vectors in order to determine R. Since B has rank two, this condition is fulfilled. One can utilise all the columns of B and U and set up an equation for each column, giving an overdetermined system. This is much simpler than Stefanovic's method.

Two decisions are required:

(1) The sign of U is arbitrarily chosen.
(2) The sign of B is also not fixed, following equation (5.59).

There are therefore four candidates for the solution:

$$\left. \begin{aligned} &(1) \ (b, R_1) \\ &(2) \ (b, R_2) \\ &(3) \ (-b, R_1) \\ &(4) \ (-b, R_2) \end{aligned} \right\} \tag{5.61}$$

where $b = {}^t(a_1, a_2, a_3)$ and R_1, R_2 are different rotation matrices. The number of candidates is reduced to two because the scaling factors are greater than zero. To decide which of the remaining alternatives is the right one, one looks at the degree of similarity of the model and control point triangles.

Absolute orientation

The computed model which is obtained by relative orientation calculation can be transformed to ground coordinates by a change of the coordinate system given by Wong (79):

$$\begin{pmatrix} X_j \\ Y_j \\ Z_j \end{pmatrix} = \begin{pmatrix} b_1 \\ b_2 \\ b_3 \end{pmatrix} + sR \begin{pmatrix} x_j \\ y_j \\ z_j \end{pmatrix} \tag{5.62}$$

where ${}^t(x_j, y_j, z_j)$ are the model coordinates, ${}^t(X_j, Y_j, Z_j)$ are the ground coordinates, ${}^t(b_1, b_2, b_3)$ is the translation vector, s is the scaling factor, and R is the rotation matrix. Every point in the model whose ground coordinates are known leads to three equations. Seven unknowns (three translations, three rotations, and one scaling parameter) are sought. Therefore one needs at least three model points with known ground coordinates in an image pair in order to determine the unknowns. The equations are non-linear, so that approximations for the solution are required when a Newton or similar method is chosen. As in all previous methods, this is not a problem in conventional photogrammetry, but it can be quite difficult for oblique archaeological aerial photographs.

Direct determination of translation and scaling parameters

It is relatively simple to determine the translation vector and scaling factor in equation (5.62). When one displaces two corresponding points in both coordinate systems to the origin, then the model only has to be scaled and rotated to obtain the correspondence between model and ground coordinates. Both the ground and model coordinates have errors. Since the errors which are made in the calculation of the translation parameter propagate and increase the errors in the following calculations of all the other parameters of exterior orientation, one must determine the translation parameter with the highest possible accuracy. It may be supposed that the X_i are independent $N(\mu_i, \sigma^2)$ distributed random variables, and that the x_i are $N(\varepsilon_i, \gamma^2)$ distributed. Similar suppositions are possible for the Y and Z coordinates.

If one computes the centre of gravity of both point systems:

$$\begin{pmatrix} \bar{X} \\ \bar{Y} \\ \bar{Z} \end{pmatrix} = \frac{1}{n} \sum_{j=1}^{n} \begin{pmatrix} X_j \\ Y_j \\ Z_j \end{pmatrix}, \qquad \begin{pmatrix} \bar{x} \\ \bar{y} \\ \bar{z} \end{pmatrix} = \frac{1}{n} \sum_{j=1}^{n} \begin{pmatrix} x_j \\ y_j \\ z_j \end{pmatrix} \qquad (5.63)$$

where $(\bar{x}, \bar{y}, \bar{z})$ in the model correspond to $(\bar{X}, \bar{Y}, \bar{Z})$ on the ground, then $\bar{X}$ is a normally distributed random variable with variance σ^2/n, and $\bar{x}$ has similarly a variance γ^2/n. One sets:

$$\begin{pmatrix} \hat{X}_j \\ \hat{Y}_j \\ \hat{Z}_j \end{pmatrix} := \begin{pmatrix} X_j \\ Y_j \\ Z_j \end{pmatrix} - \begin{pmatrix} \bar{X} \\ \bar{Y} \\ \bar{Z} \end{pmatrix} \qquad \text{and} \qquad \begin{pmatrix} \hat{x}_j \\ \hat{y}_j \\ \hat{z}_j \end{pmatrix} := \begin{pmatrix} x_j \\ y_j \\ z_j \end{pmatrix} - \begin{pmatrix} \bar{x} \\ \bar{y} \\ \bar{z} \end{pmatrix} \qquad (5.64)$$

and continues to compute with the 'centred' vectors.

The scaling factor can be calculated by comparing lengths:

$$s = \frac{\text{dist}(\mathbf{P}_j, \mathbf{P}_i)}{\text{dist}(\mathbf{p}_j, \mathbf{p}_i)}$$

with:

$$\mathbf{P}_j = \begin{pmatrix} \hat{X}_j \\ \hat{Y}_j \\ \hat{Z}_j \end{pmatrix}, \qquad \mathbf{p}_j = \begin{pmatrix} \hat{x}_j \\ \hat{y}_j \\ \hat{z}_j \end{pmatrix}, \qquad j \neq i \qquad (5.65)$$

Here too, the precision is raised if the mean value of all distances is used, whereby it is sensible to give the larger distance ratios higher weights. When the distance ratios are weighted by the distances between $\mathbf{p}_j$ and $\mathbf{p}_i$, then:

$$s = \sum_{i=1}^{n-1} \sum_{j=i+1}^{n} w_{ij} \frac{\text{dist}(\mathbf{P}_j, \mathbf{P}_j)}{\text{dist}(\mathbf{p}_j, \mathbf{p}_i)}$$

with:

$$\qquad (5.66)$$

$$w_{ij} = \frac{\text{dist}(\mathbf{p}_j, \mathbf{p}_i)}{\sum_{k=1}^{n-1} \sum_{l=k+1}^{n} \text{dist}(\mathbf{p}_k, \mathbf{p}_l)}$$

This equation can be simplified to:

$$s = \frac{\sum_{i=1}^{n-1} \sum_{j=i+1}^{n} \text{dist}(\mathbf{P}_j, \mathbf{P}_i)}{\sum_{i=1}^{n-1} \sum_{j=i+1}^{n} \text{dist}(\mathbf{p}_j, \mathbf{p}_i)} \qquad (5.67)$$

Finally, the task remains to compute the rotation matrix R, that is, to find a solution of:

$$\begin{pmatrix} \hat{X}_j \\ \hat{Y}_j \\ \hat{Z}_j \end{pmatrix} = R \begin{pmatrix} \tilde{x}_j \\ \tilde{y}_j \\ \tilde{z}_j \end{pmatrix}$$

with:

$$\begin{pmatrix} \tilde{x}_j \\ \tilde{y}_j \\ \tilde{z}_j \end{pmatrix} = s \begin{pmatrix} \hat{x}_j \\ \hat{y}_j \\ \hat{z}_j \end{pmatrix}, \qquad j = 1, \ldots, n$$

$$(5.68)$$

The rotation matrix

Rotation matrices are orthogonal. This means that:

$$^tRR = I, \qquad \det(R) = 1 \tag{5.69}$$

where I is the identity matrix.

Such matrices can be defined by three independent parameters. There are many different methods for parameterising orthogonal matrices. Tienstra (76) lists nine of them. In photogrammetry it is customary to regard the rotation matrix as the sequential application of three rotations around the coordinate axes:

$$R = \begin{pmatrix} 1 & 0 & 0 \\ 0 & \cos(\alpha) & \sin(\alpha) \\ 0 & -\sin(\alpha) & \cos(\alpha) \end{pmatrix} \begin{pmatrix} \cos(\beta) & 0 & \sin(\beta) \\ 0 & 1 & 0 \\ -\sin(\beta) & 0 & \cos(\beta) \end{pmatrix} \begin{pmatrix} \cos(\tau) & \sin(\tau) & 0 \\ -\sin(\tau) & \cos(\tau) & 0 \\ 0 & 0 & 1 \end{pmatrix}$$

$$= \begin{pmatrix} \cos(\beta)\cos(\tau) & \cos(\beta)\sin(\tau) & \sin(\beta) \\ -\sin(\alpha)\sin(\beta)\cos(\tau)+\sin(\alpha)\sin(\tau) & -\sin(\alpha)\sin(\beta)\sin(\tau)+\cos(\alpha)\cos(\tau) & \sin(\alpha)\cos(\beta) \\ -\cos(\alpha)\sin(\beta)\cos(\tau)+\sin(\alpha)\sin(\tau) & -\cos(\alpha)\sin(\beta)\sin(\tau)-\sin(\alpha)\cos(\tau) & \cos(\alpha)\cos(\beta) \end{pmatrix}$$

$$(5.70)$$

Other sequences in the rotations about the X, Y, Z axes are sometimes used, and these lead to different parameterisations, since such rotations do not obey the commutative law. Two other parameterisations are important for what follows, and they are closely related to each other. Many years ago Cayley showed that a matrix which has eigenvalues $\neq -1$ is orthogonal when it can be expressed in terms of a skew symmetric matrix S such that:

$$R = (I - S)(I + S)^{-1} \tag{5.71}$$

From the geometry of the problem it is easily seen that the rotation matrix to be computed has no negative eigenvalues, and that therefore the parameterisation above does not lead to restrictions. As shown earlier, a skew symmetric 3×3 matrix can be expressed with three parameters as:

$$S = \begin{pmatrix} 0 & -c & b \\ c & 0 & -a \\ -b & a & 0 \end{pmatrix} \tag{5.72}$$

If (5.71) is multiplied out:

$$R = \frac{1}{a^2 + b^2 + c^2 + 1} \begin{pmatrix} 1 + a^2 - b^2 - c^2 & 2ab - 2c & 2ac + 2b \\ 2ab + 2c & 1 - a^2 + b^2 - c^2 & 2bc - 2a \\ 2ac - 2b & 2bc + 2a & 1 - a^2 - b^2 + c^2 \end{pmatrix} \tag{5.73}$$

is obtained.

Further, an additional parameter d can be introduced. Then with:

$$s^2 = a^2 + b^2 + c^2 + d^2 \tag{5.74}$$

a scale factor is obtained which is set to 1 when dealing with matrices with determinant $= 1$. An orthogonal matrix is then given by:

$$R = (1/s^2)(dI + S)(dI - S)^{-1} \tag{5.75}$$

which when multiplied out gives:

$$R = \frac{1}{a^2 + b^2 + c^2 + d^2}$$
$$\times \begin{pmatrix} d^2 + a^2 - b^2 - c^2 & 2ab - 2cd & 2ac + 2bd \\ 2ab + 2cd & d^2 - a^2 + b^2 - c^2 & 2bc - 2ad \\ 2ac - 2bd & 2bc + 2ad & d^2 - a^2 - b^2 + c^2 \end{pmatrix} \tag{5.76}$$

A rotation matrix parameterised in this way is called a Rodriguez matrix. As can be seen from these examples, the elements of R are not simple functions of the parameters, and that is why it took until 1960 for a direct solution of problem (5.68) to be obtained.

The method of Thompson and Schut
This solution was found by Thompson (75) and improved upon by Schut (59). Let there be n pairs of vectors $P_i \in \mathbb{R}^3$ and $p_i \in \mathbb{R}^3$, which differ by rotation, that is:

$$P_i = Rp_i, \qquad i = 1, \ldots, n \tag{5.77}$$

whereby the unknown matrix R is parameterised as in (5.76). The parameters a, b, c, d are obtained as the solution of the following homogeneous equation system:

$$
\begin{pmatrix}
0 & -(Z_1+z_1) & (Y_1+y_1) & (X_1-x_1) \\
(Z_1+z_1) & 0 & -(X_1+x_1) & (Y_1-y_1) \\
-(Y_1+y_1) & (X_1+x_1) & 0 & (Z_1-z_1) \\
\vdots & \vdots & \vdots & \vdots \\
0 & -(Z_n+z_n) & (Y_n+y_n) & (X_n-x_n) \\
(Z_n+z_n) & 0 & -(X_n+x_n) & (Y_n+y_n) \\
-(Y_n+y_n) & (X_n+x_n) & 0 & (Z_n-z_n)
\end{pmatrix}
\begin{pmatrix} a \\ b \\ c \\ d \end{pmatrix}
=
\begin{pmatrix} 0 \\ 0 \\ 0 \\ \vdots \\ 0 \\ 0 \\ 0 \end{pmatrix}
\quad (5.78)
$$

whereby:

$$
\mathbf{P}_i = \begin{pmatrix} X_i \\ Y_i \\ Z_i \end{pmatrix}, \qquad
\mathbf{p}_i = \begin{pmatrix} x_i \\ y_i \\ z_i \end{pmatrix}
\left.\vphantom{\begin{pmatrix} X_i \\ Y_i \\ Z_i \end{pmatrix}}\right\} \quad (5.79)
$$

$$
a^2 + b^2 + c^2 + d^2 > 0
$$

At least two linear independent vector pairs $(\mathbf{P}_i, \mathbf{p}_i)$ must be given componentwise in order to determine R uniquely.

In Schut's method, the equations are first rearranged for a and b, then these are substituted in the others. Setting $d = 1$, c is calculated. If c is very large, then $c = 1$ and d is calculated. Then a and b can be calculated giving the parameters of (5.76). The advantage of this method is that three corresponding vector pairs are sufficient for the determination of the external orientation and the calculations are simple. Least squares methods have also been used by Blais (7) using four, and by Tienstra (76) using three control points to obtain R. Another method is due to Sansò (58), which can be derived in a simpler way from the method of Thompson and Schut.

Combining the relative and absolute orientation parameters

Let the parameters for the relative orientation of two photos be calculated (the model). Then it is possible to compute the model (x_a, y_a, z_a) coordinates for a ground point (X_a, Y_a, Z_a) which is visible in both images as:

$$\begin{pmatrix} x_a \\ y_a \\ z_a \end{pmatrix} = \eta'_a R_1 \begin{pmatrix} x'_a - x'_p \\ y'_a - y'_p \\ -f' \end{pmatrix} + \begin{pmatrix} a_1 \\ a_2 \\ a_3 \end{pmatrix} \tag{5.80}$$

where R_1 is the rotation matrix, and $'(a_1, a_2, a_3)$ the translation vector of the relative orientation. Furthermore:

$$\begin{pmatrix} x_a \\ y_a \\ z_a \end{pmatrix} = \eta''_a \begin{pmatrix} x''_a - x''_p \\ y''_a - y''_p \\ -f'' \end{pmatrix} \tag{5.81}$$

Equating the right hand sides of (5.80) and (5.81) gives:

$$\eta'_a v'_a + \begin{pmatrix} a_1 \\ a_2 \\ a_3 \end{pmatrix} = \eta''_a v''_a \tag{5.82}$$

whereby:

$$\left. \begin{matrix} v'_a = R_1 \begin{pmatrix} x'_a - x'_p \\ y'_a - y'_p \\ -f' \end{pmatrix} \\[20pt] v''_a = \begin{pmatrix} x''_a - x''_p \\ y''_a - y''_p \\ -f'' \end{pmatrix} \end{matrix} \right\} \tag{5.83}$$

The unknown η'_a and η''_a are obtained through the solution of:

$$(v'_a, -v''_a) \begin{pmatrix} \eta'_a \\ \eta''_a \end{pmatrix} = \begin{pmatrix} a_1 \\ a_2 \\ a_3 \end{pmatrix} \tag{5.84}$$

and thus the model coordinates are computed from (5.80) or (5.81). The parameters of the absolute orientation tell us how the model coordinates are transformed into ground coordinates, thus:

$$\begin{pmatrix} X_a \\ Y_a \\ Z_a \end{pmatrix} = sR \begin{pmatrix} x_a \\ y_a \\ z_a \end{pmatrix} + \begin{pmatrix} b_1 \\ b_2 \\ b_3 \end{pmatrix} \tag{5.85}$$

From this it follows for the first photograph:

$$\begin{pmatrix} X_a \\ Y_a \\ Z_a \end{pmatrix} = sR \left[\eta'_a R_1 \begin{pmatrix} x'_a - x'_p \\ y'_a - y'_p \\ -f' \end{pmatrix} + \begin{pmatrix} a_1 \\ a_2 \\ a_3 \end{pmatrix} \right] + \begin{pmatrix} b_1 \\ b_2 \\ b_3 \end{pmatrix} \tag{5.86}$$

and from this we obtain the following which is equivalent to equation (5.16):

$$\begin{pmatrix} X_a \\ Y_a \\ Z_a \end{pmatrix} = s\eta'_a RR_1 \begin{pmatrix} x'_a - x'_p \\ y'_a - y'_p \\ -f' \end{pmatrix} + sR \begin{pmatrix} a_1 \\ a_2 \\ a_3 \end{pmatrix} + \begin{pmatrix} b_1 \\ b_2 \\ b_3 \end{pmatrix} \tag{5.87}$$

and for the second image:

$$\begin{pmatrix} X_a \\ Y_a \\ Z_a \end{pmatrix} = s\eta''_a R \begin{pmatrix} x''_a - x''_p \\ y''_a - y''_p \\ -f'' \end{pmatrix} + \begin{pmatrix} b_1 \\ b_2 \\ b_3 \end{pmatrix} \tag{5.88}$$

With the help of these formulae, the parameters of the three point method for two images can be obtained.

5.3.9. Solution of the non-linear equation systems

In the previous section, the actual technique for calculating the sought-after parameters of the three and multipoint methods was left open, for here a non-linear equation system must be solved. This is not as simple as for the other methods where linear equations are involved. In this section, two analytic methods for determining the parameters of the three point method will be described, both of which assume that the control points have no errors, and the second of which assumes that the control points lie in a plane. The result of this computation can then be used as a starting point for an iterative technique which improves the accuracy and can take the errors into account. One such method will also be described here. It is also useful for the multipoint method, whereby the parameters from a four point technique are used as starting values.

Determining the parameters of the three point method
The method of Fischler and Bolles. In this section, we will assume that
the camera constants are known. If there are only two control points
in the image, it is still possible to compute the surface on which the
camera position must lie. Through the two control points P and Q and
the angle θ between the corresponding picture vectors $\mathbf{p}$ and $\mathbf{q}$, the radius
of the circumscribing circle around the triangle given by P, Q and the
camera position is determined as:

$$r = \frac{1}{2} \frac{\overline{PQ}}{\sin(\theta)} \tag{5.89}$$

The totality of all possible camera positions is obtained when the circle
is rotated on the axis $\mathbf{PQ}$, producing a degenerate torus. If any number
of control points lie on a circle on which the camera position is also
located, then it is not possible to determine the camera position with
this information, for it may lie anywhere on the circle. It is even simpler
to check for picture points in a line. Then as figure 5.18 shows, the
picture points also lie on the plane of the circle. Since they also lie in
the film plane, it is proven that they lie in a line.

Fischler and Bolles compute the parameters of the three point case
analytically, thereby improving upon Church. The computation has
three stages: first, the distances from the control points to the position
of the camera are calculated, then the camera position is obtained, and
finally the rotational parameters are computed. This was relegated to
an appendix by the authors. It appears to be unknown in the photogrammetric world, and because of its importance for our problem, we repeat
some of the details here.

Fig. 5.18 Two possible camera positions when these and ground control
points lie on a line.

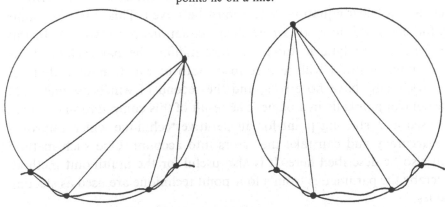

Step 1: Calculation of the distances between the control points and the camera position

Let $\mathbf{O}$ be the sought-for camera position and $\mathbf{P}_1, \mathbf{P}_2, \mathbf{P}_3$ three ground control points. Let:

$$a = \overline{OP}_1, \qquad b = \overline{OP}_2, \qquad c = \overline{OP}_3$$

$$\left. \begin{array}{l} W_1 = \cos[\angle(\mathbf{OP}_1, \mathbf{OP}_2)], \qquad W_2 = \cos[\angle(\mathbf{OP}_1, \mathbf{OP}_3)], \\ W_3 = \cos[\angle(\mathbf{OP}_2, \mathbf{OP}_3)] \end{array} \right\} \quad (5.90)$$

Then from the cosine law it follows that:

$$\left. \begin{array}{l} \overline{\mathbf{P}_1\mathbf{P}_2}^2 = a_2 + b^2 - 2ab W_1 \\ \overline{\mathbf{P}_1\mathbf{P}_3}^2 = a_2 + c^2 - 2ac W_2 \\ \overline{\mathbf{P}_2\mathbf{P}_3}^2 = b^2 + c^2 - 2bc W_3 \end{array} \right\} \quad (5.91)$$

To solve these equations one sets:

$$b = ax, \qquad c = ay \qquad \text{with } x, y \in \mathbb{R}^+. \quad (5.92)$$

then:

$$\left. \begin{array}{l} \overline{\mathbf{P}_1\mathbf{P}_2}^2 = a^2(1 + x^2 - 2x W_1) \\ \overline{\mathbf{P}_1\mathbf{P}_3}^2 = a^2(1 + y^2 - 2y W_2) \\ \overline{\mathbf{P}_2\mathbf{P}_3}^2 = a^2(x^2 + y^2 - 2xy W_3) \end{array} \right\} \quad (5.93)$$

are valid. Since $a \neq 0$ with $K_1 = \overline{\mathbf{P}_2\mathbf{P}_3}^2 / \overline{\mathbf{P}_1\mathbf{P}_3}^2$ and $K_2 = \overline{\mathbf{P}_2\mathbf{P}_3}^2 / \overline{\mathbf{P}_1\mathbf{P}_2}^2$:

$$\left. \begin{array}{l} K_1(1 + y^2 - 2y W_2) = x^2 + y^2 - 2xy W_3 \\ K_2(1 + x^2 - 2x W_1) = x^2 + y^2 - 2xy W_3 \end{array} \right\} \quad (5.94)$$

emerge.

If one rewrites the above equations as polynomials of the second degree in y then:

$$\left. \begin{array}{l} s_1 y^2 + t_1 y + r_1 = 0 \\ s_2 y^2 + t_2 y + r_2 = 0 \end{array} \right\} \quad (5.95)$$

with:

$$s_1 = 1 - K_1 \qquad t_1 = 2K_1 W_2 - 2x W_3 \qquad r_1 = x^2 - K_1$$

$$s_2 = 1 \qquad t_2 = -2x W_3 \qquad r_2 = x^2(1 - K_2) + 2K_2 W_1 x - K_2$$

Multiplying the second equation by s_1 and the first by s_2, and subtracting gives:

$$(t_1 s_2 - t_2 s_1)y + (r_1 s_2 - r_2 s_1) = 0 \quad (5.96)$$

Multiplying the first equation of (5.95) by r_2 and the second by r_1 and subtracting, eliminates the constant member. Since $y > 0$, one may divide by y to obtain:

$$(s_1 r_2 - s_2 r_1)y + (t_1 r_2 - t_2 r_1) = 0 \qquad (5.97)$$

Multiplying equation (5.96) with $(s_1 r_2 - s_2 r_1)$ and equation (5.97) by $(t_1 s_2 - t_2 s_1)$ and subtracting gives one equation in the unknown x:

$$(r_1 s_2 - r_2 s_1)^2 + (t_1 r_2 - t_2 r_1)(t_1 s_2 - t_2 s_1) = 0 \qquad (5.98)$$

If (5.94) is inserted into this equation one obtains:

$$p_4 x^4 + p_3 x^3 + p_2 x^2 + p_1 x + p_0 = 0 \qquad (5.99)$$

with:

$$p_0 = (K_1 K_2 + K_1 - K_2)^2 + 4K_1^2 K_2 W_2^2$$

$$p_1 = 4(K_1 K_2 + K_1 - K_2)K_2(1 - K_1)W_1$$
$$\quad + 4K_1[(K_1 K_2 - K_1 + K_2)W_2 W_3 + 2K_1 K_2 W_1 W_2^2]$$

$$p_2 = [2K_2(1 - K_1)W_1]^2$$
$$\quad + 2(K_1 K_2 + K_1 - K_2)(K_1 K_2 - K_1 - K_2)$$
$$\quad + 4K_1[(K_1 - K_2)W_3^2 + (1 - K_2)K_1 W_2^2 - 2K_2(1 + K_1)W_1 W_2 W_3]$$

$$p_3 = 4(K_1 K_2 - K_1 - K_2)K_2(1 - K_1)W_1$$
$$\quad + 4K_1 W_3[(K_1 K_2 + K_2 - K_1)W_2 + 2K_2 W_1 W_3]$$

$$p_4 = (K_1 K_2 - K_1 - K_2)^2 - 4K_1 K_2 W_3^2$$

Equation (5.99) has four real solutions at most; x is the distance relation between two lines which join the camera position and the ground control points and therefore one may assume that $0.3 < x < 3$. Only solutions of (5.99) which are in this interval will be considered. The roots of the polynomial of the fourth degree can be found analytically but there is a danger of instability. Since all the parameters of the three point method are computed from the roots of x, it is very important that these be obtained with high precision and stability. The method of Bairstow was used giving both the real and complex roots. The roots which lie in the set:

$$\{x \in \mathbb{C} : 0.3 < \mathrm{Re}(x) < 3, \ -\delta < \mathrm{Im}(x) < \delta\} \qquad (5.100)$$

are refined by Newton iteration of real roots of polynomials.

The distances a, b and c from the control points to the camera position are computed as:

$$\left. \begin{aligned} a &= \overline{P_1 P_2}/(1 + x^2 - 2xW_1)^{\frac{1}{2}} \\ b &= ax \\ y &= (t_2 r_1 - t_1 r_2)/(s_1 r_2 - s_2 r_1) \\ c &= ay \end{aligned} \right\} \qquad (5.101)$$

Step 2: Computing the camera position

The projection centre is computed as the intersection of three spheres around the ground control points P_1, P_2, P_3. Required are X_0, Y_0, Z_0 with:

$$(X_1 - X_0)^2 + (Y_1 - Y_0)^2 + (Z_1 - Z_0)^2 = a^2 \qquad (5.102a)$$
$$(X_2 - X_0)^2 + (Y_2 - Y_0)^2 + (Z_2 - Z_0)^2 = b^2 \qquad (5.102b)$$
$$(X_3 - X_0)^2 + (Y_3 - Y_0)^2 + (Z_3 - Z_0)^2 = c^2 \qquad (5.102c)$$

The plane which contains the intersections of spheres (5.102a) and (5.102b) can be calculated by subtracting the first equation from the second giving:

$$2(X_1 - X_2)X_0 + 2(Y_1 - Y_2)Y_0 + 2(Z_1 - Z_2)Z_0$$
$$= b^2 - a^2 + X_1^2 + Y_1^2 + Z_1^2 - X_2^2 - Y_2^2 - Z_2^2 \quad (5.103)$$

The same is done for the intersection between (5.102a) and (5.102c) which gives:

$$2(X_1 - X_3)X_0 + 2(Y_1 - Y_3)Y_0 + 2(Z_1 - Z_3)Z_0$$
$$= c^2 - a^2 + X_1^2 + Y_1^2 + Z_1^2 - X_3^2 - Y_3^2 - Z_3^2 \quad (5.104)$$

The equation for the intersection of the spheres (5.102b) and (5.102c) is linearly dependent on (5.103) and (5.104). One possibility for the calculation of the intersection of the three spheres is to determine the intersection line between the planes (5.104) and (5.103). This line is nearly vertical to the ground for aerial photographs. Then one calculates the intersection of the line with one of the spheres. As a rule, one obtains points with negative and positive Z_0 coordinates as a result. The one with positive Z_0 coordinate is sensibly assumed to be the camera position. If all the ground points lie on a line, then there is no unique intersection of the three spheres, and the camera position cannot be calculated.

Step 3: Calculating the rotation of the picture coordinate system compared with the ground coordinate system

First one calculates the scale of the individual points with:

$$\eta_i = \frac{\|(X_i - X_0, Y_i - Y_0, Z_i - Z_0)\|_2}{\|(x_i, y_i, -f)\|_2} \qquad (5.105)$$

The rotation matrix can then be computed from:

$$\eta_i \begin{pmatrix} x_i \\ y_i \\ -f \end{pmatrix} = R \begin{pmatrix} X_i - X_0 \\ Y_i - Y_0 \\ Z_i - Z_0 \end{pmatrix} \qquad (5.106)$$

The method of Thompson and Schut described above is used to compute R. In this manner, up to four solutions of the three point problem can be obtained. In order to find the correct solution, the computed parameters and the control points are inserted into equation (5.18) and the residual error is calculated. It is assumed that the result with the least error is the correct one. This initial solution can be improved iteratively. If only three control points are available, then plausibility criteria can be of help. In rare cases, the user must decide which is the best solution by trying them in turn to see which gives the best rectification of all the features in the image which can be seen on the map.

Parameters of the three point method from the four point method
The three point method has the advantage over the four point method in that with it one can deal with hilly terrain. It has the disadvantage that one must know the principal point and focal length of the lens. In this section we will describe a method for obtaining the parameters of the three point method from the four point method, by which the probable focal length will emerge from the computation. This technique can therefore be employed on a much greater number of archival images having as little as four control points in hilly terrain. The aim of the method is to determine the parameters and focal length by approximation, so that with only a few iterations, a non-linear overdetermined equation system can be solved by the method of least squares.

First assume that one has four or more control points (X_i, Y_i, Z_i) which lie roughly in a plane, and compute the parameters for the plane using the following overdetermined system of linear equations:

$$\begin{pmatrix} 1 & X_1 & Y_1 \\ 1 & X_2 & Y_2 \\ \vdots & \vdots & \vdots \\ 1 & X_n & Y_n \end{pmatrix} \begin{pmatrix} a \\ b \\ c \end{pmatrix} = \begin{pmatrix} Z_1 \\ Z_2 \\ \vdots \\ Z_n \end{pmatrix} \tag{5.107}$$

Now a coordinate transformation is carried out which carries the plane $Z = aX + bY + c$ into the plane $Z = 0$. A rotation matrix $\hat{R}$ and a translation vector $^t(T_1, T_2, T_3)$ is required such that:

$$\hat{R} \begin{pmatrix} X \\ Y \\ aX + bY + c \end{pmatrix} = \begin{pmatrix} X^* \\ Y^* \\ 0 \end{pmatrix} - \begin{pmatrix} T_1 \\ T_2 \\ T_3 \end{pmatrix} \tag{5.108}$$

$$\hat{R} = (r_{ij}), \; i, j = 1, \ldots, 3$$

$\hat{R}$ must satisfy the following condition:

$$r_{31}X + r_{32}Y + r_{33}(aX + bY + c) = -T_3 \qquad \forall X, Y \in \mathbb{R} \quad (5.109)$$

This condition is equivalent to:

$$(r_{31} + ar_{33})X + (r_{32} + br_{33})Y + r_{33}c = -T_3 \qquad \forall X, Y \in \mathbb{R} \quad (5.110)$$

from which follows:

$$r_{31} = -ar_{33}, \qquad r_{32} = -br_{33}, \qquad T_3 = -cr_{33} \qquad (5.111)$$

Since the Euclidean norm of every row of the rotation matrix is equal to 1, it follows that:

$$(-ar_{33})^2 + (-br_{33})^2 + r_{33}^2 = 1$$
$$\Leftrightarrow \qquad (a^2 + b^2 + 1)r_{33}^2 = 1 \qquad (5.112)$$

Since the rotations about the X and Y axes are less than $\pm 90°$, then $r_{33} > 0$ and from (5.112) it follows that:

$$\left. \begin{array}{cc} r_{33} = 1/(a^2 + b^2 + 1)^{\frac{1}{2}} & r_{31} = -a/(a^2 + b^2 + 1)^{\frac{1}{2}} \\ r_{32} = -b/(a^2 + b^2 + 1)^{\frac{1}{2}} & T_3 = -c/(a^2 + b^2 + 1)^{\frac{1}{2}} \end{array} \right\} \quad (5.113)$$

This determines the lowest row of $\hat{R}$ and one of the translation parameters. If one parameterises $'\hat{R}$ as in (5.70), one sees that the rotation parameters α and β are fixed thereby and τ can be freely chosen. It is intuitively obvious that rotations about the X and Y axes alone are enough to bring a slightly inclined plane into coincidence with a plane with $Z = 0$. Therefore we set $\tau = 0$ and derive $\hat{R}$ from:

$$\left. \begin{array}{c} \beta = \arcsin(r_{31}), \qquad \alpha = \arcsin(r_{32}/\cos(\beta)) \\ \hat{R} = \begin{pmatrix} \cos(\beta) & -\sin(\alpha)\sin(\beta) & -\cos(\alpha)\sin(\beta) \\ 0 & \cos(\alpha) & -\sin(\alpha) \\ r_{31} & r_{32} & r_{33} \end{pmatrix} \end{array} \right\} \quad (5.114)$$

T_1 and T_2 can be freely chosen, and they are set equal to zero.

When the ground control points lie in the $Z = 0$ plane and the picture coordinates are measured relative to the principal point, then the following relationships between the s_{ij} parameters of the four point and those of the three point methods apply:

$$\left. \begin{array}{lll} s_{11} = -(f/A_3)m_{11} & s_{21} = -(f/A_3)m_{21} & s_{31} = m_{31}/A_3 \\ s_{12} = -(f/A_3)m_{12} & s_{22} = -(f/A_3)m_{22} & s_{32} = m_{32}/A_3 \\ s_{13} = -fA_1/A_3 & s_{23} = -fA_2/A_3 \end{array} \right\} \quad (5.115)$$

with:

$$A_j = -(m_{j1}X_0 + m_{j2}Y_0 + m_{j3}Z_0), \qquad j = 1, 2, 3$$

When s_{ij} are known, then one can compute m_{ij}, A_j and f. From:

$$m_{11}^2 + m_{21}^2 + m_{31}^2 = 1 \tag{5.116}$$

it follows that:

$$(A_3^2/f^2)s_{11}^2 + (A_3^2/f^2)s_{21}^2 + A_3^2 s_{31}^2 = 1 \tag{5.117}$$

and thus:

$$(1/f^2)(s_{11}^2 + s_{21}^2) + s_{31}^2 = 1/A_3^2 \tag{5.118}$$

Similarly, from the sum of the second column of R:

$$(1/f^2)(s_{12}^2 + s_{22}^2) + s_{32}^2 = 1/A_3^2 \tag{5.119}$$

Setting (5.118) equal to (5.119) and solving for f gives:

$$f = \left(\frac{s_{11}^2 + s_{21}^2 - s_{12}^2 - s_{22}^2}{s_{32}^2 - s_{31}^2}\right)^{\frac{1}{2}} \tag{5.120}$$

for:

$$|s_{32}| \neq |s_{31}|$$

Then the absolute value of A_3 can be computed from (5.118):

$$|A_3| = f/(s_{11}^2 + s_{21}^2 + f^2 s_{31}^2)^{\frac{1}{2}} \tag{5.121}$$

With the help of equation (5.115) one can compute the first two columns of R and A_1 and A_2 except for the sign. The remaining three elements of R can be derived by assuming that $m_{33} > 0$, and that the rotations around the X and Y axes are less than $\pm 90°$ so:

$$m_{33} = (1 - m_{31}^2 - m_{32}^2)^{\frac{1}{2}} \tag{5.122}$$

Since the rows of R are orthogonal to each other:

$$\left.\begin{aligned} m_{23} &= (1/m_{33})(-m_{21}m_{31} - m_{22}m_{32}) \\ m_{13} &= (1/m_{33})(-m_{11}m_{31} - m_{12}m_{32}) \end{aligned}\right\} \tag{5.123}$$

after computing these, the third column of R is known with sign.

When A_i and R are known with sign, then the camera position is given by:

$$\begin{pmatrix} X_0 \\ Y_0 \\ Z_0 \end{pmatrix} = -{}^t R \cdot \begin{pmatrix} A_1 \\ A_2 \\ A_3 \end{pmatrix} \tag{5.124}$$

with $R = (m_{ij})$, $i = 1, 2, 3$, $j = 1, 2, 3$. The third equation is, *in extenso*:

$$Z_0 = -(m_{13}A_1 + m_{23}A_2 + m_{33}A_3) \tag{5.125}$$

and with its help the final decision as to sign can be made, since the sign of m_{i3} is already fixed. If the flight altitude turns out negative ($Z_0 < 0$) then the signs of A_i and m_{ij}, $i = 1, 2, 3$, $j = 1, 2$ must be inverted.

It may happen that the R thus computed is not orthogonal because of errors in the central perspective model. Nonetheless it fulfills five of the six independent conditions required for the elements of an orthogonal matrix. In order to find the orthogonal matrix M which is nearest to R, the following expression is minimised:

$$Q = \sum_{i=1}^{3} \sum_{j=1}^{3} (r_{ij} - m_{ij})^2$$

$$+ \eta_{11}(m_{11}^2 + m_{12}^2 + m_{13}^2 - 1) + \eta_{22}(m_{21}^2 + m_{22}^2 + m_{23}^2 - 1)$$
$$+ \eta_{33}(m_{31}^2 + m_{32}^2 + m_{33}^2 - 1) + 2\eta_{12}(m_{11}m_{21} + m_{12}m_{22} + m_{13}m_{23})$$
$$+ 2\eta_{13}(m_{11}m_{31} + m_{12}m_{32} + m_{13}m_{33})$$
$$+ 2\eta_{23}(m_{31}m_{21} + m_{32}m_{22} + m_{33}m_{23}) \qquad (5.126)$$

where $R = (r_{ij})$, $M = (m_{ij})$, η_{ij} are the Lagrange multipliers. As can be readily calculated:

$$\partial Q / \partial m_{ij} = -2(r_{ij} - m_{ij}) + 2\eta_{ii}m_{ij} + 2\eta_{ik}m_{kj} + 2\eta_{in}m_{nj} \qquad (5.127)$$

where i, k, n are pairwise different and $\eta_{ij} = \eta_{ji}$, $i = 1, \ldots, 3, j = 1, \ldots, 3$.
At the minimum of Q:

$$m_{ij} + \sum_{k=1}^{3} \eta_{ik}m_{kj} = r_{ij}, \qquad i = 1, \ldots, 3, j = 1, \ldots, 3 \qquad (5.128)$$

which written in matrix form is:

$$M + LM = R \qquad (5.129)$$

with $L = (\eta_{ij})$ symmetric. If this is transformed to:

$$(I + L)M = R$$

and both sides of the equation multiplied from the right by their transposes, we obtain:

$$(I + L)^2 = R'R$$

If the data contains no errors, then L is the null matrix. Therefore the matrix $(I + L)$ is always positive definite when errors are present. $(I + L)$ can be computed with the help of the eigenvalues and eigenvectors of $R'R$. Then M can be computed from (5.129) using the method of Thompson and Schut or since $(I + L)$ is positive definite and regular, from $M = (I + L)^{-1}R$.

When the eigenvalues and eigenvectors of $(I + L)$ are given, then the inverse of $(I + L)$ can be computed simply from:

$$I + L = {}'CSC \Rightarrow (I + L)^{-1} = {}'CS^{-1}C \qquad (5.130)$$

where C is the orthogonal matrix made up of the eigenvectors of $(I + L)$

and S is the diagonal matrix of the corresponding eigenvalues. Thus we have a method for calculating the orthogonal matrix M which is closest to R. The approximation to an orthogonal matrix is perhaps not as important as the possibility of obtaining a solution even when the denominator of (5.120) is zero, that is, $|s_{32}| = |s_{31}|$, by rotation about the Z axis.

Let s_{ij} be the parameters which transform the ground point (X, Y) to the corresponding image point. Further let:

$$\begin{pmatrix} X' \\ Y' \end{pmatrix} = \begin{pmatrix} \cos(\gamma) & -\sin(\gamma) \\ \sin(\gamma) & \cos(\gamma) \end{pmatrix} \cdot \begin{pmatrix} X \\ Y \end{pmatrix} \qquad (5.131)$$

Then the transformation parameters s'_{ij} for (X', Y') are given by:

$$\left. \begin{aligned} s'_{i1} &= s_{i1} \cos(\gamma) - s_{i2} \sin(\gamma), & i &= 1, \ldots, 3 \\ s'_{i2} &= s_{i1} \sin(\gamma) + s_{i2} \cos(\gamma), & i &= 1, \ldots, 3 \\ s'_{13} &= s_{13} \qquad s'_{23} = s_{23} \end{aligned} \right\} \qquad (5.132)$$

For $\gamma = \pi/4$ and $|s_{32}| = |s_{31}|$ we obtain:

	Case 1	Case 2
	$s_{31} = s_{32}$	$s_{31} = -s_{32}$
s'_{31}	0	$\sqrt{2}s_{31}$
s'_{32}	$\sqrt{2}s_{31}$	0

$$(5.133)$$

This shows that the case $|s_{32}| = |s_{31}|$ can be excluded by coordinate transformation unless both values are zero. This is the case of linear vertical projection, a perfectly vertical image which almost never occurs in aerial archaeology.

In order to get the parameters of the three point method for sloping terrain in the form of equation (5.17), examine:

$$\begin{pmatrix} x_i \\ y_i \\ 0 \end{pmatrix} - \begin{pmatrix} x_p \\ y_p \\ f \end{pmatrix} = \mu_i R \left[\hat{R} \begin{pmatrix} X_i \\ Y_i \\ Z_i \end{pmatrix} + \begin{pmatrix} 0 \\ 0 \\ T_3 \end{pmatrix} - \begin{pmatrix} X_0 \\ Y_0 \\ Z_0 \end{pmatrix} \right] \qquad (5.134)$$

where $\hat{R}$ is the rotation matrix which transforms the control point to the zero plane, possibly combined with a rotation about the Z axis. From (5.134) it follows that:

$$\begin{pmatrix} x_i \\ y_i \\ 0 \end{pmatrix} - \begin{pmatrix} x_p \\ y_p \\ f \end{pmatrix} = \mu_i R \hat{R} \left[\begin{pmatrix} X_i \\ Y_i \\ Z_i \end{pmatrix} - \begin{pmatrix} B_1 \\ B_2 \\ B_3 \end{pmatrix} \right] \qquad (5.135)$$

with:

$$\begin{pmatrix} B_1 \\ B_2 \\ B_3 \end{pmatrix} = {}^t\hat{R} \begin{pmatrix} X_0 \\ Y_0 \\ Z_0 - T_3 \end{pmatrix}$$

This result can be improved by iteration on the non-linear equation system.

Iterative solutions

In the preceding sections, methods for obtaining approximations for the parameters of the three point method have been described. A method for improving these results is required. For the non-linear multipoint method the four point technique gives a good starting approximation, so that here too an iterative improvement technique will be useful. The problem can be stated generally as: given m functions $F_j(x)$, $j = 1, \ldots, m$, $x \in \mathbb{R}^n$, and a vector $\mathbf{b} = (b_1, \ldots, b_m)$ with $m \geq n$, find a vector x, such that $F_j(x) = b_j$, $j = 1, \ldots, m$ as nearly as possible. This 'as nearly as possible' can be interpreted in a least squares sense by minimising:

$$Q(\mathbf{x}) = \sum_{j=1}^{m} [F_j(\mathbf{x}) - b_j]^2 \qquad (5.136)$$

and when $\hat{x}$ is the minimum:

$$\frac{\partial Q}{\partial x_k}(\hat{x}) = 0 \qquad \text{for } k = 1, \ldots, n \qquad (5.137)$$

is valid.

The first idea which emerges is that we can use a Newton method to solve the system of equations in n unknowns. For this we need not only the first, but also the second derivatives of F. Since most of the problems discussed here have at least six unknowns and two transformations for the x and y coordinates, this would require at least 12 for the first and 42 derivative formulae for the second. Not only does this require a major expenditure of effort, it is also likely to cause many errors. Therefore, we prefer a method which needs only the first derivatives. An iterative transformation formula for minimising a function has the general form:

$$x^{(k+1)} = x^{(k)} + \mu^{(k)} s^{(k)}, \qquad k = 0, 1 \ldots \qquad (5.138)$$

where $s^{(k)}$ gives the search direction for the minimum (steepest descent) and $\mu^{(k)}$ gives the step size.

If we have a good approximation $x^{(0)}$ for the minimum $\hat{x}$ of Q and F

is continuously differentiable then:

$$F(\hat{\mathbf{x}}) = F(\mathbf{x}^{(0)}) + DF(\mathbf{x}^{(0)})(\hat{\mathbf{x}} - \mathbf{x}^{(0)}) + \mathbf{h}$$

and

$$\|\mathbf{h}\| = o(\|\hat{\mathbf{x}} - \mathbf{x}^{(0)}\|) \tag{5.139}$$

where $DF(\mathbf{x}^{(0)})$ is the Jacobian matrix of F at the point $\mathbf{x}^{(0)}$.

Therefore it is sensible to find the direction of steepest descent by solving the overdetermined equation system:

$$DF(\mathbf{x}^{(k)})\mathbf{s}^{(k)} = \mathbf{b} - F(\mathbf{x}^{(k)}) \tag{5.140}$$

The step size $\mu^{(k)}$ is chosen such that:

$$Q(\mathbf{x}^{(k)} + \mu^{(k)}\mathbf{s}^{(k)}) < Q(\mathbf{x}^{(k)}) \tag{5.141}$$

If $\mathbf{s}^{(k)}$ is chosen as in (5.140), then it can be shown that an interval $[0, p^{(k)}]$ exists such that (5.141) is satisfied for all $\mu^{(k)}$ in this interval. In practice we often choose:

$$\mu^{(k)} = \max\{2^{-j}: j \in \mathbb{N}_0, Q(\mathbf{x}^{(k)} + 2^{-j}\mathbf{s}^{(k)}) < Q(\mathbf{x}^{(k)})\} \tag{5.142}$$

From this we obtain the following algorithm.

Algorithm: Iterative solution of an overdetermined non-linear system of equations.
  (1) Compute the starting value $\mathbf{x}^{(0)}$.
  (2) For $k = 0, 1, \ldots$, compute $\mathbf{x}^{(k)}$ by:
    (*a*) Compute solution $\mathbf{s}^{(k)}$ of $DF(\mathbf{x}^{(k)})\mathbf{s}^{(k)} = \mathbf{b} - F(\mathbf{x}^{(k)})$ by least squares.
    (*b*) Determine $\mu^{(k)}$ as in (5.142).
    (*c*) Set $\mathbf{x}^{(k+1)} = \mathbf{x}^{(k)} + \mu^{(k)}\mathbf{s}^{(k)}$.

This algorithm without step size control ($\mu^{(k)} \equiv 1$) is the Gauss–Newton method. The first derivatives of F can also be approximated by difference quotients. Experience has shown that three iterations are usually sufficient to refine the parameters of the three point and multipoint methods adequately.

5.3.10. Determination of ground coordinates for a point in two images

For the three and six point methods, a digital terrain model is needed to obtain height information. Pentland (52) showed that fractals are very suitable for approximating natural surfaces, and he succeeded in computing the orientation angles for several aerial photographs using an estimate of the fractal dimension. With the same technique he also managed to compute a very exact digital model of the terrain. The application of this method to archaeological aerial photographs appears to be a promising line of future research. But in this section we will use the much simpler assumption that the ground surface is almost everywhere continuously differentiable. This rule is violated at cliffs, houses, pylons, etc., but these will be ignored since they are not usually of archaeological interest.

Under this simplifying assumption, a digital terrain model is a function:

$$z:(x, y) \rightarrow z(x, y) \tag{5.143}$$

which is continuously differentiable at least once. The terrain model is computed from given support points (X_i, Y_i, Z_i), $i = 1, \ldots, n$.

Accurate contour maps at large scales exist in some parts of the world, so that many support points can be accurately identified. But for older pictures, maps of this type for the epoch in which the photograph was taken are usually not available. If the landscape has not changed much since, one can level the needed points by conventional surveying. If considerable change has taken place then older field maps may be of help, even though these do not contain height information, using methods to be described below.

When three point parameters for two images are available
If the parameters for two images using the three point method described in section 5.3.9 have been obtained, then the ground coordinates of any identifiable point visible in both images can be obtained. This is the commonest technique in conventional photogrammetry.

Given two photos with a point (X, Y, Z) visible at point (x', y') in one and (x'', y'') in the other, we write equation (5.16) as:

$$\begin{pmatrix} X \\ Y \\ Z \end{pmatrix} = \begin{pmatrix} X'_0 \\ Y'_0 \\ Z'_0 \end{pmatrix} + \mu''^t R' \left(\begin{pmatrix} x' \\ y' \\ 0 \end{pmatrix} - \begin{pmatrix} x'_p \\ y'_p \\ f' \end{pmatrix} \right) \tag{5.144}$$

$$\begin{pmatrix} X \\ Y \\ Z \end{pmatrix} = \begin{pmatrix} X''_0 \\ Y''_0 \\ Z''_0 \end{pmatrix} + \mu'''^t R'' \left(\begin{pmatrix} x'' \\ y'' \\ 0 \end{pmatrix} - \begin{pmatrix} x''_p \\ y''_p \\ f'' \end{pmatrix} \right) \tag{5.145}$$

Setting the right hand sides equal to each other to reflect the identity of the point seen gives three equations. Taking the first two and solving for a scale gives:

$$\mu'' = \frac{\hat{y}'(X_0' - X_0'') - \hat{x}'(Y_0' - Y_0'')}{\hat{x}''\hat{y}' - \hat{x}'\hat{y}''} \tag{5.146}$$

whereby:

$$\begin{pmatrix} \hat{x}' \\ \hat{y}' \\ \hat{z}' \end{pmatrix} = {}^t R' \left(\begin{pmatrix} x' \\ y' \\ 0 \end{pmatrix} - \begin{pmatrix} x_p' \\ y_p' \\ f' \end{pmatrix} \right) \tag{5.147}$$

and:

$$\begin{pmatrix} \hat{x}'' \\ \hat{y}'' \\ \hat{z}'' \end{pmatrix} = {}^t R'' \left(\begin{pmatrix} x'' \\ y'' \\ 0 \end{pmatrix} - \begin{pmatrix} x_p'' \\ y_p'' \\ f'' \end{pmatrix} \right) \tag{5.148}$$

Inserting this in (5.145) gives the solution for (X, Y, Z).

When the data contains errors, then it is sensible to consider all the equations which enter into the determination of (X, Y, Z) and this is done when one solves the overdetermined equation system:

$$\begin{pmatrix} -1 & 0 & 0 & \hat{x}' & 0 \\ 0 & -1 & 0 & \hat{y}' & 0 \\ 0 & 0 & -1 & \hat{z}' & 0 \\ -1 & 0 & 0 & 0 & \hat{x}'' \\ 0 & -1 & 0 & 0 & \hat{y}'' \\ 0 & 0 & -1 & 0 & \hat{z}'' \end{pmatrix} \begin{pmatrix} X \\ Y \\ Z \\ \mu' \\ \mu'' \end{pmatrix} = - \begin{pmatrix} X_0' \\ Y_0' \\ Z_0' \\ X_0'' \\ Y_0'' \\ Z_0'' \end{pmatrix} \tag{5.149}$$

When the six point parameters for two images are available
When the six point method is used, then following Rogers and Adams (56), given two images with known parameters t_{ij}' and t_{ij}'', with (x', y') and (x'', y'') the film points as above, we solve:

$$\begin{pmatrix} x't_{31}' - t_{11}' & x't_{32}' - t_{12}' & x't_{33}' - t_{13}' \\ y't_{31}' - t_{21}' & y't_{32}' - t_{22}' & y't_{33}' - t_{23}' \\ x''t_{31}'' - t_{11}'' & x''t_{32}'' - t_{12}'' & x''t_{33}'' - t_{13}'' \\ y''t_{31}'' - t_{21}'' & y''t_{32}'' - t_{22}'' & y''t_{33}'' - t_{23}'' \end{pmatrix} \begin{pmatrix} X \\ Y \\ Z \end{pmatrix} = \begin{pmatrix} t_{14}' - x' \\ t_{24}' - y' \\ t_{14}'' - x'' \\ t_{24}'' - y'' \end{pmatrix} \tag{5.150}$$

Here too, we have an overdetermined system of equations for computing (X, Y, Z).

5.4. Errors

5.4.1. Sources of error

In all of the foregoing, we have assumed a mathematically perfect perspective projection, which in reality is never true. In conventional photogrammetry, factors such as the curvature of the earth and atmospheric refraction are of significance but these can be shown to be insignificant in archaeological images. If a sufficiently high shutter speed and suitable film are used, then unsharpness due to camera movement can be held to insignificant values. This is seldom the case in older pictures or in pictures made under poor lighting conditions in winter or on cloudy days, and here control point information may be blurred. Although the methods of chapter 4 will help, they cannot deal with very badly blurred details.

If a slow focal plane shutter was used, then skew distortion of the image in the direction of the vector resultant of aircraft and shutter movement may be present. The four point, six point, and multipoint or polynomial methods correct automatically for this type of error, but the three point method cannot compensate for it. Therefore, even though the latter method is often the best under other circumstances, images from older cameras like the K 24 and F 24 which had large slow focal plane shutters should probably be processed with one of the self-compensating methods for best results with images made on cloudy days.

Older films, prior to the introduction of polyester based materials, are not dimensionally stable with temperature and humidity, and this leads to considerable changes after development and storage as described in chapter 3. A major source of trouble with older pictures is the poor quality of the lens used. Radial and tangential distortion, pin-cushioning, and barrelling, as well as coma and astigmatism afflict the pictures. Some methods for dealing with this were described in chapter 4. Good quality modern cameras are far better, but Hatzopoulos (30) has shown that even the best 35 mm products may have up to 60 μm radially symmetric distortion which is quite significant for control points in the picture background.

Many of the methods require knowledge of the focal length and the principal point. If, as is usual, the latter is not known, it can be approximated by computing a frame around the image by least squares and taking the intersection of diagonals. In some single lens reflex 35 mm cameras, the true principal point may be far from the centre so computed.

For these cameras, methods which do not need the inner orientation may give better results. The nominal value of the focal length can be read off the lens if the camera is still at hand. If this is not the case, a good starting guess for a normal lens may be the length of the film diagonal. If one image with lots of control points from an unknown camera is at hand, then it may be possible to reconstruct the focal length within reason using the iterative three point method with adjustment of that length. This value can then be used as a starting point in less favourable images. A similar technique for estimating the principal point can be imagined but this has not been investigated. If the camera still exists, then the best thing is to have it calibrated, an operation which manufacturers of photogrammetric equipment will usually undertake if they are feeling charitable.

If large numbers of control points are available, Mauelshagen (44) has shown that the errors can be largely compensated with polynomial or spline functions, although this may result in poor conditioning of the resulting equation systems. Weidner's multipoint method can also compensate for this type of error when many control points are at hand.

Gross error in control point identification is a frequent mistake. The standard methods for identifying outliers in data treated with least squares techniques are utterly unsatisfactory when the number of control points is small. No sure method for dealing automatically with this type of mistake is known. Fischler and Bolles as well as Haralick and Chu (28) make suggestions which may be of value in other contexts, but they are not useful for this problem. This is because there are not enough control points to produce stable statistics, and the required error bounds are not known.

Förstner (20) studied the problem of determining the extent to which an error in the input data is visible in the output. The average value for this is obtained when one computes r/n, where n is the number of equations, and $r = n - u$ where u are the number of parameters to be determined. Only r/n of the input error is maximal in the residue of the corresponding equations. This means that, for relative orientation with seven corresponding points $n = 28$ (two pictures with two coordinates each per picture), $u = 26$ for the five parameters of the relative orientation plus three times the seven model coordinates of the control points. From this $r/n = 2/28$ or 0.071 approximately. This means that only 7% of the error is on the average visible in the residue.

This very pessimistic result corresponds to the experimental results of

Kubik, Lyons, and Merchant (39). With Förstner's method it is possible to determine how much an input error affects the rth equation. In a concrete example, the effect of a gross error on the model coordinates of the relative orientation can increase the standard deviation of the input data by a factor of 13! Therefore, in computer vision problems which are not very relevant here, Förster recommends the use of at least three images to obtain an acceptable model. This result tends to support the viewpoint expressed here that, adding a second image may actually produce a poorer result than using one picture alone. Even if the error detection heuristic or robust statistical approach described by Kubik, Lyons, and Merchant is used with a small number of points, there is considerable danger of not detecting two control points which are in error, or one point which is not grossly in error. Experience has shown that the probability of gross error when using field boundaries obtained from property maps for control is very great, since the utilisation of the field visible in the photograph may differ subtly but significantly from the ownership boundaries seen in a cadastral survey map.

Errors due to map inaccuracies have, in practice, proven to be far less significant than those due to other causes, but it is often quite difficult to identify the exact correspondence between the feature in the image and its representation in a map. This problem is exacerbated by all the other sources of errors described qualitatively above. If we assume that only the picture coordinates are in error, and that these errors are of the same order of magnitude for all the control points, then, for example, in the case of the four point method, we have to minimise:

$$\sum_{i=1}^{n} \left(\frac{s_{11}X_i + s_{12}Y_i + s_{13}}{s_{31}X_i + s_{32}Y_i + 1} - x_i \right)^2 + \left(\frac{s_{21}X_i + s_{22}Y_i + s_{23}}{s_{31}X_i + s_{32}Y_i + 1} - y_i \right)^2 \quad (5.151)$$

In the solution of equations (5.25) the following expression is minimised:

$$\sum_{i=1}^{n} w_i \left[\left(\frac{s_{11}X_i + s_{12}Y_i + s_{13}}{s_{31}X_i + s_{32}Y_i + 1} - x_i \right)^2 + \left(\frac{s_{21}X_i + s_{22}Y_i + s_{23}}{s_{31}X_i + s_{32}Y_i + 1} - y_i \right)^2 \right]$$

$$(5.152)$$

with:

$$w_i = (s_{31}X_i + s_{32}Y_i + 1)^2$$

so that a weighted optimising is carried out. But the weights are not always appropriate for the problem. The greater the obliquity of the image, the more the weights become important. It is possible to determine the weights approximately, a posteriori and divide equations (5.25) by them.

5.4.2. Numerically unfavourable cases

Poor conditioning with control points on a line
This problem occurs frequently in archaeological images. Houses along a street used as control points are a typical example. A few non-aligned extra points if available, will ameliorate the problem considerably. If this is not the case, then there is no satisfactory method for rectifying the image.

Numerical instabilities in the four, six and multipoint methods
Numerical instabilities in these methods have been observed. A Monte Carlo experiment has been devised to test this. Using the picture in figure 5.10, we first tried a four point rectification using seven control points, and observed that the average residual error was 12.78 pixels. To test the conditioning of the problem the following simple experiments was devised:

(1) Compute the parameters of the four point method with the control points. Compute from these the corners of the rectification area to be projected into the map, and denote these by M_j, $j = 1, \ldots, 4$. Further, determine the local scale s_j at each of these corners.

(2) Add a random number with values $-1, 0, +1$ pixels to each of the control points and recompute the four point parameters. Then using these parameters, recompute the corner points again, and compare the distances normalised with the local scale s_j. In this way four corner errors θ_j at the scale of the map are obtained.

(3) Repeat (2) a hundred times and select the maximum computed corner error θ_j.

  The result was 139.46 pixels!

If the origin of the map coordinate system is now displaced to the centre of gravity of the control points, then the average residual error drops to 6.18 pixels and the maximum corner error is 22.77 pixels. Figure 5.19 shows the dependence of the residual error in the picture of figure 5.10 for 8500 shifts of the map coordinates relative to the centre of gravity of the control points. It has a zone of very high residual error, which can be shown to be related to poor conditioning of the equation system. This experiment shows that it is not a simple matter to tell in advance if a given set of control points is favourable or not. How to do this remains an as yet unsolved research problem.

  With the six point and multipoint methods, similar problems appear. Here too, a zone of very high error occurs, which often is reduced when the mid-point of the control points is made the centre of the map coordinate system. This discovery calls into question the uncritical use of the four point method when the map coordinate shift is not made. The three point and polynomial methods do not suffer from this difficulty.

Fig. 5.19 The dependence of residual error for the image of figure 5.10 for 8500 shifts of the map coordinates relative to the centre of gravity of the control points. High values correspond to high errors.

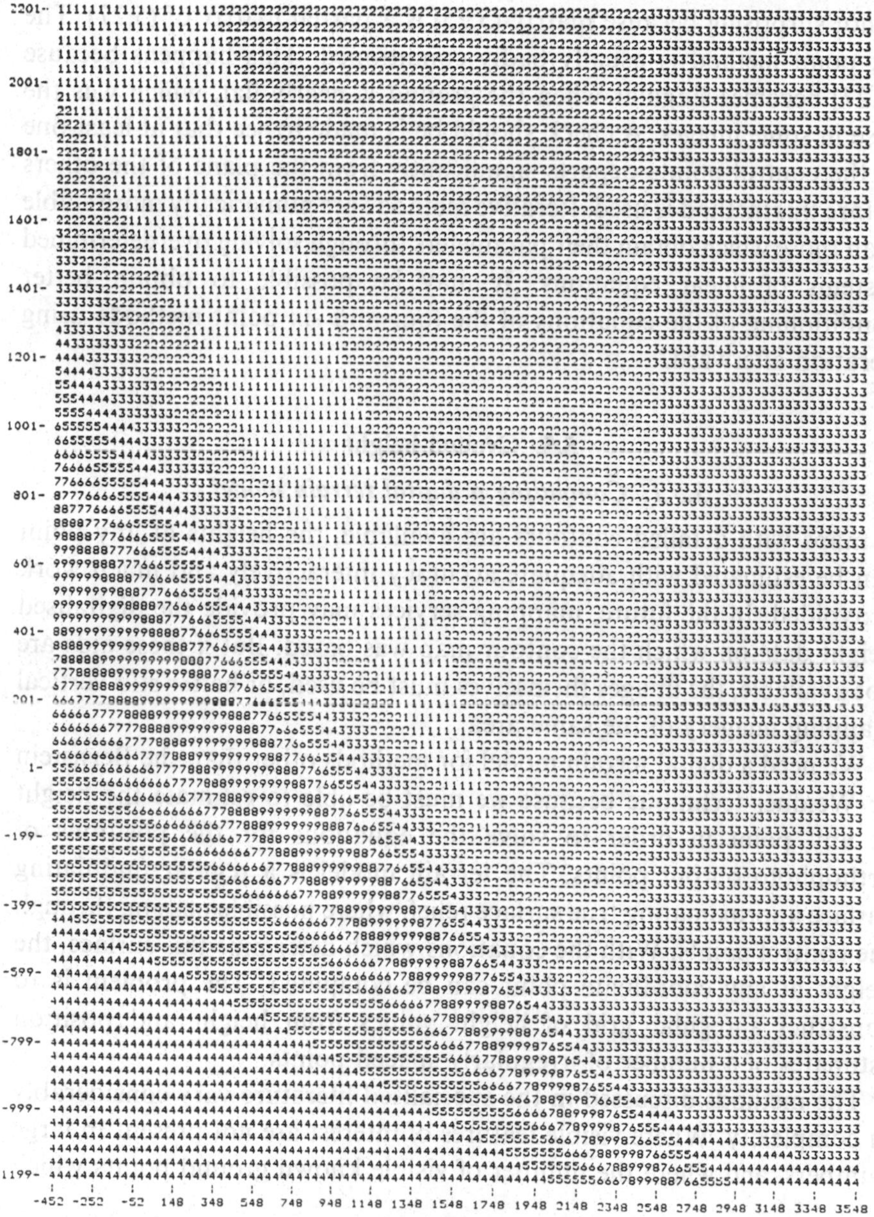

```
2201- 11111111111111111111222222222222222222222222222222222222222222222222222333333333333333333333333
       11111111111111111111222222222222222222222222222222222222222222222222222333333333333333333333333
       11111111111111111111222222222222222222222222222222222222222222222222222333333333333333333333333
       11111111111111111111222222222222222222222222222222222222222222222222222333333333333333333333333
       11111111111111111111222222222222222222222222222222222222222222222222222333333333333333333333333
2001-  11111111111111111111222222222222222222222222222222222222222222222222222333333333333333333333333
       21111111111111111111222222222222222222222222222222222222222222222222222333333333333333333333333
       22111111111111111111222222222222222222222222222222222222222222222222222333333333333333333333333
       22211111111111111111222222222222222222222222222222222222222222222222222333333333333333333333333
       22221111111111111111222222222222222222222222222222222222222222222222222333333333333333333333333
1801-  22222111111111111111222222222222222222222222222222222222222222222222222333333333333333333333333
       22222211111111111111122222222222222222222222222222222222222222222222222333333333333333333333333
       22222221111111111111112222222222222222222222222222222222222222222222222333333333333333333333333
       22222221111111111111112222222222222222222222222222222222222222222222222333333333333333333333333
       22222222111111111111112222222222222222222222222222222222222222222222222333333333333333333333333
1601-  32222222211111111111111222222222222222222222222222222222222222222222222333333333333333333333333
       32222222221111111111111222222222222222222222222222222222222222222222222333333333333333333333333
       33222222222111111111111122222222222222222222222222222222222222222222222333333333333333333333333
       33333222222221111111111111222222222222222222222222222222222222222222222333333333333333333333333
       33333322222221111111111111122222222222222222222222222222222222222222222333333333333333333333333
1401-  33333332222222211111111111111222222222222222222222222222222222222222222333333333333333333333333
       33333332222222211111111111111222222222222222222222222222222222222222222333333333333333333333333
       33333333222222221111111111111122222222222222222222222222222222222222222333333333333333333333333
       43333333332222222111111111111112222222222222222222222222222222222222222333333333333333333333333
1201-  44443333333222222221111111111111112222222222222222222222222222333333333333333333333333333333333
       54444333333332222222211111111111111122222222222222222222222222233333333333333333333333333333333
       55444433333332222222211111111111111112222222222222222222222222223333333333333333333333333333333
       55544443333333222222221111111111111112222222222222222222222222223333333333333333333333333333333
       55554444333333332222222211111111111111122222222222222222222222222333333333333333333333333333333
1001-  65555544444333333222222221111111111111112222222222222222222222222333333333333333333333333333333
       66555554444433333332222222111111111111111222222222222222222222222333333333333333333333333333333
       66665555444433333332222222111111111111111222222222222222222222222333333333333333333333333333333
       76666555544443333332222221111111111111111222222222222222222222222333333333333333333333333333333
       77666655554443333333222222111111111111111122222222222222222222222333333333333333333333333333333
801-   87776666555544443333332222221111111111111111122222222222222222222333333333333333333333333333333
       88776666555554444333333222222111111111111111122222222222222222222333333333333333333333333333333
       88887776666555544433333332222221111111111111111222222222222222222333333333333333333333333333333
       98888777666655554443333332222221111111111111111222222222222222222333333333333333333333333333333
       99988887776665555444433333222222111111111111111222222222222222222333333333333333333333333333333
601-   999998887776665555444333332222221111111111111222222222222222222222333333333333333333333333333333
       9999999888776665555444333332222221111111111111222222222222222222222333333333333333333333333333333
       99999999888776666555444333332222221111111111111222222222222222222333333333333333333333333333333
       99999999980887766666555444333332222221111111111111222222222222222333333333333333333333333333333
       999999999988877766665555444333332222221111111111111222222222222222333333333333333333333333333333
401-   8899999999999888776665555444333332222221111111111112222222222222222333333333333333333333333333333
       888999999999988877766655554443333322222211111111111222222222222222333333333333333333333333333333
       78888899999999990007766655544433333222221111111111122222222222222223333333333333333333333333333
       77788888999999999888776665554443333322222111111111122222222222222223333333333333333333333333333
       67777888899999999988877666555444333322222111111111122222222222222223333333333333333333333333333
201-   6666777788889999999999888774445551144333322222111111112222222222222333333333333333333333333333333
       66666777788889999999999888776655554443333222221111111222222222222223333333333333333333333333333
       66666667777888899999999988877656555443333322222111111122222222222223333333333333333333333333333
       666666666777788889999999988877665554433333222221111111222222222222223333333333333333333333333333
       56666666666777788899999999988776655444333322222111111222222222222223333333333333333333333333333
1-     55566666666677778889999999998887766555443333322222111111222222222222333333333333333333333333333333
       55555566666667777888999999990887766555443333322222211111222222222222333333333333333333333333333333
       555555555666666666777888999999988876655544333332222211111222222222222333333333333333333333333333333
       555555555666666666677788889999998876655544333322222211112222222222222333333333333333333333333333333
       5555555555556666666677788899999998887665554433322222212222222222222223333333333333333333333333333333
-199-  55555555555555666666667778889999999988876655544333222222222222222222222333333333333333333333333333333
       555555555555555556666666677889999999988765554333332222222222222222222233333333333333333333333333333
       5555555555555555555556666667789999999987765543333222222222222222333333333333333333333333333333333
       5555555555555555555555556666667778889999998765544333322222222222233333333333333333333333333333333
       5555555555555555555555555555666666777889999987765543333222222222333333333333333333333333333333333
-399-  55555555555555555555555555555660066677789999998765544333322222222223333333333333333333333333333333
       44455555555555555555555555555556666667778899999988765544333322222222223333333333333333333333333333
       444444445555555555555555555555556666666778889999887655443332222222222333333333333333333333333333333
       444444444445555555555555555555555556666666778899998876554433322222222223333333333333333333333333333
       44444444444445555555555555555555555556666667788999998765443333222222222233333333333333333333333333
-599-  44444444444444444555555555555555555555556666667788999998765443333222222333333333333333333333333333333
       444444444444444444455555555555555555555556666667788999887655433322222223333333333333333333333333333
       4444444444444444444455555555555555555556666667788999876554333322222233333333333333333333333333333333
       44444444444444444444444455555555555555555556666677889999876554333333333333333333333333333333333333333
       4444444444444444444444444455555555555555556666677889999876554333333333333333333333333333333333333333
-799-  44444444444444444444444444445555555555555556666678899998765544333333333333333333333333333333333333333
       444444444444444444444444444444455555555555556666667889999876554433333333333333333333333333333333333
       44444444444444444444444444444444455555555555556666677889999876554433333333333333333333333333333333
       444444444444444444444444444444444455555555555556666678899998765443333333333333333333333333333333333
       44444444444444444444444444444444444455555555555566667789999876544333333333333333333333333333333333
-999-  444444444444444444444444444444444444444555555555566667889998765444433333333333333333333333333333333
       4444444444444444444444444444444444444444445555555556666778899998765554333333333333333333333333333333
       444444444444444444444444444444444444444444445555555566677899987655444444433333333333333333333333333
       4444444444444444444444444444444444444444444444555555556666778999876555444444443333333333333333333333
       44444444444444444444444444444444444444444444444455555556666778899876555444444444333333333333333333
-1199- 4444444444444444444444444444444444444444444444444555555566667899998765555444444444444444444444
        :   :    :    :    :    :    :    :    :    :    :    :    :    :    :    :    :    :    :    :    :
      -452 -252  -52  148  348  548  748  948 1148 1348 1548 1748 1948 2148 2348 2548 2748 2948 3148 3348 3548
```

5.4.3. Comparison of the methods with regard to error

The polynomial method for $N = 2$ and $N = 3$ is very sensitive to perturbation of the control points, and is therefore not very useful. The three point method shows the greatest numerical stability in most cases. When the focal length is obtained by iteration, the stability decreases, but the maximum residual error also decreases. The four point method is almost always better conditioned than the six point technique. This is not surprising because matrix (5.25) is a partial matrix of (5.27). The six point method is usually badly conditioned. This happens because (5.27) becomes singular when the ground is nearly flat, which was the case in some images. A check should be is made to see that at least one of the control points departs significantly from the plane of the others before this method is used. Iterative methods are numerically more stable than those which obtain their parameters through solving overdetermined systems of linear equations. It may be possible to obtain better conditioning of the equations of the four and six point methods using iteration afterward.

5.5. Ground height

5.5.1. Computing a digital terrain model

If maps with detailed contours are available, the heights at any point can be estimated with adequate accuracy interactively and much work is avoided. In the future, mapping services may provide the digitalised height data on which the contour map was based. But if these aids are not available, the images themselves must be used, or better still vertical mapping imagery of the same area.

We need a digital terrain model for methods other than the Bernstein or Weidner technique because we must be able to estimate the height at any point in order to solve the equations for the positions of orthophoto output points. The terrain model is a regular undulating surface approximating true surface relief, where possible passing through the control points. With the equations which determine this surface, the height at any coordinate point can be computed. The problem is to construct a regular grid of height values from height and position estimates of a number of irregularly spaced points.

The methods discussed in the specialised literature vary considerably in complexity. Where visible features of interest are not present in large numbers, and where the number of visible known ground points which can be used for constructing the digital terrain model is small, elaborate techniques do not produce much better results than simple ones. One can appreciate the terrain model problem in terms of a simple physical analogy which has an extremely complex analytical representation discussed in detail by Terzopoulos (73).

Suppose we have a set of isolated points, think of them as tent poles, whose heights are known and drape a tent over them. If the points do not differ too much in height, all of them will poke into the tent and produce a cusped surface. This surface is not very smooth and has a discontinuity of slope at the points where the poles touch the canvas. Now imagine that we make the substance of the tent stiffer until it becomes a thin metal plate which bends so that it touches most of the points or at least comes quite close to them. We connect the plate to the points with a spring at each point whose pulling force is proportional to the accuracy with which the point is assumed to be known. Thus well-defined points will pull the plate more than poorly defined points, figure 5.20.

If we don't know anything about the quality of the points we make all the springs equal in force. We allow the plate to be pulled toward the points so that the energy stored by the springs and in the plate is minimum. This makes a nice smooth undulating surface which has no unexpected bumps in it and no discontinuities. The calculations required

Fig. 5.20 A physical model for surface reconstruction using a thin stiff plate with restoring forces. From ref. 73.

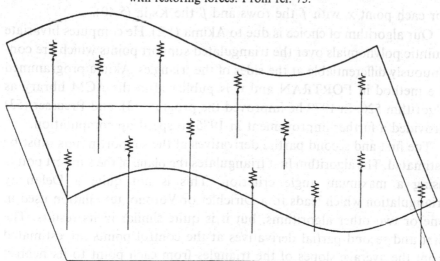

are formidable. We then use the mathematical description of the surface of the plate to find the height at any point on the plate and thus on our regular grid.

There are other methods discussed in great detail by Terzopoulos which allow redistribution of the spring forces to obtain a best fit of the plate to the points with the plate held under tension and with the redistributed forces giving an estimate of the quality of the point in terms of its neighbours. When a strong force is needed to pull the plate to the point, then it is evident that the point is not compatible with its neighbours and may be in error, with the strength of the pulling force a measure of the error.

A very much simpler approach is the use of a weighted average of the data points, where the weights are chosen as a function of distance:

$$\hat{Z} = \sum_{i=1}^{N} \lambda_i Z(x_i) \qquad \sum \lambda_i = 1 \qquad \lambda_i \propto \omega(d(x, x_i)) \qquad (5.153)$$

Typical choices for ω are d^{-r}, $e^{-\alpha d}$, whereby the first gives more weight to distant points. The weights are usually made equal to zero beyond a certain distance. This approach is also used in many contouring programs. The technique is badly influenced by clustered data points. McLain (46) therefore proposed a distance weighted least squares approach which requires solution for parameters β of a function as in (5.35):

$$\min \sum_{i=1}^{N} \omega(d(x, x_i))[Z(x_i) - {}^tf(x_i)\beta]^2 \qquad (5.154)$$

for each point x_i with f the rows and β the Ks in (5.40).

Our algorithm of choice is due to Akima (1,2). He computes bivariate quintic polynomials over the triangulated support points which are continuously differentiable at the sides of the triangles. Akima programmed the method in FORTRAN and it is published in the ACM library as algorithm 526. In 1979 he improved the program (3), and Preusser (53) provided a further improvement in 1985 to speed up computation.

The first and second partial derivatives at the support points must be estimated. The algorithm first triangulates the plane of the support points using a maximum angle criterion. This is not quite a Delaunay triangulation which leads to a Dirichlet or Voronoi tessellation used in one or two other algorithms, but it is quite similar in its results. The first and second partial derivatives at the control points are estimated from the average slopes of the triangles from each point to its nearest neighbour points. When there are relatively few control points, this computation is very rapid. In the triangles, a skew coordinate system is used to compute the weights required for interpolation of the quintic polynomial which results from the system of equations used. For a clearer explanation of the details of the algorithm than that in Akima's original publication, see Preusser.

The technique used for fitting gives a continuously differentiable surface within the triangulation. This method has been used in practical programs, and it has proven to be quick and stable for the small number of points usually available. It is invariant under rotation around the Z axis, the support points may be randomly distributed, and if the support points lie in a plane, then the computed terrain function is this plane. The method allows extrapolation, that is, for points outside the convex hull of the support points, the height value can be estimated. The technique is also local in that only the nearest support points influence the interpolation value of a given point. This has the enormous advantage over global methods that the technique does not become computationally intractable with an increase in the number of support points and that oscillations between the support points are avoided. The algorithm does not take errors at the support points into account. Since these are usually not known in this application, this does not seem to be a serious objection to its use.

5.5.2. Other terrain models

There are many other ways of constructing a three-dimensional surface to fit random support points. For the sake of completeness, a few of these are mentioned here, since some are widely used in conventional photogrammetry. Other algorithms, like the weighted distance averaging used in contouring packages, for example, SYMAP might also be considered. The statistics of fitting surfaces when the data at the support points contain errors is discussed in detail by Ripley (55) with criticism of the simpler techniques. Sibson (63–8) uses a Dirichlet tessellation, the dual of Akima's near-Delaunay triangulation, with weights assigned to each point proportional to the area of the tile in which it lies. This compensates well for clustered data points. The method is also very robust with respect to errors at the data points. Ebner (16) interpolates to a regular grid from irregularly distributed support points using bilinear interpolation. This is not differentiable over the entire surface. Kapokakis (36) uses a deterministic model for the large scale ground form and uses a stochastic model for local variation. Kriging, elaborated by Matheron (43), uses error information at the support points to produce a maximum likelihood surface. The added precision which this gives is probably not worth the very large and complex program required when very few support points are available and little is known about the magnitude of errors. Haas and Viallix (25) give a much more concise and clear explanation of the method than that given by Matheron. Leberl, Kropatsch, and Lipp (42) describe a method for obtaining the heights of a regular grid from the digitalised contours of a map. Inoue (34) gives a very extensive survey of the literature for this kind of computation, and suggests an approach which approximates the surface using cubic *B*-splines and least squares which allows for errors at the support and data points.

The subject is a favourite with photogrammetrists, geophysicists, quantitative geographers, and all those having to deal with irregularly spaced data. Only a few of the published techniques have been mentioned here. Many more methods will undoubtedly be published in the future, but for a small number of points whose error parameters are not known, the technique described here is adequate and rapid enough.

5.6. Digital photomosaics

An archaeological site is usually photographed from a number of viewpoints in order to obtain best contrast for details which are optimal from different angles and altitudes. In addition, it is good practice to make several pictures from a relatively high viewpoint in order to have as many landmarks as possible which are also visible in a map. For large complex sites, this process will probably be repeated over many years. Archive pictures will have been taken with a variety of cameras whose properties are frequently unknown, and different film types and formats may have been used. Up to several hundred images may be on hand, all of which contain information of varied quality. Over a long period, constant features in the landscape will have changed as roads are relocated, field utilisation is modified and ownership boundaries change, and houses or other structures are built or torn down. Crops differ in appearance even when the picture sequence has been made within a given season. The problem is, given all these factors, how to combine the information in a single global image which contains all the archaeological detail at once. A photomosaic is required.

By hand one would cut pieces out of verticals to paste together. Cutting along high contrast features like roads and avoiding cuts across them minimises mismatches. Where high contrast features are not present, the image edges are torn so as to produce an irregular edge, which when pasted together minimises apparent gray level differences. It is quite easy to construct a set of interactive computer assisted techniques which 'paste' an image together, with the user guiding the 'scissors'. Such a method is not faster than doing the job by hand, but the image formed is in digital form and can be readily manipulated. Simulation of tearing in low contrast regions is not quite so readily accomplished digitally.

The matching problem can be avoided if the grayscale imagery is transformed to black and white by use of an edge extraction algorithm. This is an objective alternative to the subjective alternative of interactive hand tracing of the archaeological features. However, edge extraction algorithms are scale sensitive, and features of different line thicknesses will not be transformed in quite the same fashion as a skilled observer would perform. A further problem lies in the fact that the geometric correction to an orthophoto always contains some element of error, depending on the quality of control information and camera, so that extracted features from multiple images cannot be precisely aligned. This will produce unsightly multiples of features in areas of overlap which will have to be removed by hand.

As shown in figure 5.21 two orthophotos made from low obliques can have an overlapping area which may have from three to eight sides. If

Fig. 5.21 The six possible types of overlapping areas for two orthophotos made from low obliques.

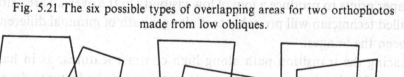
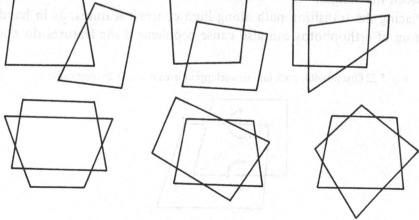

the terrain is not flat, then these sides may be curved rather than straight. Methods given by Milgram (47,48) for making mosaics from satellite imagery assume that the overlap area is rectangular. In this approach, the magnitude of the gray level difference across a boundary is reduced by a number of methods. The grayscale histograms in the overlap area are shifted so that the average values correspond. Göpfert (21) brings the grayscale transfer functions for both images in the overlapping area into nearly complete alignment. This method cannot be used for archaeological aerial photographs since fields appear to be structured differently when seen from different angles, and artificial edges would be produced when the grayscale histograms are brought into register. The method is absolutely useless when the pictures are made at different times of the year. This also applies to Milgram's techniques. In Milgram's method the points of minimal difference between the images chosen as the path of the join in each line in the overlap zone. Finally, the pixel values in the overlap area are averaged using a linear ramp weighting function whose mid-point is centred on the path. If the beginning and end of the path is given as shown in figure 5.22 then gray level artefacts may be expected at the points marked with arrows where the transition occurs between information from one or both pictures. This situation can be improved upon by using the path shown in figure 5.23. In any case, due to the averaging of the images in the transition zone, sharpness will be reduced. The method is of no use for pictures made at different seasons, where gray level differences are considerable. There is one similarity between Milgram's methods and the hand tearing and pasting of paper prints to produce a join of low visibility in the overlap area since a skilled technician will probably tear along a path of minimal difference between the images.

Placing the transition path along high contrast features, as in hand pasting of orthophotos can also cause problems if the features do not

Fig. 5.22 One possible path for an overlapping area in a two image mosaic.

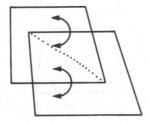

lie in favourable positions as shown in figure 5.24. Abrupt transitions of this type can be smoothed using the technique described by Peleg (51) but the computational cost of his approach is so great as to make it impractical for all but the largest machines. Peleg uses an iterative relaxation technique which leads to a smooth transition from one picture to the other, without blurring of details in the overlap zone. The method

Fig. 5.23 A better path for the overlapping area.

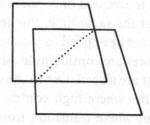

Fig. 5.24 The transition path can only be placed along high contrast features in an unfavourable position. At the transitions, the difference in a field gray level for images taken at different dates appears sharply.

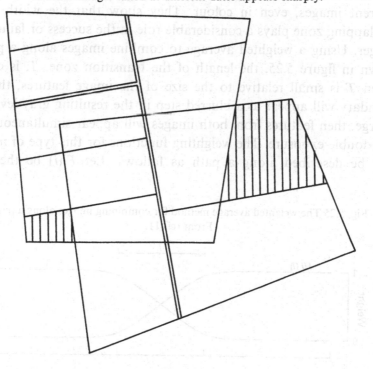

requires several hundred passes through the data before convergence is achieved.

One possible but untested approach might be to use a path which takes the pixel from the picture which was closest to the camera under the assumption that this pixel contains the most information. Some kind of smoothing would be imperative in this case. By extension, it would be sensible not to construct orthophotos and then merge, but rather combine the operations, so that in the zone of overlap, pixel interpolation in the orthophoto calculation is carried out on only one image along the path. This would have the considerable advantage that the orthophoto interpolation operation is executed only once in the overlapping region. For pictures made all at the same time, this technique is probably the most economical. It cannot be applied to pictures made at different times where gray level differences are considerable. Since it is unlikely that the edges of highest contrast are also closest to the camera, the method may produce visible anomalies where high contrast linear features are not well aligned at the point where transition from one image to the next takes place.

A most sophisticated technique for mosaic construction is due to Burt and Adelson (11), the only method which demonstrably merges highly different images, even in colour. They show that the width of the overlapping zone plays a considerable role in the success or failure of a merger. Using a weighted average to combine images along a path as shown in figure 5.25, the length of the transition zone, T, is critical. When T is small relative to the size of the image features, then the boundary will appear as a blurred step in the resulting gray level. If it is large, then features from both images will appear simultaneously, as in a double exposure. The weighting functions for this type of merging may be described along a path as follows: Let $Fl(i)$ be the pixels

Fig. 5.25 The weighted average method for combining images along a path. From ref. 11.

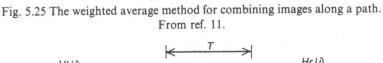
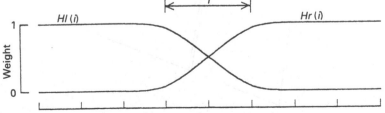

of the image on the left, $Fr(i)$ those on the right. Let $Hl(i)$ be a weighting function which decreases monotonically from left to right and $Hr(i) = 1 - Hl(i)$. Then the pixels of the merged image are given by:

$$F(i) = Hl(i - \hat{i})Fl(i) + Hr(i - \hat{i})Fr(i) \qquad (5.155)$$

To overcome the difficulties due to the width of the transition zone, Burt and Adelson propose that the image be decomposed into a set of bandpass component images, and that a separate weighting and transition size be used to combine each component. Then the components are recombined to create the final image. They call this method a 'multi-resolution' spline. The component images use the hierarchical data structure proposed by Tanimoto and Pavlidis (72) created using Burt's algorithm (9,10) for constructing fast near-Gaussian, low pass filtered data discussed in chapter 4. The hierarchy is diagrammed in figure 5.26 for one dimension. Five points are used at each level and are convolved with a set of weights to obtain the next level. Burt shows that weights of 0.4 for a, 0.25 for b, and 0.05 for c produce the desired near-Gaussian property. The convolution is repeated at every second point, so that the reduced data of the next higher level in the hierarchy contain only a quarter of the number of points from the lower level. The convolution is separable and partially redundant so that with appropriate programming, only seven multiply–add operations per original pixel are required to produce the full low pass set of images as described by Burt.

The bandpass images required are constructed by subtracting each level in the hierarchy of low pass images from the next higher level. Since the number of points at this level is only half that of the lower level, linear interpolation is used to obtain the missing values. The resulting bandpass component hierarchy is called a Laplacian pyramid by Burt and Adelson. Since the convolution extends beyond the edges of the image at each level, artificial values are assigned by reflection and

Fig. 5.26 A one-dimensional representation of the multiresolution operation. From ref. 11.

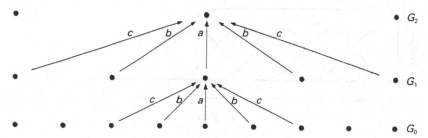

inversion such that if $G_1(0, j)$ is an edge pixel on the left edge, then the two additional values are:

$$G_1(-1, j) = 2G_1(0, j) - G_1(1, j)$$
$$G_1(-2, j) = 2G_1(0, j) - G_1(2, j)$$

(5.156)

at the edge. The actual merging or splining operation is governed by the weighting function at each level implicit in the process as shown in figure 5.27. These weighting functions are implicit in the reduction operation and need not be computed explicitly.

The final merge operation for the two images follows a simple series of steps:

(1) Laplacian pyramids *LA* and *LB* are constructed for images *A* and *B*.

(2) Laplacian pyramid *LS* is constructed by copying all points from the non-overlapping parts of *LA* and *LB* to *LS*. Nodes along the centre of the overlapping area are averaged, and along the two lines adjacent to the centre, weights of $\frac{1}{4}$ and $\frac{3}{4}$ for the respective images are used. If colour images are involved, the operations are carried out separately for the red, green, and blue components.

(3) The final result image is created by adding the levels of *LS* with interpolation at each next higher level to obtain an equal number of points at each level.

This procedure introduces no distortion in the images, and since as shown in figure 5.27 the equivalent weighting function is twice the sample distance on each side of each sample point, this is a suitable transition distance for the spatial frequencies at each level in the pyramid. The method is not as computationally intensive as Peleg's technique. It has not been tested on archaeological imagery so that possible disadvantages if any, are at present not known. But since highly differing unrelated images can be merged successfully, it is probably the method of choice.

Fig. 5.27 Equivalent weighting functions at a given level which sum to unity on the left and zero on the right. From ref. 11.

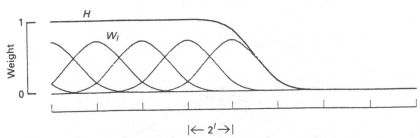

5.7. Miscellaneous practical recipes

5.7.1. Control point methods

The coordinates of common points in a picture series can be measured using conventional photogrammetric instruments, but since the pictures must be scanned anyway when one has an image processing system, it is more sensible to do this interactively on the display. The interactive devices which allow one to position a cursor directly on objects of interest on the screen most precisely are either the graphic tablet, the trackball, or the mouse. The displayed image must have reasonably high resolution if positional accuracy is to be obtained, but even with low resolution (512×512) screens, there are some simple tricks which allow one to obtain the required measurements. The scanned image provides a definite reference in each image line to each image point. When the screen cursor is properly positioned this point is identified precisely. If the picture is scanned at much higher resolution than can be displayed on the screen, then positioning can first be done in a roughly sampled image showing the whole picture at reduced resolution. Then a window in the full resolution image is displayed showing the area around the control point which can be measured to full scan resolution. If desired, the pixels and lines in the window can be doubled, so that effectively the cursor points to half a pixel in the original. This does not raise fundamental accuracy, but it simplifies positioning a cursor on small objects or sharp edges.

A control point entry program must contain facilities for fitting lines to visible linear features and constructing intersections by calculation with estimates of error. Field roads and boundaries are not always very straight and crossings of field roads ill-defined in images, but sharply drawn in maps. Hence it is usually useful to include a means for connecting a set of points along a road with the best fitting spline and constructing intersections with several lines, rather than trying to identify the exact centre or corners of road crossings in pictures directly. A considerable improvement in control point determination is possible compared with purely interactive placement of a cursor over a supposed point.

5.7.2. Hand tracing of archaeological features

Tracing of site features either from pictures on a display or with a graphic tablet suffers from a number of problems. Unless one is very skilled, minor random hand movements are repeated in the output, often magnified in the background parts of the image where small details correspond to rather large distances on the ground. If the features are known to be regular, then it is better to mark salient points and pass lines or smooth curves through them to draw the plan. Methods for passing smooth lines and curves through random points are treated widely in the literature, with a survey by Dierckx (15). Hand tracing or point determination is subjective in that the choice of lines or points depends on the user's judgement, but it eliminates any need for feature extraction algorithms and may often be more satisfactory. One problem is that no two users will produce exactly the same results.

5.7.3. *Trade-offs between scanning and computation times*

In a small machine it may be advantageous to trade storage space and scanning time against calculation time. If the image is scanned at higher resolution than needed, then bilinear interpolation gives adequate results, and if resolution is higher still, then nearest neighbour choice is good enough. Since scanning times and storage requirements rise as the square of resolution, the compromise achieved with scanning and computing time has to be weighed quite carefully against the alternatives before deciding on which method gives the lowest overall time. Costs of computer memory and fast central processors have fallen more rapidly than those for high resolution picture scanners, and since the latter cannot be speeded up very much, there is a great advantage in having as much of the picture as possible in a large central memory and using a fast processor to treat it even if the number of calculations is large. With a fixed budget, a great deal of thought must be given to questions of storage space, computing times, resolution of scanning and display systems, and useful algorithms before any decisions about acquiring equipment and writing programs for geometric correction are made.

5.8. Maps and picture processing

5.8.1. *Map availability and scales*

Picture processing operations on aerial or excavation photographs are usually meaningful only if afterwards the geographic context is established relative to a map base. Highly accurate photographic reproductions of existing maps can usually be obtained on film from government mapping agencies. If map negatives can be had, simple contact printing on film suffices to produce a transparent overlay which can be placed over geometrically corrected archaeological data reproduced at the same scale. This approach allows a 'quick look' at data in an area and doesn't tie the archaeology to a given map update.

A typical setup uses a 25 × 25 cm film for the overlay which reproduces a positive transparency of a 1:5000 base map at 1:10000. Enlargements up to 1:2500 can be used if desired. Most parts of the world are not mapped at scales better than 1:25000 and less developed regions can be counted as fortunate when maps at 1:50000 are available. Operations using pictures together with maps at scales smaller than 1:10000 are not usually meaningful, since the image data will be too small to be seen. In this case, it is better to select symbols for archaeological sites and enter these in the maps. Most archaeological maps made at scales of 1:25000 or less use symbols to designate sites of different kinds. Sometimes only a number is used which must then be looked up on an index to the map sheet to find out what kind of site is located at a given place. However, a symbol should be chosen such that a well-defined centre point is visible in the map print, and this point should be placed as accurately as possible at the centre of the site coordinates. When large scale maps are not available, the only alternative is to obtain vertical aerial cover and construct orthophotos from it by conventional photogrammetric techniques when accurate site recording is required.

5.8.2. *Processing scanned maps*

Scan resolution and fine line width
Maps may be analysed as if they are special kinds of images. The map negative can be scanned in the scanner at high resolution. Here, a trade-off is required between the fineness of the lines to be reproduced, the storage requirements for scanned maps, and the time needed for scanning. In typical maps the finest lines in the originals are about 200 μm wide. At a typical 1:2 reduction supplied for scanning, these become 100 μm. Scanning at 50 μm will adequately reproduce these lines without undersampling. It is evident that image processing of maps can only be carried out using a high resolution scanner unless work is confined to a small area of the original. At high magnification, two picture elements per line width may produce a slightly jagged appearance on diagonals, but this is not visible at normal viewing distances.

If large scale maps are available, the same scanner should be used for maps and pictures so that any geometric error in the scanner affects both to the same extent. Scanning the map produces a grayscale image, even though the original data are usually only black and white. Scanning of colour maps requires equipment which is at the limit of the current state of the art. It is more feasible to have black and white negatives made from the original map printing plates by the mapping agency. Colour copies on paper suffer from all of the problems discussed earlier and cannot be used quantitatively.

A typical map scan at the author's installation takes about 20 min due to the physical rotation time of the drum scanner used. The map negatives contain all pertinent information within a 225×225 mm area. Each scan line from a 1:5000 map normally used contains 4608 picture elements at 50 μm each, or 23.04 cm. The border information which contains the map number and the coordinate grid data is included in the scan to allow visual identification on the display afterward and to help in correcting errors in map alignment relative to the scanner axes. It also aids in correcting distortion in the reproduction of the original map for scanning, if present.

The number of pixels required in a scan line is installation and map dependent. The magic number 4608 is used on the author's machine because it is the next largest multiple of 512, the size of the smallest disk block in bytes which will cover a map of the type used from border to border. Large scale maps elsewhere will have different requirements. Increasing the scan resolution to 25 μm quadruples the scan time in a mechanical scanner. The storage requirement rises to an excessive 85 megabytes unless the data were compressed during the scan. But it is not sensible to compress the raw scanned data until they have been checked by observation on a high resolution display.

Map border information, and especially the straight lines which show the map edges should always be included in the scan if data are to be recorded which may run over the map boundary. In this case, the scan must be accurately oriented so as to be parallel and at right angles to the map sides. If a drum scanner with punched alignment holes for the scanned transparency is available, then the alignment can be most readily carried out when punching the holes. Scanners made for the printing industry usually have a transparent cylindrical drum or are flatbed designs. Scanners of this type or those using CCD arrays require modification for alignment if maps are to be scanned, or software compensation is needed. It will be necessary to rotate the map a small amount by calculation. The rotation is a simple sine–cosine coordinate transformation. For small angles, it can be carried out by table look-up with little loss of precision.

It has been observed that photographic reduction of maps made by a state mapping service sometimes is not carefully carried out. In particular, apart from blemishes (see below), the reduction camera is not always carefully adjusted, so that the map image itself apparently suffers a slightly projective transformation. Furthermore, variations in film base thickness produce slightly different length results in axial and circumferential directions when using a drum scanner. If the scanner light source is inside the drum, the film must be scanned emulsion side outward so that diffusion effects in the film base do not affect sharpness. However, the scanner circumferential digital shaft encoder calibration usually assumes that the scanned surface lies in contact with the drum. Thus the film thickness adds to the length of the scan line for the same number of shaft encoder impulses, and the circumferential size of each pixel is slightly greater than the axial. It is difficult to convince the mapping services to use a single type of film or adjust their reproduction camera carefully when reordering.

These errors can be compensated for using data obtained from an interactive map coordinate entry program. The compensation is only necessary if information from more than one map must be used in rectification, such as when the picture or control points lie partly outside the area of a single map. As long as all the information lies within a single map, and important parts of the rectified picture do not extend beyond its borders, it does not matter if control information is slightly distorted, for the picture will also be correspondingly distorted, and errors in the final result invisible. In the special case of the maps available in the author's laboratory, the fiducial marks at the edges of the map have known geographic coordinates. They can be used in a least squares calculation of the coefficients of a projective transformation. These, in turn, can be used to obtain the true coordinates of any point within the map area to subpixel accuracy. A similar approach is possible if one is forced to use paper maps, although compensation of local stretching is not possible with this simple technique.

The grayscale map output from the scanner must be inspected and treated if necessary to remove dirt spots, scratches, etc. using the algorithms discussed in chapter 4. If fine line structure is not to be degraded, these operations must be performed on the grayscale image rather than on a black and white version.

The scanned map is best displayed on a high resolution 1024×1024 pixel display system for inspection. At the same time the map number and other relevant data are read off the display and entered into the map catalogue, in a standard database system. A resolution of 512×512 is adequate for displaying parts of a map or a reduced overview. If the original scan has a resolution which is from four to ten times the resolution of the display, then a sampled image must be used for orientation when picking out a window. Creation of this by selection of every nth pixel and scan line will produce a very unsatisfactory appearance since the finer lines which will be grossly undersampled. Scale reduction using the geometric transformation resampling technique described above is ideal but time consuming, given the large number of points. An unpublished algorithm due to Weidner for making a small scale file from a large map image has been used for many years with great satisfaction.

The high resolution input map is processed checkerboard-wise with non-overlapping square sizes equal to the desired reduction factor. For each pixel in the original image above a threshold, where n is the reduction factor, the next lowest integer to $225/n^2$, is added to make up an output pixel which will represent the square. Thus, for example, if a map has been scanned at 4608×4608 and a 512×512 output picture is desired, then 3 is added for each pixel in the original above the threshold for each of the 81 pixels in the 9×9 square which is to appear as a single pixel in the output. The grayscale output pixel is thus representative of the pixels in the input, and even the finest lines make some contribution.

The full grayscale map image is far too wasteful of space to use for further processing if eight-bit pixels are retained. A map scanned according to the standards described above results in a grayscale image containing 4608×4608 pixels or 21 megabytes. Many archaeological sites are unhappily located at the intersection of four maps, and when adding information to a given map it is usually necessary to have at least three neighbouring maps in mass storage. At 21 megabytes per map, this would represent 84 megabytes. Therefore it is of interest if the grayscale maps are coded and compressed for permanent as well as on-line storage. If it is desired to keep all of the map sheets in an area on fast access mass storage, this can be a formidable requirement even for the best equipped installations. Since most maps contain only black and white information, converting the grayscale scanned map to a binary image reduces storage by a factor of 8 at little computational cost. The map images are thresholded so that only black and white information results after dirt removal and minor rotation if needed. A thresholding scheme is used which works quickly and adapts to slight variations in background gray levels:

$$O(x, y) = \quad 0 \text{ for } \alpha \left[I(x, y) - (1/(N^2)) \sum_{i=1}^{N} \sum_{j=1}^{N} I(i, j) \right] + \beta > 1$$

$$O(x, y) = 255 \text{ otherwise} \qquad \qquad \qquad (5.157)$$

where $O(x, y)$ is a black or white output pixel, $I(x, y)$ is a gray valued input pixel, and $I(i, j)$ is a pixel in an $N \times N$ window centred on $I(x, y)$, α is a gain factor, and β is a threshold level chosen by the user.

The result is a single bit of information per original picture element. These are stored eight at a time in a new byte, so that 576 bytes represent a bit-coded line from the original map at 4608 pixels per line in the sample used. These are grouped together and read or rewritten with a single disk access. The bit representation gives compression factor of 8 over the grayscale image, but still allows for rapid access to any area in the map. Unless the map negatives are unusually free of defects, it is not sensible to carry out this conversion to binary during the scan process, as is usually done in document scanners based on modified laser printers.

5.8.3. Map storage reduction through coding
Coding theory
Data storage requirements can be reduced by appropriate coding. Coding for compression of binary data by removing redundancy is the subject of a large literature, with surveys by Huang (32) and Hou (31).

If S is a source of binary data with N independent states, and the *a priori* probabilities of each of these states are $P_1, P_2, \ldots, P_N$, then according to classical Shannon theory, the information content of the source is:

$$H(S) = - \sum_{i=1}^{N} P_i \log_2(P_i) \qquad (5.158)$$

with the units in bits per state, so $\log_2$ is used. These states are encoded with code words having lengths $l_1, l_2, \ldots, l_N$. The average code word length is thus:

$$T = \sum_{i=1}^{N} P_i l_i \qquad (5.159)$$

The efficiency E of the code is defined as the ratio of the information content to the average word length:

$$E = H(S)/\bar{l} \tag{5.160}$$

and its reciprocal is the compression ratio. The most efficient codes have lengths inversely proportional to the probability of the occurrence of a state. One such code is the Huffman code (31). Such codes are of variable length which makes decoding a computationally time consuming operation.

Other codes which are suboptimal are easier to compute and decode. Some are based on lengths of runs of pixels in a scan line with unchanging polarity, and others based on short two-dimensional blocks ranging over several scan lines. In the runlength code, the binary image is considered as having alternating statistically independent runs of black (B) and white (W) pixels. If $P_i(W)$ is the probability of a white runlength $l_i(W)$, and N the number of white runs, then the average white run length is:

$$\bar{l}(W) = \sum_{i=1}^{N} l_i(W)P_i(W) \tag{5.161}$$

The information in a white run is:

$$H(W) = - \sum_{i=1}^{N} P_i(W) \log_2[P_i(W)] \tag{5.162}$$

and similarly for black runs. If black and white runs are statistically independent, the average runlength is:

$$\bar{l}_{BW} = \frac{\bar{l}(W) + \bar{l}(B)}{2} \tag{5.163}$$

and the information per run is:

$$H(B, W) = [H(W) + H(B)]/2 \tag{5.164}$$

and the information per pixel in a run is:

$$H_{\text{pixel}} = [H(W) + H(B)]/[\bar{l}(W) + \bar{l}(B)] \tag{5.165}$$

The maximum limit for compression with a given set of statistics is then:

$$E = 1/H_{\text{pixel}} \tag{5.166}$$

Practical experience has shown, however, that the probabilities of black and white runs in maps are not completely independent, and therefore this theory only holds approximately. It can nonetheless be used to guide our thinking.

Runlength coding

A very simple runlength compression scheme used for long term archiving of maps has been described in ref. 62. Statistical analysis of typical 1:5000 maps has shown that in the neighbourhood of built-up areas, a compression of about three, compared with the storage required for the binary image can be obtained by the use of a simple 16-bit runlength code. A small improvement on this figure can be obtained by dividing the 16-bit word into two bytes and using the first bit of the first byte as a flag to indicate whether the following byte is to be used together with the first or not. Runlengths of up to 128 decimal or 32 767 decimal are thus recordable in one or two bytes. Because of this simple structure, coding and decoding are very rapid operations for almost any modern computer and the coding efficiency tends to be about 3.2–3.5 times for typical maps, a significant improvement on a pure 16 bit runlength code. Typically it takes about two minutes of processing to code or decode a 21 megabit map.

However, this simple two byte method does not take the joint probability of black and white runlengths occurring together into account, and it ignores the fact that white runlengths for map negatives are much shorter than black runlengths. Statistics for a typical 1:5000 map scanned at 50 μm resolution from a 1:2 reduction negative show that the histograms of the logarithms of the frequencies of black and white runlengths are different as shown in figure 5.28. The logarithms (to the base two) are used because this corresponds to the significance of the bits in a storage word. For white, about 97% of all runlengths are less than 16 picture elements long and for black, about 88% of all lengths are less than 127, with the decrease in log runlength frequency nearly exponential. Histograms of maps in agricultural and built-up areas are similar. A few large peaks in the histograms are due to either the frame around the map or the field boundaries which have a common line thickness.

To obtain a reasonably efficient code for this case, it seems sensible to use the joint black and white probability distribution. This allows the construction of a somewhat more complicated code which can still be processed rapidly giving a further improvement over simpler methods. The code algorithm is based on the fact that a single byte can be used to count up to 127, reserving one bit as a flag to signal the possible presence of a second byte for longer lengths to follow. This is sufficient for most of the black runlengths. Since the white runlengths are shorter when using map negatives (the data lines being fine and not parallel to the direction of scanning), four bits are sufficient to describe most of them (runlength up to 16). A very efficient code takes into account the fact that for black and white runs taken together, three bits for white and five bits for black account for well over 70% of all occurrences, and four bits for white and nine for black for nearly all the rest (62). Using a 16-bit word length, the first three bits in the first byte may then be used as a flag to signal the possible presence of a following byte which in turn contains the rest of the nine bits for black. The rare case of a very long combined black and white runlength with up to 24 bits uses five bytes with the first 16 bits turned off. Most black and white runs in busy parts of a map are recorded in one byte and the remainder in two. The rare five byte long runs are usually present only on the upper and lower borders. The improvement over the code without joint probability is marked, giving a total compression between 4.5 and 6.1 for some of the maps used.

Compression is less successful if the map contains contour lines. In this case, the efficient code gives compression ratios of 3.8–4, while the

Fig. 5.28 The statistics for the $\log_2$ frequencies of (*a*) a black and (*b*) a white runlength for a 1:5000 German base map in a mixed agricultural and village area.

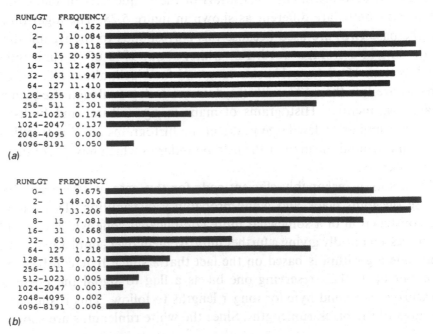

```
RUNLGT    FREQUENCY
  0-     1   4.162
  2-     3  10.084
  4-     7  18.118
  8-    15  20.935
 16-    31  12.487
 32-    63  11.947
 64-   127  11.410
128-   255   8.164
256-   511   2.301
512-  1023   0.174
1024- 2047   0.137
2048- 4095   0.030
4096- 8191   0.050
```
(*a*)

```
RUNLGT    FREQUENCY
  0-     1   9.675
  2-     3  48.016
  4-     7  33.206
  8-    15   7.081
 16-    31   0.668
 32-    63   0.103
 64-   127   1.218
128-   255   0.012
256-   511   0.006
512-  1023   0.005
1024- 2047   0.003
2048- 4095   0.002
4096- 8191   0.006
```
(*b*)

simpler code gives about 2.7. This means that a typical 1:5000 map can usually be stored in less than 600 kilobytes. That is of considerable practical importance. For smaller systems which do not have magnetic tape units, it allows storage of a map on a single, high density, double sided, industry standard floppy disk. It means that map processing is possible on one of the larger personal computers, provided that a display and scanner with sufficiently high resolution are available. A geographically organised database system can thus be implemented at relatively modest cost in hardware provided that a high resolution scanner and output device are available.

Problems with runlength codes

Runlength codes of all types place no upper limit on the maximum number of bytes of coded data which may represent a line of a scanned map. In 'busy' city areas, where the runlengths may be very short, coded lines may actually be longer than an uncoded bit representation. There are several possible solutions to this problem. The simplest is to keep track of the number of coded bytes in the output line and when it rises to equal that of a bit coded line, switch over to straight bit representation for that line and use an additional two bytes at the beginning of every code line to tell the decoding program the number of code bytes to expect. For example, using a map scanned at a resolution of 4608 pixels, reduced to binary by thresholding, a scan line can be contained in 576 bytes. When, during the coding process, the number of coded bytes reaches 576, the coding program must switch over to straight bit representation or the coded line will actually be longer than the decoded input. This happens only a few times in a typical map. This strategy gives a significant improvement in city areas surrounded by open countryside.

Another strategy makes use of the fact that for very 'busy' areas a block coding scheme is much more efficient than a runlength code. With two data lines in storage, Kunt's code (40) seems to be a suitable simple choice. Three by two bit blocks are observed. For map negatives if all are black, the most frequent case, a zero is recorded (compression 6:1), if all are white, two ones are recorded (compression 3:1) and if mixed pixels are present, a one and a zero prefix is followed by the actual bit pattern in six bit (loss of compression 6:8). This code requires little computation if implemented via table look-up for there are only 2^6 possible bit arrangements in the small blocks. Decoding can be similarly carried out. The initial value at the beginning of a code line tells the decoding program whether or not a runlength or a pair of Kunt coded lines follows.

Decoding and map handling utilities

The decoding program is used to convert back to black or white pixels so that maps can be reproduced on a film writer which has the same resolution as the scanner. The decoded grayscale maps are needed for merging true picture information, corrected and superimposed on the map. They are also needed for the display system. There must also be conversion programs for going from bit coding to runlength coding and the reverse. Programs are also required for merging up to four maps by removing unwanted edge information to produce a new map at their junction so that sites which extend beyond a given map sheet can be properly placed on up to three neighbouring sheets. Experience has shown that the maps may vary in size by a few pixels, and that it is necessary to compute ideal map sides by measurement and least squares fitting of lines to a number of points on each side if maps are to be joined, or if relative pixel coordinates are needed outside the limits of a single map.

5.9. Incorporating non-pictorial information in images

A very large portion of information about archaeological sites comes from sources which are quite conventional in nature. These include stray finds, orderly excavations, field walking, observation of large construction sites etc. They are recorded in a large number of archive sources in most establishments, with no concordance between them. In order to use this information in the picture and map system, it must not only be entered into a database system, it must be standardised and compressed. Then the database must be coupled to the picture processing system.

5.9.1. Storing and finding political boundaries and place names

A particular problem is raised by the place names used in almost all of the sources. Communal boundaries have changed repeatedly during the course of time, and older records often use names which are no longer current. This problem can be solved by obtaining a digitised version of the current communal boundaries together with the currently valid names. A special set of programs transforms these into a database which can be queried by the data entry program. When the user enters the geographical coordinates of a site, the program automatically computes the name of the current political unit in which it is found. If the site occupies an area extending over several such units, then the one where the centre lies is recorded, together with a flag signifying that the boundary has been approached. The search algorithm for the communes makes use of modern techniques in computer graphics for determining whether or not a point lies inside or outside a complicated reentrant polygon as described by Kalay (35). The published version of the algorithm makes use of a long known principle due to Euler. If a ray is extended to infinity from a point, the ray will intersect the boundary of an enclosing polygon an odd number of times. If the number of intersections is even, then the point is outside the polygon.

The practical implementation of this technique is quite straightforward. The digitised version of the communal boundaries must contain the coordinates of all the vertices, the names of the political units associated with them and the number of vertices in a given unit (which may be added after digitalisation). The coordinates for a given polygon must be sorted in ascending order. This data can be stored in an indexed database file. In addition, the lower left and upper right hand coordinates of the rectangles which contain each communal polygon are also stored, and two of the file's indices are sorted by these coordinates. Given a coordinate pair, this allows a quick rough search to be made among candidate polygons by choosing those which contain the point. Usually there are less than half a dozen candidates.

The actual implementation of the Euler algorithm must take a number of degenerate cases into account, namely polygons with coinciding edges, self-intersecting polygons and points exactly on the border of two polygons. A Fortran implementation is given in section 5.10.

5.9.2. The geographic data header

The result from the data entry program is a short record which contains coordinate information, coded data on the type and age of a site, communal names, older names if referred to in the sources, archive information, dates of recording, and dates in the sources etc. Preprocessor codes are decoded automatically and clear text is presented to the user. A suite of programs for servicing and changing the codes as required is also extant. The report generation feature of the database management system is used to produce catalogues classified as desired for administrative purposes. A further suite of programs extracts data from the reports and prints labels which can be stripped from the carrier and pasted onto conventional index cards for re-incorporation in the source files, or on to the reverses of air photos, so that the data in the database and on the pictures is the same.

5.9.3. Scanned symbols and fonts

In addition to the monument database and the on-line corrected pictures which have been merged into decoded grayscale maps, symbols and text must be included. Suitable fonts or symbols may be photographed and scanned. The scanned font image is processed interactively to select the letters. Each letter is recorded as a pattern of bits in an indexed file with the index corresponding to the letter. The bit pattern is recalled by a program which searches the data-base file for text keys at a given coordinate position, and the letter written into the image. The merged data is passed to the film writer for recording. Fonts do not scale well, becoming blocky in appearance when size is reduced by sampling. More subtle methods use mathematically generated fonts which compensate for the reduction process, as suggested by Knuth (37).

Sometimes the map information with lettering is kept separately to be used as an overlay for the picture data. There are cases where a large quantity of symbolic data would obscure picture details, or where the map base is frequently being changed. Therefore symbols cannot be placed automatically in their final position, since human judgement is required to prevent obscuring significant map details. The placement is done approximately by the program, and the user must have the freedom to override or change this by using suitable graphic entry devices. Text must stand out from conventional map text, and suitable fonts must be chosen for this purpose.

5.10. Fortran function for rapid determination of presence of a coordinate point inside or outside a polygon

IPPOLY – DETERMINE POINT TO POLYGON RELATION

CALLING SEQUENCE

IPPOLY(X,Y,P,N)   Where an integer function reference is legal
    IPPOLY   Function value:
          1   Point is interior point of polygon
          0   Point is boundary point of polygon
       −1   Point is exterior point of polygon
    X   Integer x coordinate of test point
    Y   Integer y coordinate of test point
    P   Integer array representing the polygon of
          DIMENSION (2,N).
            P(1,I)   X Coordinate of vertex
            P(2,I)   Y Coordinate of vertex
    N   Number in vertices, second dimension of P

Notes:
The program uses the ray intersection method to determine the topological relation of point and polygon. A point is defined to be an interior one, if a ray coming from infinity and ending at the test intersects the polygon boundary an odd number of times.

This algorithm works independently of the choice of orientation of coordinate system and allows not only proper convex and concave polygons but also degenerate (coinciding edges) and self-intersecting polygons.

```
      INTEGER FUNCTION IPPOLY(X,Y,P,N)
      INTEGER X,Y,P(2,0:N−1)
      INTEGER XIJ,YIJ,XJ,YJ
      BYTE ACTION(−40:+40)
      DATA ACTION/ 3, 3, 3, 3, 1, 3, 2, 2, 0,
     X             3, 3, 3, 3, 1, 3, 2, 1, 3,
     X             3, 3, 3, 3, 1, 3, 0, 3, 3,
     X             3, 3, 3, 3, 1, 1, 2, 2, 2,
     X             1, 1, 1, 1, 1, 1, 1, 1, 1,
     X             3, 3, 3, 1, 1, 3, 3, 3, 3,
     X             2, 2, 0, 2, 1, 3, 3, 3, 3,
     X             2, 1, 3, 2, 1, 3, 3, 3, 3,
     X             0, 3, 3, 2, 1, 3, 3, 3, 3/
      IPAR=0
      DO 90 I=0,N−1
      J=I+1
```

IF (J.EQ.N) J=0

Get polygon edge. Convert it to a standard frame.

XIJ = P(1,I) − P(1,J)
XJ = X − P(1,J)

YIJ = P(2,I) − P(2,J)
YJ = Y − P(2,J)

Get position orientation of polygon edge. The possible cases are coded in the action table with the states:

0   determine intersection relation by calculation,

1   boundary incidence;

2   edge intersects test ray;

3   edge does not intersect test ray.

The index calculation is optimized to require only four additions on average. The code is equivalent to addressing an array dimensioned $(-1:1, -1:1, -1:1, -1:1)$

```
       KA = −40
       IF (XJ − XIJ) 11,12,20
11        KA = KA + 1
12        KA = KA + 1
20        IF (YJ − YIJ) 21,22,30
21        KA = KA + 3
22        KA = KA + 3
30        IF (XJ) 31,32,40
31        KA = KA + 9
32        KA = KA + 9
40        IF (YJ) 41,42,50
41        KA = KA + 27
42        KA = KA + 27
50        CONTINUE

          GOTO (60,70,90),ACTION(KA)
```

Falling through when slope calculation is necessary.

```
       IF (XJ − XIJ*(YJ/FLOAT(YIJ))) 90,60,70
```

Point is on boundary.

```
60        IPPOLY = 0
          RETURN
```

Increment parity counter

70   IPAR = IPAR + 1

Do nothing at all.

90   CONTINUE

Return function value − 1,0,1

        IPPOLY = 2*MOD(IPAR,2) − 1
        RETURN
        END

Notes

(1) Akima, H. 1978a, A method for bivariate interpolation and smooth surface fitting for irregularly distributed data points, *ACM Transactions on Mathematical Software*, **4**, 148–59.

(2) Akima, H. 1978b, Algorithm 526: Bivariate interpolation and smooth surface fitting for irregularly distributed data points, *ACM Transactions on Mathematical Software*, **4**, 160–4.

(3) Akima, H. 1979, Remark on algorithm, 526, *ACM Transactions on Mathematical Software*, **5**, 242–3.

(4) Bakker, L. 1977, Alpen, römisches Lager, *Bonner Jahrbuch, Jahresbericht 1975*, **177**, 703–4.

(5) Barnea, D. I., Silverman, H. F. 1972, A class of algorithms for fast digital image registration, *IEEE Transactions on Computers*, **21**, 179–86.

(6) Bernstein, R. 1976, Digital image processing of earth observation data, *IBM Journal of Research and Development*, **20**, 40–57.

(7) Blais, J. A. R. 1972, Three dimensional similarity, *Canadian Surveyor*, **26**, 71–6.

(8) Burnside, C. D. *et al.* 1983, A digital single photograph technique for archaeological mapping and its application to map revision, *Photogrammetric Record*, **11**, 59–68.

(9) Burt, P. J. 1981, Fast filter transforms for image processing, *Computer Graphics and Image Processing*, **16**, 20–51.

(10) Burt, P. J. 1983, Fast algorithms for estimating local image properties. *Computer Vision, Graphics and Image Processing*, **21**, 368–82.

(11) Burt, P. J., Adelson, E. H. 1983, A multiresolution spline with application to image mosaics, *ACM Transactions on Graphics*, **2**, 217–36.

(12) Carlbom, I., Paciorek, J. 1978, Geometric projection and viewing transformations, *Computing Survey*, **1**, 465–502.

(13) Church, E. 1945, *Revised Geometry of the Aerial Photography*, Syracuse University Bulletin 15, Syracuse University Press, Syracuse, NY.

(14) Cyganski, D., Orr, J. A. 1985, Application of tensor theory to object recognition and orientation determination, *IEEE Transactions Pattern Analysis and Machine Intelligence*, **7**, 662–73.

(15) Dierckx, P. 1982, Algorithms for smoothing data with periodic and parametric splines, *Computer Graphics and Image Processing*, **20**, 171–84.

(16) Ebner, H. 1983, Berücksichtigung der lokalen Geländeform bei der Höheninterpolation mit finiten Elementen, *Bildmessung und Luftbildwesen*, **51**, 3–8.

(17) Fischler, M. A., Bolles, R. C. 1981, Random sample consensus: a paradigm for model fitting with applications to image analysis and automated cartography, *Association for Computing Machinery Communications*, **24**, 381–95.

(18) Foley, J. D., Van Dam, A. 1984, *Fundamentals of Interactive Computer Graphics*, Addison-Wesley, Reading, Ma.

(19) Frederiksen, P. Kubik, K., Weng, W. 1985, Ah, robust estimation, *The Australian Journal of Geodesy Photogrammetry & Surveying*, **42**, 1–18.

(20) Förstner, W. 1987, Reliability analysis of parameter estimation in linear models with applications to mensuration problems in computer vision, *Computer Vision, Graphics and Image Processing*, **40**, 273–310.

(21) Göpfert, W. 1981, Ein Entzerrungsverfahren zur Herstellung digitaler Orthophotos, *Bildmessung und Luftbildwesen*, **49**, 117–25.

(22) Goshtasby, A. 1986, Piecewise linear mapping functions for image registration, *Pattern Recognition*, **19**, 459–66.

(23) Goshtasby, A. 1987, Piecewise cubic mapping functions for image registration, *Pattern Recognition*, **20**, 525–34.

(24) Green, P. J., Sibson, R. 1978, Computing Dirichlet tesselations in the plane, *Computer Journal*, **21**, 168–73.

(25) Haas, A. G., Viallix, J. R. 1976, Krigeage applied to geophysics. The answer to the problem of estimates and contouring, *Geophysical Prospecting*, **24**, 49–69.

(26) Hall, E. L. 1979, *Computer Image Processing and Recognition*, Academic Press, New York, NY.

(27) Hallert, B. 1960, *Photogrammetry*, McGraw-Hill, New York, NY.

(28) Haralick, R. M., Chu, Y. H. 1984, Solving camera parameters from the perspective projection of a parameterized curve, *Pattern Recognition*, **17**, 637–45.

(29) Haralick, R. M. 1986, Computer vision theory: the lack thereof, *Computer Vision, Graphics and Image Processing*, **36**, 372–86.

(30) Hatzopoulos, J. N. 1985, An analytical system for close range photogrammetry, *Photogrammetric Engineering & Remote Sensing*, **51**, 1583–8.

(31) Hou, H. S. 1983, *Digital Document Processing*, John Wiley & Sons, New York, NY.

(32) Huang, T. S. 1977, Coding of two-tone images, *Technical Report TR-EE 77-10*, Purdue University, West Lafayette, Ind.

(33) Huang, T. S., Scollar, I., Weidner, B. 1977, An installation for interactive transfer of information from oblique aerial photos to maps, in *Informatik Fachberichte*, W. Brauer, ed., **8**, 198–211 Springer Verlag, Heidelberg.

(34) Inoue, H. 1986, A least squares smooth fitting for irregularly spaced data. Finite element approach using the cubic *B*-spline basis, *Geophysics*, **51**, 2051–66.

(35) Kalay, Y. E. 1982, Determining the spatial containment of a point in general polyhedra, *Computer Graphics and Image Processing*, **19**, 303–34.

(36) Kapokakis, E. 1984, Least squares co-location in digital terrain modelling, *Photogrammetric Record*, **11**, 303–9.

(37) Knuth, D. E. 1979, *TEX and Metafont, New Directions in Typesetting*, Digital Press, Bedford, Ma.

(38) Konecny, G. 1980, *Luftbildvermessung und Kartographie, in Erfassung und maschinelle Verarbeitung von Bilddaten*, H. Kazmierczak, ed., Springer Verlag, Vienna.

(39) Kubik, K., Lyons, K., Merchant, M. 1988, Photogrammetric work without blunders, *Photogrammetric Engineering & Remote Sensing*, **54**, 51–5.

(40) Kunt, M., Johnsen, O. 1980, Block coding of graphics: a tutorial review, *Proceedings of the IEEE*, **68**, 770–86.

(41) Lawson, C. L., Hanson, R. J. 1974, *Solving Least Squares Problems*, Prentice-Hall, Englewood Cliffs, NJ.

(42) Leberl, F. Kropatsch, W. Lipp, V. 1980, Interpolation of raster heights from digitized contour lines, paper presented, Congress of the International Society for Photogrammetry, 14, Hamburg.

(43) Matheron, G. F. 1965, *Les variables régionalisées et leur estimation*,Thesis, Université de Paris 6, Masson et Cie., Paris.

(44) Mauelshagen, L. 1977, Teilkalibrierung eines photogrammetrischen Systems mit variabler Paßpunktanordnung und unterschiedlichen deterministischen Ansätzen, *Deutsche Geodätische Kommission C*, **236**.

(45) McLain, D. H. 1974, Drawing contours from arbitrary data points, *Computer Journal*, **17**, 318–24.

(46) McLain, D. H. 1976, Two dimensional interpolation from random data, *Computer Journal*, **19**, 178–81.

(47) Milgram, D. L. 1975, Computer methods for creating photomosaics, *IEEE Transactions on Computers*, **24**, 1113–19.

(48) Milgram, D. L. 1977, Adaptive techniques for photomosaicing, *IEEE Transactions on Computers*, **26**, 1175–80.

(49) Palmer, R. 1977, A computer method for transcribing information graphically from oblique aerial photographs to maps, *Journal of Archaeological Science*, **4**, 283–90.

(50) Park, K., Schowengerdt, R. A. 1983, Image reconstruction by parametric cubic convolution, *Computer Vision, Graphics and Image Processing*, **23**, 258–72.

(51) Peleg, S. 1981, Elimination of seams from photomosaics, *Computer Graphics and Image Processing*, **16**, 90–4.

(52) Pentland, A. 1984, Fractal based description of natural scenes, *IEEE Transactions on Pattern Analysis and Machine Intelligence*, **6**, 666–74.

(53) Preußer, A. 1984, Bivariate Interpolation über Dreieckselementen durch Polynome 5. Ordnung mit C1 Kontinuität, *Zeitschrift für Vermessung*, **6**, 292–301.

(54) Preußer, A. 1985, Remark on Algorithm 526, *ACM Transactions on Mathematical Software*, **11**, 186–7.

(55) Ripley, B. D. 1981, *Spatial Statistics*, John Wiley & Sons, New York, NY.

(56) Rogers, D. F., Adams, J. A. 1976, *Mathematical Elements for Computer Graphics*, McGraw-Hill, New York, NY.

(57) Rosenfeld, A. Kak, A. C. 1982, *Digital Image Processing*, John Wiley & Sons, New York, NY.

(58) Sansò, F. 1973, An exact solution of the roto-translation problem, *Photogrammetria*, **29**, 163–78.

(59) Schut, G. H. 1960, On exact linear equations for the computation of the rotational elements of absolute orientation, *Photogrammetria*, **15**, 34–7.

(60) Scollar, I. 1975, Transformation of extreme oblique aerial photographs to maps or plans by conventional means or by computer, *Aerial Reconnaissance for Archaeology, Research Report*, 12, 52–8 Council for British Archaeology, London.

(61) Scollar, I., Weidner, B. 1979, Computer production of orthophotos from single oblique images or from rotating mirror scanners, *Aerial Archaeology*, **3**, 17–28.

(62) Scollar, I. 1984, Coding schemes for high resolution maps used in an archaeological database with superimposed air photographs, *Proceedings of the Symposium on Archaeometry, Washington, DC*, 217–23, Smithsonian Institution Press, Washington, DC.

(63) Sibson, R. 1978, Locally equiangular triangulations, *Computer Journal*, **21**, 243–5.

(64) Sibson, R. 1980a, The Dirichlet tessellation as an aid in data analysis, *Scandinavian Journal of Statistics*, **7**, 14–20.

(65) Sibson, R. 1980b, A vector identity for the Dirichlet tessellation, *Mathematical Proceedings Cambridge Philosophical Society*, **87**, 151–5.

(66) Sibson, R. 1980c, A brief description of natural neighbor interpolation, Privately printed, R. Sibson, Bath.

(67) Sibson, R. 1980d, Natural neighbourhood interpolation, *Graphical Methods for Multivariate Data*, V. D. Barnett, ed., John Wiley & Sons, Chichester.

(68) Sibson, R. 1986, Terrain modelling using quadratics, in: *State of the Art in Stereo and Terrain Modelling*, R. A. Earnshaw, ed., British Computer Society.

(69) Simon, K. W. 1975, Digital image reconstruction and resampling for geometric manipulation, reprinted in *Digital Image Processing for Remote Sensing*, R. Bernstein, ed. 1978, Proceedings of the IEEE Symposium Machine Processing Remotely Sensed Data, 3A1–3A11, Institute of Electrical and Electronic Engineers, New York, NY.

(70) Slama, C. C., ed. 1980, *Manual of Photogrammetry*, 4th edn, American Society of Photogrammetry, Falls Church, Va.

(71) Stefanovic, P. 1973, Relative orientation – a new approach, *ITC Journal*, International Training Center, Delft.

(72) Tanimoto, S. L., Pavlidis, T. 1975, A hierarchical data structure for picture processing, *Computer Graphics and Image Processing*, **4**, 104–19.

(73) Terzopoulos, D. 1983, Multilevel computational processes for visual surface reconstruction, *Computer Graphics and Image Processing*, **24**, 52–96.

(74) Tewinkel, G. C. et al. 1966, Basic mathematics of photogrammetry, *Manual of Photogrammetry*, 3rd edn, M. M. Thompson, ed., 17–65, American Society of Photogrammetry, Falls Church, Va.

(75) Thompson, E. H. 1958, An exact linear solution to the problem of absolute orientation, *Photogrammetria*, **14**, 163–78.

(76) Tienstra, M. 1974, A method for the calculation of orthogonal transformation matrices and its application to photogrammetry and other disciplines, *ITC Publication Series A*, **48**, International Training Center, Enschede.

(77) Welch, R. et al. 1985, Comparative evaluations of the geodetic accuracy and cartographic potential of Landsat-4 and Landsat-5 thematic mapper image data, *Photogrammetric Engineering & Remote Sensing*, **51**, 1249–62.

(78) Wiesel, W. J. 1981, Paßpunktbestimmung und geometrische Genauigkeit bei der relativen Enzerrung von Abstastdaten, *Deutsche Geodätische Kommission C*, **268**, 1981.

(79) Wong, K. W. 1980. Basic mathematics of photogrammetry, *Manual of Photogrammetry*, 4th edn, C. C. Slama, ed., 37–101, American Society of Photogrammetry, Falls Church, Va.

6

Resistivity prospecting

6.1. Electric currents and soil resistivity

6.1.1. Theoretical introduction

All materials allow movements of electric charge to a certain extent. In some, conductors like metals and electrolytes, the mobility is very large, in others, insulators like glass, plastics, air and ice, it is very weak or almost zero. Between these extreme cases, soil materials have intermediate properties which depend in a rather specific manner on their structural and chemical properties which permits them to be characterised rather well (see chapter 2). This ability to allow electric currents to circulate is expressed by the notion of electrical conductivity or its inverse, resistivity, and serves as the basis of an archaeological prospection method via recognition of the contrast between a sought-for structure and its milieu. Historically it is the oldest of the archaeological prospecting techniques, having been introduced in the mid-1940s by Atkinson (3,4).

Distribution of a current in the ground
Following Cagniard (6) we have Ohm's law: In conductors, electric charges move under the effect of an electric field **E** which may be expressed in the form:

$$\mathbf{E} = -\operatorname{grad} V \tag{6.1}$$

where V is the electric potential. If one defines intensity **I**, as the quantity of electric charge which flows through a surface S in unit time with the vector i:

$$i = dI/dS \tag{6.2}$$

expressing the density of the current at a point, it can be shown that:

$$i = \sigma E = \frac{1}{\rho} E = -\sigma \, grad \, V = -\left(\frac{1}{\rho}\right) grad \, V \tag{6.3}$$

where σ and ρ are the electrical conductivity and resistivity. The intensity of the current can be expressed in the form:

$$I = \iint_S i \cos \theta \, dS \tag{6.4}$$

where θ is the angle formed by the vector i with the normal to the surface S at the point considered. If S is any closed surface, and since there can be no accumulation of charge within the volume v limited by the surface S, it follows that:

$$I = \iint_S i \cos \theta \, dS = \iiint_V (div \, i) \, dv = 0 \tag{6.5}$$

Therefore it is necessary that at all points of the conductor one has:

$$div \, i = \frac{\partial i_x}{\partial x} + \frac{\partial i_y}{\partial y} + \frac{\partial i_z}{\partial z} = 0 \tag{6.6}$$

and as a consequence of Ohm's law:

$$\frac{\partial^2 V}{\partial x^2} + \frac{\partial^2 V}{\partial y^2} + \frac{\partial^2 V}{\partial z^2} = \nabla^2 V = 0 \tag{6.7}$$

The potential V is therefore a harmonic function.

If the trajectory of an electric charge (field or current line) traverses a surface which separates two conductors of different resistivities ρ_1 and ρ_2, one can show that the trajectory is broken and obeys the relationship:

$$\frac{1}{\rho_1}\left(\frac{\partial V}{\partial n}\right)_1 = \frac{1}{\rho_2}\left(\frac{\partial V}{\partial n}\right)_2 \tag{6.8}$$

along n, the normal to S or:

$$\rho_1 \cot \theta_1 = \rho_2 \cot \theta_2 \tag{6.9}$$

where θ_1 and θ_2 are the angles formed in each of the media by the lines of current with the normal to the surface of partition, figure 6.1. It follows that a line of current approaches the vertical when it enters a more resistive environment. In particular, at the surface of separation between a conducting and an insulating body with infinite contrast in resistivity, the lines of current are tangent to the surface of separation whereas the equipotential surfaces are orthogonal to it. This is reciprocally true when the second milieu has zero resistivity.

In summary, all solutions to the problem of the distribution of potential must obey the following conditions:

(1) V is a harmonic function with:

$$\nabla^2 V = 0$$

(2) V is continuous overall, especially at the surfaces of the discontinuity. The derivatives of V are continuous in homogeneous media and only the derivative with respect to the tangent to the surface is continuous at the surface.

(3) Over the whole surface of the discontinuity:

$$\frac{1}{\rho}\left(\frac{\partial V}{\partial n}\right)$$

is continuous.

(4) V is considered to be zero at infinity.

Distribution of a point source at the surface of the ground

Assume that A, a point source electrode which carries electric charge is located in the interior of a space with homogeneous resistivity ρ. By symmetry, at a non-zero distance r from the electrode, the equipotential surface V is a sphere which, following Ohm's law, one may write at any

Fig. 6.1 Refraction of a line of current at the interface between two media.

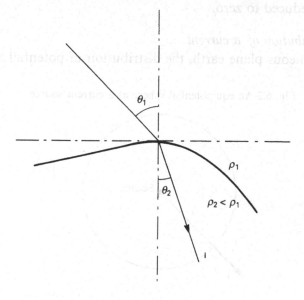

point:

$$|i| = \frac{I}{4\pi r^2} = -\frac{1}{\rho}\frac{dV}{dr} \tag{6.10}$$

i being radial and thus perpendicular to the equipotential sphere. I represents the current which passes through the electrode, figure 6.2. Therefore by integration:

$$V = \frac{\rho I}{4\pi}\frac{1}{r} + C \tag{6.11}$$

where C is a constant.

If the resistivity of the half-space divided by a plane passing through the electrode becomes infinite, nothing is changed at the surface of separation inasmuch as the orthogonality of the equipotential surfaces and the tangentiality of the lines of current are respected. However, the current only passes through the equipotential hemispheres in such a way that one obtains, for the potential created by a point source electrode at the surface of the ground:

$$V = \frac{\rho I}{2\pi}\frac{1}{r} + C \tag{6.12}$$

In the real world of experiment, there are no strictly point source electrodes. The condition $r \to 0$ which causes the potential to be infinite is respected. The formula is only valid at a distance from the electrode which is sufficient such that its actual dimensions may be considered negligible relative to r. Furthermore, if one sets V to zero at infinity, the constant is reduced to zero.

Normal distribution of a current
In a homogeneous plane earth, the distribution of potential at a point

Fig. 6.2 An equipotential sphere as a current source.

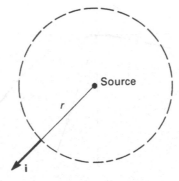

M results in the circulation of a current between electrodes A and B connected to a generator of direct or alternating current. Because the current enters via one of these electrodes and leaves by the other, the potential at a point in the ground results from the superposition of the potentials attributable to each of the electrodes, thus from the relationship of equation (6.12) and taking into account the signs:

$$V_M = \frac{\rho I}{2\pi}\left(\frac{1}{MA} - \frac{1}{MB}\right) \tag{6.13}$$

The potential is set to zero at infinity, but since we can only measure differences in potential, and since it is frequently more practical to make measurements with short cables, we consider a second point N for which:

$$V_N = \frac{\rho I}{2\pi}\left(\frac{1}{NA} - \frac{1}{NB}\right) \tag{6.14}$$

such that if we set $\Delta V = V_M - V_N$ the resistivity of the ground is expressed for an electrode quadripole by the relationship:

$$\rho = 2\pi\frac{\Delta V}{I}\frac{1}{(1/MA - 1/MB + 1/NB - 1/NA)} = 2\pi k\frac{\Delta V}{I} \tag{6.15}$$

This equation permits us to determine the resistivity of the ground from four electrodes placed in any arbitrary arrangement on the surface, two of which serve to inject current and the other two to measure potential. The constant k is called the geometric coefficient of the quadripole, figure 6.3.

One of the interesting properties of the current distribution in a homogeneous ground results from the symmetric role of the dipoles, A, B and M, N in the preceding formula. One may exchange current and potential electrodes without modifying the value of $\Delta V/I$. Even though this remark is not applicable in the case of an inhomogeneous terrain,

Fig. 6.3 Various electrode arrangements: (a) random, (b) in-line quadripole.

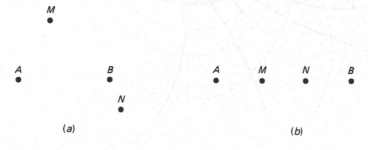

(a)   (b)

it remains true in practice, and can be applied to obtain a rapid technique
for moving the electrode array.

6.1.2. Properties of quadripoles

In a homogeneous terrain, the current is distributed in the ground
between the points A and B with a regular geometric shape as shown
in figure 6.4. In this distribution, the lines of current linking A and B,
and the equipotential surfaces which are close to hemispherical form
near A and B cross each other. The shape obtained in the vertical section
between A and B is reproduced in all the other planes passing through
this line and the surface is symmetric. It is valid apart from a scaling
factor for all distances AB, all resistivities, and all values of current, and
the potential obtained varies proportionally as a consequence. Therefore
any arrangement of the electrodes MN, figure 6.3, will give the same
resistivity. Two special configurations have been most widely used in
practice, one with the MN electrodes on the line AB, an arrangement
which has many subvarieties, and one, which will be discussed later, in
which M and N are at the corners of a square, one of whose sides is AB.

If one now considers the current densities, the preceding formulae
show that this density is not equal at all points in the figure and that
most of the intensity I emitted between A and B is concentrated in the
neighbourhood of the segment AB which constitutes the shortest path,

Fig. 6.4 Distribution of current lines (dashed) and equipotential surfaces
between two electrodes A and B.

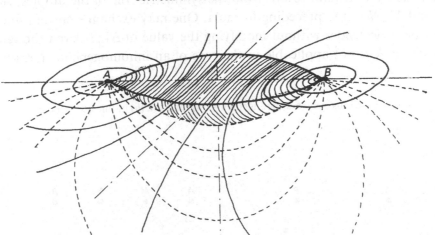

that is, in a volume which has approximately the shape of half a rugby football with pointed extremities. We shall see later that the important consideration of a 'volume of influence' of the quadripole intervenes in the estimation of the depth of investigation and the lateral sensitivity during prospection.

Apparent resistivity

Now suppose that the ground considered is no longer homogeneous, but rather presents different resistivities from one point or zone to another. Figure 6.4 no longer represents reality. The lines of current, figure 6.1, will refract at the boundary of the surfaces of discontinuity or will inflect around perturbing bodies. The theoretical definition of resistivity no longer makes sense, but it is convenient to call the value of ρ_a obtained from ΔV and I the 'apparent resistivity', as if the ground were homogeneous. One thus obtains a 'kind of average of the true resistivities of the various media actually affected by the current distribution' (6).

The inhomogeneities of the ground perturb the normal current distribution, and it is clear that the measurements obtained will vary with at least two parameters:

(1) for identical electrode configurations, depending on the orientation of the quadripole relative to the body. The perturbation is maximum, when this presents the largest cross-section to the volume of influence of the quadripole, figure 6.5;

(2) depending on the positions of M and N relative to A and B, since each type of quadripole will have a different response to identical subsurface structures, figure 6.6.

Calculation of the effect of an inhomogeneity

The determination of these anomalies as a function of the quadripole configuration and their positions with respect to the subsurface structures normally requires solving the fundamental current distribution equations in a case which is much more complex than that of the homogeneous terrain which has been discussed up to now. In practice, we only know how to solve the equations for the simple cases of a layered ground or certain cylindrical structures with simple geometry such as linear ditches with semicircular cross-sections (5). In the case of more complex three-dimensional finite length structures we do not know how to find analytic solutions to the differential equations, with the exceptions of a hemisphere at the surface (9) or a buried sphere below the surface in a uniform primary field (35).

For the latter case, the calculation is made within the reference frame of a spherical coordinate system, figure 6.7, a sphere of radius a and resistivity ρ_2 placed in a medium of resistivity ρ_1 in which a uniform electric field $\mathbf{E}$ exists parallel to the x axis. In this system, the equation for the potential V, where V_1 and V_2 are respectively the potentials at the interior and exterior of the sphere, can be written as:

$$\nabla^2 V = \frac{1}{r^2}\frac{\partial}{\partial r}\left(r^2\frac{\partial V}{\partial r}\right) + \frac{1}{r^2\sin\theta}\frac{\partial}{\partial\theta}\left[\sin(\theta)\frac{\partial V}{\partial\theta}\right] = 0 \qquad (6.16)$$

Fig. 6.5 Apparent resistivity profile at right angles to a ditch filled with stones. A Wenner quadripole, electrode spacing $a = 1$ m, (a) parallel, (b) perpendicular to the ditch. From A. Hesse, Geophysical prospecting for archaeology in France, *Archaeometry*, **5**, 1962.

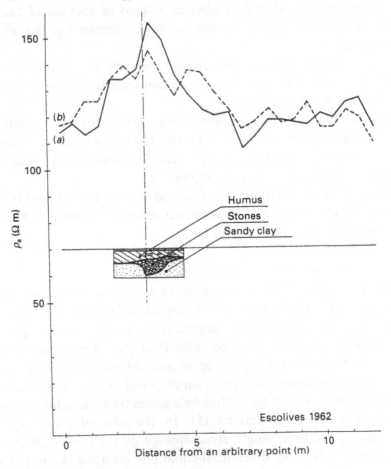

Humus

Stones

Sandy clay

Escolives 1962

and for $r = a$, then $V_1 = V_2$ and:

$$\frac{1}{\rho_1} \frac{\partial V_1}{\partial r} = \frac{1}{\rho_2} \frac{\partial V_2}{\partial r} \tag{6.17}$$

and for large values of x or r, V_1 must tend toward $-Ex = Er \cos(\theta)$.

The solution of these equations is given by a series of Legendre polynomials $P_n(\mu)$, where $P_1(\mu) = \mu = \cos(\theta)$, in the following form:

$$\left. \begin{aligned} V_1 &= -Er \cos(\theta) + \sum_{n=0}^{\infty} b_n P_n(\mu)/r^{n+1}, & r > a \\ V_2 &= \sum_{n=0}^{\infty} a_n r^n P_n(\mu), & r < a \end{aligned} \right\} \tag{6.18}$$

Expansion of these polynomials, taking into account the second and

Fig. 6.6 Three resistivity profiles with two different electrode configurations, normal Wenner, dipole–dipole Wenner (dashed), $a = 2$ m. The arrow marks the edge of the buried resistive surface. (Maya remains of the pre-classical period at Copan, Honduras.) From ref. 6.

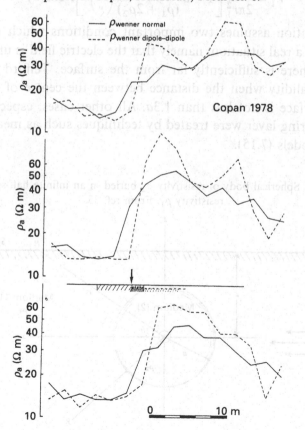

third conditions above, allows calculation of the constants a_n and b_n which are all zero with the exception of:

$$b_1 = \frac{E(\rho_1 - \rho_2)a^3}{(\rho_1 + 2\rho_2)}, \qquad a_1 = -E \frac{3\rho_2}{(\rho_1 + 2\rho_2)} \qquad (6.19)$$

The potential at the exterior of the sphere can therefore be written as:

$$V_1 = -Er\cos(\theta) + \frac{E(\rho_1 - \rho_2)a^3\cos(\theta)}{r^2(\rho_1 + 2\rho_2)} \qquad (6.20)$$

In the particular case which is of interest to the prospector, the sphere is located in a half-space of resistivity ρ_1 in which the electric field is created by a source of current of intensity I at a distance r_1. The same line of reasoning which was used to establish equation (6.13), namely the electrical image of the sphere with respect to a plane given by the surface of the ground, allows one to write the potential at any point in the half-space as:

$$V_1 = \frac{I\rho_1}{2\pi r_1^2}\left[1 - 2\frac{(\rho_1 - \rho_2)}{(\rho_1 + 2\rho_2)}\left(\frac{a}{r}\right)^3\right]r\cos(\theta) \qquad (6.21)$$

This calculation assumes two important conditions which are poorly observed in a real situation, namely that the electric field is uniform and that the sphere is sufficiently far from the surface. Telford *et al.* (35) admit its validity when the distance between the centre of the sphere and the surface is not less than $1.3a$. All other cases, especially those with a covering layer were treated by techniques such as measurements on scale models (7,15).

Fig. 6.7 Spherical body of resistivity ρ_2 buried in an infinite half-space of resistivity ρ_1. From ref. 35.

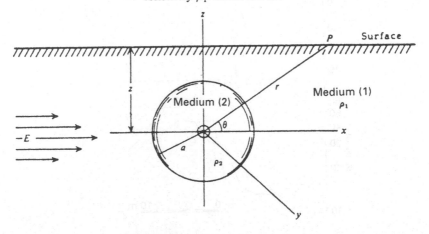

Other theoretical cases, in particular including the model of a vertical fault, a hemisphere, and a half-cylinder touching the surface have been treated and are presented by Cook and Van Nostrand (9). They have been compared with measurements on scale models and have contributed important data concerning two points: the lateral influence of these kinds of structures and the existence of sudden jumps which occur when an electrode passes over the inhomogeneity. These jumps correspond to negative or positive peaks in the apparent resistivity curves which, in consequence, take on a highly disturbed appearance for structures which are close to the surface. From the end of the 1960s onwards, the availability of computers has made possible the analysis of the general case of the response of a structure of arbitrary shape with numerical methods, as described by Keller and Frischknecht or Dieter, Patterson and Grant (26,12).

The principle of numerical calculation (32) consists in using a Green function, that is, in supposing that the presence of a body of resistivity ρ_2 is equivalent to a certain distribution of electrostatic elementary sources on its surface, and adding the effects of all of these sources, with ρ_1 being the resistivity of the surrounding medium. In the case of perturbing body C buried in a half-space, the potential at a point $P(r)$ in the half-space at the exterior of C may be written:

$$V_1(P) = \frac{\rho_1 I}{2\pi} \frac{1}{|r_1 - r|} + \iint_C \frac{q(r')}{|r - r'|} \, dr' + \iint_{\bar{C}} \frac{q(\bar{r}')}{|r - \bar{r}'|} \, d\bar{r}' \quad (6.22)$$

in which the first term represents the potential due to the source A (r_1 and r being the position of A and P respectively). The second and third terms correspond to two terms of the Green function. They represent the potential due to the distribution at the surface of C of charges of superficial density $q(r')$ at the position r'. To realise the theoretical conditions of the distribution of potential at the earth's surface, one must not only consider C, but also its electrical image $\bar{C}$ which is symmetrical with respect to the surface and is characterised by a moving point at position $\bar{r}'$.

The external electric field normal to a body at a point r' on its surface can be expressed as:

$$\left(\frac{\partial V_1}{\partial n}\right)_{r'} = 4\pi q \left(\frac{\rho_1}{\rho_2 - \rho_1}\right) \quad (6.23)$$

inasmuch as:

$$\frac{1}{\rho_1} \frac{\partial V_1}{\partial n} = \frac{1}{\rho_2} \frac{\partial V_2}{\partial n} \quad (6.24)$$

and:

$$\frac{\partial V_1}{\partial n} - \frac{\partial V_2}{\partial n} = -4\pi q \tag{6.25}$$

Elsewhere, one has, using the general expression for potential:

$$\left(\frac{\partial V_1}{\partial n}\right)_{r'} = \frac{\rho_2 I}{2\pi} \frac{\partial}{\partial n}\left(\frac{1}{|r_1 - r'|}\right) + \oint\!\!\oint_C \frac{\partial}{\partial n} \frac{q(r'')}{|r' - r''|}\, dr'' + \int\!\!\int \frac{\partial}{\partial n} \frac{q(r'')}{|r' - r''|}\, dr'' \tag{6.26}$$

Taking into account the singularity of the integral over C when $r'' = r'$, we consider the contribution to the normal field $-2\pi q$ of the charge q carried by a circle of infinitesimal radius around the point $r'' = r'$, and obtain:

$$4\pi q_{(r')} = \left(\frac{\rho_1}{\rho_2 - \rho_1}\right) = \frac{\rho_1 I}{2\pi} \frac{\partial}{\partial n}\left(\frac{1}{r_1 - r'}\right) - 2\pi q_{(r')} + \int\!\!\int_{C - S_r} \frac{\partial}{\partial n} \frac{q(r'')}{|r' - r''|}\, dr''$$

$$+ \oint\!\!\oint_{\bar{C}} \frac{\partial}{\partial n} \frac{q(r'')}{|r' - r''|}\, dr'' \tag{6.27}$$

The surface of C has therefore been decomposed into a series of small facets on each of which one may consider the charge density to be constant, and after simplification in setting:

$$K = \frac{\rho_2 - \rho_1}{\rho_2 + \rho_1} \tag{6.28}$$

equation (6.27) allows the calculation of the density q with the aid of a system of linear equations. Summing the effect of the charges thus calculated and adding the effect of injection, one obtains the potential at a point on the surface where:

$$V(P) = \frac{\rho_1 I}{2\pi} \frac{1}{|r_1 - r_P|} + \sum_i \frac{q_i \,\Delta S}{|r_i' - r_P|} + \sum_i \frac{\bar{q}_i \,\Delta S}{|r_i' - r_P|} \tag{6.29}$$

in which r_P, q and ΔS represent the position of the point P, the charge density and the surface of each facet respectively.

   The apparent resistivity observed at the surface affected by a perturbing body is therefore expressed via the potentials V_M and V_N calculated at two points M and N of the quadripole following the fundamental formula given above:

$$\rho_a = 2\pi k \frac{V_M - V_N}{I} \tag{6.30}$$

The resistivity anomaly can be expressed conventionally in the form of the ratio ρ_a/ρ_1 which is equal to 1 when the quadripole is sufficiently distant from the cause of the anomaly, figure 6.8.

Anisotropy

It is known that certain bodies or materials show different properties depending on whether the current passes through them in one direction or another. This is particularly the case for bodies which are formed of alternating conductive and resistive layers. In this case the current does not encounter the same resistance when it passes perpendicular or parallel to the layers. Thus for a layered structure with alternating thicknesses e_1 and e_2 and resistivities ρ_1 and ρ_2, the opposing resistances

Fig. 6.8 Calculated conductivity (inverse of resistivity) anomaly for two Wenner configurations, normal and dipole–dipole and two electrode separations. The structure is of finite dimensions under a conductive surface layer. The model characteristics are:

$$e_1 = 0.2 \text{ m} \qquad x = 1.2 \text{ m} \qquad \sigma_1 = 0.02 \text{ S m}^{-1}$$
$$e_2 = 0.1 \text{ m} \qquad y = 1.2 \text{ m} \qquad \sigma_2 = 0.02 \text{ S m}^{-1}$$
$$e_3 = 0.6 \text{ m} \qquad\qquad\qquad\qquad \sigma_3 = 0.01 \text{ S m}^{-1}$$

From ref. 34.

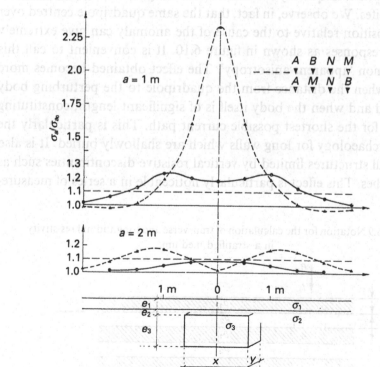

of an elementary cube of side length $e_1 + e_2$ may be expressed transversely in the form, figure 6.9:

$$R_T = \frac{\rho_1 e_1 + \rho_2 e_2}{(e_1 + e_2)^2} \qquad (6.31)$$

and in parallel in the form:

$$R_P = \frac{\rho_1 \rho_2}{e_1 \rho_1 + e_2 \rho_2} \qquad (6.32)$$

for which it can be easily shown that $R_T \geqslant R_P$ is always satisfied. The medium is considered to be electrically anisotropic because it does not possess the same properties in all directions. It is evident that the measurements of apparent resistivity as defined above on terrain presenting these properties (notably schists) will be affected by the orientation of the lines of current. That is, it will vary according to the angle formed between the current injection AB with the plane of the layers such that an arbitrary single measurement of resistivity of such a milieu does not have intrinsic significance.

Although it is relatively rare in soils to find areas which have real anisotropy, one frequently encounters an analogous situation in inhomogeneous sites. We observe, in fact, that the same quadripole centred over a fixed position relative to the cause of the anomaly can give extremely variable responses as shown in figure 6.10. It is convenient to call this phenomenon 'apparent anisotropy'. The effect obtained becomes more marked when the distance from the quadripole to the perturbing body is reduced and when the body itself is of significant length, constituting a barrier for the shortest possible current path. This is particularly the case in archaeology for long walls which are shallowly buried. It is also true for all structures limited by vertical resistive discontinuities such as filled ditches. This effect is particularly noticeable in a series of measure-

Fig. 6.9 Notation for the calculation of transverse and longitudinal resistivity in a stratified medium.

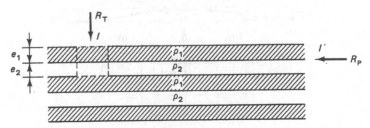

ments made along a set of profiles. It can readily be observed in practice, figure 6.5, on scale models, figure 6.11 or by calculation, figure 6.12.

Depth of investigation and lateral sensitivity

As a consequence of the resemblance stated above, the greater the distance *AB*, the greater the volume traversed by the current and the greater the depth at which one may hope that structures to be detected perturb the normal current distribution under the restriction that their volume and resistivity contrast is sufficient. One can only give an empirical definition to this depth which is very dependent on the relative geometric configuration of the structure to be detected and the quadripole. For a given electrode configuration, it is that for which the anomaly rises sufficiently above the noise level to be detected. In theoretical calculations on models, a 10% level is considered to be significant (34) as shown in figure 6.8.

The quadripoles most frequently used, figure 6.24 have four in-line

Fig. 6.10 Anisotropy ellipses obtained over Merovingian inhumation pits containing sarcophagi. From ref. 15.

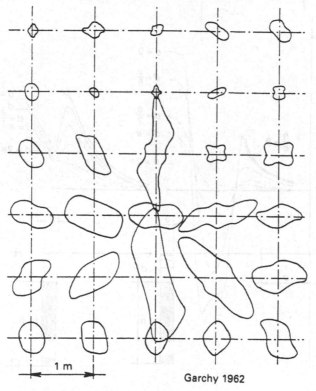

1 m

Garchy 1962

electrodes, with the current injectors *A, B* at the ends and the potential electrodes in the middle. The name Wenner has been given to the arrangement in which the spacing between the electrodes is equal, and Schlumberger to that where *MN* is small relative to *AB* (*ca* 1/10). In practice, investigators voluntarily adopt the approximation due to

Fig. 6.11 Anomalies obtained on two scale models of an infinitely resistive wall at depth *d*, of height *h*, and thickness *e*, for electrode spacing *a*, ($h/e = 1$, $h/e = 8$) at different depths. (*a*) Quadripole perpendicular to the structure. (*b*) Quadripole parallel to it. Width of the wall $e = 3a/10$. From ref. 15.

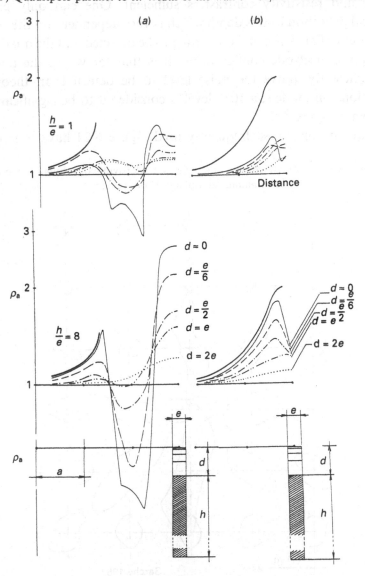

Schlumberger according to which the depth of investigation is equal to $AB/4$ for the arrangement which bears his name. Later we shall see for the special case of the detection of layered strata (electrical sounding) that this approximation is optimistic concerning the length AB which just allows detection of a very resistive or conductive layer, and that it doesn't always allow a precise interpretation of the resistivity and the exact depth of the top of such a layer.

As far as lateral effects go, they can be readily calculated for the case of an infinitely resistive medium bounded by a vertical plane for a normal Wenner linear quadripole. The effect on the measured resistivity is characterised by functions K' and K'' in which α is the distance to the boundary expressed in units a, figure 6.13. For a quadripole

Fig. 6.12 Calculated anomalies for a square quadripole with current injection AB parallel or perpendicular to a structure. Dimensions: $2.6 \times 0.4 \times 0.6a$ under a surface layer of $0.5a$. From ref. 32.

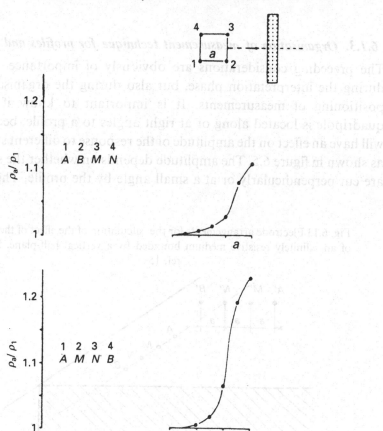

perpendicular to the boundary:

$$K' = 1 + \frac{1}{2\alpha(2\alpha - 1)} - \frac{1}{2(2\alpha^2 + 5\alpha + 3)} \qquad (6.33)$$

and for one parallel to it:

$$K'' = 1 + \frac{1}{(1 + 4\alpha^2)^{\frac{1}{2}}} - \frac{1}{1 + \alpha^2} \qquad (6.34)$$

From the curves obtained, it may be deduced that an effect equal to or greater than 10% is obtained when the centre of the quadripole approaches the boundary for a value of at least $2.2a$ in the first case and $1.4a$ in the second. This calculation not only permits the fixing of the extreme limits of lateral detection for one of the linear quadripoles most frequently used, but also allows the evaluation of the perturbing effect of working near the boundaries of an excavation (15).

6.1.3. *Organisation of measurement technique for profiles and mapping*

The preceding considerations are obviously of importance, not only during the interpretation phase, but also during the organisation and positioning of measurements. It is important to know if a linear quadripole is located along or at right angles to a profile, because this will have an effect on the amplitude of the response to different structures, as shown in figure 6.5. The amplitude depends on whether the structures are cut perpendicularly or at a small angle by the profile. The effect of

Fig. 6.13 Electrode arrangements for the calculation of the effect of the edge of an infinitely resistive medium bounded by a vertical half-plane. From ref. 15.

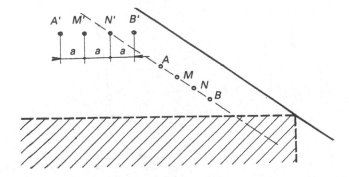

the orientation of a linear quadripole relative to the profile also helps determine the optimal distance between measurements. Due to the lengthening of the volume of influence, the percentage overlap is better for in-line than for transverse measurements.

As soon as one is interested in the recognition of the shapes of the structures in the horizontal plane, the problem arises of creating a map from a series of parallel profiles whose distances must be defined under these restrictions. The distances between measurements which define a step along a profile lead to the choice of a grid, which for the evident reason of homogeneity in both directions in the plane, should be kept square. The phenomenon of apparent anisotropy is therefore of fundamental importance in archaeology for networks of walls which have a relative perpendicular orientation. However, compromises are inevitable, as shown in figure 6.14 and described in ref. 16, in order to limit the distortions which create artefacts during interpretation. To these will be added the effects of deformations in the distribution of resistivities which are collinear or perpendicular to the profiles (topographic and climatological effects or those due to numerical treatment). The compromise to be adopted also has to take into account logistical problems related to the measurement at each station, the global cost of the prospecting campaign, and the administration of the stored measurements when automatic recording is used. This explains the current interest in the use of square quadripoles which, apart from some theoretical advantages (32) has the further merit of a greater isotropy which can be augmented by calculating the average of two measurements at the same point, one along the line of the profile, the other perpendicular to it. The ratio of these two values also constitutes an index of apparent anisotropy, figure 6.15, which can emphasise the more marked resistivity discontinuities.

Fig. 6.14 Fish-tail orientation of quadripoles on an offset grid to reduce the effects of apparent anisotropy in a resistivity map. From ref. 16.

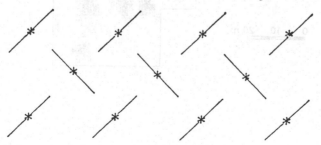

Fig. 6.15 Map of the anisotropy index calculated for a square quadripole to enhance the most important resistivity inhomogeneities at a peat bog in the Airette valley in the department of Lozère, France. The darker areas correspond to water saturated hollows and subsidence zones in the bog as opposed to the central areas which are probably more stratified.

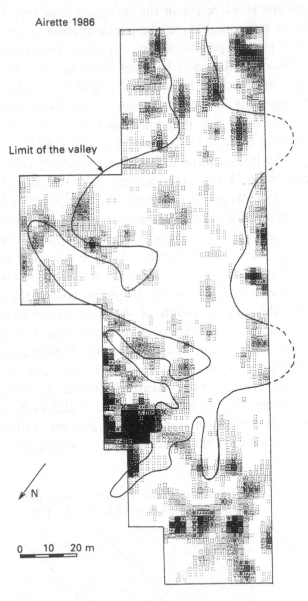

6.1.4. Electrical sounding and interpretation of stratification

Current distribution for a point electrode in a layered earth

This fundamental calculation was carried out exactly for the first time by Schlumberger, Stefanescu and Schlumberger in 1930 (33). Later it was taken up by a large number of others, including Cagniard (6) whose development has been followed in this treatment. This approach results in a mathematical expression which contains a Hankel transform, whose calculation represents the principal difficulty in resolving the problem. The Hankel transform is equivalent to a series of functions of the type $1/R$, but such a series converges slowly. The computation was carried out rapidly and in a satisfactory fashion when Ghosh (13) introduced a method allowing the transformation of the expression into a convolution which led to numerical linear filtering. Other authors such as Niwas and Israil (29) have proposed various other procedures to accelerate the convergence of the series.

At the surface of an infinite half-space which is composed of a superficial layer of thickness h and resistivity ρ_1, resting on a substratum of unlimited depth with resistivity ρ_2, we consider the case of a point electrode A, B being placed at infinity, figure 6.16. Taking into account rotational symmetry, an arbitrary point in the half-space can be characterised by its cylindrical coordinates:

$$R = (x^2 + y^2)^{\frac{1}{2}}, z \tag{6.35}$$

and the equation:

$$\nabla^2 V = \frac{\partial^2 V}{\partial R^2} + \frac{1}{R}\frac{\partial V}{\partial R} + \frac{\partial^2 V}{\partial z^2} = 0 \tag{6.36}$$

expresses the fact that the potential $V(R, z)$ at point M is harmonic. The method of integration due to Riemann consists in looking for a particular solution of equation (6.36) depending on parameters or arbitrary functions (which are specified later) and which satisfy initial and boundary conditions. This solution will have the form $V = F(R)\Phi(z)$ the product of a function of R and z only.

Equation (6.36) can be satisfied if F and Φ fulfil the following conditions:

$$d^2\Phi/dz^2 - \lambda^2\Phi = 0 \tag{6.37}$$

$$\frac{d^2 F}{dR^2} + \frac{1}{R}\frac{dF}{dR} + \lambda^2 F = 0 \tag{6.38}$$

where λ is an arbitrary parameter. The general integral of equation (6.37) is:

$$\Phi = A \exp(\lambda z) + B \exp(-\lambda z) \tag{6.39}$$

in which A and B are constants.

The general integral equation (6.38) serves as the definition of Bessel functions:

$$F(\xi) = CJ_0(\xi) + DN_0(\xi) \tag{6.40}$$

where C and D are constants, J_0 is a Bessel function of the first kind, and N_0 is a Bessel or Neumann function of the second kind. By setting $\varepsilon = \lambda R$, we observe that N_0 introduces an infinite term along the vertical axis. The coefficient D must therefore be zero. A solution of the form:

$$V = J_0(\lambda R)[A \exp(\lambda z) + B \exp(-\lambda z)] \tag{6.41}$$

is obtained. Since the choice of λ is arbitrary, a more general solution is obtained by taking:

$$V = \int_0^\infty J_0(\lambda R)[A(\lambda) \exp(\lambda z) + B(\lambda) \exp(-\lambda z)] \, d\lambda \tag{6.42}$$

This is equivalent to expressing the solution as a Hankel transform. For a layered earth with two strata one has:

(1) in the first medium ($0 < z < h$) where the injection point A is located:

$$V = V_1 = \frac{\rho_1 I}{2\pi} \left[\int_0^\infty J_0(\lambda R) \exp(-\lambda|z|) \, d\lambda + \int_0^\infty J_0(\lambda R) \exp(-\lambda z)\alpha(\lambda) \, d\lambda \right.$$

$$\left. + \int_0^\infty J_0(\lambda R) \exp(\lambda z)\zeta(\lambda) \, d\lambda \right] \tag{6.43}$$

Fig. 6.16 Notation for the calculation of potential created by a current source in a two layered medium.

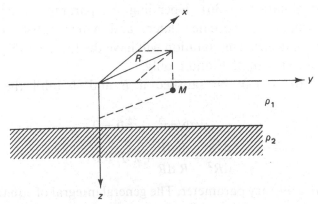

in which the first term is chosen such that in the vicinity of A, the function becomes infinite with:

$$\rho_1 I/[2\pi(R^2 + z^2)^{\frac{1}{2}}]$$

following the identity:

$$1/(R^2 + z^2)^{\frac{1}{2}} = \int_0^\infty J_0(\lambda R) \exp(-\lambda|z|) \, d\lambda \qquad (6.44)$$

(2) in the second medium $(h < z < \infty)$:

$$V = V_2 = \frac{\rho_2 I}{2\pi} \int_0^\infty J_0(\lambda R) \exp(-\lambda z)\psi(\lambda) \, d\lambda \qquad (6.45)$$

in which the first term is no longer necessary because there is no source and the third must be zero, since it causes the potential to become infinite with z.

Therefore three arbitrary functions $\alpha(\lambda)$, $\zeta(\lambda)$, and $\psi(\lambda)$ must be found which satisfy the following boundary conditions:
(a) for $z = 0$:

$$\partial V_1/\partial z = 0 \qquad (6.46)$$

(orthogonality of the equipotential surfaces at the ground surface);
(b) for $z = h$:

$$V_1 = V_2 \qquad (6.47)$$

(at the boundary of each medium);
(c) for $z = h$:

$$\frac{1}{\rho_1}\frac{\partial V_1}{\partial z} = \frac{1}{\rho_2}\frac{\partial V_2}{\partial z} \qquad (6.48)$$

(refraction condition of the current lines at the surface of separation).

The first condition leads from the derivation of V_1 and for $z = 0$ the first derivative of (6.44) being null to impose:

$$\int_0^\infty \lambda J_0(\lambda R)[-\alpha(\lambda) + \zeta(\lambda)] \, d\lambda = 0 \qquad (6.49)$$

from which $\alpha = \zeta$. The second and third conditions impose:

$$\rho_1[(1 + \alpha)\exp(-\lambda h) + \zeta \exp(\lambda h)] = \rho_2 \psi \exp(-\lambda h) \qquad (6.50)$$

and:

$$(1 + \alpha)\exp(-\lambda h) - \zeta \exp(\lambda h) = \psi \exp(-\lambda h) \qquad (6.51)$$

respectively. Equations (6.49)–(6.51), being linear with respect to three unknowns, can be solved as:

$$\alpha = \zeta = -\frac{1 - \rho_2/\rho_1}{(1 - \rho_2/\rho_1) + (1 + \rho_2/\rho_1)\exp(2\lambda h)} \quad (6.52)$$

$$\psi = \frac{2\exp(2\lambda h)}{(1 - \rho_2/\rho_1) + (1 + \rho_2/\rho_1)\exp(2\lambda h)} \quad (6.53)$$

The potential at the surface of the ground, $z = 0$ which we are looking for is thus:

$$V = \frac{\rho_1}{2\pi}\frac{1}{R} - \frac{\rho_1 I}{\pi}\frac{\rho_1 - \rho_2}{\rho_1 + \rho_2}\int_0^\infty \frac{J_0(\lambda R)\exp(-2\lambda h)}{1 + [(\rho_1 - \rho_2)/(\rho_1 + \rho_2)]\exp(-2\lambda h)} \quad (6.54)$$

This expression reduces in practice to the potential in a medium of resistivity ρ_1 when R is small compared with h, as if the lower layer did not exist. When R is large, it reduces to that in a medium of resistivity ρ_2 as if the superficial cover does not exist. For a layered ground of n strata, there is a more complex kernel under the integral but it can be described by a recursion relation in which the influence of each layer intervenes via contrast coefficients of the type $(\rho_2 - \rho_1)/(\rho_2 + \rho_1)$ and via exponentials in which the thickness h intervenes of the type $\exp(-\lambda h)$.

Numerical calculation of the Hankel transform
The calculation of the potential at a point implies the calculation of a Hankel transform $K(R)$ of a function k according to a general expression of the type:

$$K(R) = \int_0^\infty k(\lambda)J_n(\lambda R)\,d\lambda, \quad R > 0 \quad (6.55)$$

where n equals 0 or 1 depending on the type of quadripole. The numerical integration of this expression is possible but tedious because the Bessel functions are oscillatory. It is therefore preferable to utilise a convolution method whose principle was proposed first by Ghosh (13). First a change of variable is carried out whereby the distances are replaced by their logarithms giving:

$$x = \ln(R), \quad y = \ln(1/\lambda) \quad (6.56)$$

This same change of variables is utilised in the representation of electrical sounding curves which are log–log (see below). Expression (6.55) can therefore be written in the form:

$$\exp(x)K(\exp(x)) = \int_{-\infty}^{+\infty} k(\exp(-y))[\exp(x - y)J_n(\exp(x - y))]\,dy$$

$$(6.57)$$

which is the convolution of the function $k(\exp(-y))$ by a filter with an impulse response:

$$\delta = \exp(x - y)J_n(\exp(x - y)) \tag{6.58}$$

To calculate this impulse response, that is, to determine the weights to be attached to each lag $(x - y)$, the easiest thing is to utilise functions whose Hankel transforms are known, for example:

$$\int_0^\infty \lambda \exp(-a\lambda^2)J_0(\lambda R) = \frac{\exp(-R^2/4a)}{2a} \tag{6.59}$$

The functions are quantised with a step $\Delta\psi$ of $0.20 = \ln(10)/11.513$ following Anderson (2). The numerical Fourier transform is taken, and by dividing the spectrum of the output function by the spectrum of the input function one obtains the spectrum of the filter. The inverse Fourier transform of this spectrum is thus the desired sinc response or filter weight. The calculation of equation (6.55) reduces thus to the calculation of:

$$K(R) = \left[\sum_{i=N_1}^{N_2} W_i k(\exp(A_i - x)) \right] \frac{1}{R} \tag{6.60}$$

where N_1 and N_2 are the limits adopted and $(A_i - x)$ the shifted abcissa, with W_i the weights determined as above.

Electrical sounding over a two layer earth

The preceding calculations allow the determination of the potential resulting from the injection of current into a stratified medium at every point. The most common application of this technique leads to the establishment of the graph of variation of apparent resistivity over a milieu with two layers with increasing separation between the electrodes. It can be observed, in conformity with the preceding remarks, that for short distances (a in the Wenner, $AB/2$ in the Schlumberger configurations serving as variables), the curve departs from resistivity ρ_1, of the first layer to approach asymptotically the value of the second, ρ_2, for values which are lesser or greater depending on whether or not ρ_2 is greater or less than ρ_1. This curve is called an electrical sounding. In effect it gives the apparent resistivity as a function of a quantity which varies with the depth of investigation. The curve can be calculated for all values of the parameters h, ρ_1 and ρ_2, but the law of symmetry allows reduction of all these cases to a limited series of models for which the thickness and resistivity of the first layer are normalised, allowing evaluation in terms of the dimension of the quadripole and ρ_2 respectively. Any two

layered case can thus be evaluated from a set of standard curves. One
may proceed by multiplication or by simple translation along the axis,
if, as is the most common usage, one takes the precaution of representing
the curves on log–log axes, as shown in figure 6.17. The most important
application of these standard curves consists in solving the inverse
problem, that is, from real measurements made on a supposedly layered
earth, to determine the depth of the surface of separation and the
electrical characteristics of both layers by comparison with theoretical
curves. The layers may be, as shown in figure 6.18 a water table, the
depth of a base rock under alluvium, or, in archaeology, the study of
stratigraphic layers with sufficiently contrasting resistivities. One shifts
the real curve which is drawn on transparent paper along the standard
curves until satisfactory coincidence is obtained in order to identify a

Fig. 6.17 Master curves for a two layer medium using a normal Wenner
quadripole.

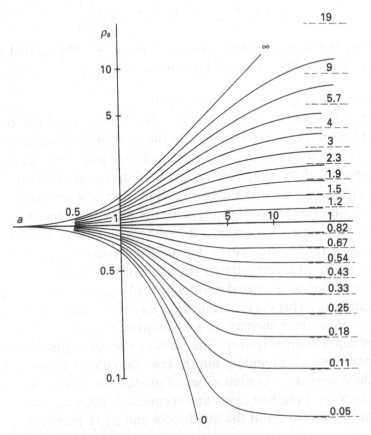

model which is representative of a real situation in which the parameters h, ρ_1 and ρ_2 are thus determined.

Interpretation of complex layered media
Since we know how to calculate theoretically all cases for n layers, one might envisage making a complete catalogue of sounding curves. It can readily be understood that the number of parameters which have to be taken into account with the introduction of a third layer leads to the construction of several families of standard curves which rapidly become unwieldly because of their number. At the most one can construct a new set of curves which Cagniard christened 'crossover points' (6) from the standard curves for three layers. These new guide curves can be used to position simple two layer curves with respect to each other such that the real measured curve is approached, as shown in figure 6.19. When the existence of more than three layers is recognised, one proceeds from point to point on the whole curve and considers successive groups of three contiguous layers.

This method only allows a first approximation to the desired solution and modern methods of calculation put a set of programs at the disposal of the investigator which are very efficient and extremely rapid for determining the exact sounding curve which corresponds to a hypothetical stratigraphic sequence. It is even possible to obtain an automatic improvement in the interpretation by an average least squares error reduction to a given threshold between the observed and calculated points as shown in figure 6.20.

In reality, the greatest limitation to the utilisation of soundings in archaeology comes from the fact that the effective layering of the medium is not assured and the exploitation of more complex models runs up

Fig. 6.18 Example showing the limits of the base rock or the surface of an aquifer at the Forum of Paestum, Italy.

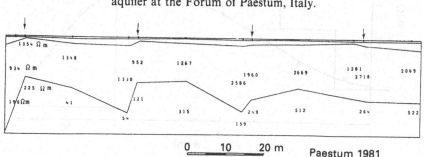

Fig. 6.19 Interpretation of an electrical sondage at the Forum of Paestum compared with two theoretical curves for a two layer medium.

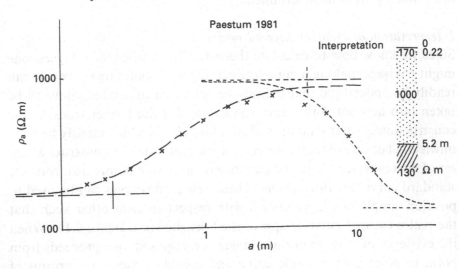

Fig. 6.20 Results of the automatic interpretation of the data (crosses) of the sondage shown in figure 6.19. The continuous curve corresponds to the resistivity and thickness of four layers.

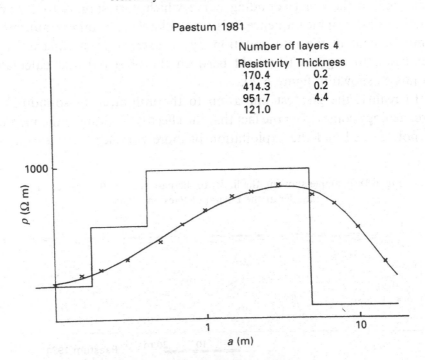

against practical difficulties. Therefore it is convenient to go ahead as if the terrain is layered, but several precautions are taken. For example, the axes of the current electrodes are oriented in the direction of the structures (if known) to reduce the effects of apparent anistropy. In those cases where doubt about the layering exists, two soundings are carried out at right angles and the differences between them allow an appreciation of the limits of interpretation.

Without going necessarily to a complete interpretation, electrical sounding can be a valuable aid prior to a resistivity mapping in order to determine the order of size and the sequence of resistivities which will be encountered, and to adjust the quadripole which will be utilised. The same is true after a mapping in which a series of soundings can be placed at the centre of the principal anomalies encountered, that is, where the layering is likely to be greatest. From the curves obtained, it may be possible to deduce qualitatively the superficial or deep-lying character of the cause of the anomaly, an essential piece of information for archaeological interpretation (16).

6.2. Practical aspects of electrical prospecting

6.2.1. Speed of measurement

The measurement of resistivities in the field with the electrical method suffered from the very beginning from lack of speed. The handicap which results from the need to move four cables, one for each electrode, is very obvious for non-archaeological prospecting because the cables can be hundreds or thousands of metres in length. In the case considered here the time for moving cables remains important but it can be reduced sufficiently so that the measurement time itself becomes non-negligible. If one considers the equation which governs the total measurement time T for a given surface regardless of method:

$$T = N(t_d + t_m) \tag{6.61}$$

where N is the number of readings, t_d the time required for movement from one station to another, and t_m the time for the measurement itself, all efforts directed toward increasing efficiency in the field have tried to reduce each of the terms of the second part of the equation. This effort is particularly necessary in electrical prospecting which suffered considerably despite its irreplaceable virtues from competition which was suddenly introduced at the end of the 1950s in the form of the proton magnetometer, which permits measurements to be made at 4–5 s time intervals or less.

6.2.2 Measuring instruments

Balance measurements using direct currents

There is a clear evolutionary trend with respect to measurement time due to progress in instrument design which is linked to the state of the art in electronics at a given moment. The direct current method which has been inherited from the earliest geophysical prospecting days can be rapidly reviewed, as in figure 6.21. It requires a source of direct current (batteries or generator) whose power is sufficient for the planned depth of investigation to inject a current of intensity I which is measured using an ammeter in series with the line supplying the electrodes A and B. It also needs a compensating potentiometer which can oppose the difference in potential ΔV which appears between the electrodes M and N as a consequence of I. This technique which follows the theory of the method is particularly slow, with 10–20 seconds per measurement needed. It also requires training of the operator and suffers considerably from imprecision due to the imperfect compensation of the self-potential of the electrodes. This phenomenon is the result of the discharge in the potential circuit of ionic charges accumulated at the contact between the metallic electrodes and the solutes in the soil. The effect is accentuated by the passage of a direct current and the changing role of current and potential electrodes along a profile. It is, however, the only possible approach when prospecting to a depth greater than 50 m. Modern devices store the value of the self-potential and repeat several measurement cycles automatically with and without current injection to overcome this problem.

Alternating current potentiometers

The utilisation of alternating current brings with it several advantages.

Fig. 6.21 Schematic of the principle of the measurement of resistivity with direct current. From ref. 16.

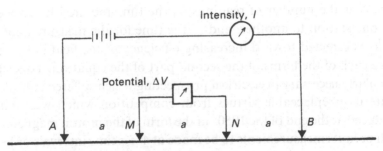

On the one hand the effects of spontaneous polarisation of the electrodes are suppressed. On the other hand the possibility exists for using a potentiometer system which directly cancels the voltage which appears at the potential electrodes as shown in figure 6.22. The annulling voltage is taken from the variable secondary of a transformer whose primary is in series with the current electrode circuit, and this voltage is evidently proportional to I. It permits the calibration of the annulling potentiometer $R = \Delta V/I$ is independent of the current and corresponds directly to the apparent resistivity within a factor k. The measurement is obviously much quicker than that described above, taking about 5 s and it is much more repeatable and precise. The disadvantage of alternating current lies in the limitation on the depth of investigation due to the skin effect (chapter 9) and in rarely encountered disturbances due to noise sources at frequencies close to that used in the measurement. Such disturbances are largely eliminated in most instruments by the use of phase coherent detection of the signal to be annulled. Numerous commercial versions of compact resistivity meters have been produced for the market since the 1950s. They usually use batteries as a current source and a frequency between 100 and 150 Hz, examples being those manufactured by Nash and Thompson, Gossen, Norma, Chauvin Arnoux etc. The use of the classical 'Megger' described first by Atkinson (3) and later by Aitken (1) remains an historical curiosity because of its purely electromechanical construction.

Direct measurement of ΔV using a regular alternating current
Progress in the measurement of potential differences currently allows us to go beyond the limits of balance measurements. These were necessary earlier in order to avoid disturbing the distribution of current in the soil which would have been produced by using part of the current in the

Fig. 6.22 Schematic of the principle of the measurement of resistivity with alternating current. From ref. 16.

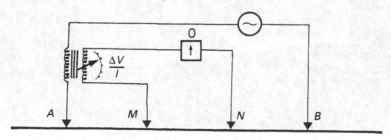

potential circuit. Most modern devices measure the potential difference ΔV directly using an input impedance which is sufficiently high so that the condition of non-disturbance is fulfilled. In order to obtain a reading which is directly proportional to the resistivity, the current I is maintained at a constant level. The reading therefore requires no manual intervention and is much more rapid. The display is either a moving coil meter (Bradphys) or digital (Gossen, Septa). The latter device, figure 6.23, has been especially designed for use with automatic recording which is compatible with goals to be described below. Thus, the efforts aimed at reducing the measurement and station time have joined to replace the obsolete procedures which have previously been used for electrical prospecting.

Electromagnetic techniques
The improvements in resistivity measurement techniques by classical electrical methods cannot mask an incontestable advantage available in electromagnetic approaches to the same problem as discussed in chapter 9. This eliminates the electrodes and the cables. However, the electrical technique has several important advantages: a great flexibility in the adaptability of different electrode configurations which can be used to distinguish responses from different depths, a superior ability to detect

Fig. 6.23 Resistivity meter using regulated alternating current, RMCA3-SEPTA made under licence from the CNRS-ANVAR, France.

highly resistive structures in less resistive media, and finally the fact that metallic objects or structures nearby do not disturb the measurements significantly.

6.2.3. Electrode systems

We have seen above that the electrode system imposed by the theory of resistivity measurement requires four points in contact with the ground in order to overcome the effect of contact resistance. A large variety of arrangements has been suggested since the method was first used in archaeology, without being able to say that one is definitely better than another. In order to judge the matter we must take into consideration a series of heterogeneous factors, such as the nature of the targets sought, the dimensions of the surfaces to be explored, the type of measuring equipment etc. In current practice, without automatic recording, the choice must essentially be made of a system of electrodes, their construction, and the methods for connecting to them in I and V. These factors have a direct influence on measuring time and on the quality with which subsurface structures are represented by the resistivity anomalies which they provoke.

In-line quadripoles

Independently of the above mentioned constraints, investigators rapidly discovered that it was practical to place the electrodes in line and Schlumberger proposed the configuration which carries his name, for which MN is small ($\leqslant \frac{1}{10}a$), figure 6.24(a). Even if one obtains an investigation depth which sometimes appears to be remarkable, several disadvantages result, especially the parasitic effects which occur when the potential electrodes encounter small superficial structures with high resistivity contrast. In addition, it is difficult to move the quadripole easily.

These problems disappear with the use of the Wenner configuration, figure 6.24(b) in which the in-line electrodes are equidistant. The movement of a single electrode is sufficient to obtain the same geometric configuration displaced by one unit, a. It is merely necessary to reestablish the order of connection to the electrodes to recover the initial Wenner arrangement. Several types of switches using four or five prewired positions have been proposed (3,31). It may also be observed that the connection obtained by the first electrode displacement, namely $MN\ BA$ is of the dipole–dipole type in which the measurement of potential is not made in the interval which separates the two current electrodes, but rather is external to them. This configuration can also be obtained from the initial one by changing the lines B and M at the contacts of the resistivity meter. By combining these two observations with the possibility of exchanging current and potential electrodes without modifying the measured value, as demonstrated theoretically for homogeneous media, one can rapidly realise a linear moving system having two different responses, namely the normal Wenner and the dipole–dipole Wenner.

Among the inconveniences which are inherent in these procedures for moving the electrode system is the lengthening of the quadripole along the profile axis which constitutes a factor leading to apparent anisotropy. In addition, a certain amount of unnecessary remeasurement of the intervening ground takes place in two successive readings. As a corollary, two successive profiles on surface S cannot be separated too widely for fear of accentuating this effect if the distance is greater than a. It follows that this procedure practically forces a number of measurements $N = S/a^2$ for a given investigation depth. For extended shallow structures, this results in a large number of unnecessary measurements.

Fig. 6.24 Various electrode configurations: (a) Schlumberger; (b) Wenner; (c) twin electrodes.

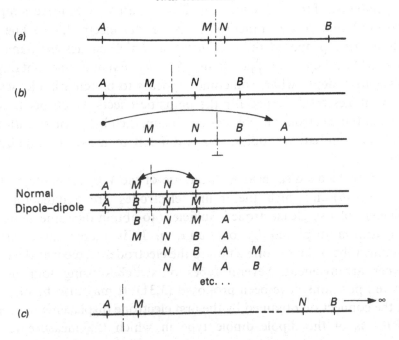

Fixed electrode arrangements

If the quadripole utilised is small in size, a few metres or less, then it is possible to attach the electrodes to a fixed or articulated frame and to move the whole assembly at once for each reading. The in-line arrangement is no longer required and Clark invented an arrangement using the four legs of a table carrying the resistivity meter and the sheet of paper for recording the results, figure 6.25, which can easily be moved from one point to another (8).

The advantage of using a square array with respect to apparent anisotropy has already been described. If the electrodes are in-line, it is preferable that the frame is articulated (15) in order to follow terrain undulations better. Therefore one has complete latitude in the choice of the position of the quadripole with respect to the profile, the interval between measurement and between the profiles, and consequently the superficial measurement density *N/S*.

Fig. 6.25 Portable measuring table the legs of which form a square quadripole, following Clark. From ref. 8.

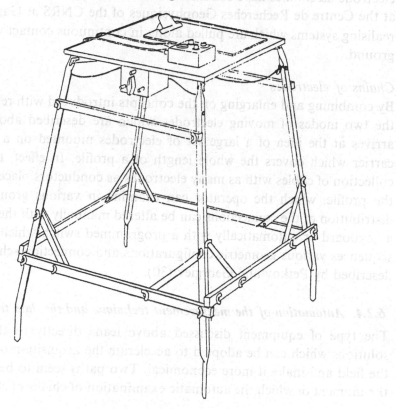

The same latitude is offered by an arrangement favoured by the Anglo-saxon school of archaeological prospecting. The quadripole is reduced to a twin electrode which is made up of a single current and a single potential electrode spaced from 0.5 to 1 m apart. Another electrode pair is are placed at a great distance, in theory at infinity relative to the area to be investigated. The electrode carrier which is light and easily handled is designed so that one can carry the measurement device from point to point. This method slightly resembles the mapping of potential, and apart from the lightness of the apparatus has the same properties as the previously discussed arrangements. It is well adapted for the exploration of small surfaces but requires the merging of a number of survey grids when the surfaces are extensive. In terms of apparent resistivity this is difficult and it may lead to false anomalies being detected.

All of the techniques mentioned are useful on light soils. A hard, stony, compact surface always requires a return to the use of steel electrodes which are hammered into the ground separately. The trials with fixed electrode assemblies have culminated in recent experiments carried out at the Centre de Recherches Geophysiques of the CNRS at Garchy for realising systems which are pulled along in continuous contact with the ground.

Chains of electrodes

By combining and enlarging on the concepts introduced with respect to the two modes of moving electrodes which are described above, one arrives at the idea of a large set of electrodes mounted on a flexible carrier which covers the whole length of a profile. In effect, this is a collection of cables with as many electrodes as conductors placed along the profile, which the operator can examine in various groups. The distribution of the connections can be altered manually with the help of a keyboard or automatically with a programmed switch which rapidly sequences various geometric configurations and connection schemes as described by Petkov and Georgiev (30).

6.2.4. *Automation of the measurement technique and the data treatment*

The type of equipment discussed above leads directly to the latest solutions which can be adopted to accelerate the acquisition of data in the field and make it more economical. Two paths seem to be open at the moment of which the automatic examination of chains of electrodes has the incontestable advantage of lightness as compared with motorised quadripoles mounted on frames. The fact that one has electrodes along the whole length of the profile allows the examination of electrodes according to the way they are connected and with very different dimensions. A project, which is in the course of realisation as this is being written, aims at obtaining a sequence of measurements over a line of 30 electrodes spaced at metre intervals in a short period of time (of the order of a second). The arrangement which will allow the use of a normal Wenner and a dipole–dipole scheme gives 30 readings at 1 m, is at 2 m, etc., figure 6.26. By comparing successive profiles a resistivity

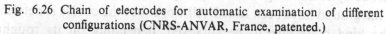

Fig. 6.26 Chain of electrodes for automatic examination of different configurations (CNRS-ANVAR, France, patented.)

sounding series is obtained, so that the whole gives a three-dimensional cover of the underlying medium.

The other technique which has already been tested using the so-called RATEAU (Résistivimètre Auto-Tracté à Enregistrement AUtomatique), an acronym which is a play on words for a harrow, which the device resembles (23,10). It is used in recording the resistivities obtained by towing a quadripole in contact with the ground along a profile. The idea which was suggested quite a long time ago (14) did not then lead to the construction of an operational prototype because a number of problems remained to be solved for the whole system, figure 6.27. Electrode systems using either liquid jets, points, or discs for contact with the ground have been tested. All work, but the choice is determined by the kind of terrain to be surveyed, depending on its roughness, hardness, humidity, and the state of surface vegetation. The resistivity meter has been the object of a special study, since no commercial instrument is applicable for this kind of system. Great attention has been given to obtaining the shortest possible measurement time (24) and correct operation at high contact resistance.

The data acquisition system is not the least of the problems to be

Fig. 6.27 The RATEAU (harrow) with disc electrodes in continuous contact with the ground.

resolved. The recording of position is assured by means of the wheel which carries the quadripole. At each partial turn corresponding to an advance of 10 cm, a reading is started and this is recorded in a microprocessor after analogue–digital conversion. The recorded data are then transferred to floppy disk, whose analysis allows either the visualisation of the raw data or the application of various treatments. We can thus obtain an almost continuous profile, as shown in figure 6.28 in which the imperceptible discontinuities due to sampling only affect the small resistivity anomalies recorded. That these are real and not noise is clearly proved by the repeatability of the profile, with the small differences observed attributable to the impossibility of placing the quadripoles on the same line exactly. The map shown in figure 6.29 constitutes one of the first examples of electrical prospecting carried out with this technique.

6.3. Problems in electrical prospecting

It is not evident that the factors which perturb measurements and which are likely to cause erroneous interpretation are more numerous in resistivity than in other prospecting techniques. Undoubtedly experience and a certain ease in identifying them allows us to catalogue the pitfalls. Although this is not directly applicable to other methods, it allows us

Fig. 6.28 Continuous resistivity profile obtained with the RATEAU (square quadripole of $a = 1$ m) at the medieval metal working site at Minot, Côte d'Or, France. The large sharp anomaly probably represents a wall, while the other strong values are accumulations of slag and waste. From ref. 10.

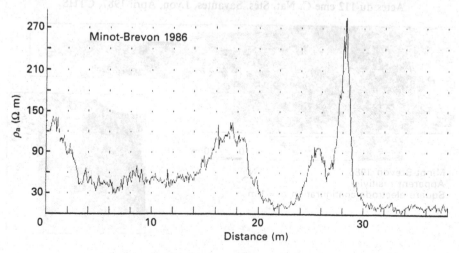

to suggest a number and variety of hazards which await the user at all phases of the operation.

6.3.1. Instrumental errors

These errors come from the measuring instrument itself. It is, however, the most reliable link in the measurement chain, and given the great diversity of types manufactured, it is not realistic to list their respective merits and defects here. The errors which exist in the coverage of different measurement ranges present in nearly all instruments are a consequence of the very large range of values encountered in soil resistivities. Absolute calibration is frequently inadequate, and this may be a source of important parasitic anomalies if several instruments are used in the same survey. An elementary precaution at least consists in taking a joint profile at the boundary of two subareas surveyed with different devices. The main source of error or imprecision comes from the electrode system itself.

Loss of precision through loss of sensitivity
Several factors reduce sensitivity:
(1) Contact resistance which is too large between the electrodes and the ground limits the magnitude of injected current and consequently

Fig. 6.29 Apparent resistivity map of nearly 2 ha at the metal working site of Minot using a square array of $a = 1$ m, one measurement per square metre, obtained in two working days. From M. Dabas, A. Hesse and A. Jolivet, Prospection geophysique de site metallurgique anciens à Minot Actes du 112 eme C. Nat. Stés. Savantes, Lyon, April 1987, CTHS.

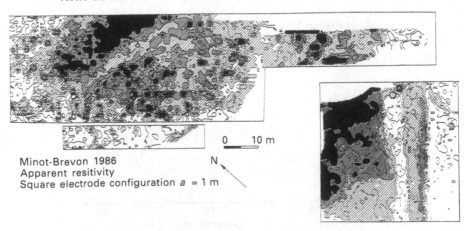

0　　10 m

Minot-Brevon 1986
Apparent resistivity
Square electrode configuration a = 1 m

N

reduces the voltage which appears at the potential electrodes. This is reflected in a reduction of unbalanced current and a difficulty in obtaining a precise zero in instruments which have manual balancing. In more modern devices with regulated constant current the more complex effects are often masked by the direct readout circuitry. The RCMA3, for example, remedies these difficulties within certain limits by having very great tolerance to high contact resistance.

(2) Difficulties appear at large electrode spacings. In a homogeneous earth with constant contact resistance, raising the electrode separation AB causes a reduction in ΔV in the formula $\rho = 2\pi k \, \Delta V/I$ not only because of the increased soil resistance between the two electrodes but also because of the coefficient k. If a Schlumberger arrangement is used, the effect is increased because of the small size of MN. All of this becomes more important when the resistivity contrasts to be detected are small, the sought-after bodies less voluminous, and when they lie at greater depth.

Errors in electrode positioning
Electrodes are never theoretical point sources, since they penetrate several centimetres into the ground. An acceptable limit for the depth of insertion appears to be 10% of the difference between the electrodes. Beyond that, variations can provoke parasitic anomalies. An error in the distance between electrodes can be more significant in that it modifies the coefficient k. This error (15) varies depending on whether the current or the potential electrodes are affected. In the first case:

$$\left|\frac{d\rho_a}{\rho_a}\right| = \frac{3}{4}\frac{da}{a} \qquad (6.62)$$

for a Wenner configuration and in the second:

$$\left|\frac{d\rho_a}{\rho_a}\right| = \frac{5}{4}\frac{da}{a} \qquad (6.63)$$

The parasitic anomaly obtained with an incorrectly placed electrode has the characteristic shape shown in figure 6.30. It should be kept in mind that the perturbation of the readings is of the same order of magnitude as that introduced by errors in electrode spacing for a Wenner configuration. Obviously if a fixed electrode configuration mounted on a jointed or towed frame is used, the source of error vanishes. The error in the absolute position of the quadripole itself remains. This does not create

parasitic anomalies, it merely distorts their shape, especially in areas
where the variation in resistivity is abruptly changed between two
measured points in the vicinity of an anomaly. A similar phenomenon
is frequently observed in magnetic prospecting (see chapter 8).

6.3.2. Parasitic electrical phenomena

It is not possible to escape entirely from the effects of parasitic currents
which disturb the measurements, especially when using direct current.
However, experience shows that they are negligible, due to the very small
voltages encountered with short electrode spacings. With alternating
current, the filtering effect of the circuits which respond only to the
injected frequency makes this even more certain. The only case where
such disturbances have been observed is when two resistivity meters
are used near each other simultaneously. From 20 m onwards, the
disturbance usually vanishes.

   The problem of electrode self-potential can be treated by use of
alternating current instrumentation. If this is not done, it appears as a
sawtooth curve rather like that shown in figure 6.30, in which a parasitic
anomaly is superimposed on the true variation whose measurement is
desired. This is why alternating current devices are almost universally
used for shallow depth archaeological prospecting.

Fig. 6.30 Parasitic anomaly caused by incorrect positioning of an electrode
in a Wenner quadripole. From ref. 15.

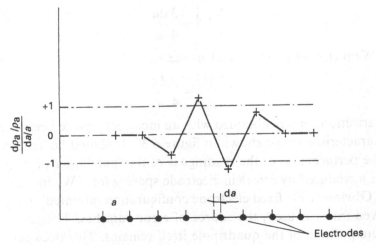

6.3.3. *Topographic and surface effects*

Surface relief

The user frequently observes that passage over a hillock or a ditch provokes an anomaly. The simple examination of the distribution of the lines of current in the ground shows very clearly that in certain electrode orientations which are particularly unfavourable with respect to variations in topography, the current density rises when it is obliged to approach the surface, as is the case with an open ditch. This causes an increase in ΔV at the surface even if the soil is totally homogeneous and a consequent positive anomaly, figure 6.31(a). Conversely, under a hillock, the lines of current are free to spread, the current density diminishes and a minimum is observed, figure 6.31(b). This can be a serious difficulty when prospecting on uneven terrain.

Effects due to agriculture

Previous treatment and the state of the surface can affect measurements. Working the soil to conserve its humidity is a well-known process. The mechanical modification of the mobility of the first few decimetres of soil changes its cohesion. Plants modify the equilibrium by transpiration. All of these factors, which are rather poorly understood at present, influence the surface resistivity strongly. This necessarily affects the measurements even if these are made with quadripoles several metres long designed to detect deeper structures. Parasitic anomalies are frequently observed when passing over field boundaries and have to be

Fig. 6.31 Anomalies caused by topographic disturbance of a homogeneous terrain: (a) a ditch, (b) a bank. From ref. 16.

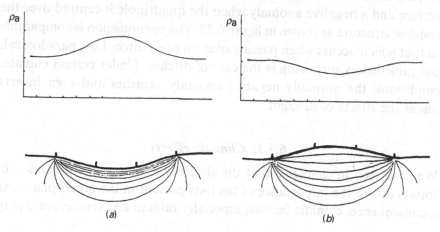

(a)   (b)

eliminated from the geophysical interpretation by accurate observation
of the surface topography in the area surveyed. Archaeologically,
these phenomena should never be eliminated without serious reflection,
inasmuch as they may be due to the survival of older field boundaries
or structures reflected in the contemporary boundaries.

Archaeological excavations
The increase in resistivity produced by proximity to an excavation cutting
must be classified along with other members of the family of disturbances.
The cutting constitutes an infinitely resistant volume in the ground which
is detectable like any other cavity. It is obviously impossible to make
measurements over the excavation, and the disturbance is evident in the
form of a lateral effect on all profiles which run parallel to it. The
theoretical effect can be expressed through the formulae already given
(equations (6.31), (6.32)). The resistivity anomalies are accentuated by
the drying out of the section which is exposed to the atmosphere. In
general, all terrain changes – filled in cuttings, spoil heaps even after
their removal, etc. can cause problems. This is behind the failure or lack
of significance of all attempts to evaluate anomalies on structures which
have been reburied after excavation.

6.3.4. *Resistivity paradoxes*

Although examples are rare, the case of a highly resistant wall which
outcrops at the surface, masked by a very slight cover of earth or grass,
is not an academic hypothesis. In this case the lines of current flow
cannot pass between the surface and the structure and are constrained
to pass under the wall. The result is a drop in current density at the
surface and a negative anomaly when the quadripole is centred over the
resistive structure as shown in figure 6.32. The phenomenon is comparable
to that which occurs when passing over an open ditch. Less paradoxical,
but nonetheless surprising is the case of ditches. Under certain climatic
conditions, the normally negative anomaly vanishes and even inverts
under the effects of drought.

6.3.5. *Climatic effects*

We have seen in chapter 2 that the electrolytic nature of conduction in
soils is due to the importance of the ions present in the liquid phase. As
a consequence, climatic factors, especially rainfall and temperature play

a fundamental role in the electrical properties of the uppermost soil layers which are directly subject to strong and abrupt changes. Two aspects must be considered.

Variations in the resistivity of a homogeneous earth
Experiments conducted during the 1960s have shown that it is convenient to distinguish two periods in the year for middle latitudes in Europe (15): (1) a humid period which extends from the beginning of autumn to the end of spring during which we may consider the soil to be permanently saturated with water, from which it follows that the predominant effects are due to temperature; (2) a dry period from the end of spring to the beginning of autumn during which alternating rainy and dry spells lead to a slow drying out and sudden rehumidification of the soil which is superimposed on the thermal effects and which exceeds them in amplitude.

Temperature effects. The conductivity of a saturated soil is directly linked to the conductivity of the interstitial water whose variation with temperature is approximately linear at ambient temperatures. The way in which climatic variation affects prospecting is not simple, however, because the thermal variations act primarily on the surface rather than on the bulk of the soil. In addition, the user does not measure resistivity

Fig. 6.32 Paradoxical anomaly produced by a resistive wall which rises to the surface, with *a* less than or equal to the thickness *e* of the wall. To be compared with figure 6.11, $a > e$. From ref. 16.

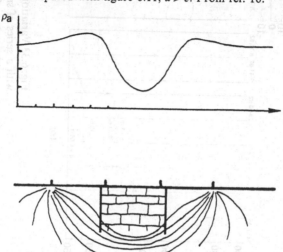

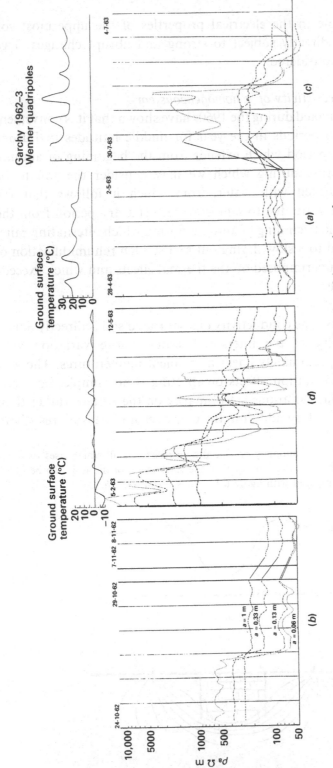

Fig. 6.33 Variation of the surface resistivity as a function of time and weather with a series of small quadripoles. Visible are: (*a*) diurnal oscillations; (*b*) rapid inversion of the order of resistivities between the surface and at depth following rain on the 27 October after a long dry period; (*c*) slow reduction of the resistivity of the largest quadripole (*a* = 1 m); (*d*) the effects of surface freezing. From ref. 15.

at a point but rather apparent resistivity which integrates the thermal effects from the surface right down to the maximum depth of investigation.

A basic experiment has been performed by following the apparent resistivity during a period for a set of small quadripoles with spacings of 6, 13, 33 and 100 cm as shown in figure 6.33. A diurnal variation is observed with a minimum in the middle of the day corresponding to the maximum solar energy received. The effect is obviously more marked for the smallest quadripoles, but it remains visible even for a 3 m total length and reaches 5–10% in less favourable instances. Obviously, for surveys in which the anomalies can attain an order of magnitude which is much greater, these relatively small perturbations can be neglected. The particular case of frost must be considered, however. It is sufficient to observe that the diurnal variation can reach ratios of 1:50, figure 6.33, for the smallest quadripole when there is alternate freezing and thawing. The effect on larger quadripoles is not negligible and can be as much as 50%. If these oscillations are superimposed on variations of greater duration due to heating or cooling over several days, the thermal effects are transmitted at depth, following laws discussed in chapter 10. The variations in apparent resistivity are very much greater and must be monitored during the whole campaign in order to make eventual corrections to the measurements.

By observing electrical soundings for the four fixed quadripoles described above and measuring the surface temperature with a thermometer barely inserted into the soil, an experimental relationship between the resistivity of the uppermost layer and temperature can be established. Figure 6.34 shows, in logarithmic scale, the existence of two mathematical laws within the limits of experimental error with a sharp anomaly at 0°C, which can be written in the following form:

(1) above 0°C:

$$\rho = \exp(-A\theta + C) \tag{6.64}$$

(2) at 0°C the resistivity jumps sharply from a value $\exp(C)$ to a value C' which probably corresponds to the freezing of free water;

(3) below 0°C:

$$\rho = C' - B\theta \tag{6.65}$$

Humidity effects. Observations from a set of fixed quadripoles during the spring and summer show a rise in resistivity for the smallest spacing which may exceed 10:1. If the resistivity of this layer was less than that of the subsurface during the winter (in a clay soil on limestone, for

example), it may be much greater during the course of the summer.
A Wenner configuration with $a = 1$ m is evidently very sensitive to this
change, and it must be monitored during a long survey in order to
correct the readings. Obviously, in this case, the effect is attributable to
a drying out of the ground. Later rainfall, in particular in the autumn
after a dry summer, quickly restores the apparent resistivities to their
former values as shown in figure 6.33.

Variation of anomalies with time

Instead of recording the data from a set of fixed quadripoles of various
lengths, a profile can be repeated at regular intervals during the course
of a year. Evidently it will be observed that all values increase when
drying or freezing takes place. During heating or after rainfall, the values
decrease. If the profile contains an anomaly, that is, if the measured
values are distributed over a large range, it is also observed that two
profiles made under different climatic conditions are related by a simple
coefficient of proportionality. This is manifest as a simple translation of
the profile along a logarithmic resistivity ordinate. It can also be observed
as a characteristic alignment of the cloud of points which is representative
of the measurements made at two different times, figures 6.35 and 6.36.
This situation occurs only when variations are relatively weak. If
observations are extended over a long period, the variations in resistivity

Fig. 6.34 Variation of the resistivity of a water-saturated clay as a function
of temperature. The open circles correspond to a dry period, so that
depending on the humidity of the soil, they do not follow the same law.
From ref. 15.

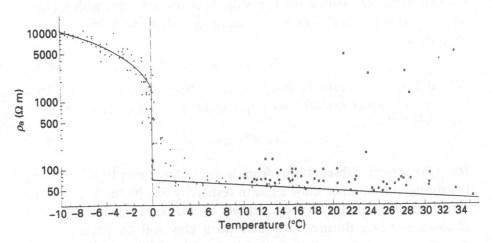

may be important and a new phenomenon appears. The relative values of the anomalies are not conserved and this is reflected as a significant difference in the logarithmic ordinate, figure 6.37.

There are at least two explanations which can be offered for this phenomenon, and both mechanisms may occur at the same time. On the one hand, when the chance variations in temperature and humidity last long enough, their effects may exceed the thickness of the relatively homogeneous surface plough layer. They therefore reach the archaeological layer, the site of the sought-after anomalies, whose various materials may react differently to climatic change. The modification of the ratio of their resistivities may modify the amplitudes of the anomalies. On the other hand, an important change in the superficial resistivity may seriously modify the depth of investigation for a given quadripole. The contribution from the materials of the archaeological layer toward the formation of an anomaly may therefore be more or less important depending on whether the lines of current are attracted toward the surface (surface layer more conductive), or pushed downward (surface layer more resistive). Theoretical calculation (32) shows that the anomalies increase in a logarithmic fashion as a function of the resistivity of the first layer, starting from a model which assumes a surface layer resistivity ρ_1. When ρ_1 is very high, the anomalies tend to vanish, as one may observe in the field especially in the case of freezing temperatures.

In an experiment on a Merovingian cemetery with anomalies obtained

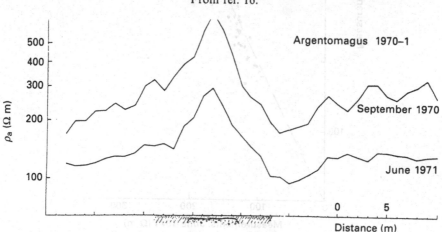

Fig. 6.35 Comparison between two resistivity profiles measured over a Roman road at Argentomagus (Indre) under different climatic conditions. From ref. 16.

Argentomagus 1970–1

September 1970

June 1971

ρ_a (Ω m)

500
400
300
200
100

0   5

Distance (m)

over pits containing sarcophagi cut into a Jurassic limestone and covered with a superficial clay layer, figure 6.37, several phases can be distinguished.

Dry period. At the beginning of the dry period, the profile contains low values. The anomalies, two peaks, are weak. When the soil dries out, the resistivity of the superficial structures rises and the peaks increase along with the small anomalies at the extremities of the profile. As the

Fig. 6.36 Comparison of the values obtained at each point in the resistivity profile of figure 6.35 at the two periods under consideration. Linearity is not fully respected for the observed slope of about 2. From ref. 16.

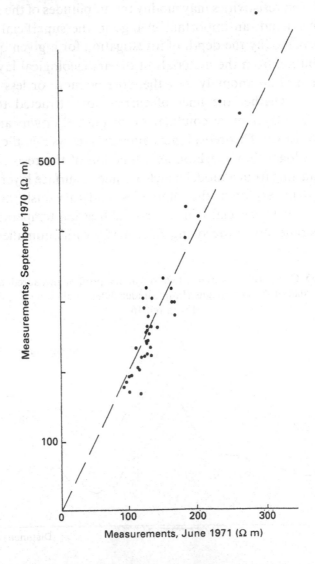

depth of drying increases, the average resistivity rises, increasing anomalies along with it. If moderate rainfall follows, the water does not reach the deeper dry layers, but does modify the depth of investigation. Therefore the anomaly is reduced, as seen in a profile made on 13 September, 1962. This takes place without disturbing the irrevocable rise in average resistivity. The phenomenon of anomaly fluctuation may obviously be repeated on several occasions during the dry spell. At the end of the summer, the apparent resistivity has tripled.

Humid period. With the first rains, the anomalies fall rapidly as if a dry period fluctuation had occurred, with a slight reduction of the average apparent resistivity. If the rains continue, the anomalies do not rise again, the drying is no longer sufficient, and in a dozen days or so, the deeper layers are also saturated. Therefore the average resistivity falls rapidly, and this evolution continues until the end of the humid period. Through heating of the ground and perhaps through seasonal evolution of the electrolytes in the soil, the resistivity drops to 50 or 60 Ω m. The anomalies weaken and even tend to disappear.

Freezing weather. We have noted above that freezing produces a considerable rise in soil resistivity. However, a light frost like moderate

Fig. 6.37 Climatic variation of anomalies over a test body (a sarcophagus pit in the Merovingian cemetery at Garchy) using a normal Wenner quadripole with $a = 1$ m. Compare with the dates of figure 6.33. (a) Dry phase, (b) humid phase, and (c) freezing phase. From ref. 15

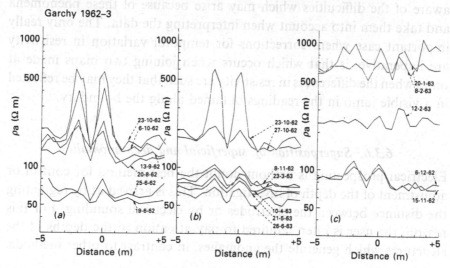

drying only gives rise to a weak increase in resistivity without significant modification of the anomalies. If the cold weather continues, parasitic anomalies of an amplitude comparable with those of true anomalies appear, all the resistivity values rise considerably and the profile becomes unreadable.

Correction of readings for temperature and humidity
From the preceding observations, it is evident that no *a priori* consideration allows one to predict or evaluate the amplitude of the temperature and humidity dependent effects in order to make corrections. The experienced user can frequently appreciate subjectively the importance of the perturbations and neglect them. But when in doubt it is prudent to verify things, either with a fixed quadripole or better still, with a profile which contains a good sample of the observed resistivity values which is remeasured at regular intervals. The average of the ratios of the readings obtained on two occasions, or better still the slope of the linear regression between readings, figure 6.36, can be used as a correction coefficient.

If variation is severe, the correlation may no longer be linear, as is the case with the variation of the amplitudes of the anomalies. One may therefore introduce finer empirical corrections based on the curve obtained. It is important to be conscious of the limits of this method, since it is never certain that the variations in resistivity are the same at two points in the same area or that the anomalies, which may be of different origin, are subject to the same modifications. The user must be aware of the difficulties which may arise because of these phenomena and take them into account when interpreting the data. The only really important case where corrections for temporal variation in resistivity are imperative, is that which occurs when joining two maps made at times when the differences in resistivity are such that they may be reflected in a visible jump in the readings obtained along the boundary.

6.3.6. Superposition of superficial and deep anomalies

Electrical prospecting is without doubt the best method for control or adjustment of the depth of investigation. This is arrived at by regulating the distance between the electrodes or by electrical sounding. For this reason, the user is often required to pay attention to the depths of the structures which generate the anomalies, in contrast to other methods.

However, causes of anomalies of archaeological origin are most frequently characterised by their presence at intermediate depths between:
  (1) A topsoil layer whose depth of 30–50 cm is defined by ploughing which has mixed it thoroughly and thus made it quite uniform. The tops of buried structures have been removed by the process.
  (2) A subsoil layer of geological origin on which the ancient structures sought were constructed.

It is an illusion to think of these two milieu as being homogeneous. The geological subsoil displays strong electrical contrasts between clays, loams, sands, and gravels in alluvial terrain, between limestones and marls in a sedimentary context, between weathered and unweathered zones on hard rocks etc. If the archaeological layer is sufficiently thick, one may hope to use an electrode spacing which is sufficiently short so as to be free of deeper effects. But this may be in conflict with logistic constraints, such as the desired rapidity of exploration. Larger electrode spacings give faster coverage of an area, of course. In any case, it is difficult to be free of the effects of the surface layer whose contribution to the apparent resistivity is always important as a consequence of the strong current density flowing through it. All of the parameters which characterise this layer (inhomogeneity, water content, temperature, thickness, etc.) are likely to affect the measurements.

In the final analysis, the measurements represent the superposition of the perturbing effects of the two layers on the actual anomalies of the archaeological structures sought. It is the role of the interpreter to separate and ignore those anomalies which are not of archaeological interest. Experience shows that the situation is less desperate than this account might lead one to believe (20) inasmuch as certain information relative to the geological substructure can contribute to a better geographic understanding of the environment of the archaeological site under study.

6.4. Structures of different types

6.4.1. Conductive structures

These are essentially archaeological structures excavated at a remote period in a resistant material such as limestone, gravel, or alluvial sands. These structures, mainly pits and ditches, were either filled up during the course of time, or filled deliberately at the time with a loose material which was either finer or contained more clay than the surrounding rock. These features therefore show characteristic resistivity minima. The first surveys by Atkinson (4) were carried out on this type of site.

Long structures which are made up of rectilinear ditches are rather suitable for prospecting by means of widely spaced profiles, figure 6.38, but the risks inherent in the identification of an anomaly via such a technique are well known (31). It is therefore preferable to measure a series of closely spaced profiles so that the anomaly does not disappear between two of them. With complete resistivity mapping in cases such as that of the circular enclosure at Moneteau (15), the identity between the response obtained in electrical and magnetic prospecting over a ditch filled with black earth having considerable magnetic susceptibility was shown. This explains the economics of the long standing competition between the two methods for rapidity.

In another example shown here, figure 6.39, the conductive structures are remarkably clear. They are due to shallow foundation trenches in volcanic tufa in the Forum of Paestum in southern Italy. The tufa is covered with 30–50 cm of earth and a Wenner quadripole with $a = 1$ m on a 1 m grid shows fine detail. The anomaly aligned SW–NE is not a

Fig. 6.38 Successive resistivity profiles at 3 m intervals, $a = 1$ m, which show a straight ditch at the Gaulish site of La Croix des Sables at Mainxe (Charente). From C. Burnez and A. Hesse, Le site gaulois de la Croix de Sables, *Bulletin SPF*, **68**, 1971, Etudes et travaux, fasc. 1.

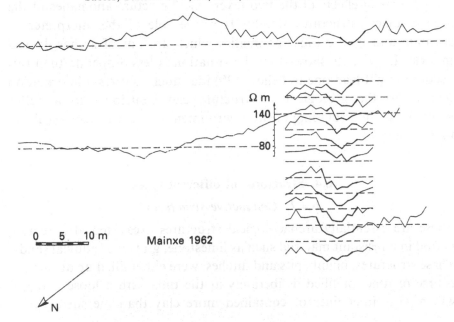

Ω m
140
—80

0   5   10 m   Mainxe 1962

N

foundation trench, but rather a drain of which only the upper conductive part was detectable.

Pits (dwelling pits, wells, graves) can also be identified through resistivity prospecting. However, magnetic prospecting is more effective if their dimensions are small, such as seen in the inhumation graves at Garchy (15). They are particularly difficult to identify in a disturbed geological context without appropriate data treatment (19).

Fig. 6.39 Resistivity map of the Forum at Paestum. The darker anomalies correspond to ancient ditches in the lower volcanic tufa (Wenner dipole–dipole, $a = 1$ m).

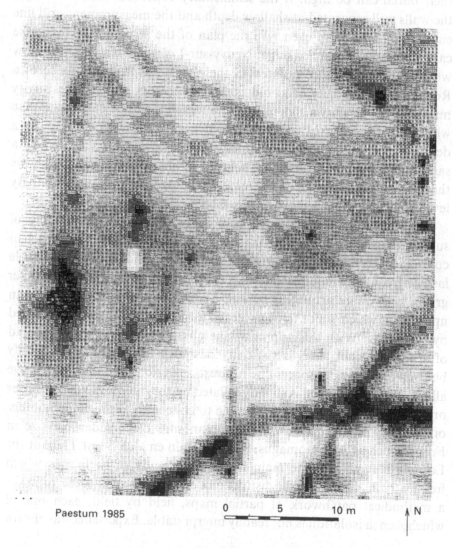

Paestum 1985

0   5   10 m

↑ N

6.4.2. Resistive structures

This is the incontestably privileged domain of application of resistivity methods.

Walls and urban structures

Stone, which is used in the construction of walls and in a large number of dwellings is frequently by its nature a body of high electrical resistance. Put together with sand mortar or dry built, walls constitute bodies which pose a strong barrier to the passage of current and the contrast with detrital material which results from their destruction or which led to their burial can be high. If the sedimentary context is simple enough, the walls well separated at shallow depth and the measurement grid fine enough, mapping can even give the plan of the structures. This is the case for the survey carried out at Neuvy-sur-Loire in a large flat meadow where various evidence pointed to the poorly localised presence of a Roman site which may be as old as the first century AD. The resistivity map, figure 6.40, shows high resistivity anomalies with a rectangular plan which subsequent excavation identified as walling with its base at a depth of 1 m, made of flint or limestone blocks solidly cemented by a sand-lime mortar. The figure, displayed in shades of gray also shows that the degree of blackening is more advantageous for some resistivity levels and that each is more or less favourable for certain walls.

The stratigraphic situation does not always permit the making of such a map, but above all, the necessity of identifying large urban complexes occupying several hectares in an economically acceptable fashion frequently requires that the measurements be made with a coarser grid. The Wenner arrangement with $a = 2$ m seems, however, to be an upper limit which is not to be exceeded for sites which have been erased by ploughing. With this approach one obtains a rough outline instead of a detailed plan. This is sufficient inasmuch as it is easy to identify built-up areas, empty spaces due to courts, exterior areas etc., and above all the street plan which is slightly inflated by the buildings which border on it, figure 6.41. It has been possible to collect an appreciable number of images which represent Gallo-Roman sites of considerable size in France, such as Argentomagus, Saint Romain en Gal, Saint Thibaut sur Loire, Glanum, etc. The main rule for the study of such sites which frequently lie under several distinct modern fields seems to be to make a methodical patchwork of partial maps, field by field, each one of which seen in isolation is not readily interpretable. Experience has shown on several occasions that even if important areas remain unmeasured, as in figure 6.41, the final collection of maps will allow identification of the major features of large organised urban complexes.

The survey of some large sites is equivalent to the case described above

Fig. 6.40 Resistivity map of the Gallo-Roman site at Neuvy-sur-Loire (Nièvre). Wenner dipole–dipole $a = 1$ m. Different features appear depending on the coarseness of quantisation of resistivities to produce a given gray value.

neuvydd 50 120

neuvydd 65 120

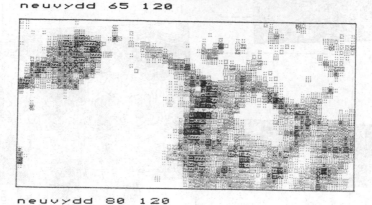

neuvydd 80 120

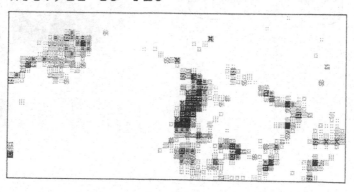

Fig. 6.41 Resistivity map of the Gallo-Roman site at Saint Thibaut (Cher). The high resistivity values reveal a network of perpendicular roads at approximately 70 m intervals in spite of the fact that large areas were not measured (normal Wenner quadripole, $a = 2$ m).

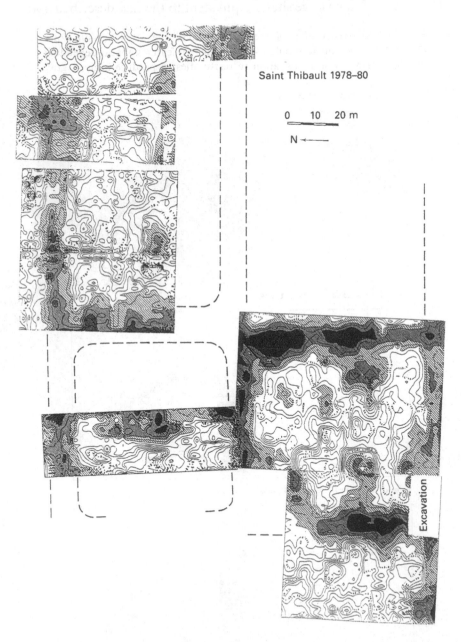

even if, properly speaking, detection of walls is not the aim. Very fine anomalies have been obtained at the Achaeminian palaces at Susa in southern Iran, even though the walls which were constructed in unbaked brick and surrounded by the clayey matter which resulted from their collapse are practically undetectable. These walls were placed on artificial pebble packing of a rather complex and variable structure. In some cases, the packing which is rather deep and narrow serves as the foundation of an individual facade wall, figure 6.42. In other cases, the pebble extends under the whole surface of a building, a courtyard or an enclosure wall defining its outline. Two preliminary electrical soundings, one carried out in a known area of pebble packing, the other outside it, allowed the setting of 2.5 m as the most appropriate spacing for the survey (17).

Isolated structures and cavities
This type of structure is typified by graves, a frequent preoccupation of archaeologists. Unfortunately, in general, individual graves are of modest size, made up of a few stones which offer only a weak contrast with their surroundings. Above all the volume, a cubic metre or so, is too small for the anomaly to appear in more than one measurement in a 1 m grid. Therefore the identification of such anomalies is difficult, and if the structures are very isolated, the percentage of non-productive measurement is high. This explains why examples of surveys of this kind are comparatively infrequent, even though the detection of sarcophagi or groups of Merovingian sarcophagi or graves under stone slabs has been successful, figures 6.10 and 6.37. This no longer applies when the structure is of considerable size and a tumulus made up of a significant quantity of stones, even when flattened, can give rise to a quite spectacular anomaly, as at Saint Martin, figure 9.9(a). If the barrow is preserved, the survey permits a glimpse of the details of the internal structure and the clear characterisation of the plan of the buried substructure, when the object is a medieval motte (22).

Graves are also encountered in the form of a natural or artificial cavity. The latter have given the best results, and the work of the Fondazione Lerici at the cemeteries of Tarquinia and Cerveteri especially, figure 6.43, led to the discovery of numerous painted tombs with rich inventories (28). It is evident in this case that the empty volume of the tomb which acts as a perfect insulator opposing the passage of current gives a strong resistive anomaly. Paradoxically, this case which appears to be ideal isn't as perfect as it would appear at first sight. Cavities frequently have an effective volume which is relatively too small compared with the depth of their roofs to give an observable anomaly. Other effects, like the wetness of the walls, and a more or less complete conductive filling are perturbing factors. From these effects, the detection of the tomb can just as well result from the identification of a low

Fig. 6.42 Portion of the facade wall of the palace of Darius I at Susa, southern Iran, made visible by high values of resistivity (normal Wenner quadripole, $a = 2.5$ m).

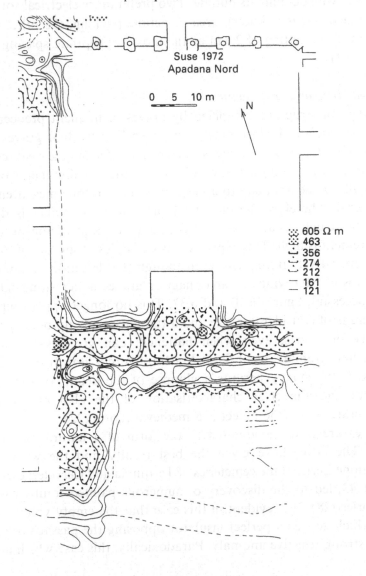

Suse 1972
Apadana Nord

0   5   10 m

N

605 Ω m
463
356
274
212
161
121

resistance anomaly coming from the loose filling of the corridor or access shaft which, when it outcrops at the surface gives an anomaly which is larger than that of the tomb itself.

The same difficulties appear in the study of natural cavities which were occupied in prehistoric times (15) or which served as refuges in more recent epochs. Experience shows that the very large variety of fill material encountered (clay, dry or wet gravel etc.) corresponds to an almost continuous scale of resistivities of extreme contrast, which viewed

Fig. 6.43 Contour map of resistivities of part of the cemetery at Tarquinia, Italy. The chambered tombs later excavated are superimposed on the map. From ref. 28.

via the integrating and deforming effect of the apparent resistivity will never permit precise identification of the constituents of the subsurface. There again, one can rarely hope for direct detection of the cavity. Frequently, secondary effects are observed such as diaclases, aquifers, entry fill, etc.

6.4.3. *Unusual structures and unconventional procedures*

In the case of natural cavities, one touches on an area where the measurement of resistivities together with geomorphology contributes to the description of the sedimentary and geological context in which archaeological structures are found. Most frequently it is a question of large regional anomalies which have been more or less considered as perturbations due to the deeper subsurface. An interesting example was described by Parchas and Spahos at Amathonte in Cyprus. In a series of survey grids designed to detect the remains of stone construction, all the anomalies with high resistivities disappear along a common line. The fact that this line is parallel with the modern beach located about 100 m away suggests that the phenomenon may correspond to the line of an old bank, an ancient wharf filled with sediment, the limit of the destruction of the settlement or the line of an enclosure wall. Whatever the cause may be, the resistivity map only furnishes a silent schematic document which must be clarified through excavation.

More generally, one must keep in mind that resistivity is an intrinsic physical property of the materials of the subsoil. Therefore, even an isolated measurement is directly significant and resistivity like magnetic susceptibility allows surveying with a coarse grid. The survey of the site of Saint Romain en Gal (21) also shows that the mapping of the average resistivity with a Wenner configuration with $a = 2$ m at intervals of 30 m or more, figure 6.44, clearly indicates a raised area with high resistivity which not only corresponds to the best drainage but is also the region occupied in the Roman period. The high resistivities are also a consequence of the presence of walling which cannot be mapped with such a coarse grid, scattered building stone and rubble of all kinds such as bricks and shards, all of the elements of an archaeological site (24).

Figure 6.45 shows a Roman villa in Switzerland mapped by Leckebusch and treated with the methods described in chapters 4 and 8. In this case the high resistivities shown in white define the areas of collapsed walling and parts of buildings quite well. Clearly, the display procedure allows

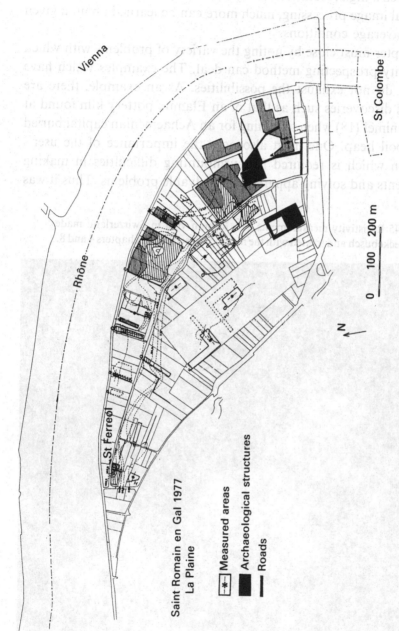

Saint Romain en Gal 1977
La Plaine

* Measured areas

■ Archaeological structures

— Roads

Vienna

St Colombe

Rhône

St Ferreol

N

0   100   200 m

Fig. 6.44 Saint Romain en Gal (Rhône). Large anomalies are revealed on a very coarse grid at 30 m intervals, which permit the identification of the Gallo-Roman occupation zone shown with strong cross-hatching over an area of about 40 ha. From ref. 17.

a much more satisfactory interpretation of the data than older techniques permit. Given a high resolution survey over a large area and presentation using digital image processing, much more can be learned about a given site under average conditions.

This chapter must close by noting the variety of problems with which the resistivity prospecting method can deal. The examples which have been given do not exhaust the possibilities. As an example, there are unexpected discoveries such as that of an Elamite pottery kiln found at Susa by Renimel (18) when searching for an Achaemenian capital buried under a spoil heap. One must also note the importance of the user's imagination which is required for surmounting difficulties in making measurements and solving apparently unsolvable problems. Thus it was

Fig. 6.45 Resistivity measurements at a Roman villa in Switzerland made by J. Leckebusch and treated with the methods described in chapters 4 and 8.

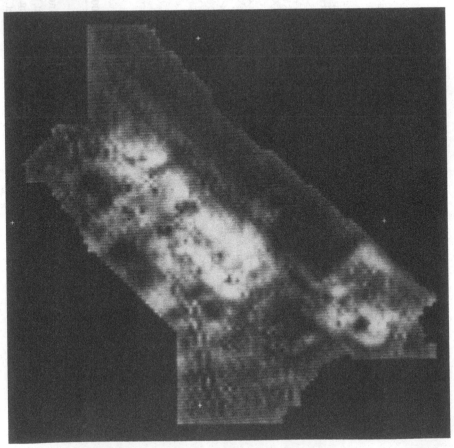

enough to put electrodes into a grid of holes punched through the asphalt surface of a parking lot adjacent to the cathedral of Beauvais with the aid of an iron bar and fill them with water to permit the detection of a medieval building underneath the otherwise perfectly insulating layer.

6.5. History of the application of resistivity methods

With the exception of the empirical use of mine detectors in the period immediately following Second World War, resistivity measurements made in England in 1946 by Atkinson (3) and by de Terra (11) in Mexico in 1949 were the first conscious attempts at utilising a geophysical method in archaeology. For several years electrical prospecting remained the sole technique employed. It developed primarily in England, where several manufacturers produced alternating current resistivity meters which were relatively fast, as well as in Italy, under the influence of the Fondazione Lerici (27,28). The reasons appear to be connected with the low price of equipment which was readily available, and the relative simplicity of interpretation of the anomalies. The variety of archaeological targets which were detectable, principally stone construction and hollows filled with earth, were also favourable points.

After the first announcements of the proton magnetometer by Aitken of the Research Laboratory for Archaeology at Oxford, a device which permitted measurements at intervals of several seconds with a precision in measurement of the earth's magnetic field an order of magnitude better than that of the magnetic balances previously available, this device looked as though it would replace the resistivity meter. The relative slowness of electrical measurements was seen as a very serious handicap and almost all researchers tried to reduce the measurement time. In practice, several tests of magnetic prospecting (10) and systematic experiments conducted at the Centre de Recherches Geophysiques at Garchy showed that each of the methods had its own area of application, with resistivity particularly efficient for use on remains constructed in stone. The search for a method to speed up the measurements was, in the beginning, directed toward reducing the time required for moving the electrode system, and highly diverse ingenious schemes were developed by various workers in the field. Clark used rigid quadripoles (8), Hesse (15) used articulated sets, Scollar and Hesse both used a switching arrangement for a linear electrode array (31,16) and several other workers used arrays of cables.

As far as the time required to make a measurement, the alternating current bridges commercially available required 5–10 seconds per reading, and this represented a considerable improvement over the direct current devices which required 30–40 seconds per reading. They are still used by some, but they have been improved so that a direct readout of the radio $\Delta V/I$ allows competition with the speed of free precession proton magnetometers.

Devices with automatic recording constitute the ultimate stage of evolution in measurement technology in that they follow the evolution in all of the detection methods for structures which are near the surface, which necessitate the taking of a large number of readings at short distances. Suggested several years ago by Hesse, automatic recording of resistivity promises considerable developments. It opens the way to a rich technique for the interpretation of prospecting results via automatic commutation of the electrode system, giving indices of anisotropy and simultaneous measurements at several depths.

Notes

(1) Aitken, M. J. 1961, *Physics and Archaeology*, Wiley-Interscience, London.
(2) Anderson, W. L. 1979, Computer program numerical integration of related Hankel transform of order 0 and 1 by adaptive digital filtering, *Geophysics*, **44**, 1287–1305.
(3) Atkinson, R. J. C. 1953, *Field Archaeology*, Methuen, London.
(4) Atkinson, R. J. C. 1952, Méthodes électriques de prospection en archéologie, in *La Découverte du Passé*, A. Laming, ed., Picard, Paris.
(5) Bertrand, Y. 1972, *La Prospection Électrique Appliquée aux Problèmes des Ponts et Chaussées*, July, p. 172, Laboratoire Central de Ponts and Chaussées, Paris.
(6) Cagniard, L. 1959, Cours de géophysique appliqués, *Electricité*, **5**, 3, Université de Paris (Sorbonne), Paris.
(7) Cagniard, L., Neale, R. N. 1957, Technique nouvelle de modèles réduits pour la prospection électrique, *Geophysical Prospecting*, **5**, 3.
(8) Clark, A. J. 1968, A square array for resistivity surveying, *Prospezioni Archaeologiche*, **3**, 111–14.
(9) Cook, K. L., Van Nostrand, R. G. 1954, Interpretation of resistivity data over filled sinks, *Geophysics*, **19**, 761–90.
(10) Dabas, M., Hesse, A., Jolivet, A. 1987, Prospection électrique de subsurface automatisés, *Photographie aérienne et prospection géophysique*, Symposium , Centre Interdisciplinaire de Recherche Archéologique, Bruxelles.
(11) De Terra, H., Rowers, J., Stewart, T. D. *The Tepexpan Man*, *Viking Fund Publications in Archaeology*, **11**, Axel Werner-Grenn Foundation, New York, NY.
(12) Dieter, K., Patterson, N. R., Grant, F. S. 1969, IP and resistivity type curves for three dimensional bodies, *Geophysics*, **34**, 615–32.
(13) Ghosh, D. P. 1971, Inverse filter coefficients for the computation of apparent resistivity standard curves for a horizontal stratified earth, *Geophysical Prospecting*, **19**, 769–75.

(14) Hesse, A. 1966a, Perfectionnement des applications archéologiques de la prospection électrique, *Comptes Rendus mensuels Société Préhistorique Française*, **1**, 15–19.

(15) Hesse, A. 1966b, *Prospections géophysiques à faible profondeur*, Dunod, Paris.

(16) Hesse, A. 1978, Manuel de prospection géophysique appliquée à la reconnaissance archéologique, *Centre de Recherches sur les Techniques Gréco-Romaines*, **8**, 127, Université de Dijon, Dijon.

(17) Hesse, A. 1979, Reconnaissance d'ensemble du palais du Chaour par la méthode des résistivités électriques, *Cahiers de la DAFI*, **10**, 137–44, Paléorient, Paris.

(18) Hesse, A. 1980, La prospection des vestiges préhistoriques en milieu proche-oriental: une douzaine d'années d'expériences géophysiques, *Paléorient*, **6**, 45–54.

(19) Hesse, A. 1985, La reconnaissance archéologique sur le terrain, in *L'Archéologie et Ses Méthodes*, A. Pelletier, ed., Horwath, Roanne.

(20) Hesse, A. 1987, Méthodes géophysiques de la prospection, in *Géologie de la Préhistoire*, J. C. Miskovsky, ed., Géopré, Paris.

(21) Hesse, A. *et al.* 1978, Succès de l'archéologie prospective à Saint Romain en Gal, *Archéologia*, **122**, 7–17.

(22) Hesse, A. *et al.* 1979, Des contrastes physiques à la détection dans le sol, *Dossiers de l'Archéologie*, **39**, 26–41.

(23) Hesse, A., Jolivet, A. 1981, Réalisation et expérimentation d'un résistivimètre autotracté – enregistreur, *DGRST*, **17**, Direction Générale à la Recherche Scientifique et Technique, Paris.

(24) Hesse, A., Jolivet, A., Tabbagh, A. 1986, New prospects in shallow depth electrical surveying for archaeological and pedological applications, *Geophysics*, **51**, 585–94.

(25) Hesse, A., Spahos, Y. 1979, The evaluation of Wenner and dipole–dipole resistivity measurements and the use of a new switch for archaeological field work, *Archaeo-Physika*, **10**, 647–55.

(26) Keller, G. V., Frischknecht, F. C. 1966, *Electrical Methods in Geophysical Prospecting*, Pergamon Press, New York, NY.

(27) Lerici, C. M., Carabelli, E., Segre, E. 1958, Prospezioni geofisiche nella zona archaeologica de Vulci, *Quaderni de geofisica applicata*, **20**, 47.

(28) Lerici, C. M., Linington, R. E. 1969, *Nuovi Sviluppi delle Applicazioni della Scienza e della Tecnica alla Ricerca Archeologica*, Consiglio Nazionale delle Ricerche, Roma.

(29) Niwas, S., Israil, M. 1986, Computation of apparent resistivities using an exponential approximation of kernel functions, *Geophysics*, **51**, 1594–602.

(30) Petkov, A., Georgiev, M. 1986, Automatic and electromechanical electro-prospecting equipment, *Proceedings Symposium on Archaeometry, Athens*, **150**, Demokritos, Athens.

(31) Scollar, I. 1959, Einführung in die Widerstandsmessung, *Bonner Jahrbücher*, **159**, 284–313.

(32) Spahos, Y. 1979, Calculs sur modèles et role du quadripole en prospection électrique de subsurface. Application à la détection archéologique, Thesis, Université de Paris 6, Paris.

(33) Stefanescu, S., Schlumberger, C., Schlumberger, M. 1930, Sur la distribution électrique autour d'une prise de terre ponctuelle dans un terrain à couches horizontales, homogènes et isotropes, *Le Journal de Physique et le Radium*, **7**, 132–40.

(34) Tabbagh, A. 1985, The response of a three dimensional magnetic and conductive body in shallow depth electromagnetic prospecting, *Geophysical Journal of the Royal Astronomical Society*, **81**, 215–30.
(35) Telford, W. M. *et al*. 1976, *Geophysical Prospection*, Cambridge University Press, Cambridge.

7

Magnetic properties of soils

7.1. Introduction

Magnetic prospecting requires careful measurement of the magnitude or of differences in magnitudes of the earth's magnetic field at many points over a large area if fine details of buried structures are to be revealed after processing. Often, such prospecting campaigns have proven disappointing. It is the purpose of this chapter to discuss the foundations and limitations of the technique. Some of the information is also applicable to electromagnetic prospecting methods.

An empirical approach to the problem of the detectability of archaeological remains is sometimes possible if excavation has partly exposed a feature and measurements can be made on parts of it which are still buried. When a clear magnetic contrast exists, this should encourage further survey. This is rarely the case in soils which are only weakly magnetic. A negative result does not necessarily imply that sophisticated interpretation methods or a more sensitive instrument will fail to yield useful information. Some of the methods described in this chapter may be of considerable help in deciding whether or not to carry out a survey of a large area. They may also be useful in deciding that entire geologically similar areas are unlikely to be productive when using magnetic prospecting techniques.

7.2. Some magnetic preliminaries
7.2.1. Magnets and magnetic fields

Magnetism is one of the fundamental fields in nature, one of the first to be perceived as action at a distance. Magnetism is observed as the force exerted between magnetic poles. In this short review, the notation of Chikazumi (3) will be followed. In the MKS system, two magnetic poles whose strengths are defined in webers m_1 and m_2, exert a force in newtons at distance r on each other of:

$$F = m_1 m_2 / (4\pi\mu_0 r^2) \qquad (7.1)$$

This is Coulomb's Law which has a similar form with electric charges; μ_0, a constant, is called the permeability of a vacuum and in the rationalised system used here is given in units of henry per metre:

$$\mu_0 = 4\pi 10^{-7} \, \text{H m}^{-1} \qquad (7.2)$$

This definition has the advantage that all further equations contain only rational numbers. Moving electric charges which constitute an electric current produce a magnetic field. A wire in which a current flows is surrounded by such a field. When the wire is wound into a long coil or solenoid, a nearly uniform magnetic field is produced at the centre. With a current of i A (amperes) in the coil with n turns per metre, the intensity of the field at the centre is:

$$H = ni \qquad (7.3)$$

in the rationalised system. In this system $1 \, \text{A m}^{-1} = 0.0126$ Oe (oersted) in the non-rationalised system which is still widely used in geophysical practice.

A magnetic pole of strength m weber experiences a force of:

$$F = mH \qquad (7.4)$$

in a field of intensity H. Single magnetic poles have never been observed, although their existence is predicted by certain theories in modern physics. What is observed is a magnet of length l (a bar magnet), which has magnetic poles of opposite strengths m and $-m$ near each end. When such a magnet is placed in uniform field of strength H, each pole experiences a force, figure 7.1(*a*), which gives rise to a turning moment or couple L which is related to the angle of the bar magnet relative to the field by:

$$L = -mlH \, \sin(\theta) \qquad (7.5)$$

where θ is the angle between the magnet and the field.

If the field varies in intensity from place to place, it is said to have a gradient, which is locally defined as:

$$\nabla = \partial H_x / \partial x \qquad (7.6)$$

and a bar magnet experiences a translational force, figure 7.1(*b*), as well as the rotational force described above of:

$$F_x = ml\, \partial H_x / \partial x \tag{7.7}$$

in the direction of the field. Both the rotational and translational forces have in common the product ml, and this is called the magnetic moment:

$$M = ml \tag{7.8}$$

which is defined in the rationalised system in units of weber metres. The rotational torque which a magnet in a uniform field experiences is therefore:

$$L = -MH \sin(\theta) \tag{7.9}$$

The energy of the system being conserved if there are no losses, this gives rise to a potential energy of:

$$U = -mH \cos(\theta) \tag{7.10}$$

Small magnetic particles such as those encountered in soils behave like small bar magnets. The interaction between two such magnets of moments M_1 and M_2 in the absence of an external field, and separated by a distance r_{12} can be calculated easily if it is assumed that the length of each magnet is small compared with the distance between them. The magnetic field components H_{1x} and H_{1y} experienced by one magnet produced by the second, figure 7.2, are:

$$H_{1x} = \frac{M_2}{4\pi\mu_0} \frac{2\cos(\theta_2)}{r_{12}^3}$$

$$H_{1y} = -\frac{M_2}{4\pi\mu_0} \frac{\sin(\theta_2)}{r_{12}^3} \tag{7.11}$$

Fig. 7.1 (a) A magnet under the action of a couple of forces in a uniform magnetic field. (b) A magnet under the action of a translational force in a gradient field. From ref. 3.

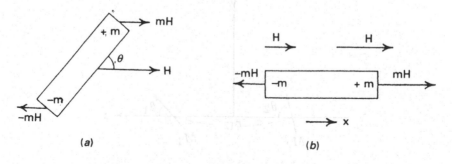

(a)   (b)

and the potential energy of the two magnet system is:

$$U = -\frac{M_1 M_2}{4\pi\mu_0 r_{12}^3} [2\cos(\theta_1)\cos(\theta_2) - \sin(\theta_1)\sin(\theta_2)] \qquad (7.12)$$

When the magnets have the same moments and are parallel to each other the potential energy is:

$$U = -\frac{3M^2}{4\pi\mu_0 r_{12}^3} (\cos^2(\theta) - \tfrac{1}{3}) \qquad (7.13)$$

which is obviously minimum for $\theta = 0$, when the magnets are aligned pole to opposite pole on the same axis.

7.2.2. *Kinds of magnetism*

The intensity of magnetisation, the magnetic moment of a substance per unit volume, is a vector quantity I_s. The MKS unit is the weber per square metre and is related to the older cgs system, being equal to 7.96×10^2 G (gauss). The intensity of magnetisation I is related to the strength of the applied magnetic field by a proportionality constant χ, the magnetic susceptibility:

$$\chi = I/H \qquad \text{and} \qquad \bar{\chi} = \chi/\mu_0 \qquad (7.14)$$

which is dimensionless because I and H have the same dimensions, but its value depends on the system of units. In the MKS or International system units it is SIu. Susceptibilities are observed in natural substances with values ranging between 10^{-5} and 10^6 or eleven orders of magnitude, and weak negative values also occur. The relationship between I and H in some substances is not linear.

The source of magnetic behaviour in materials is the orbital motion and spin of the electrons in the material. In a classical model of an atom,

Fig. 7.2 Magnetic interaction between two dipoles. From ref. 3.

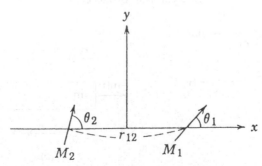

an electron moves in a circular orbit around the nucleus at radius r, at an angular velocity ω. The electron rotates therefore with $\omega/(2\pi)$ revolutions per second, and since motion of an electric charge constitutes a current, this is therefore:

$$i = -e\omega/2\pi A \tag{7.15}$$

where $-e$ is the charge of the electron. From the relationships above, the magnetic moment of a circular electric current i, with cross-section S m², is $\mu_0 i S$ Wb m. Thus the magnetic moment of the electron in circular orbit according to this classical model is:

$$M = -\mu_0 \frac{e\omega}{2\pi} (\pi r^2) = -\tfrac{1}{2}\mu_0 e\omega r^2 \tag{7.16}$$

The angular momentum of the motion is:

$$P = m\omega r^2 \tag{7.17}$$

where m is the electron mass, and the magnetic moment is therefore:

$$M = -(\mu_0 e/2m)P \tag{7.18}$$

But quantum theory teaches that electrons may only move in orbits such that the angular momentum is an integer multiple of Planck's constant h divided by 2π such that:

$$P = nh \qquad h = h/2\pi \tag{7.19}$$

and the quantised magnetic moment thus becomes:

$$M = -n(\mu_0 h e/2m) \tag{7.20}$$

and therefore the magnetic moment can vary only by the smallest magnetic unit, the Bohr magneton:

$$M_B = \mu_0 h e/2m \tag{7.21}$$

The relationship between angular momentum and the spinning motion of the electron differs from above by the ratio 1:2, the gyromagnetic ratio, and is expressed as:

$$M = -(\mu_0 e/m)P \tag{7.22}$$

so that the orbital ($g = 1$) and spin ($g = 2$) motion contributions to magnetic moment can be written generally as:

$$M = -g(\mu_0 e/2m)P \tag{7.23}$$

The spin angular momentum changes by units of $h/2$, so that the spin magnetic moment is also in Bohr magnetons. In the materials of interest here, spin effects are predominant.

Diamagnetism

Orbital rotation of the electrons about the nucleus with the application of an external field produces weak magnetism opposite in direction to that of the applied field following Lenz's Law. The susceptibility, $\bar{\chi}$, is of the order of -10^{-5}, and such substances are called diamagnetic. If the substance is mixed with other more magnetic materials, the effect will be masked by much stronger positive susceptibilities.

Paramagnetism

If a substance contains atoms whose electron spins are not coupled with each other which allows them to change direction freely, the magnetisation is proportional to the applied field. The susceptibility, $\bar{\chi}$, is weakly positive and ranges from 10^{-5} to 10^{-3}. Such substances contain randomly oriented spins, which upon application of an external field tend to orient parallel to the field direction. These materials are called paramagnetic. Since thermal agitation randomises the spin order, susceptibility is inversely proportional to temperature. Some soil minerals behave paramagnetically.

Ferrimagnetism

Néel (26,27) proposed the term ferrimagnetism to describe the magnetism observed in certain very special substances which occur in nature and which can be produced in the laboratory. These materials, the ferrites, are of prime importance in soil magnetism. They are crystalline, and have well-defined lattices. The magnetic ions are placed at two kinds of lattice sites, designated A and B as shown in figure 7.3. The spins in the A sites point in one direction usually called plus, the spins in the B sites point in the other, called minus. There is a strong negative interaction between the spin systems at A and B sites. The number of ions and the magnitudes of the spins of the single ions are not the same at the A and B sites, so that a net magnetisation results. This magnetisation is present without any external field. Heating introduces randomness, and at a very well-defined temperature, the spin arrangement breaks down completely and the substance is demagnetised, becoming merely paramagnetic. This is the Curie temperature. In metals like iron, the spins are completely parallel to each other, and there is strong interaction between neighbouring spins. Such materials are ferromagnetic, but they are not of interest here although very important in other contexts. The force which determines alignment is an interactive one, governed by the probabilistic behaviour of the electrons. The mutual electrostatic energy of two atoms is reduced when the unbalanced magnetic moments of the electrons in the $3d$ shells are parallel. Thus the probability that the magnetic moment in one atom is parallel or anti-parallel to its neighbour's is determined by the orientation of the neighbouring moments, and this requires the solution of all the mutual interactions. This remains one of the unsolved problems (Ising) in physics.

A simplified explanation which gives satisfactory results at normal temperatures is possible however. This is the quantum corrected Weiss theory based on the hypothesis of a molecular field (42). It assumes that an atom in a lattice has z nearest neighbours and that there exists an exchange interaction between them such that when the magnetic moments of a pair are parallel, the mutual energy is $-U$, and when anti-parallel $+U$. These interactions are equivalent to a molecular field for the whole substance. It is equivalent to assuming that the interactions correspond to neighbours with an average alignment typical of all the atoms. In crystals of interest here, the structure is either body centred cubic, where $z = 8$, or face centred cubic or hexagonal close-packed, in which case $z = 12$. Stacey and Banerjee (34) state that these numbers

Fig. 7.3 Spin rotation and lattice sites in a magnetite crystal. From ref. 34.

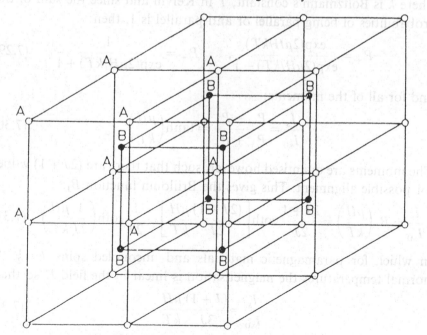

are just sufficient to make the intermolecular field a reasonable approximation. The sum of the parallel and anti-parallel alignments $p + a$, is z. The magnetic moment per unit volume, relative to complete magnetisation is:

$$\frac{I_s}{I_{s0}} = \frac{(p - a)}{(p + a)} \tag{7.24}$$

the energy of the central atom is:

$$E = \pm (p - a)U = \pm zUI_s/I_{s0} \tag{7.25}$$

and the intermolecular field coefficient is:

$$E = \pm F\mu = \pm \lambda I_s \mu \tag{7.26}$$

with:

$$\lambda = zU/\mu I_{s0} \tag{7.27}$$

Thus, the magnetic moments are subject to a force which is proportional to the degree of alignment. For uncoupled electron spins of moment μ in a field H, the energies are $\pm \mu H$, and the parallel direction has a preferred energy difference of $2\mu H$. This state is more probable by a factor f:

$$f = \exp(2\mu H/kT) \tag{7.28}$$

where k is Boltzmann's constant, T in Kelvin and since the sum of the probabilities of being parallel or anti-parallel is 1, then:

$$P_+ = \frac{\exp(2\mu H/kT)}{\exp(2\mu H/kT) + 1}, \qquad P_- = \frac{1}{\exp(2\mu H/kT) + 1} \tag{7.29}$$

and for all of the moments:

$$\frac{I_s}{I_{s0}} = \frac{P_+ - P_-}{P_+ + P_-} = \tanh\left(\frac{\mu H}{kT}\right) \tag{7.30}$$

The moments are quantised however, such that there are $(2J + 1)$ values for possible alignment. This gives the Brillouin function B_J:

$$\frac{I_s}{I_{s0}} = B_J\left(\frac{\mu H}{kT}\right) = \frac{2J + 1}{2J} \coth\left[\frac{(2J + 1)}{2J}\frac{\mu H}{kT}\right] - \frac{1}{2J}\coth\left(\frac{1}{2J}\frac{\mu H}{kT}\right) \tag{7.31}$$

in which for paramagnetic materials and uncoupled spins $J = \frac{1}{2}$. At normal temperatures the magnetisation is linear in the field H so that:

$$\frac{I_s}{I_{s0}} = \frac{(J + 1)}{3J}\frac{\mu H}{kT} \tag{7.32}$$

and the susceptibility is:

$$\chi = \frac{I}{H} = \frac{(J+1)\,\mu I_{so}}{3J} \frac{\mu I_{so}}{kT} = \frac{(J+1)\,N\mu^2}{3J} \frac{N\mu^2}{kT} \tag{7.33}$$

where N is the number of moments per unit volume (Curie's Law). For ferrimagnetic materials an intermolecular field λI must be added which is usually stronger than H:

$$\frac{I}{I_{so}} = \tanh\left[\frac{\mu}{kT}(H + \lambda I)\right] \tag{7.34}$$

and for I small, $I/I_s \ll 1$:

$$\chi = \frac{\mu I_{so}/kT}{1 - \lambda\mu I_{so}/kT} = \frac{C}{T - \theta}, \qquad C = \mu I_{so}/k, \qquad \theta = \lambda C \tag{7.35}$$

This theory gives a good approximation to experimental observations on magnetite and ferrimagnetic materials at normal temperatures. In fact, reversals of individual atomic moments are not involved, but rather spin waves and the energies of such waves in the spin structures. These much more complex quantum mechanical considerations are significant only at low temperatures which are not of concern here.

7.2.3. Dilution factor

It is evident that the surface material disturbed during the construction of a structure is returned in part to a lower level along with some of the lower level material itself when the structure collapses (chapter 1). The quick and subsequent fills of a feature thus contain a mixture of materials and one may speak of a dilution of surface material at depth. The dilution factor is defined by Mullins (23) as the percentage of surface material present in the fill. We shall see in a moment why this is important. Conventionally, archaeologists describe surface material as the biologically active layer, usually between 30 and 50 cm deep. It is characterised by root penetration of grasses, grain crops, and underbrush. There is high earthworm activity in temperate climates and correspondingly large amounts of organic material. This zone, the A horizon of the pedologist is sometimes differentiated into further subhorizons. In an excavated section, this layer sometimes goes deeper than does contemporary ploughing. Soil horizons below the surface layer are considerably poorer in organic remains, and the mineral composition has usually been more affected by climatic than by biological processes.

7.2.4. Parent material

The 'parent' rock minerals on which soil develops have remained in place since their deposition at a remote geological epoch. There follows a gradual exposure due to water and wind erosion. If wind transport predominated, then the material is quite uniform, as is also the case if initial deposits are of marine or lake origin. If the origin is glacial, then the material is quite heterogeneous. Soils on igneous or metamorphic rocks weathered in place are likely to vary widely even over short distances, as do soils transported by rivers. The wind transported loesses which are quite important in wide areas of northern Europe, Asia, and north America are mineralogically very uniform over wide areas. A sorting process occurs whereby rock particles of very small sizes and similar composition are deposited at different distances from their sources depending on their settling times in air.

7.2.5. Estimates of dilution factors

Although examinations of experimental archaeological structures are too recent for conclusive results, it seems reasonable to suppose that the dilution factor is more dependent on how a feature is constructed and what happens afterwards than on the kind of soil present. More mobile soils are nonetheless more likely to be less diluted at depth, since more material can be washed in from the surface with the quick fill. For ditches of moderate size, 1–2 m in depth and 3–5 m wide, dilution factors can be estimated roughly from the appearance of the ditch in section and from the depth of visible surface material. If the sides of a ditch are nearly vertical and the surface material is uniformly returned to the fill, then the dilution factor is simply the proportion given by the depth of the A horizon to the total depth. From observations on experimental features, the factor appears to range between 10 and 30%. For pits and ditches which were deliberately filled, the percentage is probably higher. Estimates of dilution factors are quite important when attempting to predict possible magnetic anomalies from measurements made on surface and subsurface samples especially those obtained by coring.

7.3. Soil magnetism research

The study of the magnetic properties of soil has a history going back many decades. It is closely linked with the research into rock magnetism and the magnetism of burned clays, both important archaeo- and paleomagnetic chronological methods. Less attention has been paid to soil magnetism as related to magnetic and electromagnetic prospecting in archaeology. Rock magnetism research is well summarised in Nagata (24), Stacey and Banerjee (34), and O'Reilly (28). Soils, being made up of rock particle mixtures have properties in common with the parent materials, but due to the changes brought about by weather, plants, and man, they have some properties which are quite different. An outline theory of soil magnetism was sketched by Le Borgne in a series of papers published between 1955 and 1960 (16,17,18) and in his doctoral thesis which was unfortunately published in Turkey and is not readily available. Earlier investigators noted some of the curious aspects of the magnetic behaviour of soils, but these did not constitute a systematic study. Of historic interest are the contributions of Belluigi (2), Tucker, (40) and Stavrou (35). After the appearance of Le Borgne's classic papers, an independent study by Cook and Carts (4) introduced the important concept of local surface variation in magnetic properties of soil, which Scollar (29) rechristened 'soil noise'.

The importance of soil magnetism in archaeological prospecting was first recognised by Aitken, Webster, and Reeds (1). Belshé suggested to Aitken that a sensitive magnetometer might be useful in detecting pottery kilns (see chapter 8), and Aitken noted that the magnetic properties of the soil in ditches and pits at the same site made them detectable. It was known as far back as 1896 that the firing of clay used in making pottery enhanced the measurable magnetism of the piece (8). Thellier provided the first systematic study of the phenomenon in his thesis (36). Mullins later provided a well-founded physical explanation of many features of soil magnetism (23) expanding the results published in a preliminary paper by Tite and Mullins (37). Other studies of interest during the late 60s and early 70s are the papers of Neumeister and Peschl (25) and Vadunina and Babanin (41) on the magnetism of soils in the GDR and the Soviet Union. The latter paper summarises most of the Russian literature on the subject. A later summary paper by Le Borgne (19) was written for archaeologists. The magnetic properties of some soils in the Rhineland were also reported in the same publication by Scollar (29). An extensive non-mathematical treatment is given by Graham and Scollar (11). Longworth and Tite provided very important experimental confirmation of some of the hypotheses expressed in earlier studies (20). Tite and Linington published results from an extensive survey of Italian soils and discussed a possible climatic dependence of observed susceptibilities (38,39).

7.4. Minerals with magnetic properties in soils

The principal source of the magnetic properties of a soil is its iron content. With the exception of rocks of organic origin like limestone and chalk, most rocks contain from 1 to 10% by weight of iron oxides like Fe_2O_3 or FeO. The chalks and limestones contain far less, but since the bodies of certain marine organisms and bacteria appear to concentrate small magnetite particles (possibly for navigational purposes) the result is highly magnetic. Most of the principal soil forming minerals are silicates, but most of the iron which is of interest in archaeological prospecting is in non-silicate form. In temperate wet climates, most of the iron is encountered in some complex ferric hydroxide gel, geothite (FeOOH), and haematite (αFe_2O_3). This is especially true of soils formed on rocks of the Devonian and Triassic periods. FitzPatrick (7) describes ferric hydroxide gel as being amorphous and of a yellowish brown colour, whereas geothite is crystalline and reddish brown, changing to yellowish brown with increasing hydration. Extremely hydrated forms are sometimes called limonite. Lepidocrocite ($\gamma Fe_2O_3 \cdot H_2O$) is bright orange and is found in soils subject to waterlogging.

Within soils, ferric hydroxide and ferric oxides occur as discrete particles, usually in coatings or microaggregates on the quartz feldspar grains which make up the bulk of the material. The iron minerals are responsible for much of the observed colour in soils, and are implicitly given considerable attention by archaeologists who pay great attention to minor colour differences in excavated features.

Iron is the principal element in soils affected by the natural processes of oxidation and reduction, and it is one of the few substances which is found in a reduced state in some primary minerals. When iron is released by hydrolysis and enters an oxygen-rich atmosphere, it is quickly oxidised to the ferric state and precipitated as ferric hydroxide. When the iron is released in an oxygen-free state, it remains ferrous and has a gray-blue colour characteristic of some constantly waterlogged soils. The change from red to gray-blue soil colour due to the action of microorganisms in the prolonged absence of oxygen has been observed frequently by pedologists. Fluctuation in the water table reverses the effect, producing the elaborate gray-blue and orange banding typical of a gley soil.

Colour in soil is also affected by the presence of organic matter, and the darker browns and blacks of the upper layers are attributable to this, rather than to the state of the iron oxides, as almost every gardener knows. This effect is enhanced in soils which are rich in calcium and sodium like those developed on rocks of marine origin. Therefore colour alone is not a guide to the state or quantity of iron oxides which a given soil may contain. One should not attempt to infer the practicability of magnetic survey from the presence of a very dark brown soil layer in a section.

Highly hydrated iron oxides are usually weakly paramagnetic. Muscovite, dolomite, lepidocrocite, geothite, etc. are examples of such minerals. There are no published data on the magnetic susceptibility of ferric hydroxide amorphous gels. According to Selwood (32) susceptibility of the materials is around 480×10^{-6} SIu. In areas of extensive human occupation, it is not possible to separate the effects due to the natural paramagnetic minerals present in significant amounts from those due to the ferrimagnetics which, even if present in small quantities, have a very great effect on the magnetic properties of the soil. The paramagnetics may be responsible for the weak magnetism observed in soil formations in areas far removed from civilisation, and almost certainly for most of the magnetism observed in fossil soil horizons. Stronger magnetic properties are almost certainly due to ferrimagnetics, principally maghaemite, which Longworth and Tite (20) demonstrated through Mössbauer spectroscopy to be the only significant contributor as Le Borgne proposed. Magnetite was shown in these experiments to be insignificant in agricultural soils in non-volcanic areas.

Nagata (24) states that the term 'magnetite' has been used very loosely in petrology for magnetic minerals which can be separated magnetically. The magnetites in rocks have variable compositions. Chemical analysis shows that these 'magnetites' are really complex mixtures of metallic oxides in the iron–titanium series, with small quantities of manganese, magnesium, aluminium, and vanadium as minor components. If rocks with igneous origins are present, then the magnetic properties of the ferrimagnetic minerals are interpretable as characteristic of the Fe, Fe_2, Fe_3–TiO_2 ternary system. This includes most of the minerals of interest to studies in rock magnetism like wuestite (FeO), magnetite (Fe_3O_4), maghaemite (γFe_2O_3), haematite (αFe_2O_3), ilmenite ($FeTiO_3$), ulvospinel (Fe_2TiO_4) and pseudo-brookite (Fe_2TiO_5). The literature of paleo and rock magnetism is full of references to the magnetic properties of rocks containing these minerals. On the basis of measurements made by Graham and Scollar (11) and those of Longworth and Tite cited above, there is no evidence to show that these substances are of importance in the sedimentary soils of the north Eurasian or north American temperate plains. They are probably significant elsewhere, but most prospecting activity has taken place in the named areas. Soils of igneous rocks origin are far less widespread, and Mössbauer data are not available for them at the time of writing.

7.4.1. Magnetite

Although its presence has not been demonstrated in soils of archaeological interest examined to date, the substance of such is of interest because of its relation to other more significant materials. Magnetite is the most magnetic of the iron oxides. The structure of a unit cell of the basic building block of a magnetite crystal is shown in figure 7.4. This structure contains eight Fe_3O_4 molecules in a very special arrangement called an inverse spinel. The drawing represents a quarter of the smallest structure out of which magnetite crystals can be made. There are 24 iron atoms and 32 oxygen atoms arranged in a complex interacting fashion. The important feature of the crystal which endows it with its strong magnetic properties is the direction of the iron spins. The interactions between the atoms at the various sites in the crystal cause the magnetic moments of the atoms in the 8 a sites and the 16 d sites to be oppositely directed. Of the 16 d sites, half are occupied by Fe^{2+} ions, the other half by Fe^{3+} ions. All of the 8 a sites are occupied by Fe^{3+} ions. The magnetic moments of the Fe^{3+} ions in both types of sites are opposed and equal in number and thus cancel out. But the magnetic moments due to the Fe^{2+} ions are not compensated, and thus the crystal has a net magnetic moment or permanent magnetisation. A single crystal of magnetite is made up of four such unit cells and has a magnetic moment equal to the sum of the net moments of each cell.

7.4.2. Maghaemite

The most important mineral responsible for soil magnetism in areas of human occupation with agriculture, as Le Borgne, Tite, Tite and Mullins, Graham and Scollar, Longworth and Tite have demonstrated convincingly, is maghaemite. It has the same structure as magnetite, but a ninth of the iron atoms are absent. Only Fe^{3+} ions are present. Therefore the unit cell looks very much like that of magnetite, with some vacancies. These are preferentially located at 16 of the d sites. On average, there are 2.67 vacancies per unit cell. Thus in three unit cells, containing a total of 72 possible iron positions, only 64 iron atoms are actually present. The arrangement is such that, in a total of three cells, two lack three iron atoms each while two are missing from the third. This apparently unstable arrangement is thought to be held together by the presence of other atoms such as sodium which fill the vacancies left by the iron. As in the structure of magnetite, the magnetic moments of the 16 d sites

Fig. 7.4 The inverse spinel structure of magnetite. From ref. 3.

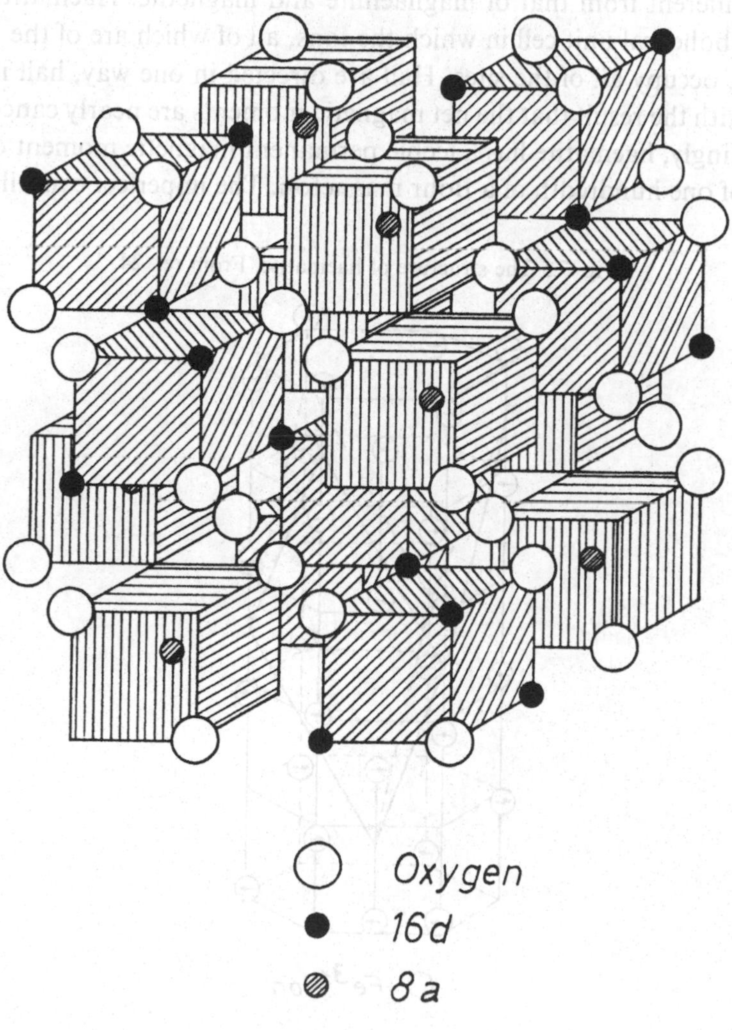

○ Oxygen

● 16d

◉ 8a

and those at the 8 a sites lie in opposite directions. On average, for each unit cell there are 8 iron ions occupying the 8 sites and 13.33 at the 16 d sites leaving a net magnetic moment slightly less than that of magnetite. Observed values are 4.07 and 2.36 Bohr magnetons per molecule of magnetite and maghaemite respectively.

7.4.3. Haematite

The most common of the iron oxides is haematite (αFe_2O_3). Its crystal arrangement is shown in figure 7.5. It is found in nearly all soils, usually in one of its hydrated forms, in amounts ranging from a few tenths of 1% to more than 10% by weight. The crystal structure of haematite is quite different from that of maghaemite and magnetite. Haematite has a rhombohedral unit cell in which the ions, all of which are of the Fe^{3+} variety, occupy all of the sites. Half are directed in one way, half in the other with the result that the net magnetic moments are nearly cancelled. Accordingly, haematite has a feeble permanent magnetic moment of the order of one hundredth of a Bohr magneton. The imperfect cancellation

Fig. 7.5 The structure of haematite. From ref. 24.

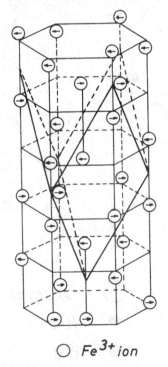

O Fe^{3+} ion

of magnetic moment is treated extensively in the literature. The fact that haematite is slightly paramagnetic in its behaviour above 250 K makes it difficult to decide whether this material or other paramagnetics are responsible for some of the weaker magnetic effects in soils. For our purposes, the importance of haematite lies in its role as a source mineral for the development of much more magnetic forms due to human or natural agencies.

7.4.4. Grain size effects

In contrast with ferrimagnetic materials, paramagnetic and diamagnetic substances retain their properties independently of grain size or dilution in a non-magnetic matrix. In ferrimagnetic substances, however, behaviour is very much more complex due to the interactions between the grains and among the ions in the crystal lattice. In most ferro and ferrimagnetic substances, the permanent magnetisation of each unit cell does not result in a total magnetic moment for a large mass of material. If this were the case, a large amount of magnetostatic energy would have to be stored. This does not arise because the large ferrimagnetic grain, composed of many unit cells, breaks up into a series of smaller volumes called domains, each of which is magnetised in a different direction, figure 7.6. Néel was responsible for a complete analysis of this very complex physical process (27).

Fig. 7.6 Magnetic domains in a polycrystalline specimen. From ref. 24.

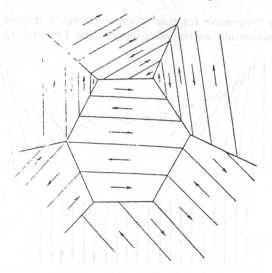

In certain directions relative to the crystal lattice, magnetism is facilitated. These are called the directions of easy magnetisation in the material. The direction of easy magnetisation refers to a result of the interaction between spin and electron orbital magnetism which creates more or less free energy in the crystal, depending on the orientation of magnetisation with respect to the crystallographic axes. Each domain is normally oriented along one of these easy directions, and the total magnetic moment of the crystal tends to be small since the domains, which are magnetised in different directions, tend to neutralise each other. It might be expected that without an external magnetic field, the grain would contain a very large number of domains giving very good cancellation of the magnetic moments. However, the domains are surrounded by so-called walls in which the direction of magnetism of the atoms changes slowly from the directon in one domain to that of the next, figure 7.7(a). Sudden changes as shown in figure 7.7(b) constitute an unfavourable energy distribution. These walls can actually be made visible under the microscope in a carefully polished section which is then coated with a fine dispersion of very small magnetite particles.

The domain walls store considerable potential energy. Total potential energy is minimised by sharing the magnetic energy due to imperfect cancellation of the total magnetic moment and the domain wall energy. This equilibrium condition is the normal state of the grain. A magnetic field applied to a collection of domains causes some of those whose magnetisation is opposite to that of the applied field to reverse. It also

Fig. 7.7 (a) Progressive rotation of spins through a domain wall. (b) Energetically unfavourable spin reversal. From ref. 32.

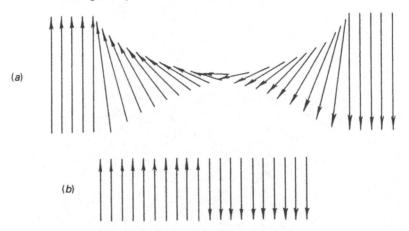

(a)

(b)

causes some domains whose magnetisation lies in the direction of the applied field to increase in size until a new equilibrium condition is achieved with minimum potential energy and a finite magnetic moment parallel to the field. A large crystal reaches equilibrium with very many magnetic domains having different directions. Its net magnetic moment is therefore low. Very small grains, with dimensions only a little larger than the domain wall size are at their lowest potential energy when they contain just one domain. Such single domain grains therefore have a much larger measurable susceptibility than do multidomain grains. Most soil magnetism is produced by grains in the mono-domain size range, which is thought to be of the order of 20–40 nm radius. Any chemical or mechanical process which results in grain size reduction to this range will significantly increase the magnetism of a sample, even if the actual composition remains unchanged.

7.4.5. Time dependence

When a magnetic field is suddenly applied, the domain size in a monodomain cannot grow to maintain energy equilibrium as it can in a multidomain grain. Reduction of the stored potential energy which results is only possible by a reversal of the direction of magnetisation. A definite amount of energy is required to effect this reversal and this can readily be supplied by thermal energy coming from outside the grain. Hence the magnetic moment of materials containing monodomains tends to increase slowly with time when they are placed in a magnetic field as more and more grains gradually receive enough thermal energy to reverse direction or 'flip'. This phenomenon of the growth of magnetism has been christened magnetic viscosity. Magnetic forces between neighbouring atoms are short range, whereas the actual magnetostatic forces in the material are effective over a much greater distance. The spin alignment which is produced by the exchange force is disturbed and a domain structure results, with regions in which alignment is present separated by regions with narrow walls in which the spins progressively rotate to new positions. The domains form closed loops within the grain of magnetic material. The grain can become magnetised by an external field which changes the total balance of internal magnetic energy. There is a probability $dP \ll 1$ that a change can occur in a short time dt relative to the activation energy E, and the absolute temperature T:

$$dP = C \exp(-E/kT)\, dt \qquad (7.36)$$

where k is Boltzmann's constant and C is a frequency factor which Stacey and Banerjee (34) give as 10^{-8} s, this being the frequency of a spin wave of half-wavelength equal to the domain wall thickness in magnetite. Thus the characteristic relaxation time is:

$$\tau = (1/C)\exp(E/kT) \tag{7.37}$$

which leads to Néel's concept of a blocking temperature. Thermal activation is a very strong function of temperature and at any temperature, monodomains undergo thermally excited magnetic reversals. After application (or removal) of a magnetic field, there will be n moments parallel to the original direction of the field and n' anti-parallel so that the magnetisation of the whole is:

$$\frac{M}{M_0} = \frac{n - n'}{n + n'} \tag{7.38}$$

and therefore:

$$\frac{d}{dt}\left(\frac{M}{M_0}\right) = \frac{2}{(n + n')}\frac{dn}{dt} \tag{7.39}$$

but since both processes proceed at a rate given by the probability equation (7.36):

$$\frac{d}{dt}\left(\frac{M}{M_0}\right) = -\frac{M}{M_0}[2C\exp(-E/kT)] \tag{7.40}$$

which upon integration gives:

$$M = M_0\exp\left(\frac{-t}{\tau}\right), \qquad \tau = \frac{1}{2C}\exp\left(\frac{E}{kT}\right) \tag{7.41}$$

Thus as Néel has shown (27), magnetisation varies with the logarithm of time and at a rate proportional to absolute temperature.

Monodomain material, mostly maghaemite, is usually in a particle range between 20 and 40 nm radius as determined by electron microscopy. The probability that a transition from any direction of magnetisation to that of the applied field is assumed to be a constant in the Néel model. Therefore there is a definite time needed at a given temperature for $1/e$ or 38% of all of the particles present to change direction. This value is a characteristic of the material and it is called the relaxation time. Néel showed that the time depends in an extremely critical way on the grain size of the substance. At room temperature, a particle of magnetite of 34 nm radius has a relaxation time of 10^9 s or about 30 y. However a particle of 26 nm radius has a relaxation time of less than

0.1 s. Nagata (24) gives a table of critical sizes. In soils which have a range of grain sizes spanning this very critical region, the superposition of the various relaxation times spans an enormous range for a very small change in the particle size distribution. An extremely sharp peak at the monodomain–multidomain transition is produced by the phenomenon as described. For grains smaller than monodomain size, behaviour is paramagnetic with relatively high values of susceptibility (super-paramagnetism).

Given a distribution of grain sizes spanning the Néel transition peak, the sum of the individual exponential relaxation times is logarithmic. It follows from this that, if grains sizes are present in equal numbers, the increase with time of the magnetic moment of a soil sample will follow a nearly logarithmic law. The moment at any given time can be described by:

$$M_t = M_0[1 + V_c \log_{10}(t/t_0)] \qquad (7.42)$$

where M_t is the magnetic moment at a given instant, M_0 the moment after some definite time t_0 and V_c a constant of proportionality, the viscosity constant. The value of t_0 depends on the method of measurement. One hundred seconds after the application of the field was used by Graham and Scollar, a value which was convenient for the measuring equipment employed (11).

The increase in measured magnetic moment of a typical soil sample is very considerable when archaeological time scales are involved. Typical values for V_c range between 3 and 6%. For example, a 5% value for V_c results in a net 45% increase in magnetic moment between 100 s and 3000 y. Thus, when calculating the magnetic anomalies to be expected from archaeological features of given shapes and containing given amounts of magnetic materials, the magnetic viscosity must be taken into account or the calculation will seriously underestimate magnitudes.

7.4.6. Dilution effects and field dependence of viscosity

In the majority of soils examined by Graham and Scollar (11), the amount of magnetic iron oxides by weight ran in the range of between 0.5 and 5% as shown in figure 7.8. Magnetic particles are uniformly distributed throughout the soil matrix. The soil matrix is usually diamagnetic since it is commonly composed of quartz, feldspar, calcium carbonate, and other non-magnetic minerals. Let all of the magnetic particles be of the same size or nearly so. Suppose further that these grains are roughly equal in size or larger than the grains of the non-magnetic matrix. Then from simple geometric considerations, the average distance of one magnetic particle from its nearest neighbour will be about four times its diameter for the iron oxide concentration in question. A 1% concentration of iron oxide in a non-magnetic matrix means that the mixed material contains 1 oxide particle with 99 non-magnetic closely packed neighbours. If this mixture is contained within the walls of a cube, then the cube will have a side length equal to the cube root of 100, so that the distance from an oxide particle to its nearest neighbour will be 4.64. This distance depends only slightly on iron oxide concentration, the values being 6.84 and 2.71 from a 5% and 0.5% concentration respectively. This was christened the 'currant bun' model by Graham because it resembles a currant bun with 1% weight in currants in cross-section.

The fields from the elementary dipoles represented by the magnetic particles fall off as the cube of the distance between them. Therefore the degree of interaction is a function which depends on the iron concentration. The effect of the interaction between isolated grains is to make single domain grains appear to be more like multidomain grains in their behaviour, partially suppressing magnetic viscosity effects at high iron concentrations. The greater the dilution, the lower the iron oxide concentration, and the greater the magnetic viscosity. For extremely low concentrations, this does not appear to be true. Another exception are basalts whose monodomains are large and of magnetite. They have low magnetic viscosity, whereas multidomains of this material have high magnetic viscosity.

Le Borgne (18) showed that magnetic viscosity is nearly independent of the field strength for fields which are of practical interest. Hence it is

Fig. 7.8 Chemically determined iron percentages by weight in a set of soil samples from the Rhineland. From ref. 11.

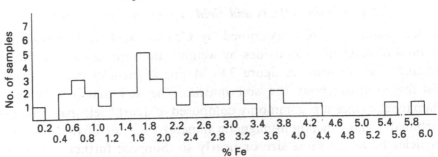

possible to measure at fields higher than those encountered in nature and still draw valid conclusions about the viscosity. This is of importance in the case of soils with low iron concentration and which are difficult to measure in fields comparable with that of the earth.

7.5. Depth dependence of magnetic properties: the Le Borgne effect

Tucker (40) was probably the first to observe anomalously high susceptibility in topsoils. Le Borgne was the first to study the phenomenon systematically for a number of soils, and offered what is still considered to be the most acceptable explanation of its origins (17). The A horizon in most soils, noted Le Borgne, is more magnetic than the lower horizons or the parent rock. The magnetic susceptibilities at different depths in a loess soil are shown in figure 7.9. The archaeological fill often contains surface material as shown. Scollar (29) later noted that certain soils which display high vertical mobility, such as washed brown earths of Dudal type B (6) sometimes have higher susceptibilities in the B than in the C or A horizons. Le Borgne offered two theories for the origin of the enhanced susceptibility of topsoil. In his 1955 paper, he suggested that bacterial action was responsible for reduction of natural haematite to magnetite, followed by reoxidation to maghaemite, with the crystal structure of the magnetite formed at the intermediate step preserved. There is no known chemical mechanism which can transform haematite directly into maghaemite.

Fig. 7.9 Dependence of magnetic susceptibility with depth in a deep, fine grained soil at three nearby locations in the Rhineland. From ref. 11.

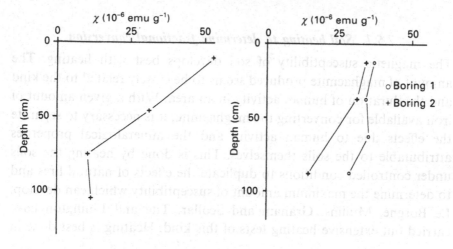

The production of maghaemite in nature is an incompletely understood process. Stacey and Banerjee summarise various references from the literature (34). They state: 'Maghaemite is formed by the low temperature (150–250°C) oxidation of magnetite. Because of the instability, single crystals larger than 1 micron can rarely be synthesized or found in nature. The oxidation is achieved by a topotactic process, so-called because the anionic (oxygen structure) is left unchanged; the Fe^{2+} ions diffuse to the surface of the grain, oxidise and form Fe^{3+} ions, leaving lattice vacancies. If the diffusion can proceed unhindered, one microcrystal of magnetite will be completely transformed to maghaemite without an additional new crystal of maghaemite being formed. Extensive work . . . has shown, however, that if a minute amount of haematite impurity is present in the initial magnetite grain, two competitive processes occur. In addition to the slow conversion from magnetite to maghaemite, there is an autocatalytic growth of haematite which becomes predominant in a mixed crystal . . . (producing) a mixture of magnetite, maghaemite and haematite'.

Le Borgne obtained inconclusive results with an attempt to grow bacteria on soils to enhance susceptibility. It is not at all evident why bacteria should instigate a reaction like the haematite–maghaemite transformation which requires the expenditure of energy. Certain higher organisms, as O'Reilly (28) states, concentrate magnetite and similar phenomena have been observed in certain very specialised micro-organisms. Perhaps the observations which Graham and Scollar made of enhanced susceptibility in some fossil soils may be attributable to this effect (11).

7.5.1. Soil heating to determine fractional conversion

The magnetic susceptibility of soil develops best with heating. The amount of maghaemite produced seems to be closely related to the kind and the duration of human activity in an area. With a given amount of iron available for converting the maghaemite, it is necessary to separate the effects due to human activity and the mineralogical properties attributable to the soils themselves. This is done by heating the soils under controlled conditions to duplicate the effects of natural fires and to determine the maximum amount of susceptibility which can develop. Le Borgne, Mullins, Graham and Scollar, Tite and Linington have carried out extensive heating tests of this kind. Heating is best done in

a tube furnace. Such an arrangement is shown in figure 7.10. First a
nitrogen atmosphere is used, then after the necessary temperature (about
550°C) has been reached, air is admitted. Graham and Scollar mixed
household flour with the soil to saturate it with organic material, and
highly repeatable results were obtained. The natural organic material in
soil produces hydrocarbon vapour at high temperatures which breaks
down at the still higher temperatures used. The actual reduction seems
to be mainly due to carbon monoxide developed at these temperatures.
The rise in susceptibility produced by heating to different temperatures
is shown in figure 7.11 for two quartz boats placed one above the other
containing the specimen soils.

Tite and Mullins (37) introduced the concept of fractional conversion.
A 100% conversion of all of the αFe_2O_3 present to γFe_2O_3 can only be
obtained in the laboratory. Fractional conversion is defined as the ratio
of the susceptibility actually observed in the field to that obtained in
the laboratory after heating:

$$C = (\chi_0/\chi_h)100\% \tag{7.43}$$

Typical conversions are usually less than 25% in Mediterranean areas,
and are as low as 1%, with 5% a typical figure in northern Europe.
Details are described in Graham and Scollar (11) and Tite and Linington
(39). They state that freshly exposed soils in Italy show considerable
enhancement of susceptibility after a few years, but the mechanism here

Fig. 7.10 An arrangement for heating soil samples. From ref. 11.

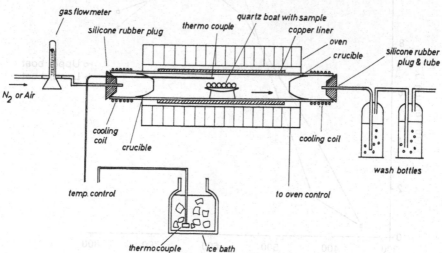

is also unclear. Extensive evidence for much higher fractional conversion in Mediterranean climates than in northern Europe is presented as evidence for the action of anaerobic bacteria in wet winters and aerobic bacteria in hot dry summers. Such reactions have never been demonstrated under controlled laboratory conditions and the energy considerations mentioned above led Graham and Scollar to scepticism as to the likelihood of this mechanism. It may be that partial laterisation which alters the grain size distribution of soils, with an increase in the numbers of monodomain particles could also account for the observed effect. However, the observed higher conversion rates in alternating wet/arid hot climates are a fact, whatever the cause may be, and this leads to higher magnetic anomalies for a given feature size and depth of burial in these regions.

The second explanation for the enhancement of topsoil susceptibility offered by Le Borgne in 1960 seems to account for most of the evidence in areas much affected by human occupation. The action of fire is the mechanism proposed. Heating haematite in the presence of organic material which in burning provides a carbon monoxide reducing atmosphere turns it into magnetite above 500°C. On cooling in air, the magnetite oxidises back to haematite in part, but other ions, perhaps

Fig. 7.11 The rise in constant field susceptibility produced by heating a soil sample to different temperatures under first reducing and then oxidising conditions at two levels in the furnace. From ref. 11.

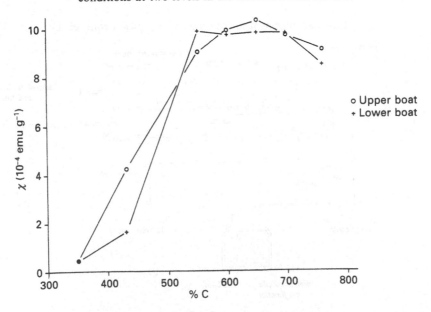

sodium or even water, as proposed by Stacey and Banerjee (34), stabilise the cubic crystal lattice and maghaemite results. Very few neutral ions are needed, since the number of vacancies which must be filled is small. Sodium is present in nearly all soils, as is aluminium and magnesium, other possible candidates. Water is almost universally abundant. That the heating mechanism accounts for the enhanced soil magnetism has been demonstrated repeatedly in laboratory experiments. In temperate climates, fire occurred frequently, either deliberately kindled for land clearance, or accidentally produced by lightning or deliberate destruction. Such a fire heats only the first few centimetres of soil, a poor conductor of heat. Kilns and hearths produce heating to depths of more than 10 cm. Archaeologists are familiar with the visible change in colour to a striking reddish-brown when such heating has occurred.

The maghaemite developed by surface burning is rapidly mixed, either by natural soil displacement due to earthworms and burrowing animals and insects or through subsequent ploughing. There is also a long term downward migration of the small maghaemite particles with subsequent deposition on non-magnetic carrier grains. This filtering action produces the occasionally observed raised susceptibility of a strong B horizon soil. Le Borgne proposed that certain bacteria may be capable of extracting energy from maghaemite and oxidising it back to haematite. This would account for the fluctuating oxidation–reduction conditions observable in gley soils. In a typical gley soil profile, a significant drop in susceptibility in the waterlogged horizon has been observed by Graham and Scollar (11).

On an archaeological site, the natural soil profile which was developed before settlement has been disturbed by construction. There are two dilution processes which can affect the magnetic particles present. The first is particle migration due to leaching, the second is mechanical mixing of the infilling of pits and ditches. Susceptibility contrasts between the pits and ditches and that of the surrounding B and C horizons are usually less than that found between the topsoil and those horizons. Such developments, if dilution has not been too pronounced, are favourable for magnetic prospecting.

7.6. Measurement of the magnetic properties of soils

Studies in soil magnetism have been based on a number of observations made on a wide variety of samples taken from different areas. Tite and

Mullins (37), Mullins (23), Tite and Linington (38,39), Graham and Scollar (11), and Longworth and Tite (20) are examples of research devoted to soil magnetism on archaeological sites. All of these studies have been devoted to soils developed in moderate temperate climates. There are no data available at the time of writing for other areas.

7.6.1. Sample taking and preparation

Soils, as seen from the agricultural standpoint which dominates in pedology, vary greatly from place to place depending on the parent rocks, local topography, climate, and plant growth history. Graham and Scollar (11) have shown that soils are magnetically much more uniform than pedological considerations imply when deep sediments are the parent material. Since areas having such soils are the most fertile and hence most densely occupied, they are of greatest interest archaeologically. Hence the sampling interval required for stable results for archaeological prospecting via enhanced susceptibility can be much greater than that common in pedological studies. Figure 7.12 shows a histogram of susceptibilities of samples taken at 50 m intervals over a large area. If no archaeological site with extensive burning is present, the histogram

Fig. 7.12 Susceptibility distributions of samples taken on a 50 m grid in an area with an archaeological site. From ref. 11.

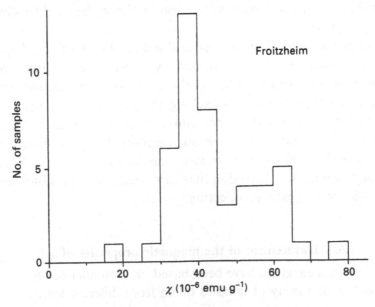

is unimodal with a relatively narrow peak. Areas which include long occupied sites tend to be multimodal with several peaks.

Sample size

Even if soils in densely occupied agricultural areas are horizontally quite uniform and well mixed, they contain a wide variety of minerals, some of which have different magnetic properties. To guard against the inclusion of a few large grains of highly magnetic material, fairly large samples must be taken when compared to those commonly used in paleomagnetic studies. A half kilogram of material is a reasonable quantity, and at least 100 g of prepared material should be measured. For samples from settlements where lumps of highly magnetic burned brick or pottery may intrude, a large sample is essential to allow for losses in sieving. The best method for sample collection is to take them from an excavated and cleaned section which shows archaeological features and the soil profile. If this is not possible, then a soil borer can be used. This is difficult to apply if the ground contains many stones. The iron fragments scraped off the borer do not cause a problem, because the few milligrams involved quickly turn to nearly non-magnetic rust during the drying process, and the grain size of any remaining metallic iron is not significant in producing effects in the range of interest. Mixing of materials from different levels is more of a problem with a borer unless care is taken when extracting the contexts of the bore chamber. For open sections, a brass shovel prevents any suspicion of contamination and is easily made. Samples must be stored in open containers to prevent growth of organisms.

Calculation of magnetic anomalies based on the geometry of features must necessarily be based on an estimate of the volume susceptibility contrast. Samples taken with a borer or shovel do not have the same volume when dried that they did *in situ*. Tests using plastic cylinders with sharpened rims pushed into profiles to extract a volumetrically correct sample (11) showed that it is possible to correct adequately for volume when using weighted samples, and such precautions proved unnecessary. An advantage of the plastic sleeve method is that any orientation dependent magnetisation is preserved, but such sample taking is difficult.

Correction for density and drying

Since the relative humidity in the laboratory is usually lower than that in the field, soil samples lose weight on drying. The loss is of the order of 10–20% and about a week is necessary before the weight of a sample stored in an open tray stabilises. This is shown in figure 7.13 for three different soils. Weighing need be only a bit more accurate than the susceptibility measurement itself. Allowing for the probable errors in estimates of density to obtain volume susceptibility, an easily achieved 0.1% is more than good enough. Hillel (14) notes that the bulk density of soils usually ranges from 1.1 to 1.6. Tests made by Graham and Scollar (11), show that 1.4 is a suitable value to use, based on comparison of volumetrically correct and dried, weighed samples. If samples are heated, as discussed below, a further density correction of 5% is needed in addition to that necessary for air dried samples.

Mechanical preparation

To remove any large lumps of pottery or brick, soils from settlement sites should be sieved. The standard 2 mm brass sieve used in pedology is suitable. A practical installation for mechanical pretreatment is shown in figure 7.14. By measuring only the fraction under 2 mm in size, comparisons are more stable. Experiments conducted by Le Borgne and others have shown (17) that almost all of the magnetic materials are contained in the smallest grains of soils, so that removal of the larger fraction does not falsify results if the weight is corrected for the rejected material when calculating the susceptibility. Soils from areas where dangerous fungi or other parasites are suspected should be heated to at least 105°C for 24 h before preparing the samples.

Fig. 7.13 Stabilisation of the weight of three different soil samples on drying to room humidity. From ref. 11.

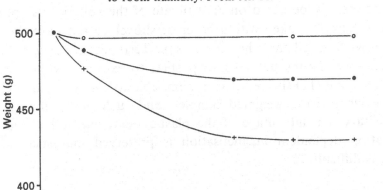

Calibration substances

Graham and Scollar (11) used an artificial soil made of laboratory grade haematite and salt for heating studies. X-ray spectroscopy indicated that maghaemite was formed. Longworth and Tite (20), using Mössbauer spectroscopy on actual soils were able to demonstrate this convincingly. For susceptibility comparison purposes, actual soils might be thought of as calibration samples but because of viscosity and grain size dependent effects, this is not very satisfactory. The material of choice as a calibration body is partly hydrated manganese sulphate, $MnSO_4 \cdot H_2O$, which is paramagnetic, of high purity, and has a very accurately known susceptibility (12). Cross-calibration using the Zahn bridge method (43), which provides an independent non-chemical check showed a value for the susceptibility of $MnSO_4 \cdot H_2O$ which was within the accuracy of the measuring equipment used (0.5%). All measurements with different instruments were referred to a standard maganese sulphate calibration body which could be transferred to instruments working on different principles. Although this technique gives very accurate results in susceptibility, a standard of this type is not available for magnetic viscosity measurements, and it is not at all clear how to make one. Measured soil samples are probably best used for this purpose.

Fig. 7.14 An installation for weighing and sieving of soil samples prior to magnetic susceptibility measurements. From ref. 11.

Determination of iron content

In standard texts on soils chemistry, iron content is usually determined colorimetrically, and this was also used by Mullins (23). The correlation between iron measured in this way and the susceptibilities developed on heating is not very good. Graham and Scollar (11) used complexometric titration which gave accuracies to better than 1 part per 1000 in iron and the correlation with measured susceptibility was good at lower iron percentages, measured in a constant field as shown in figure 7.15, but the scatter increased at higher percentages. The tendency was the same when measured in an alternating field as shown in figure 7.16. The correlation between both types of measurement is very good, figure 7.17. A chemical iron determination of this type is easy to carry out, and it allows a very useful estimate to be made of the maximum susceptibility which can be developed on a given soil without complicated equipment or special laboratory conditions.

7.6.2. Measurement of susceptibility and viscosity

There are many techniques described in the physics literature for the measurement of magnetic susceptibility. Only some of these are applicable

Fig. 7.15 Constant field (100 s) susceptibility and percentage of chemically determined iron for a collection of samples from various parts of Europe. From ref. 11.

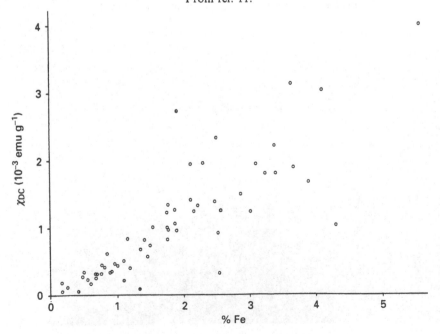

Fig. 7.16 Alternating field susceptibility for the same samples as in figure 7.15. From ref. 11.

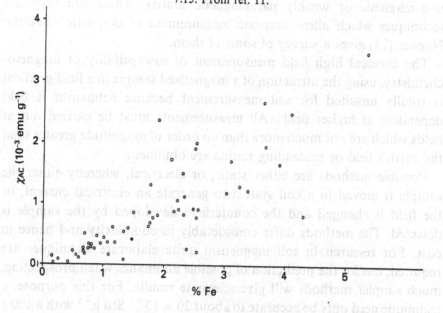

Fig. 7.17 Alternating and constant field susceptibilities for a set of samples showing a slope which approaches 1/2 for many soils (the Mullins constant). From ref. 11.

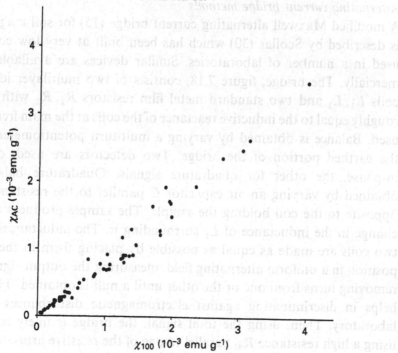

to soils because of the dispersed mixture of ferrimagnetic grains in a non-magnetic or weakly paramagnetic matrix. There are very few techniques which allow accurate measurement of magnetic viscosity. Nagata (24) gives a survey of some of them.

The classical high field measurement of susceptibility of magneto-chemistry, using the attraction of a magnetised sample in a field gradient is totally unsuited for soil measurement because behaviour is field dependent at higher fields. All measurements must be carried out at fields which are not much more than an order of magnitude greater than the earth's field or misleading results are obtained.

Possible methods are either static, or electrical, whereby either the sample is moved in a coil system to generate an electrical current, or the field is changed and the counterfield developed by the sample is detected. The methods differ considerably in complexity and hence in cost. For research in soil magnetism quite elaborate techniques are required, but for the prediction of possible anomalies when prospecting, much simpler methods will give adequate results. For this purpose, a technique need only be accurate to about 20×10^{-7} SIu g^{-1} with a 100 g sample.

Alternating current bridge methods

A modified Maxwell alternating current bridge (15) for soil magnetism is described by Scollar (30) which has been built at very low cost and used in a number of laboratories. Similar devices are available commercially. The bridge, figure 7.18, consists of two multilayer identical coils L_1, L_2 and two standard metal film resistors R_1, R_2 with values roughly equal to the inductive reactance of the coils at the main frequency used. Balance is obtained by varying a multiturn potentiometer R_4 in the earthed portion of the bridge. Two detectors are used, one for in-phase, the other for quadrature signals. Quadrature balance is obtained by varying an air capacitor C parallel to the resistance arm opposite to the coil holding the sample. The sample produces a small change in the inductance of L_1 surrounding it. The inductances of the two coils are made as equal as possible by placing them in their final position in a uniform alternating field, measuring the output signal and removing turns from one or the other until a null is obtained. This also helps in discriminating against electromagnetic disturbances in the laboratory. Then, using the total signal, the bridge is finely balanced using a high resistance R_3, parallel to one of the resistive arms, with the multiturn potentiometer at zero. Direct reading in susceptibility from this potentiometer is determined by the choice of R_6 and higher variable resistor R_5 parallel to it.

If two frequencies are used in the bridge, additional capacitative trimming to make the bridge frequency insensitive may be required, and the standard resistors should be chosen to have a value roughly equal to the mean of the inductive reactance at each frequency to maintain roughly constant sensitivity. The drive signal must be a pure sinusoidal varying either side of each potential, and an instrumentation type operational amplifier must be coupled to the bridge via capacitors and small damping resistors as shown to prevent parasitic oscillations.

The use of Garrett overwound end coils (9) to obtain a uniform field is advisable, allowing for large samples with possible inhomogeneities. If the field is uniform enough, standard plastic wide mouth bottles can be used as sample holders with no correction for the sample geometry necessary. The arrangement is readily capable of direct reading to 5 SIu or less, and with care in construction and low noise conditions in the laboratory, even higher sensitivity is obtainable.

Using manganese sulphate accurately diluted with a diamagnetic material like powdered aluminium oxide, Al_2O_3, samples of known paramagnetic susceptibility can be easily made. The bridge can be calibrated to read directly from the balance potentiometer and any minor non-linearity corrected. An alternative calibration scheme using a coil

Fig. 7.18 A Maxwell bridge for the measurement of soil susceptibilities.

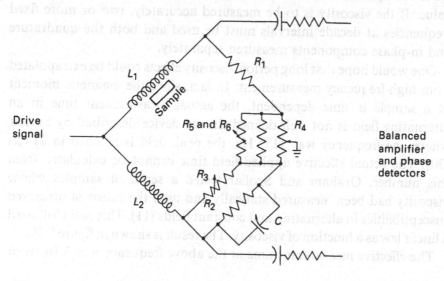

of known dimensions and a precision capacitor has been described by Zahn (43). If the field in the coil system is highly uniform relative to the calibration sample size, it does not matter which method of calibration is employed. If very high sensitivity in the presence of external fields is required, considerable care in coil construction, mounting, orientation, and trimming is required. Any remaining asymmetry in the system can be tuned out with a small variable capacitor parallel to one of the resistive bridge arms. Both coils must be at the same temperature so both are left open to the atmosphere but only one is used.

The optimum frequency of operation is chosen to be between power line harmonics to reduce problems of pickup of external disturbance in the coil system. If it is too low, then the coils must be very large in order to obtain a significant amount of change upon introducing the sample. If it is too high, then the effective measurement time will be too short. Frequencies in the low audio range are suitable. Using a balanced input amplifier, interference is further reduced. By careful coil construction and pruning, nearly perfect cancellation of external fields can be obtained, and the device can be operated in a normal laboratory environment. A synchronous phase detection scheme with a very narrow bandwidth reduces interference sensitivity and allows measurement of either the in-phase or the out-of-phase component of the signal produced when a sample is inserted into one of the coils. This allows some separation of the effects due to susceptibility and viscosity. The out-of-phase signal is also affected by the conductivity of the sample, but this effect is small for the sample sizes used so that the viscosity can be estimated from this value. If the viscosity is to be measured accurately, two or more fixed frequencies at decade intervals must be used and both the quadrature and in-phase components measured separately.

One would hope that long period viscosity effects could be extrapolated from high frequency measurement. In fact, since the magnetic moment of a sample is time dependent, the actual measurement time in an alternating field is not well defined. In the device described by Scollar where one frequency was 1175 Hz, the peak field is reached in 213 μs (30). The actual effective applied field time cannot be calculated from this number. Graham and Scollar tested a series of samples whose viscosity had been measured statically and used the ratios of measured susceptibilities in alternating and constant fields (11). This ratio followed a linear law as a function of viscosity. The result is shown in figure 7.19.

The effective measuremnt time at the above frequency was 3.16 ms or more than an order magnitude larger than the peak field time would indicate. Mullins and Tite (22) took the long term magnetic moment to be of the order of twice that predicted by the high frequency bridge measurement. Graham and Scollar confirmed this for a large number of soil samples. The differing results of measuring a sample at 1175 Hz and at 100 s are shown in figure 7.17. Significant departure was only observed for soils developed on carbonate or igneous rocks, with the latter tending to higher values, the former to lower ones. Soils in the lower range of susceptibilities, under 1500×10^{-6} SIu g^{-1} at 100 s were all very close to the Mullins constant when extrapolated to archaeological time scales. These soils comprise a very large portion of those encountered in northern Europe. Therefore, the alternating current bridge represents a very useful and cost effective device for prediction of anomaly strengths based on soil samples if allowance is made for all of the variable quantities involved in the measurement. Commercial instruments usually use two frequencies. Some measure the frequency shift produced by the sample in a coil used in an oscillator circuit. This is mainly produced by the in-phase component, and is a measure of susceptibility only at one frequency.

Fig. 7.19 The ratio of measured susceptibility in alternating and constant fields as a linear function of magnetic viscosity. From ref. 11.

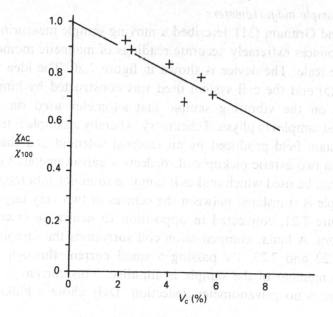

Fluxgate and astatic magnetometers

Molyneux (21) and Heye and Meyer (13) have used fluxgate magnetometers (see chapter 8) for the determination of long term magnetic
moments of rock samples. In this technique, a pair of balanced fluxgates
is mounted horizontally within a magnetically shielded enclosure. A
small sample is introduced and demagnetised by a decreasing alternating
field of high amplitude which is smoothly reduced to zero. It is then
remagnetised in a known field and the magnetic moment developed read
from the fluxgate device. Correction for sample geometry is necessary
when the samples are close to the fluxgate. The technique is the only
constant field method which can be made readily portable and taken
into the field. The results are quite readily extrapolatable to the long
archaeological time scale and no correction factors except for sample
shape are needed. The samples must be highly homogeneous, however,
something which is not always the case with archaeological material. In
this respect the technique can replace the very much more cumbersome
astatic magnetometer laboratory method used by Le Borgne (18). Costs
are much higher than for the alternating current bridge approach, but
much more realistic estimates of anomaly strengths can be made.
Measurements made directly in the field avoid problems of drying and
mechanically treating samples. Care must be taken not to have any
highly magnetic inclusions like fragments of pottery in the samples.

Moving sample magnetometers

Scollar and Graham (31) described a moving sample measuring device
which produces extremely accurate readings of magnetic moment on a
long time scale. The device is shown in figure 7.20. The idea was due
to Daly (5) and the coil system used was constructed by him. It is a
variation on the vibrating sample magnetometer used on solid or
compacted samples in physical chemistry, whereby a sample is translated
in a constant field produced by an external solenoid and the current
induced in two astatic pickup coils deflects a galvanometer. Very large
samples can be used which makes it immune to minor inhomogeneities.
The sample is translated between the centres of two very large pickup
coils, figure 7.21, connected in opposition to neutralise external field
fluctuations. A bifilar compensation coil surrounds the sample holder,
figures 7.22 and 7.23. By passing a small current through this, the
magnetic moment of the sample is annulled. This current is adjusted
until there is no galvanometer deflection. Daly chose a bifilar coil so

that a very small known magnetic surface resulted, making it relatively easy to measure compensation currents used. A large coil which allows application of an external field up to 20 Oe surrounds the pickup systems, and the entire group is mounted in the centre of a set of Garrett-Pissanetzky (10) coils to remove the earth's field over a large uniform volume. Signal detection is by means of a liquid damped

Fig. 7.20 A translating sample magnetometer mounted in a set of Garrett–Pissanetzky coils which annul the earth's magnetic field. From ref. 11.

galvanometer system with a sensitivity better than the Johnson noise due to the resistance of the pickup coils themselves. The high sensitivity of the system made it imperative to house it in a calm magnetic area, and it cannot be used in the disturbed environment of a city laboratory.

Accuracy was limited mainly by the precision of the geometry of the annulling coil and the long term stability of the compensating current

Fig. 7.21 The translation magnetometer principle: a sample is moved linearly in an applied field and the current induced in a pair of opposing pickup coils is detected.

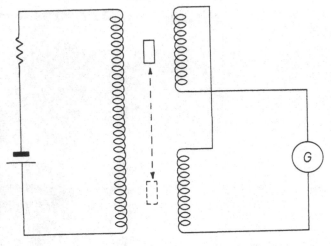

Fig. 7.22 Translation magnetometer with compensation of sample suscepti-bility and light-valve galvanometer.

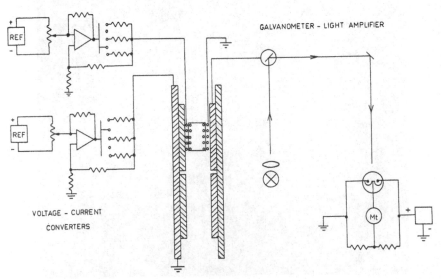

GALVANOMETER - LIGHT AMPLIFIER

VOLTAGE - CURRENT CONVERTERS

source. As constructed, 2.5×10^{-5} SIu could readily be detected in the earth's field, so that 5×10^{-6} SIu g^{-1} in a 100 g sample could be measured reliably. By raising the external field to 20 Oe, weak samples under 10×10^{-6} SIu g^{-1} and even as low as 1×10^{-6} SIu g^{-1} could be

Fig. 7.23 Detail of the standard sample holder and its bifilar coil container which compensates for the susceptibility of the sample on the detector coil box. A Plexiglas block with milled slots mounted on the coil box allows exact orientation of the fluxgates used to control the zeroing of the earth's magnetic field. From ref. 11.

reliably measured. Field dependence is linear as shown in figure 7.24. The sample's viscous magnetisation acquired during transport and storage was removed by rigorous shaking in zero field which randomised the direction of the grains. Only strongly magnetic samples with a few highly magnetic grains resisted this type of demagnetisation. A simple correction using the compensating current could be applied in these cases. Alternating current demagnetisation as used in rock magnetism studies was not necessary. Viscosity coefficients up to 8% were observed on several hundred soil samples. Typical values, as shown in the histogram of figure 7.25 average between 4 and 5%. Six–eight readings taken at half-decade intervals beginning at 100 s were used with a resulting accuracy of 0.1% for the magnetic viscosity. For a typical measurement sequence covering three decades (100–10000 s) the total change in magnetic moment is of the order of 15%. The moment can be tracked over very long periods, as shown in figure 7.26. The values are very slightly temperature dependent, as shown in figure 7.27 for Shimizu's results on magnetite (33). The time slope of the curve depends on the magnetic viscosity of the sample as shown in figure 7.28. Samples so measured were used to check magnetic viscosity effects in the alternating current bridge, and to conclude that this latter technique is

Fig. 7.24 The linear dependence of constant field measured magnetic moment of a soil sample on field strength. From ref. 11.

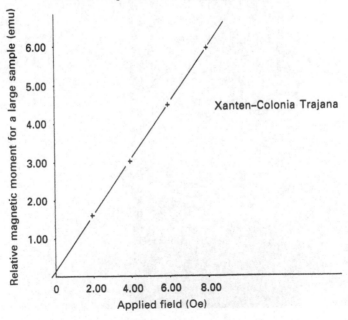

Xanten–Colonia Trajana

sufficiently accurate when using the Mullins correction for anomaly estimates. A moving sample device is probably the most accurate and reproducible method for measurement of the absolute magnetic moment over a period which allows direct extrapolation to the archaeological time scale. It has the great disadvantage of requiring very quiet magnetic conditions unless expensive shielding and compensation techniques are used.

Fig. 7.25 Magnetic viscosities of a collection of soil samples from Europe. From ref. 11.

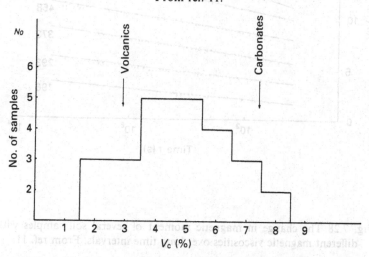

Fig. 7.26 The dependence of measured susceptibility of a soil sample on a constant field strength at a constant long time interval. From ref. 11.

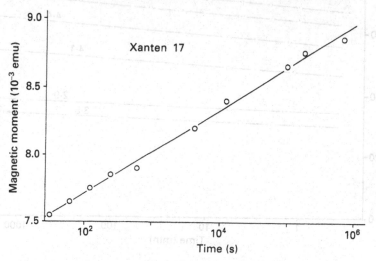

Fig. 7.27 Dependence of magnetic moment on temperature with time. From ref. 33.

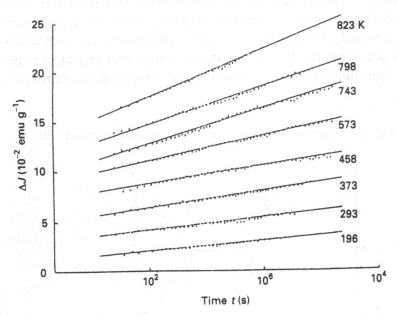

Fig. 7.28 The change in magnetic moment of several soil samples with different magnetic viscosities over long time intervals. From ref. 11.

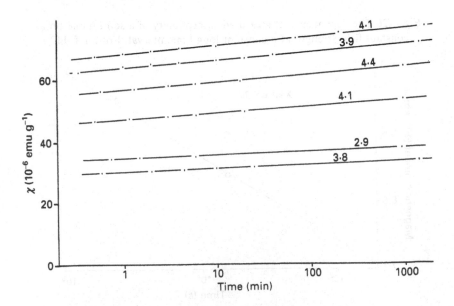

7.7. Conclusion

The magnetic properties of soils are well understood and readily measured with comparatively unsophisticated techniques. A set of soil sieves, a good scale, a kitchen or laboratory exhaust hood, and a susceptibility bridge, which can be purchased or constructed, are all that is required. In addition, a soil borer is useful for sample taking in the field if the terrain does not contain too many stones. When soil magnetic properties have been measured, it is quite simple to compute the probable strengths of anomalies to be expected on sites with typical geometries using the methods described in the next chapter. This preliminary investigation makes it possible to decide if a site is worth surveying or not with a given level of equipment. Failure to make such an investigation can lead to a considerable waste of time and resources on sites which will not produce detectable anomalies.

Notes

(1) Aitken, M. J., Webster, G., Reeds, A. 1958, Magnetic prospecting, *Antiquity*, **32**, 270–1.

(2) Belluigi, A. 1931, Present time magnetic investigation of the subsoil, *Miniere Italiana*, **15**, 7, 37–41.

(3) Chikazumi, S. 1964, *Physics of Magnetism*, John Wiley & Sons, New York, NY.

(4) Cook, J. C., Carts, S. L. 1962, Magnetic effects and properties of typical topsoils, *Journal of Geophysical Research*, **67**, 815–28.

(5) Daly, L. 1970, Études des propriétes magnétiques des roches metamorphiques ou simplement tectonisées. Thesis, Université de Paris 6, Paris.

(6) Dudal, R. 1953, Etude morphologique et génétique d'une séquence de sols sur limon loessique, *Agricultura*, **1**, 2, 119–63.

(7) FitzPatrick, E. A. 1971, *Pedology*, Oliver & Boyd, Edinburgh.

(8) Folgheraiter, A. 1896, Determinazione sperimentale della direzione di un campo magnetico uniform dalla orientazione del magnetismo da esso indotto, *Rendiconti Academia Lincei*, **5**, 127–35.

(9) Garrett, M. W. 1967, Thick cylindrical coil systems for strong magnetic fields with field or gradient homogeneities of the 6th to the 20th order, *Journal of Applied Physics*, **38**, 2563–86.

(10) Garrett, M. W., Pissanetzky, S. 1971, Polygonal coil systems for magnetic fields with homogeneity of the fourth to eighth order, *Review of Scientific Instruments*, **42**, 840–57.

(11) Graham, I. D. G., Scollar, I. 1976, Limitations on magnetic prospection in archaeology imposed by soil properties, *Archaeo-Physika*, **6**, 1–124.

(12) *Handbook of Chemistry and Physics*, 1966, 47th edition, Chemical Rubber Publishing Company, Cleveland, Oh.

(13) Heye, D., Meyer, H. 1972, Ein Messverfahren zu paläomagnetischen Messungen an Tiefseesedimentkernen an Bord eines Schiffes, *Zeitschrift für Geophysik*, **38**, 937–48.

(14) Hillel, D. 1971, *Soil and Water, Physical Principles and Processes*, Academic Press, New York, NY.

(15) Laws, F. A. 1938, *Electrical Measurement*, p. 397–400, McGraw-Hill, New York, NY.

(16) Le Borgne, E. 1955, Susceptibilité magnétique anormale du sol superficiel, *Annales de Géophysique*, 11, 399–419.

(17) Le Borgne, E. 1960a, Influence du feu sur les propriétés magnétiques du sol et sur celles du schist et du granite, *Annales de Géophysique*, 16, 159–96.

(18) Le Borgne, E. 1960b, Etude expérimentale du trainage magnétique dans le cas d'un ensemble de grains magnétiques trés fins dispersés dans une substance non-magnétique, *Annales de Géophysique*, 16, 445–93.

(19) Le Borgne, E. 1965, Les propriétés magnétiques du sol. Application à la prospection des sites archéologiques, *Archaeo-Physika*, 1, 1–20.

(20) Longworth, G., Tite, M. S. 1977, Mössbauer and magnetic susceptibility studies of iron oxides in soils from archaeological sites, *Archaeometry*, 19, 1, 3–14.

(21) Molyneux, L. 1971, A complete result magnetometer for measuring the remanent magnetization of rocks, *Geophysical Journal of the Royal Astronomical Society*, 24, 429–35.

(22) Mullins, C. E., Tite, M. S. 1973, Preisach diagrams and magnetic viscosity phenomena for soils and synthetic assemblies of iron oxide grains, *Journal of Geomagnetism and Geoelectricity*, 25, 213–29.

(23) Mullins, C. E. 1974, The magnetic properties of the soil and their application to archaeological prospecting, *Archaeo-Physika*, 5, 144–348.

(24) Nagata, T. 1961, *Rock Magnetism*, Maruzen, Tokyo.

(25) Neumeister, H., Peschl, G. 1968, Die magnetische Suszeptibilität von Böden und pleistozänen Sedimenten in der Umgebung Leipzigs, *Albrecht Thaer Archive*, 12, 12.

(26) Néel, L. 1948, Magnetic properties of ferrites: ferrimagnetism and antiferromagnetism, *Annales de Physique*, 3, 137ff.

(27) Néel, L. 1955, Some theoretical aspects of rock magnetism, *Advances in Physics*, 4, 191–243.

(28) O'Reilly, W. 1984, *Rock and Mineral Magnetism*, Blackie, Glasgow & London.

(29) Scollar, I. 1965, A contribution to magnetic prospecting in archaeology, *Archaeo-Physika*, 1, 21–92.

(30) Scollar, I. 1968, A simple direct reading susceptibility bridge, *Journal of Scientific Instruments*, 1, 2, 781–2.

(31) Scollar, I., Graham, I. D. G. 1972, A method for the determination of the total magnetic moment of soil samples in a constant field, *Prospezioni Archeologiche*, 7, 85–92.

(32) Selwood, P. W. 1956, *Magnetochemistry*, p. 308, Interscience, New York, NY.

(33) Shimizu, Y. 1960, Magnetic viscosity of magnetite, *Journal of Geomagnetism and Geoelectricity*, 11, 125.

(34) Stacey, F. D., Banerjee, S. K. 1974, *The Physical Principles of Rock Magnetism*, Elsevier Scientific Publishing Co, Amsterdam.

(35) Stavrou, A. 1957, Investigation of the magnetic properties of Greek red earth, *Gerlands Beiträge zur Geophysik*, 66, 214–55.

(36) Thellier, E. 1938, Sur l'aimantation des terres cuites et ses applications géophysiques, *Annales de l'Institut Physique du Globe*, 16, 156–302.

(37) Tite, M. S., Mullins, C. 1970, Magnetic properties of soils, *Prospezioni Archeologiche*, 5, 111–12.

(38) Tite, M. S., Linington, R. E. 1975, Effect of climate on the magnetic susceptibility of soils, *Nature*, **246**, 565–6.

(39) Tite, M. S., Linington, R. E. 1986, The magnetic susceptibility of soils from central and southern Italy, *Prospezioni Archeologiche*, **10**, 25–36.

(40) Tucker, P. M. 1952, High magnetic effect of lateritic soil in Cuba, *Geophysics*, **17**, 753–5.

(41) Vadunina, A. F., Babanin, V. F. 1972, Magnetic susceptibility of some soils in the USSR, *Soviet Soil Science (Pochvovedeniye)*, **10**, 55–66.

(42) Weiss, P. 1907, L'hypothèse du champ moléculaire et la propriété ferrimagnétique, *Journal de Physique*, **6**, 661–91.

(43) Zahn, C. T. 1963, New absolute null method for measurement of magnetic susceptibilities in weak low-frequency fields, *Review of Scientific Instruments*, **34**, 285–91.

8

Magnetic prospecting

8.1. Prediction of magnetic anomalies

8.1.1. Introduction

When the absolute magnitudes of the differences in susceptibility, magnetic viscosity, and remanent magnetisation between a feature and its surroundings are known, then the magnetic anomaly produced in the earth's field can be calculated in closed form for simple regular shapes (40). There are a number of methods given in the geophysical literature which allow calculation of theoretical anomalies from objects of arbitrary shape. Only a few of these are of interest for archaeological structures. Some methods allow for differing susceptibilities in various parts of a structure and for remanent magnetisation with direction differing from that of the contemporary field as well as for a magnetic moment of arbitrary strength. The relevant geophysical literature up to 1969 is summarised by Scollar (73) and additional references for the early 70s are given by Linington (45).

Calculated anomalies from structures of known shape serve a function similar to that of standard resistivity curves used in electrical prospecting. They can be compared with observed values in an attempt to estimate the shape, size, and depth of burial of the features of interest. They can also be used to design matched filters which can be used to search data for structures of an assumed geometry (74,76). However, such matching techniques are not reliable when soil noise and other forms of disturbance are significant. Therefore the most useful aspect of anomaly prediction lies in deciding on the feasibility of a survey campaign where soil susceptibility data have been gathered and something is known about the possible shape and depth of burial of the sought-after features. It also helps in determining required magnetometer sensitivity, accuracy, and long term stability. Another important application lies in creating realistic synthetic data for testing new display and interpretation schemes. Finally, theoretical calculation of anomalies is instructive in that it allows full appreciation of the physical principles underlying the magnetic prospecting problem. Measurements on scale models of archaeological features have confirmed the theoretical calculations quite closely (2).

Assuming a filling of uniform susceptibility without remanent magnetisation does not lead to undue error when constructing a model which is to be compared with actual measurements. The random fluctuations in the latter due to all possible causes will usually be more significant than that attributable to variable susceptibility in the fill. The inclusion of remanent magnetisation effects is readily possible in most methods, but this is rarely done because the direction of magnetisation is unknown in a real situation.

8.1.2. Calculation of the anomaly due to objects of arbitrary shape

Linington (45) gave a large selection of analytical anomaly calculations for features with various two-dimensional cross-sections extending to infinity. The first collection of shapes, figure 8.1, are idealisations of those which are typically encountered in archaeological practice. The second collection, figure 8.2, are of interest in the analysis of measurements made on irregular terrain, a problem discussed in detail by Linington (46).

Definitions
The components of the earth's magnetic field are defined as:
F the total intensity in nanotesla (nT)
H the horizontal component of the total intensity
Z the vertical component of the total intensity
X the horizontal component in a geographic north–south direction
Y the horizontal component in a geographic east–west direction
I the angle of inclination of the field direction to the horizontal
D the angle of declination between H and geographic north.

These field components and directions are related by:

$$F^2 = H^2 + Z^2, \quad H = F\cos(I), \quad Z = F\sin(I), \quad \tan(I) = Z/H \left.\right\}$$
$$H^2 = X^2 + Y^2, \quad X = H\cos(D), \quad Y = H\sin(D), \quad \tan(D) = Y/X \left.\right\} \quad (8.1)$$

To simplify the calculation, D, X, and Y are ignored by considering magnetic north to coincide with geographic north. Thus H is the north–south component and the east–west component is zero.

Field perturbation from a feature

The concentration of magnetic material which differs from that of the surrounding soil in an archaeological feature perturbs the earth's field. The amount of perturbation is:

$$\Delta F = F' - F, \qquad \Delta H = H' - H, \qquad \Delta Z = Z' - Z \qquad (8.2)$$

Fig. 8.1 Idealised cross-sections of archaeological features. From ref. 45.

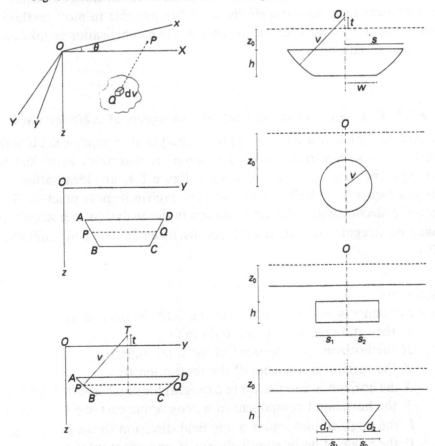

where the F, H, and Z are the unperturbed fields, and F', H' and Z' are the fields in the presence of the perturbation.

The horizontal component of the anomaly ΔH can be further resolved into two components ΔH_X and ΔH_Y in east–west and north–south directions, respectively. Then (8.2) ΔH can be replaced by:

$$\Delta H_X = H'_X - H, \qquad \Delta H_Y = H'_Y \tag{8.3}$$

From these equations it follows by simple vector algebra that the

Fig. 8.2 Irregular terrain features of interest in magnetic prospecting. From ref. 45.

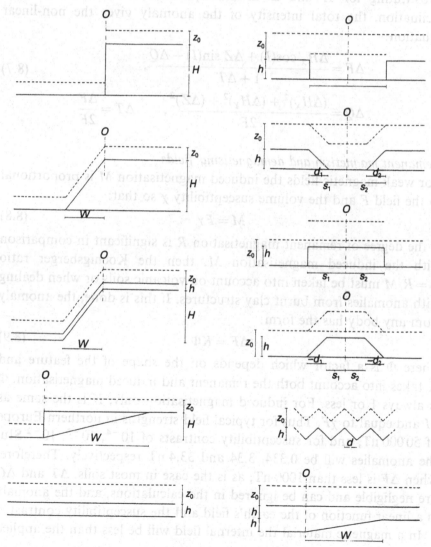

anomaly field ΔF due to the perturbation is:

$$\Delta F = F' - F, \qquad F'^2 = H_X'^2 + H_Y'^2 + H_Z'^2 \tag{8.4}$$

multiplying ΔF by $(F' + F)/(F' + F)$ then:

$$\Delta F = \frac{F'^2 - F^2}{F' + F} \tag{8.5}$$

and thus substituting and multiplying:

$$\Delta F = \frac{2H\,\Delta H_X + 2Z\,\Delta Z + (\Delta H_X)^2 + (\Delta H_Y)^2 + (\Delta Z)^2}{2F + \Delta F} \tag{8.6}$$

Substituting for H and Z in the equations above which include inclination, the total intensity of the anomaly gives the non-linear equation:

$$\Delta F = \frac{\Delta H_X \cos(I) + \Delta Z \sin(I) + \Delta Q}{1 + \Delta T}, \tag{8.7}$$

$$\Delta Q = \frac{(\Delta H_X)^2 + (\Delta H_Y)^2 + (\Delta Z)^2}{2F}, \qquad \Delta T = \frac{\Delta F}{2F}$$

Remanent magnetism and demagnetising fields
For weak magnetic fields the induced magnetisation M is proportional to the field F and the volume susceptibility χ so that:

$$M = F\chi \tag{8.8}$$

If the degree of remanent magnetisation R is significant in comparison with the induced magnetisation M, then the Koenigsberger ratio $Q = R/M$ must be taken into account on volcanic soils or when dealing with anomalies from burnt clay structures. If this is done, the anomaly from any body has the form:

$$\Delta F = K\Phi \tag{8.9}$$

where Φ is a factor which depends on the shape of the feature and K takes into account both the remanent and induced magnetisation. Φ is always 1 or less. For induced magnetisation only, K is the same as M and equal to χF. Thus for typical field strengths in northern Europe of $50\,000$ nT, and for susceptibility contrasts of 10^{-4}, 10^{-3}, 10^{-2} SIu, the anomalies will be 0.334, 3.34 and 33.4 nT respectively. Therefore, when ΔF is less than 1000 nT, as is the case in most soils, ΔT and ΔQ are negligible and can be ignored in the calculations, and the anomaly is a linear function of the earth's field and the susceptibility contrast.

In a magnetic material the internal field will be less than the applied external field because induced magnetisation opposes the external field. Thus the induced magnetisation produces a demagnetising field which is dependent on the shape of the structure. Where N is dependent on the shape of the body, the effective field is reduced by a factor:

$$\frac{M - m}{M} = \frac{\chi N M}{M} \tag{8.10}$$

(m due to demagnetisation). But since χ is always a small number the effect can be ignored for archaeological problems.

The field of a buried dipole at the surface

The magnetic intensity F_1 at point P from an object with magnetic moment m_1 at another point $P_1 = (x_1, y_1, z_1)$ at a distance r_1 from $P = (x, y, z)$ is:

$$F_1 = m_1/r_1^2, \qquad r_1^2 = (x_1 - x)^2 + (y_1 - y)^2 + (z_1 - z)^2 \tag{8.11}$$

The components of the intensity F at P due to m_1 are therefore:

$$F_{x_1} = \frac{m_1(x - x_1)}{r_1^3}, \qquad F_{y_1} = \frac{m_1(y - y_1)}{r_1^3}, \qquad F_{z_1} = \frac{m_1(z - z_1)}{r_1^3} \tag{8.12}$$

This can be written:

$$F_{x_1} = -\partial V_1/\partial x, \qquad F_{y_1} = -\partial V_1/\partial y, \qquad F_{z_1} = -\partial V_1/\partial z \tag{8.13}$$

with

$$V_1 = \mu_0/4\pi$$

where V_1 is the magnetic potential due to the moment m_1 at P_1. A small volume of magnetic material at a distance z below the surface corresponds to a dipole. Assuming that it is inclined at an angle J which may or may not be equal to the angle of dip of the earth's field, depending on whether remanent magnetism is present or not, and assuming that the line of the dipole is oriented in the x direction, then the field components at the surface can be derived from the above equations by differentiation:

$$\left.\begin{aligned}
\Delta H_x &= M \frac{(2x^2 - y^2 - z^2)\cos(J) - 3xz\,\sin(J)}{(x^2 + y^2 + z^2)^{\frac{5}{2}}} V_1 \\[2mm]
\Delta H_y &= M \frac{3yz\,\sin(J) - 3xy\,\cos(J)}{(x^2 + y^2 + z^2)^{\frac{5}{2}}} V_1 \\[2mm]
\Delta H_z &= M \frac{(2z^2 - x^2 - y^2)\sin(J) - 3xz\,\cos(J)}{(x^2 + y^2 + z^2)^{\frac{5}{2}}} V_1
\end{aligned}\right\} \tag{8.14}$$

Combining gives the total field anomaly:

$$\Delta F = V_1 \frac{M}{(x^2 + y^2 + z^2)^{\frac{5}{2}}} [(2x^2 - y^2 - z^2)\cos(E)\cos(J)\cos(I)$$
$$+ (2z^2 - x^2 - y^2)\sin(J)\sin(I) - 3xy\sin(E)\cos(J)\cos(I)$$
$$- 3xz\cos(J)\sin(I) - 3xz\cos(E)\sin(J)\cos(I)$$
$$+ 3yz\sin(E)\sin(J)\cos(I)] \tag{8.15}$$

where E is the angle between the dipole axis and magnetic north in the horizontal plane, and J the angle of dipole dip. When remanent magnetism is not present and the induced magnetisation is in the direction of the earth's field, then the expression can be rearranged and reduces to:

$$\Delta F = V_1 M \frac{x^2(3\cos^2(I) - 1) + z^2(3\sin^2(I) - 1) - 6xz\sin(I)\cos(I) - y^2}{(x^2 + y^2 + z^2)^{\frac{5}{2}}} \tag{8.16}$$

An alternative formulation for the field components H_X, H_Y, H_Z is given by Scollar (73) with $r = (x^2 + y^2 + z^2)^{-\frac{5}{2}}$ as

$$\left. \begin{array}{l} H_x = [3xz\sin(I) - 3xy\cos(I)]r \\ H_y = [\cos(I)(2y^2 - z^2 - x^2) - 3yz\sin(I)]r \\ H_z = [\sin(I)(2z^2 - x^2 - y^2) - 3yz\cos(I)]r \end{array} \right\} \tag{8.17}$$

and for the total field anomaly due to a set of dipoles of strength M and without remanence:

$$\Delta F = V_1 M \{\cos(I)\sum H_Y + \sin(I)\sum H_Z + \tfrac{1}{2}\chi[(\sum H_x)^2 + (\sum H_Y)^2 + (\sum H_Z)^2]\} \tag{8.18}$$

This formulation is useful when using the method for computing anomalies described in the following section.

For the case where the dipole direction is significantly different from that of the earth's field, the expressions are much more complex and are given by Linington's equations (33)–(36). Since the remanent magnetism of a buried structure is nearly never known, that formulation is not of practical importance in prospecting.

Calculation by summing dipole moments
With the expressions for the field of a dipole at any depth and at any geographic location known, then the anomaly for any three-dimensional shape can be readily computed by adding the fields from a set of dipoles

filling the three-dimensional shape. Scollar (73) gives a description of one possible computer method for calculating the field at the surface. The dipole field at various depths is shown in figure 8.3. Since the field is symmetrical around the axis of magnetic north, only half the values need be calculated. The field becomes insignificantly small for magnetisations of archaeological importance at a distance of 20 units from the dipole, so that only 841 field values need be calculated for each depth. If magnetisation is assumed to be non-uniform with depth, the values at each depth are multiplied by a constant and remanence can be allowed by summing separate dipole fields for the induced and remanent directions. The summing arrangement is shown in figure 8.4. Summation

Fig. 8.3 The dipole field of a spherical anomaly at various depths at 50° north magnetic latitude. From ref. 73.

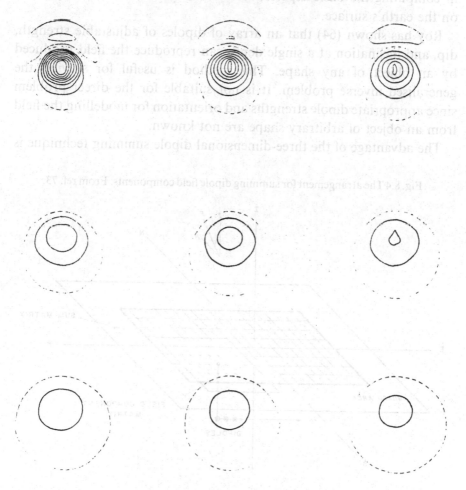

is carried out separately for each field component and then the total field is computed from the three components at each point.

The anomalous shape is assumed to be filled with closely packed spheres of magnetic material. Each sphere can be considered to be a dipole source. A correction is needed to take into account the fact that the effective volume of magnetic material is greater than the spacing of the coordinates of the dipoles. This correction is equal to the radius of the sphere of unit volume or 0.6203. It must also be subtracted from the coordinates to allow specification to the top and sides of a feature rather than to dipole centres below the top or side layers. Using standard computer graphic techniques to obtain coordinates for lines, arcs, etc, the coordinates for individual dipoles for objects of arbitrary shape can readily be computed. By changing the inclination or dip of the field used in computing the basic dipoles, results can be obtained for any point on the earth's surface.

Roy has shown (64) that an array of dipoles of adjustable strength, dip, and inclination at a single depth can reproduce the field produced by an object of any shape. This method is useful for solving the generalised inverse problem. It is not suitable for the direct problem since appropriate dipole strengths and orientation for modelling the field from an object of arbitrary shape are not known.

The advantage of the three-dimensional dipole summing technique is

Fig. 8.4 The arrangement for summing dipole field components. From ref. 73.

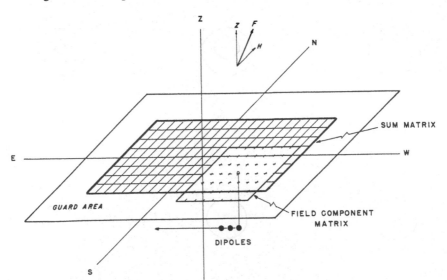

that it is mathematically straightforward, requires no complex algebraic expressions which may result in hard-to-find typing errors in the program, and if the computer has a high speed floating point add instruction it is quite fast. A disadvantage is that the accuracy of the representation of the field depends on the number of dipoles used to represent complex shapes. Use of a very fine grid to obtain good shape representation raises the calculation time as the third power of the number of dipoles employed. Experiments have shown that a $100 \times 100 \times 10$ grid is adequate for most shapes of archaeological interest, and with these values reasonable calculation times are obtained even on small machines. If higher resolution is needed, then it may be more advantageous to use Linington's or Talwani's method (83) below.

Calculation by analytical triple integration

Linington (see below) observed that the expression for a dipole field can be written generally as a function of distances and angles:

$$\Delta C = V_1 M \sum_{n=1}^{7} f(n) g(n, C) \qquad (8.19)$$

where $f(n)$ are functions of distance and $g(n, C)$ are functions of angles for any of the field components C or for the total field. The seven distance coefficients $f(1, \ldots, 7)$ all have the form of the quotient of a polynomial of the second degree at most in x, y, and z with the familiar constant denominator $(x^2 + y^2 + z^2)^{5/2}$. Linington (45) notes that all these are not independent, for the first four can be related by:

$$f(1) + f(2) + f(3) = 0, \qquad f(4) = f(1) - f(2) \qquad (8.20)$$

so that only five need be considered. The anomaly can then be expressed as the volume integral of a distance function:

$$\Delta C = V_1 M \sum_{n=1}^{7} g(n, C) \iiint f(n) \, dx \, dy \, dz \qquad (8.21)$$

The coefficients for $g(n, C)$ are given by Linington in his Table 1. The shape of the feature determines the limits of the volume integral and the formulation of the distance functions $f(n)$. Linington breaks the integration into three phases, obtaining a line integral in the first, a plane in the second and the volume in the third. The resultant distance functions for the line source are quotients of polynomials of the fourth degree. For the horizontal plane the expressions are more complex still and for three-dimensional features they contain eight terms of high order polynomial quotients for a rectangular prism. Linington lists all of these along with the case of a prism with sloping sides. If sources with a finite two-dimensional cross-section and infinite length are used, the expressions simplify considerably. A large number of practical cases are given by Linington for shapes related to surface topography whose effects can thus be taken into account if desired. He states 'The formulae . . . have all been recalculated starting from basic principles with comparisons being made with other published results where possible. One general point has become obvious in this comparison; the number of printing errors that manage to occur in many articles. It is hoped that the present study may not contain too many of these . . .'.

If this method is used, it would not be advisable to program Linington's equations as published. Rather, it would be sensible to use a symbolic mathematical processor like MACSYMA or REDUCE and carry out the integration, followed by automatic translation to a suitable programming language. This avoids transcription errors for the extremely complex expressions which may not be entirely correct as published.

Calculation by analytical double and numerical single integration
Talwani (83) published a method for obtaining the field from an object of any desired shape which is widely used in geological geophysics. In this technique, the body is represented by polygonal lamina with a number of sides sufficient to describe a section through the shape. The method assumes a uniform magnetisation M of the body, and with the three components of M at distance r, the magnetic potential is:

$$V = \frac{\mu_0}{4\pi} \frac{M_x x + M_y y + M_z z}{r^3} \, \Delta x \, \Delta y \, \Delta z \qquad (8.22)$$

Thus the magnetic anomaly is:

$$\Delta X = \iiint -\frac{\partial V}{\partial x} \, dx \, dy \, dz, \qquad \Delta Y = \iiint -\frac{\partial V}{\partial y} \, dx \, dy \, dz \qquad (8.23)$$

$$\Delta Z = \iiint -\frac{\partial V}{\partial z} \, dx \, dy \, dz$$

Substituting the magnetic potential and differentiating with respect to x, y and z:

$$\Delta X = M_x V_1 + M_y V_2 + M_z V_3$$
$$\Delta Y = M_x V_2 + M_y V_4 + M_z V_5 \qquad (8.24)$$
$$\Delta Z = M_x V_3 + M_y V_5 + M_z V_6$$

where:

$$V_1 = \iiint \frac{3x^2 - r^2}{r^5} \, dx \, dy \, dz, \qquad V_2 = \iiint \frac{3xy}{r^5} \, dx \, dy \, dz$$

$$V_3 = \iiint \frac{3xz}{r^5} \, dx \, dy \, dz, \qquad V_4 = \iiint \frac{3y^2 - r^2}{r^5} \, dx \, dy \, dz \quad (8.25)$$

$$V_5 = \iiint \frac{3yz}{r^5} \, dx \, dy \, dz, \qquad V_6 = \iiint \frac{3z^2 - r^2}{r^5} \, dx \, dy \, dz$$

The integrals over the volumes V_n are reduced to an analytical double integration performed on the surface of horizontal lamina and a numerical integration follows for the depth axis. When R is the distance to an elementary volume element in a lamina, the surface integrals $S(n)$ for a polygon sectioned from a body as shown in figure 8.5 are solved by Talwani to give:

$$S_1 = -\sum_i \frac{\cos^2(\theta_i)}{z^2 + p_i^2} \left[\frac{g_i y_{i+1} - z^2 \tan(\theta_i)}{R_{i+1}} - \frac{g_i y_i - z^2 \tan(\theta_i)}{R_i} \right]$$

$$S_2 = -\sum_i \frac{\cos^2(\theta_i)}{z^2 + p_i^2} \left[\frac{g_i y_{i+1} \tan(\theta_i) + g_i^2 + z^2}{R_{i+1}} - \frac{g_i y_i \tan(\theta_i) + g_i^2 + z^2}{R_i} \right]$$

$$S_3 = -\sum_i \frac{z \cos^2(\theta_i)}{z^2 + p_i^2} \left[\frac{y_{i+1} \sec^2(\theta_i) + g_i \tan(\theta_i)}{R_{i+1}} - \frac{y_i \sec^2(\theta_i) + g_i \tan(\theta_i)}{R_i} \right]$$

$$S_4 = -\sum_i \frac{\sin^2(\theta_i)}{z^2 + p_i^2} \left[\frac{c_i x_{i+1} - z^2 \cot(\theta_i)}{R_{i+1}} - \frac{c_i x_i - z^2 \cot(\theta_i)}{R_i} \right]$$

$$S_5 = -\sum_i \frac{z \sin^2(\theta_i)}{z^2 + p_i^2} \left[\frac{x_{i+1} \operatorname{cosec}^2(\theta_i) + c_i \cot(\theta_i)}{R_{i+1}} - \frac{x_i \operatorname{cosec}^2(\theta_i) + c_i \cot(\theta_i)}{R_i} \right]$$

$$S_6 = -\sum_i \frac{p_i}{z^2 + p_i^2} \left[\frac{r_{i+1} \cos(\gamma_i)}{R_{i+1}} - \frac{r_i \cos(\beta_i)}{R_i} \right]$$

(8.26)

in which the distances and angles are those shown in the figure repeated for each side of the polygon. This is repeated for all the other polygonal lamina. The volume integration is then performed numerically either by the technique suggested by Talwani or in another standard manner. If desired, different values of the intensity and direction of magnetisation for each of the lamina may be incorporated into the calculation to allow for remanence. The anomaly in the total field with inclination I and declination D is given as in the other methods by:

$$\Delta F = \Delta X \cos(D) \cos(I) + \Delta Y \sin(D) \cos(I) + \Delta Z \sin(I) \quad (8.27)$$

With the large computers which are now readily available, there is really no compelling reason to use any of the methods described above. A brute force numerical integration of the field components is feasible even when using a fine volume grid to describe the shape of the feature. But the methods described are of interest if personal computers are to be used for the calculations.

Fig. 8.5 Polygon approximation for a section of a body following Talwani. From ref. 83.

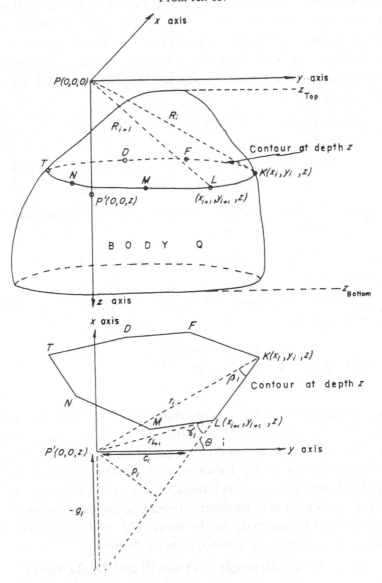

Table 8.1 *Anomaly due to a buried sphere of one unit radius at a depth expressed in units of the same size. Earth's field: 47 000 nT, inclination: 65°N, susceptibility contrast: 120×10^{-6} SIu.*

Depth (unit radii)	Anomaly (nT)
1.5	2.05
1.7	1.41
2.0	0.86
2.5	0.44
2.7	0.35
3.0	0.26
3.7	0.14
4.0	0.11

8.1.3. Feature shapes and the perturbed field

The two types of features which are of the greatest archaeological interest are the pit and the linear ditch. The former can be equated to an isolated dipole equivalent to a uniform spherical feature at a known depth, the latter to a triangular cross-section ditch of great length. There is excellent agreement between the Scollar, Linington and Talwani calculations and the experimental data obtained by Aitken and Alldred (2). A third shape, that of a buried disk was used by Sternberg (81) to approximate frequently encountered buried hearths which are sought for archaeo-magnetic measurements. The calculations used by him are based on a very simple method for the magnetic anomaly due to circular lamina.

Anomaly from a buried sphere
The total magnetic anomaly due to a buried sphere of one unit radius at a depth expressed in the same units is given by equation (8.27), and for the earth's field in northern Europe (47 000 nT, inclination 65° and a moderate susceptibility contrast) it is shown in table 8.1. A sphere with 1 m radius represents the filling of a very large pit, and with the susceptibility contrast shown in the table, it can be detected at about 1.5 m from the centre by typical commercial magnetometers. Smaller pits produce proportionally smaller anomalies and will be usually undetectable at greater depths with many instruments unless the susceptibility contrast is at least an order of magnitude higher.

Theoretical anomalies are calculated using the volume susceptibility contrast. However, it is much easier to measure susceptibility by sample weight, so that a correction for soil density of about 1.4–1.6 must be applied (see chapter 7). Typically, a weight susceptibility contrast of 630×10^{-6} SIu results in a 15 nT anomaly for a pit of 1 m radius buried just under a 30 cm thick plough layer with a measuring sonde held 20 cm above the surface. This anomaly can be detected with almost any magnetometer. If, for reasons to be discussed shortly, the sonde is kept at 50 cm from the surface, and if more typical pits with 50 cm radius are assumed having the filling described, and with a wet soil density of 1.5, the anomaly will be only 1.2 nT, which cannot be detected reliably by all instruments. For more weakly developed soil magnetism, values near or under a tenth of a nanotesla are more common. Few commercially available instruments can detect these reliably with a small number of readings. The shape of the anomaly is best judged from figure 8.6 which shows the negative values to the north of the feature on a north–south

Fig. 8.6 The anomaly from a buried sphere in mid-north latitudes. From ref. 45.

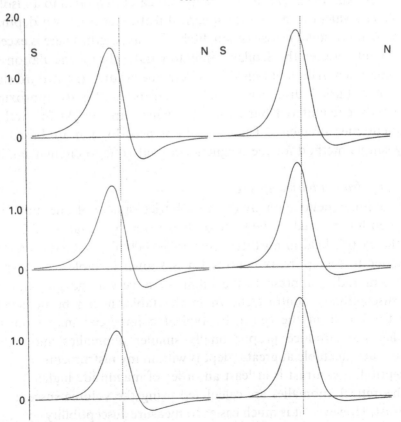

line, and the symmetry in an east–west direction when measurements are made at a fairly high magnetic latitude in the northern hemisphere. Nearer to the magnetic equator the anomalies are almost S shaped in a north–south direction and are considerably weaker.

Linear features

The situation is somewhat more favourable for long linear features, because the anomaly drops off almost linearly with depth when the measurements are made close to the upper surface. The anomaly strength depends on the orientation of the feature relative to the earth's field. It is maximum when the feature lies east–west, minimum when north–south. Long features are easier to detect than isolated pits, not only because the anomaly is slightly stronger, but also because the data are more obvious to the eye if the evaluation technique is well chosen. A 4.5 nT anomaly is observed with material having 1260×10^{-6} SIu contrast, 30 cm above a 30 cm deep ploughed layer over a 90 cm wide ditch. For greater depths, a small feature like this quickly goes below the level of detectability of a single set of measurements, but an order of magnitude in detectability can be gained by use of the proper evaluation technique.

The width of the anomaly depends on the depth of burial of the feature and on its size. For a vertical plate anomaly of various vertical sizes at constant depth to the top of the structure, the anomalies are plotted in figure 8.7. It can readily be seen that the width of the anomaly is hardly affected by the deeper lying material. The shape of the anomaly depends on the orientation of the feature as shown in figure 8.8 with negative anomalies quite pronounced even for a ditch with a north–south orientation.

If the feature is of considerable width, then the shape of the anomaly changes, as shown in figure 8.9, becoming flat-topped. The figure shows only the case for a north–south orientation. With other orientations, the amplitude of the north magnetic anomaly increases, and the flat top is distorted. The shape of the anomaly is slightly affected by the shape of the lower parts of a feature, but since the deeper layers contribute little, there is difficulty in distinguishing these effects from those due to the material nearest to the surface.

Multiple ditches are frequently encountered. A problem arises in distinguishing this case from that of the broad single feature when the multiple ditches are deeply buried. This is shown in figure 8.10 in which the double peak becomes a single peak with increasing depth of burial, flat topped at first.

Interruptions in ditches caused by entryways are also frequent. They can be readily detected only if the feature is close enough to the surface. For example, in figure 8.11, the separation is clearly visible, but in a simulated circular feature, intended to represent a barrow ditch shown in figure 8.12 there is no evidence that an opening exists, other than a very slight inflection of the innermost contours on the northeast perimeter. In a real survey this would undoubtedly be masked by soil noise.

Fig. 8.7 The anomaly from a vertical plate at various sizes at constant depth. From ref. 73.

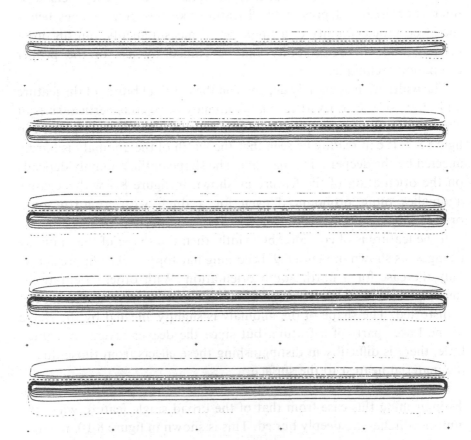

Fig. 8.8 The anomaly of a north–south oriented ditch. From ref. 45.

Fig. 8.9 The anomaly for a wide ditch. From ref. 45.

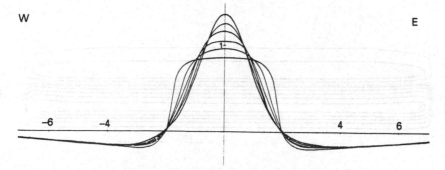

8.1.4. Soil noise and other sources of disturbance

It is convenient to consider the problem of ultimate sensitivity in archaeological magnetic prospecting via an analogy with communication theory. Buried structures of interest create magnetic anomalies of a size and strength which depend on their shapes, magnetisation, and depth of burial. These may be considered as 'signals', and all other sources of magnetic anomalies or variations in measurement are taken to be 'noise'. The aim of a measurement technique is to maximise the signal to noise

Fig. 8.10 Anomalies produced by multiple ditches: E–W oriented spaced triangular ditches, width 5.2 units, depth 3.2 units, $F = 47$ K gamma, $I = 65°$, $k = 10^{-3}$ emu: (a) separation 1.8 units, depth to top of features 0.38 units, contour interval 50 gamma; (b) separation 0 units, depth of feature 1.0 unit, contour interval 10 gamma; (c) separation 0 units, depth of feature 2.0 units, contour interval 10 gamma. From ref. 73.

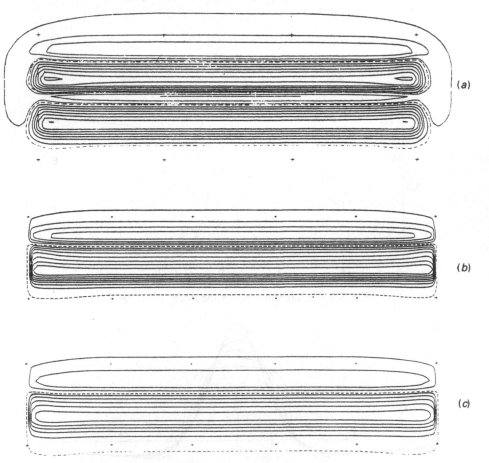

ratio, and the aim of a display and evaluation technique is to present the results in the most effective possible way for human cognition. The use of signal processing theory cannot be taken too far because the data are not infinite in extent in the same way that time varying signals and noise are taken to be. Scollar (74) distinguished between two major classes of noise, depending on the degree of correlation of the magnetic anomaly produced from place to place, as defined by the spread of the autocorrelation function (see chapter 4). A highly uncorrelated noise source usually comes from non-systematic errors in the measuring instrument itself or from rapid external magnetic disturbances like

Fig. 8.11 A visible interruption in a ditch at shallow depth of burial. From ref. 73.

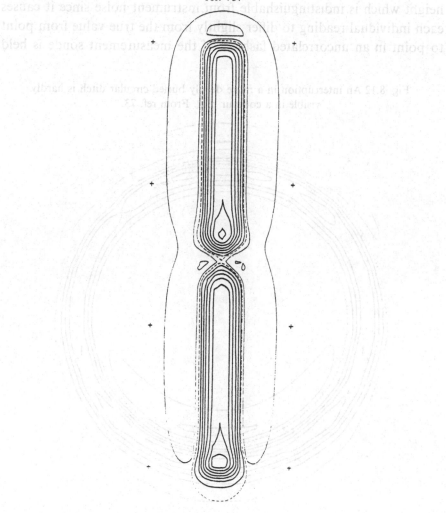

passing vehicles or uncompensated micropulsations in the magnetic field of the earth.

Errors in the position of the sonde relative to a true three-dimensional grid are always present. The extent to which this affects the accuracy of measurement of a magnetic anomaly in the x, y direction depends on the depth of burial of the feature. If the feature is far from the sonde, the anomaly will be wide and small positional errors will not matter much. Horizontal positional errors contain a correlated component which distorts the shape of the anomalies and an uncorrelated one which merely causes the edges to be less sharply defined. An example is given in figure 8.13.

There is always an unsystematic component in measurement sonde height which is indistinguishable from instrument noise since it causes each individual reading to differ slightly from the true value from point to point in an uncorrelated fashion. If the measurement sonde is held

Fig. 8.12 An interruption in a more deeply buried circular ditch is hardly visible in a contour plot. From ref. 73.

above the ground by a rod which does not sink into the surface, this source of noise is usually insignificant. If the soil is highly magnetic, the irregularities in the surface itself which are close to the sonde may introduce some random effects which are easily eliminated by raising the sonde a bit. However, if the base rock is magnetic, this will not help very much. The anomalies due to small amounts of surface iron drop off almost as the cube of the distance. Therefore a higher sonde position is useful if the archaeological anomalies are extended enough so that their fields fall off more slowly. A point is quickly reached where the uncorrelated instrument noise is of similar magnitude, and further raising the height will degrade the signal to noise ratio. A test over a known feature on a site is the only way to determine optimum sonde height, although knowledge of the surface soil susceptibility and roughness helps. Good results have been obtained for soils with surface susceptibility less than 500×10^{-6} SIu g^{-1}, sonde height 30 cm, less than 2000×10^{-6} SIu g^{-1}, 60 cm and from 2000×10^{-6} to 5000×10^{-6} SIu g^{-1}, 1 m.

Instrumental noise sources can be reduced by proper design and the

Fig. 8.13 Bomb and shell crater anoamlies at a Second World War battlefield distorted by local error in the recording of sonde position.

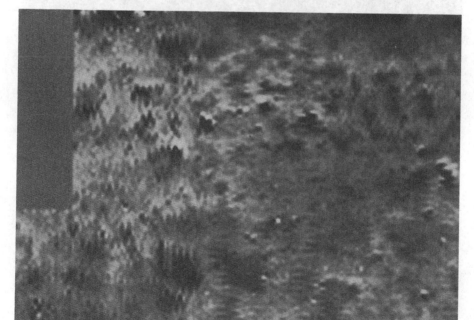

effect of surface iron ameliorated by raising the sonde higher above the surface but major anomalies produced by bomb or shell craters in former battle areas as shown in figure 8.13 cannot be eliminated in this way. The additional errors visible in this figure as a wavy outline of the strong crater anomalies are due to incorrect recording of the sonde position which is then exaggerated by the enlargement procedure used to produce the image. There is no way to reduce this kind of error.

A subtle source of noise is the heading error to which nearly all types of magnetometer, and especially optically pumped instruments, are subject. In this case variation of sonde orientation from reading to reading or line to line will produce small deviations of up to 0.5 nT if uncompensated. A similar effect is seen in proton magnetometer sondes which are insufficiently free of magnetic contamination or when the sonde carrier has not been sufficiently decontaminated prior to making the readings. This effect is shown in figure 8.14.

Correlated noise due to natural horizontal irregularities in superficial susceptibility also cannot be eliminated. Cook and Carts (20) measured

Fig. 8.14 Heading error in a sonde produces stripe anomalies parallel to the direction of measurement.

small anomalies quite close to the surface which they attributed to small stones of varying magnetic composition. The distance from sonde to surface was much smaller than usually used in archaeological prospecting. The results of measurements on samples of agricultural topsoils taken over a wide area by Graham and Scollar (26) show that the ploughed layer is quite well mixed and that local variations seldom exceed more than 10% of the mean value of susceptibility on the site. Hence the effect is not very important at typical sonde heights.

Variation in subsurface geometry of a non-archaeological kind due to the mode of deposition or formation of the soil depends on the microtopography and the mode of soil formation. Erosion or gullies which were subsequently filled in produce anomalies which can be distinguished by shape from archaeological features only if the sampling rate of the field is high and the data are properly displayed. In volcanic soils, noise is particularly annoying because the individual rock fragments present have very high thermoremanent magnetism. Ditches have been detected in such surroundings, not by looking for a positive magnetic anomaly, but rather by observing the less noisy zone produced by the mixed material in a ditch, as opposed to the highly noisy magnetic surroundings (18).

Disturbances of man-made origin such as old field boundaries, uprooted tree stumps, drainage ditches, wartime trenches, etc, can be distinguished if their shapes differ from those of archaeological features. The ability to do so depends very much on the technique chosen for displaying the data. In figure 8.15 two faint Roman ditches, one wide and one narrow, are visible in an area of bomb craters, uprooted trees, and strong surface iron disturbances only because of their characteristic shapes. This is despite the fact that the magnetic anomalies of archaeological interest are two orders of magnitude weaker than the modern disturbances. Strong local anomalies due to lightning strikes have been reported in the palaeomagnetism literature, but these have not been observed in an archaeological context.

Models for soil noise

Scollar (74) described a number of techniques for simulating noise by calculation. Two basically different approaches were tried. The first takes real measurements and transforms them to a new set of values, the second uses arrays of randomly distributed dipoles from which the perturbed magnetic field can be calculated. The former automatically includes all other instrumental and natural error sources and thus gives very realistic results, but it is mathematically intractable. The latter is less realistic, but more readily analysed. In the first technique the two-dimensional Fourier transform of a grid of real readings is taken. This gives a complex number array which is the spectrum of the data. The modulus of such a spectrum looks like figure 8.16 where the low spatial frequencies are shown in the centre of the plot. The frequency distribution falls slowly from the centre indicating high correlation of the noise features. The spectrum has real and imaginary components r and i. To obtain the new data with the same noise spectrum envelope, the angle $\theta = \arctan(r/i)$ is modulated by addition of a linear random number Δ less than 2π.

A new Fourier coefficient is calculated from this by taking:

$$F(x, y) = |r \cos(\theta + \Delta) + i \sin(\theta + \Delta)| \tag{8.28}$$

and then the inverse transform taken. The result is a set of noise anomalies with all the features in different places and having different shapes. The spectrum from an archaeological site with strong signal to noise ratio looks very different, of course, as shown in figure 8.17 for the data of

Fig. 8.15 Wide and narrow roman ditches are faintly visible despite the bomb craters when suitable range compression techniques are used.

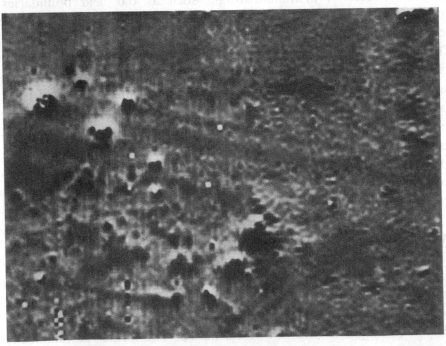

figure 8.18. Here the linear anomalies contribute to frequency components which spread over the entire plane and produce large peaks in the centre corresponding to the longer feature lengths.

The second technique uses the dipole field summing technique described in the previous section. One assigns linear random numbers to the coordinates of the dipoles present in x, y, and z. This is a stochastic process indexed on pairs of integers $Z(t)$ in the plane:

$$Z(t) \qquad t \in T \tag{8.29}$$

with T in the set of real numbers. The presence or absence of a dipole can be considered as a set of discrete binary values assigned to each location by the toss of a coin. If the strengths of the dipoles are allowed to vary, then continuous variables are assigned to each of n locations from a discrete distribution of zero mean and unit variance $N(0, 1)$.

Fig. 8.16 The modulus of the spectrum of typical soil noise with the low frequencies in the centre of the plot. From ref. 74.

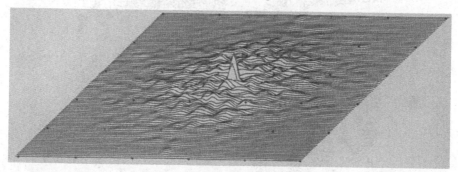

Fig. 8.17 The modulus of the spectrum of an archaeological site with good signal to noise ratio. From ref. 74.

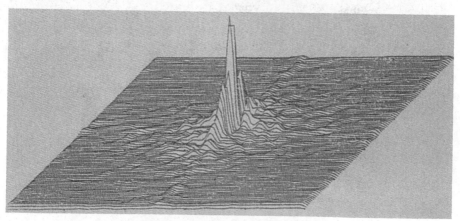

This allows for dipoles with negative polarities such as might be produced by random chunks of iron or stone whose remanent magnetism is much stronger than that induced by the earth's field. The process is stationary under translation and isotropic. Patterns are possible both in values and in locations. Instrument noise can be simulated by adding to each computed field value of the model a normally distributed random number with zero mean and a desired variance.

Fig. 8.18 The raw data used for the calculations of figure 8.17. From ref. 74.

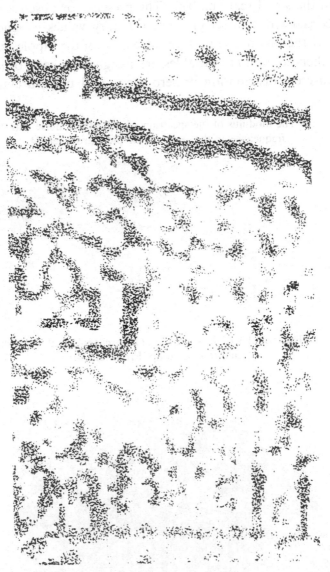

Consider first a single plane of random dipoles. The region in the plane has an area A in arbitrary units. There are λA points at random in the region. The intensity is λ per unit area. The region is divided into N subregions with N large so that there is a negligible chance that more than one dipole occurs in a subregion. Each subregion has an equal probability of containing a dipole. Then for N subregions, the probability P that the ith region contains a dipole is:

$$P(i) = \lambda A / N \tag{8.30}$$

When N goes to infinity, this gives a Poisson distribution of dipoles as the limit of the binomial distribution (62):

$$P(r) = \lambda^r \exp(-\lambda)/r!, \qquad r = 0, 1, 2, \ldots \tag{8.31}$$

where $P(r)$ is the probability of a region having r points. For a Poisson process, the mean is equal to the variance λ.

From the discussion in the previous section, we have observed many different sources of noise, some of which extend over a limited area. The suggested process does not model the clustering effects which are observed in real data. For this purpose, the Poisson process can be generalised using a parent–offspring model which leads to a Neyman Type A distribution (62) of random cluster locations with random cluster sizes such that random dipoles are considered as generating a set of daughter random dipoles. This is mathematically messy, so a nearly equivalent negative binomial approximation is used, giving:

$$P(r) = \binom{r+k-1}{k-1} p^k (1-p)^k, \qquad P(n) \propto \frac{(1-p)^n}{n} \qquad r, n = 0, 1, 2 \ldots \tag{8.32}$$

where n is the size of the dipole clusters and p and k are the parameters of the negative binomial distribution:

$$P(k) = \frac{(r+k-1)!}{k!\,(r-1)} p^r q^k, \qquad q = 1 - p \tag{8.33}$$

In general:

$$N_{\text{tot}} = N_1 + \cdots + N_m \tag{8.34}$$

where N_i is a random variable with a distribution giving the number of dipoles with probability P_2 in each cluster and m a random variable giving the number of clusters with probability P_1. This is not quite the double Poisson process of the Neyman A distribution, but it is a Poisson and log series process which is very similar but mathematically much simpler.

The distribution of dipoles with depth need not be considered since dipoles of different strengths can be used instead. If only variations in soil properties are considered, then a half-Gaussian vertical distribution of dipoles seems appropriate since the upper layers are more likely to contain highly magnetic material. If surface iron is to be taken into account, some of the dipoles must be given anomalously strong values in the first layer. This might be done by assuming a distribution which is long tailed rather than zero mean and unit variance for the dipole strengths.

The most important conclusion which emerges from the model studies is that the noise spectrum is biased towards the lower spatial frequencies but without visible geometrical properties in the Fourier plane. Signals from archaeological anomalies also have this kind of spectrum when pronounced geometric structures are not present, and linear filtering cannot separate them well from soil noise.

8.2. Magnetometers

8.2.1. Introduction

The theory governing magnetic anomalies from archaeological structures developed rather later than the techniques for measuring them. Mechanical magnetometers based on the torsion balance, whereby the torque on a small magnet which dips in the vertical component of the earth's field is restored by an opposite twist given to the suspending fibre were available for many decades. In fact the Askania torsion magnetometer was used in the first actual prospecting campaign ever undertaken in archaeology (see below). The device was used experimentally even after the development of the proton magnetometer, the instrument which provided the main impetus for magnetic prospecting in archaeology. It is probably not an exaggeration to state that the proton magnetometer has been more widely used in archaeological prospecting than all other instruments taken together. The principle upon which it is based is so elegant and simple that it retains its importance despite many decades of development of other methods.

8.2.2. The free precession proton magnetometer

Physical principles
A proton has an intrinsic spin $I\hbar$ where $\hbar$ is Planck's constant. It also has a magnetic dipole moment $\lambda I\hbar$ where λ is the gyromagnetic ratio

($\lambda = 2.67513 \times 10^4$). When materials are placed in a magnetic field H, a small magnetisation is established parallel to H in the volume v, whose moment is:

$$M_0 = N\lambda^2 \hbar H / 2kT = \chi Hv \qquad (8.35)$$

where χ is the static nuclear susceptibility, which for water at room temperature is ca 3×10^{-9} SIu and N the number of protons.

When the energy of magnetisation μH (where μ is the magnetic moment of the nucleus) is small compared with that of thermal agitation kT (k being Boltzmann's constant), the susceptibility is given by the Curie formula:

$$\chi = \frac{(j+1)}{3j} \frac{n\mu^2}{kT} \qquad (8.36)$$

where n is the number of nuclei per cubic centimetre, j is the angular momentum in units of $\hbar/2\pi$ and T is the absolute temperature.

In a proton magnetometer, the material is an organic liquid rich in protons such as methanol, kerosene, etc. Although water can be used, the danger of freezing in northern climates precludes this choice for practical instruments. A field H which is much greater than the earth's total field F is applied for a time to the liquid using a polarising coil. After some time, the nuclear magnetisation reaches the value given by equation (8.35). When H is suddenly removed, only F remains. Because of the gyromagnetic properties of the protons, the magnetisation vector precesses about F until the nuclear spins reach a new equilibrium alignment. The rotating flux of the precessing magnetisation induces a small voltage in the polarising coil. The frequency is given by the product of the gyromagnetic ratio and the earth's field:

$$f = \lambda F / 2\pi \qquad (8.37)$$

Since the gyromagnetic ratio is known to very high accuracy, the frequency is a direct measure of the strength of F. It lies between 1100 and 2700 Hz. In north European magnetic latitudes, the frequency is of the order of 2000 Hz.

The growth and decay of nuclear magnetisation in an applied field are determined by the spin lattice relaxation time T_1 and the spin–spin relaxation time T_2. The magnetisation, t seconds after applying the field H is given by:

$$M_t = M[1 - \exp(-t/T_1)] \qquad (8.38)$$

The decay is determined by the properties of the liquid and the magnetic field gradient. In typical liquids, the time is of the order of several seconds.

The precessing protons induce a voltage in the polarising coil:

$$V = 4\pi\omega_0 \alpha NAM \exp(-t/T_2) \sin(\omega_0 t) \qquad (8.39)$$

by Faraday's Law where NA is the area–turns product (total surface) of the coil, α is a filling factor (typically 0.5–0.7) which describes the degree of flux linkage between fluid and coil, and ω_0 is the initial signal amplitude at the beginning of precession.

For typical coil constructions, practical polarising currents and reasonable quantities of liquids at room temperature, the initial signal is of the order of 1–5 μV. In practice, the coil is tuned by a capacitor to the precession frequency, so that some resonant amplification of the signal results. The tuning of the system pulls the precession frequency slightly, but this effect is usually negligible for typical coil Qs of 20 or less. The signal to noise ratio of the input amplifier is most important because this determines the precision with which the precession frequency can be measured during the short time available before the signal decays.

In airborne proton magnetometers where a regular repetition of readings is required, the polarising current is usually applied in phase with the preceding decay signal so that any remaining magnetisation is augmented rather than starting afresh. This synchronous technique can double the signal strength when the polarisation times are short relative to the T_1 time of the liquid.

Most modern instruments use a pair of coils connected in opposition (astatic windings) immersed in the liquid to get maximum coupling and to cancel external electromagnetic interference. One instrument, no longer manufactured, used a toroidal container for the liquid which offers maximum coupling with great reduction of external field interference at the cost of considerable constructional complexity. Since copper sometimes contains ferromagnetic impurities, aluminium wire of square cross-section may be used to obtain maximum packing of the windings and freedom from long term magnetisation effects. But a simple copper wound thick solenoid will work well enough, and this has been used in cheaper instruments for many years.

The amplitude of the proton signal depends not only on coil construction, polarising current, the earth's field, and the local gradient, but also to some extent on the rapidity with which the polarising field collapses (54). In high gradients, there are slightly different precession frequencies in different parts of the liquid, with a consequent loss of phase coherence following the termination of the polarising field. This produces a shorter decay time. The effect is reduced in commercial instruments by using as small a sample as possible consistent with other requirements, and by suitable electronics capable of measuring the frequency quickly in the short time available.

Frequency measurement techniques

Phase locked loop multiplication. Shortly after the invention of the proton magnetometer, a plethora of frequency measuring techniques were published, the earliest due to Waters, Francis, and Phillips (86,87). These were designed using the electronics of the day to obtain as much accuracy as possible from the proton signal. Serson later patented a method which has since been widely used in many commercial instruments. Since the development of readily available integrated circuits for all functions, it is quite cheap to build an instrument using this type of processing. The scheme is shown in figure 8.19. After amplification, the proton signal is

Fig. 8.19 (*a*) Phase locked loop as a noise reduction filter. The basic phase locked loop consists of of three function blocks: a phase comparator, a low pass filter, and a voltage-controlled oscillator. (*b*) Phase locked loop with programmable divider as a frequency multiplier. From Exar Databook.

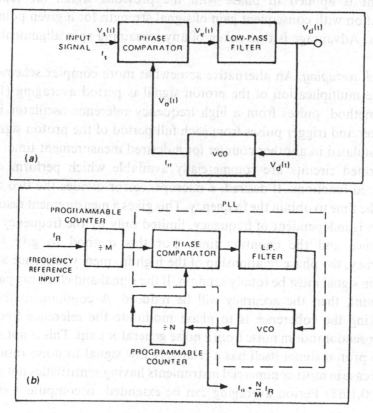

applied to a phase detector. It is compared with a divided down high frequency, voltage-controlled oscillator whose frequency is controlled by the detector output. The proton frequency is thus multiplied by a factor of several hundred or more. The high frequency is counted for a time related to the reciprocal of the gyromagnetic constant and the multiplication factor to give a direct readout in nanoteslas. Since the bandwidth of the phase locked loop can be made quite narrow, noise rejection is excellent and tuning is automatic over the lock-in range of the phase locked loop system. With a signal to noise ratio of only 2:1, 1 nT accuracy is readily attained. This allows a considerable reduction in polarising power, with consequent savings in the weight of batteries.

Excellent components are commercially available which can implement this arrangement with but a few parts. One refinement includes indication of loss of lock should the signal amplitude be insufficient. Using a very stable voltage-controlled oscillator held at constant frequency after acquisition of lock and measurement, the phase of the proton signal can be retained until the next polarisation period if desired. The polarising current is applied in phase with the previous signal for repetitive operation with consequent gain of signal strength for a given polarising power. Advantage is thus taken of any remaining spin alignment.

Period averaging. An alternative somewhat more complex scheme than simple multiplication of the proton signal is period averaging (16). In this method, pulses from a high frequency reference oscillator in one counter and trigger pulses from each full period of the proton signal are accumulated in another counter for a desired measurement time. Single integrated circuits are commercially available which perform all the required functions. If desired, a microprocessor divides the two counts and the time to obtain the frequency. This gives a measurement resolution which is independent of frequency, limited only by the frequency of the reference and the counting time. For this method to give highest accuracy, the phase relationship of the high frequency reference and the proton signal must be totally random. If the signal and clock are partially coherent, then the accuracy will be reduced. A common method for breaking the coherence is to phase modulate the reference frequency with pseudorandom noise from a noise generator chip. This is not needed if the proton signal itself has a relatively low signal to noise ratio. This is the case in most commercial instruments having sensitivities not greater than 0.1 nT. Period averaging can be extended to compute a running average of the high frequency pulses. The standard deviation can also be computed at the same time and this can be displayed or recorded as a measure of the quality of the reading.

Differential operation

The single channel proton magnetometer is affected by diurnal variations in the earth's magnetic field and by short term fluctuations which are comparable to or longer than the frequency measuring time (15). To eliminate these, two instruments can be used, with polarisation and frequency measurement synchronised. The readings are then subtracted in the recording microprocessor. Before the days of microprocessors, instruments were constructed to obtain the difference readings directly (53,68), but this is no longer necessary. If sondes are arranged vertically on a common staff, near-gradient readings can be obtained in the same fashion. Reduction of external disturbance is less when the sondes are widely spaced.

The bleeper. A highly simplified gradiometer with qualitative indication of anomalies was developed by Aitken and Tite (1). In this device, two sondes are connected in series and polarised together. Two precession signals are created and amplified in a common amplifier. They are of equal amplitude and phase initially, but of slightly different frequency. Added together the amplitude is:

$$V = V_0 \exp(-t/T_2) \cos[\pi(f_1 - f_2)]t \qquad (8.40)$$

The difference in frequency for the time to first zero, $t = t_0$ is:

$$f_1 - f_2 = 1/2t_0 \qquad (8.41)$$

and the difference in field between the two sondes is

$$\Delta F = (11.7431/t_0) \times 10^{-5} \text{ nT} \qquad (8.42)$$

The beat signal is passed through a threshold detector to an earphone, and the user estimates the anomaly strength by the length of time the signal is heard. At the magnetic latitude of northern Europe, a difference of field of 2 nT produces a time to first zero of a bit less than 6 s, which is within the range of decays possible with deoxygenated methyl alcohol.

Listening to the proton signal for any length of time is quite irritating, and a simple modification (82) improved on the original version by connecting the sondes in phase opposition after polarising. Thus, in the absence of an anomaly no signal is produced at all, but when one is present the bleeps are heard. By use of a small annulling coil placed at the centre of the staff holding the sondes (usually held vertically), the bleeps can be cancelled. The angle of the coil and its orientation are a measure of the strength and the sign of the anomaly. Although the device is quite slow in operation, it is surprisingly sensitive and accurate, and, of course, extremely cheap. Searches for objects with relatively strong anomalies like kilns can readily be carried out by an unskilled user with a simple device of this type.

8.2.3. The fluxgate gradiometer

Physical principles

The first purely electronic magnetometer, the fluxgate, was invented in the 1930s (25) and achieved prominent use during the Second World War and shortly thereafter for military and geophysical exploration purposes. It was not used in archaeological prospecting until much later. In this type of device, a highly permeable nickel–iron alloy core is magnetised by the earth's field and an alternating field applied via a primary winding simultaneously. The simplest form of device is essentially a saturated transformer with an overall secondary winding which is used to pick up the induced second harmonic component. The second harmonic voltage is proportional to the external constant field H_{ex} which produces a magnetic flux BA in the core of average cross sectional area A. When the permeability μ of the core alternates, the change of flux produces a voltage in the n turns of the secondary winding:

$$V_{sec} = nA \, dB/dt \tag{8.43}$$

The core field B is proportional to B_{ex} for small magnetisations and the proportionality factor is the effective permeability μ_a which depends on the kind of material used in the core and its geometry, so that:

$$B = \mu_a B_{ex} \tag{8.44}$$

The internal core field is:

$$B = \mu_0 (H + M) \tag{8.45}$$

and the magnetisation M is proportional to H in the core multiplied by the susceptibility, $M = \chi H$ and:

$$H = H_{ex} - DM, \qquad H_{ex} = B_{ex}/\mu_0 \tag{8.46}$$

where D is a demagnetising factor which depends on core geometry. Therefore:

$$B = \frac{\mu_r B_{ex}}{1 + D(\mu_r - 1)} \tag{8.47}$$

where $\mu_r = 1 + \chi$ is the relative permeability, and the apparent permeability is:

$$\mu_a = \frac{\mu_r}{1 + D(\mu_r - 1)} \tag{8.48}$$

Combining gives:

$$V_{sec} = \frac{nAB_{ex}(1 - D)\,d\mu_r/dt}{[1 + D(\mu_r - 1)]^2} \tag{8.49}$$

which Primdahl (60) calls the basic fluxgate equation.

Many different types of fluxgate geometries have been constructed, some of which are shown in figure 8.20. Commercial production of several of these types by a number of manufacturers makes the sensors readily available, standardised, and reliable devices. All of these are sensitive to the component of the earth's field in the sensor axis only, with the second harmonic output voltage V_{sh} related to the field by:

$$V_{sh} = kF \cos(\alpha) \tag{8.50}$$

where α is the angle between the magnetic axis of the sensor and the field F and k a constant.

Cosine law directivity has also made the devices useful for magnetic compasses in ships and aircraft. These usually use a servo system which rotates the fluxgate physically until zero output is obtained. For field strength measurement from an aircraft, three orthogonal fluxgates are used. Because of this directional property, a single fluxgate cannot be used for archaeological prospecting since it is impossible to hold it at a

Fig. 8.20 A few types of fluxgate sensors. From ref. 60.

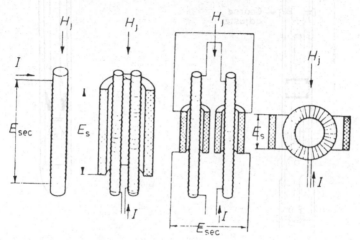

constant angle to the earth's field. Therefore, gradiometers with two fluxgates mounted in a rigid staff which is allowed to hang vertically are universally used. Slight changes in verticality affect both fluxgates equally. These are arranged so that their output signals are in opposition, and the orientation effects are very much reduced. A clever mechanical arrangement used in the gradiometer produced in the early 1960s by Alldred (3,4) and later improved upon by Philpot (57) in the commercially

Fig. 8.21 A mechanical arrangement for reducing the orientation error in a fluxgate gradiometer due to bending of the sonde. From ref. 3.

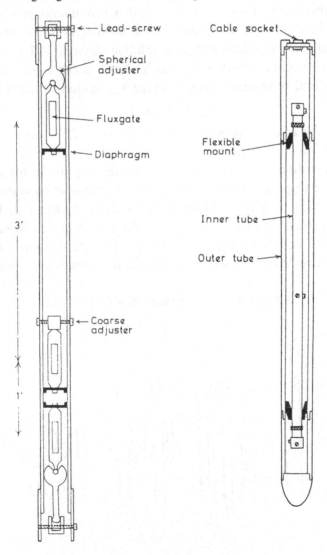

available instrument which reduces orientation error of the fluxgates to a minimum is shown in figure 8.21.

The mathematical analysis of the output signal from the fluxgate is quite complex, but Primdahl offered a simplified approach which accounts well for most observations. From equation (8.49):

$$V_{sec}/B_{ex} = nA \, d\mu_a/dt \qquad (8.51)$$

The apparent permeability is a periodic function of the driving field with a maximum at the zero crossings of the field as shown in figure 8.22. It contains the even higher harmonics of that field. It can be approximated by a cosine wave at the second harmonic with a small phase delay due to hysteresis $\Delta\phi$ where:

$$\Delta\phi = \arcsin(H_1/H_0) \qquad (8.52)$$

and H_1 is the driving field at the point of inflection of the B/H curve and H_0 the driving amplitude.

The function of apparent permeability with time is approximated by:

$$\mu_a(t) \approx \tfrac{1}{2}(\mu_{a\,max} + \mu_{a\,min}) + \tfrac{1}{2}(\mu_{a\,max} - \mu_{a\,min}) \cos(2\omega t + \Delta\phi) \qquad (8.53)$$

Substituting, the amplitude is:

$$\left.\begin{aligned} V_{sec}/B_{ex} &= -nA\omega(\mu_{a\,max} - \mu_{a\,min}) \sin(2\omega t + \Delta\phi) \\ V_{sec} &= nA\omega(\mu_{a\,max} - \mu_{a\,min})B_{ex} \end{aligned}\right\} \qquad (8.54)$$

Typically, the second harmonic is detected by the feedback method shown in figure 8.23, in which the fluxgate is used as a null detector, with a tertiary coil used to apply a field which cancels that of the earth. This arrangement, now almost universally used, has the advantage that calibration depends only on the annulling coil constant and the fluxgate operates at its highest sensitivity in a near-zero field.

Fig. 8.22 The apparent permeability of a fluxgate sonde as a function of the driving field. From ref. 60.

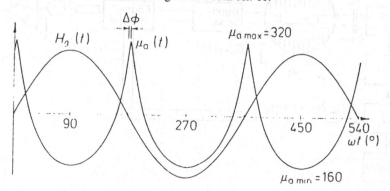

Sources of error, accuracy and sensitivity

When the secondary winding is tuned by a capacitor to the second harmonic, parametric amplification results (25). This allows a very high degree of sensitivity to be obtained. The price paid for amplification of the signal through resonance is instability due to the highly temperature dependent self-inductance of the core. This produces unwanted phase changes which alter the apparent null value in a feedback system. Since very low noise amplification is not a problem with modern circuit technology, there is no longer any reason to use resonant amplification to obtain high sensitivity.

The ability to follow rapidly changing magnetic fields is determined by the drive frequency of the device. It is modulated by the field changes, and the highest modulation frequency is about one tenth of the drive frequency. Response is also affected by integration in the detector circuitry. Typical values as given by Primdahl, are of the order of 300–700 Hz, corresponding to several milliseconds (60).

The sensitivity of a fluxgate depends on the signal to noise ratio of the sensor. Noise can be expressed in picotesla, and carefully designed fluxgates for space research have achieved a value as low as 8 pT, with

Fig. 8.23 Direct current feedback annulling of the external field and second harmonic detection in a fluxgate gradiometer. From ref. 60.

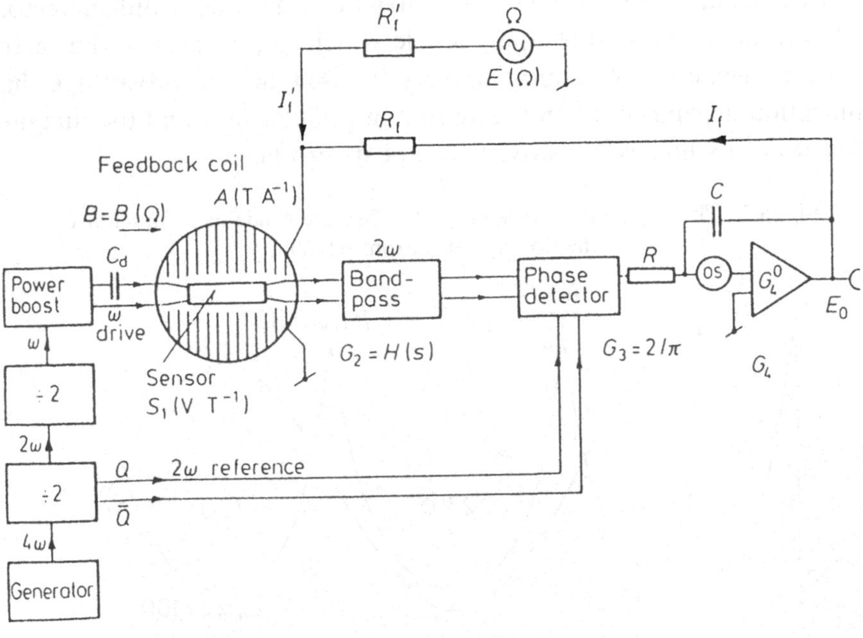

20–50 pT typical (42). Low field instruments for satellite use have been made with a stability of 0.1 nT over a wide temperature and time range. Sensor noise is caused mainly by the Barkhausen effect. Each magnetic grain in the core material has a slightly different magnetisation threshold. Hence all are not magnetised equally at the same time, and the sudden magnetisation of a hard to magnetise domain produces a noise pulse. Barkhausen noise is minimised by careful choice and treatment of the core material and especially by use of the highly uniform substances which can be created by modern methods (52). The design and construction of a highly sensitive ultra low noise thermally stable fluxgate is an art and few details about it are available in the open literature.

Practical gradiometers for archaeological prospecting have been published with a stability of 1 nT per hour and a sensitivity of better than 0.5 nT. Using modern materials and techniques, these figures could be substantially improved. Nonetheless, the mechanical design of a fluxgate gradiometer for rugged field use remains difficult. Since the competing proton magnetometer is inherently free of mechanical problems, it can be built at much lower cost.

The primary advantage of the fluxgate gradiometer is its inherent immunity to external magnetic or electromagnetic disturbance and high local gradients allowing it to be used in built-up areas or even inside a building. An example of the latter are some of the walk-through security detectors used at airports which often have a combined fluxgate and electromagnetic detection system working at high sensitivity in a very perturbed environment. A second advantage is the inherent filtering which a gradient device gives to eliminate deep-lying features. Other advantages are the inherent speed of operation due to the very fast response time to magnetic field changes and extremely low power consumption. For large scale surveys carried out over months or years, correction for long term drift of the instrument from time to time is a nuisance, but not an insurmountable one. If the anomaly strengths sought are greater than 0.5 nT, speed of operation important, and disturbance levels high, then the fluxgate gradiometer is probably the best instrumental choice.

8.2.4. *The optically pumped magnetometer*

Physical principles

As discussed in chapter 7, an electron in orbit has an angular momentum and a magnetic moment. An angular momentum of $\frac{1}{2}$ is assigned to the electron spin. When monochromatic light passes through a magnetic field in an appropriate material, there is interaction between the spins of the substance and the electromagnetic properties of the light. The monochromatic light breaks up into several lines whose spectrum depends on field intensity and the direction from which the source is viewed. A single spectral line from a simple element gives rise to two components at modest fields with equal frequency differences from the original line. This line splitting is the Zeeman effect (8) and the necessity for explaining it was one of the primary stimuli to the development of quantum electrodynamics at the beginning of the twentieth century.

The light components are circularly polarised in opposite directions with electric vectors at right angles to the applied field. For more complex elements, larger numbers of splittings occur at different field strengths. Monochromatic light of frequency v is emitted when an electron moves from one stable energy state W_1 to another W_2 such that:

$$v = (1/c\hbar)(W_2 - W_1) \tag{8.55}$$

where c is the velocity of light and $\hbar$ is Planck's constant. When a magnetic field is applied, the energy of each state is changed because of the Larmor precession around the field direction. An atom in state W_2 has a magnetic moment $M_2 g_2 \mu_B$ parallel to H, where g_2 is the Landé splitting factor (8) with magnetic energy:

$$\Delta W_2 = -M_2 g_2 \mu_B H \tag{8.56}$$

and μ_B is in Bohr magnetons. The changes in magnetic energy $\Delta W_2 - \Delta W_1$ correspond to frequency changes:

$$\Delta v = \frac{1}{c\hbar}(\Delta W_2 - \Delta W_1) = \frac{(M_1 g_1 - M_2 g_2)\mu_B H}{c\hbar} \tag{8.57}$$

which accounts for the split line spectra quite precisely.

Use is made of this phenomenon in the optically pumped alkali vapour magnetometer first proposed by Dehmelt and perfected by Bloom and Bender (28,41). In contemporary instruments (50) caesium 133 is used because it has only one isotope, but in the past rubidium 85 or 87 was favoured. The energy level diagram for caesium is shown in figure 8.24. The simplest form of magnetometer is shown in figure 8.25. A glass cell containing metallic caesium is heated slightly to vaporise the material, a temperature of 22°C sufficing to produce a pressure of 10^{-6} mm Hg. This low temperature (and hence low power) requirement is one of several reasons why caesium is now preferred. Helium is used in a military

Fig. 8.24 The energy levels of caesium. From ref. 28.

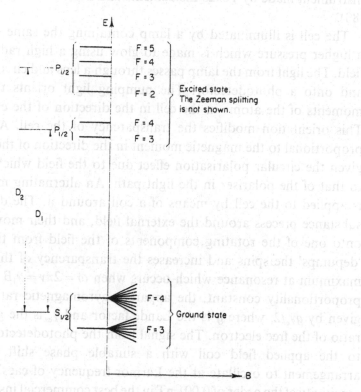

Fig. 8.25 A simple form of optically pumped magnetometer. From ref. 28.

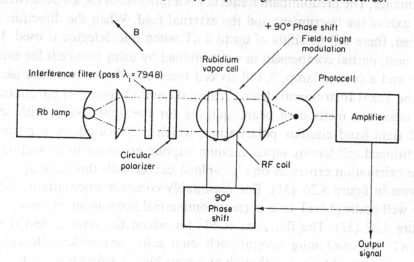

instrument made by Texas Instruments and described briefly by Slocum (80).

The cell is illuminated by a lamp containing the same substance at a higher pressure which is made to glow using a high radio frequency field. The light from the lamp passes through a lens and circular polariser and onto a photodetector. The pumping light orients the magnetic moments of the atoms in the cell in the direction of the external field. This orientation modifies the transparency of the cell. Absorption is proportional to the magnetic moment in the direction of the light beam, given the circular polarisation effect due to the field which is opposite to that of the polariser in the light path. An alternating magnetic field is applied to the cell by means of a coil around it. The dipoles of the substance precess around the external field, and their movement locks onto one of the rotating components of the field from the coil. This 'depumps' the spins and increases the transparency of the cell with a maximum at resonance which occurs when $\omega = 2\pi v = \gamma_i B$ where γ_i is a proportionality constant, the apparent gyromagnetic ratio of a level given by $g\gamma_e/2$, where g is the Landé factor and γ_e is the gyromagnetic ratio of the free electron. The signal from the photodetector is fed back to the applied field coil with a suitable phase shift, causing the arrangement to oscillate at the Larmor frequency of caesium, giving a sensitivity of the order of 0.001 nT in the best commercial instruments.

Since there are a number of hyperfine components, the oscillation frequency depends on the eight possible spin states in caesium for example, giving eight possible Larmor frequencies and one composite resonance. The predominant frequency is a function of the angle between the axis of the instrument and the external field. When the direction is varied, there will be shifts of up to 8 nT when one detector is used. In the past, partial compensation was obtained by using two cells for each axis and a central lamp. A full six cell three axis device has been used in the Texas Instruments device. Yabuzaki and Ogawa (90) introduced an ingenious method of beam splitting for the light source with left and right-hand circular polarisation, which are focused on a porous partitioned cell having equal caesium vapour pressures in its sections. The orientation errors in one half-plane cancel with this technique as shown in figure 8.26 (31). For reasonably constant orientations, they are well under 0.1 nT in a current commercial instrument as shown in figure 8.27 (32). The flat zone is 65° for which the error is less than 0.5 nT. By combining several such dual cells, near-independence of orientation is obtained, although at a very high construction cost.

Gradient operation

Commercially, the multicell split light arrangement is used only for expensive airborne installations. The simpler non-compensated oscillator is used in a gradiometer arrangement on a rigid staff in the Scintrex (formerly Varian) device. If the two sensors are reasonably identical in

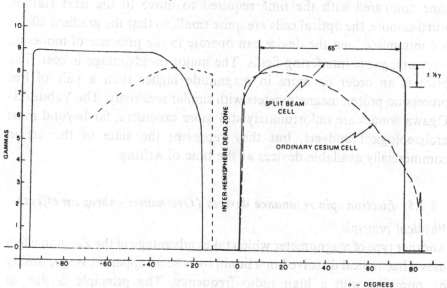

Fig. 8.26 Orientation error of a caesium magnetometer with and without a split beam. From ref. 31.

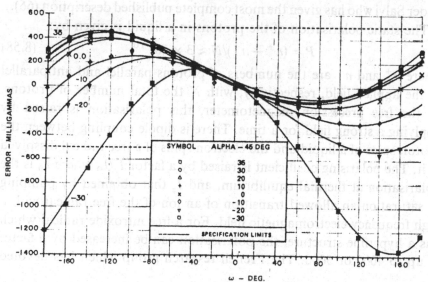

Fig. 8.27 Orientation error of an airborne caesium magnetometer. From ref. 32.

construction, then heading errors under 0.5 nT for the combination can be obtained. If the arrangement is held at a reasonably constant orientation during measurement, the residual heading error will be small enough on most archaeological sites so as not to be troublesome. The great advantage of this type of magnetometer is the very high Larmor frequency and good signal to noise ratio which allows an accurate reading to be made with relatively primitive signal processing in a very short time compared with the time required to move to the next station. Furthermore, the optical cells are quite small, so that the gradient effects are minimised, and the device can operate in the presence of moderate electromagnetic interfering fields. The major disadvantage is cost. The price is an order or more in magnitude higher than a pair of free precession proton magnetometers with similar sensitivity. The Yabuzaki–Ogawa sondes are unfortunately still more expensive, far beyond most archaeological budgets, but they represent the state of the art in commercially available devices at the time of writing.

8.2.5. Electron spin resonance devices (Overhauser–Abragam effect)

Physical principle

Another type of magmometer which takes advantage of the Zeeman effect uses a free radical dissolved in a liquid to raise its apparent susceptibility by pumping with a high radio frequency. The principle is due to Overhauser and Abragam and a production instrument is based on ideas due to Combrisson and Solomon with the operational design by a team under Salvi who has given the most complete published description (66).

The static polarisation of the protons in a liquid is given by:

$$P = (n^+ - n^-)/N \approx 3 \times 10^{-10} \tag{8.58}$$

where n^+ and n^- are the numbers of protons parallel and antiparallel to the applied field, respectively, with N the total number of protons. In the free precession magnetometer, this polarisation is raised by applying a strong field for a time. There is dipole coupling between the proton spins of a liquid and the electron spins of a free radical dissolved in it. The polarising coefficient is raised by a factor I_z/I_0 where I_0 is the polarisation at thermal equilibrium, and I_z that obtained by pumping to saturation an allowed transition of an ion of the free radical with a high frequency electromagnetic field. For a free nitroxide radical which has a hyperfine structure, the polarisation can be increased by a factor of up to 4000 or 5000. The French research group which has devoted nearly a generation to perfecting this type of instrument has settled on a deuterium substituted version of triacetone-amine-nitroxide which has proven to be soluble in many liquids, and possesses long term stability.

The hyperfine transition which can be used for measurement purposes with a 0.001 molar solution lies at 62.6 MHz which is easily produced. The freqency for pumping depends on the choice of solvent. There are $2(2K + 1)$ energy levels for the coupling of the electron spin of $\frac{1}{2}$ with the nuclear spin of nitrogen, and the Zeeman diagram looks like figure 8.28. There are two energy transitions for the material which are displaced in phase by 180° corresponding to the inversion of the macroscopic resultant in the solvent. This phase dependency, which is shown in figure 8.29, is exploited in the dual cell instrument by using methanol and dimethoxyethane, both stable with low freezing and high boiling points as well as low viscosity, all desirable properties. A container with a septum separating the fluids is used.

As shown in figure 8.30(a) the sample is placed between two coils. This system is decoupled as far as possible, so that no oscillations take place despite the high gain of the amplifier in the loop. When the pumping frequency is applied, the coils are coupled by the excited spins, and continuous oscillation at the Larmor frequency results. By using low Q coils, the Larmor frequency is not pulled, and a large oscillating range results. In a practical instrument, as shown in figure 8.30(b), four coils and two cells using different solvents are used. The signal to noise ratio

Fig. 8.28 The Zeeman diagram of a free radical suitable for magnetometry. From ref. 66.

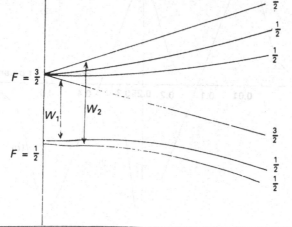

is extremely good, and due to the cancelling properties of the coil system, the arrangement is largely immune to external disturbance. Because of the very great increase in polarisation, very small amounts of fluid can be used, which makes the sondes quite small and therefore also highly resistant to gradients. The electromagnetic pumping energy is applied via a coaxial cable to the metallised glass walls of the container which acts as a cavity resonator. Quite low pumping power of the order of half a watt is all that is needed, so that total power consumption is very reasonable. The arrangement is in principle totally insensitive to orientation, although small stray magnetic fields from the wiring of the electronics produce a small orientation error. Sensitivities of the order of 0.01 nT are readily obtained in practice. Because a continuous signal with high signal to noise ratio is available, sophisticated signal processing is not required but may be used to obtain better sensitivity if desired.

With all of these advantages, one may well enquire as to why the device is not widely used in archaeological prospecting. The major reason is that it is only produced for sale to the French military establishment. Although rental to others is said to be possible, the price is well beyond most archaeological budgets. Duplication of the instrument, despite its superficially simple concept is not easy. Production of the free radical, especially in its deuterated form (which gives a 30% improvement in

Fig. 8.29 Phase/frequency dependence on the solvent in a dual cell Salvi magnetometer: (a) DME; (b) methanol. From ref. 66.

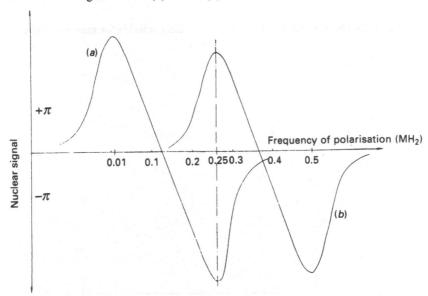

signal to noise ratio due to the larger number of protons in a given
volume of liquid), is an exceedingly complex task and full details have
never been disclosed in the open literature. Other substances which are
more readily available do not have the long term stability with respect
to time and temperature required of a practical instrument. It is hoped
that when the patents expire, commercial production may be possible
outside of France. Although inherently much more costly than the
free precession proton magnetometer, it has a number of significant
advantages compared with optically pumped devices, given the inherent
freedom from orientation error at comparable sensitivity.

Fig. 8.30 An Overhauser–Abragam magnetometer as perfected by Salvi and
coworkers. (*a*) The principle of a spin coupled oscillator. From ref. 28. (*b*)
Implementation in a dual cell arrangement with a noise cancelling bridge.
From ref. 66.

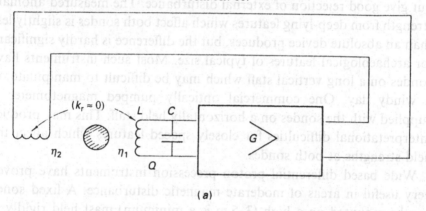

(*a*)

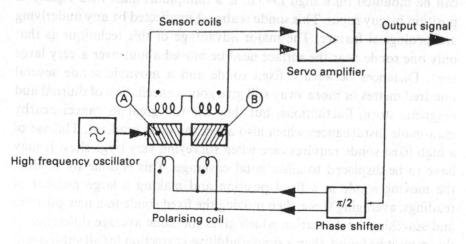

(*b*)

8.3. Some practical considerations for magnetic measurements

8.3.1. *Use of absolute, differential, and gradient magnetometers*

Commercially produced proton magnetometers for geological prospecting are usually capable of measuring and recording only the absolute value of the earth's magnetic field. Diurnal variations and short term magnetic fluctuations due to magnetic forms or man-made sources are often larger in magnitude than archaeological anomalies, so that differential operation of two instruments is highly recommended. The use of one absolute instrument is reasonable if a very small area is surveyed quickly and sensitivities of 10 nT are acceptable when searching for large anomalies like kilns.

Short base differential proton devices with sonde spacing of the order of several metres are also available. They do not measure the true gradient but give good rejection of external disturbance. The measured anomaly strength from deep-lying features which affect both sondes is slightly less than an absolute device produces, but the difference is hardly significant for archaeological features of typical size. Most such instruments have sondes on a long vertical staff which may be difficult to manipulate on a windy day. One commercial optically pumped magnetometer is supplied with the sondes on a horizontally held staff. This may produce interpretational difficulties for closely spaced features which affect the field strengths at both sondes.

Wide based differential proton precession instruments have proven very useful in areas of moderate magnetic disturbance. A fixed sonde can be mounted on a high (3–5 m is a minimum) mast held rigidly in position by guy ropes. This sonde is almost unaffected by any underlying archaeological feature. The major advantage of this technique is that only one sonde near the surface need be moved about over a very large area. Distances between a fixed sonde and a movable sonde several hundred metres or more away still give good cancellation of diurnal and magnetic storm fluctuations, but they are too great to cancel nearby man-made disturbances which also affect the local gradient. The use of a high fixed sonde requires care when surveying very large sites. It may have to be displaced to allow total coverage. This is done by holding the moving sonde in a fixed position and making a large number of readings, averaging these, then moving the fixed sonde to a new position and searching for a location which gives the same average difference. If this cannot be found, then a simple additive correction for all subsequent readings can also be used. With microcomputer field recording of results, this is the simpler technique and it must be provided for in the program.

If a magnometer with internal storage is available, then it can be used as a base station. Measurements must be made automatically at the highest rate of which the instrument is capable. The time for each reading must also be recorded. This record can be used for pseudodifferential operation if the reading times for the moving sonde are also recorded and a simple linear time interpolation correction used when subtracting the results from both devices. The method is very convenient because long cables connecting two instruments are not required and considerable separation between the reference and measurement stations is possible. Weymouth and Lessard (89) show that the correction is good to about 1 nT or so when the cycle time of the reference instrument is short relative to the level of magnetic disturbance. In highly disturbed areas or for sensitivity better than 1 nT, this technique is inadequate.

'Continuous' vs. discrete measurement and sampling
All magnetometers require a definite amount of time to process signals and reach a stable reading. In a fluxgate gradiometer, some form of analogue integration for the second harmonic signal is used, and this integration must have a time constant equal to a significant number of periods so that noise becomes negligible. The integration time needed depends on the excitation frequency, and is of the order of milliseconds. With proton magnetometers, much of the time required for a measurement is taken up with the polarisation of the fluid. An additional time is then needed for processing the proton decay signal. Polarisation times in commercial instruments are usually between 0.5 and 3 s, and signal processing times usually less than 0.75 s. Therefore a single reading takes something of the order of 2–3 s. Readings can be speeded up by the choice of a proton fluid with higher viscosity and consequent rapid decay or by doping it with a soluble salt like ferric nitrate to reduce decay time. Polarisation power can be increased to obtain a higher signal to noise ratio for the resulting loss of signal processing time. Since most instruments are battery powered, weight considerations quickly set a limitation to this approach. Optically pumped magnetometers, being continuous oscillators at a relatively high frequency usually require less than 100 ms for signal processing of a reading. The same is true of Overhauser–Abragam devices.

All instruments other than proton magnetometers give the impression of yielding continuous readings. This leads to the thought that higher horizontal resolution and detection of higher spatial frequencies for magnetic anomalies are possible with such devices. It is indeed true that a larger number of readings is obtained in one direction, but since the spacing of the measurements in the other is substantially constant for all types of instruments, this means that the sampling of the magnetic field is not the same in both directions. The highest spatial frequencies are limited by the sampling interval of the line spacing, not the sampling along the line itself. Thus, there is no great difference in the horizontal resolution of data obtained with the slower proton magnetometer compared with the others for typical line spacings of the order of a metre. With walking speeds of the order of $1.5 \mathrm{~m~s}^{-1}$ most instruments yield comparable spatial high frequencies. However, for more closely spaced survey lines of 50 cm, the fluxgate gradiometer or optically pumped device can produce higher resolution for the same total time. Since a decrease in line spacing is accompanied by a squaring of the number of readings, a practical limit to an improvement in resolution is quickly reached. Small surface anomalies which are usually responsible for most of the high frequency information are seldom of archaeological interest anyway. The time required to cover a given area adequately is more important.

   If there is a serious discrepancy between the high spatial frequency content of the data in one direction compared with the other, computational artefacts due to aliasing may be introduced when low pass filtering techniques are used in the evaluation phase. Higher measurement speeds can be put to use by the averaging of readings in the instrument itself to obtain greater measurement precision at each point, if relevant. At least one commercial magnetometer provides an arrangement for averaging a given number of readings automatically. On the whole, it is advantageous to have an instrument which can measure accurately in a time which is smaller than that required to move a sonde from one point to the next. Automatic positioning and recording are needed if maximum possible speed is desired. The time required for moving from one station to another, as discussed in chapter 6, is always a limiting factor.

Diurnal variation correction, magnetic storms
The earth's magnetic field fluctuates constantly about a local average value (15) due to variations in the stream of charged particles from the

sun, the solar wind. If only absolute measurements of the field are made, these fluctuations will mask the sought-after anomalies and produce artefacts in displayed results which are unacceptable.

The preferred method for removing these effects is the use of a differential or gradient magnetometer, but if one is not available, then some amelioration is possible during the mathematical evaluation phase. The information which allows this correction is obtained from a repetition of measurement at defined points at frequent intervals. In aeromagnetic and marine surveys, the measurements are usually made along long nearly parallel lines, and the repetition obtained with cross-tie intersecting the main measurements as shown in figure 8.31. In archaeological measurements based on regular squares, readings can be duplicated along the square boundaries or at an arbitrarily chosen fixed point. A second sonde is usually set up permanently here, and the associated field recorded at regular intervals. A second magnetometer is usually not needed, a simple shielded switch to connect the magnetometer alternatively to the fixed and moving sonde will suffice if rapid field changes are absent.

Linear interpolation. The simplest methods for diurnal correction of measurement lines used by Aitken and Linington in the earliest archaeological surveys consisted in a return to a starting point every ten or twenty readings. The observed difference at this point if any, is linearly distributed among the intervening readings. This is equivalent to

Fig. 8.31 Cross tie measurements for reduction of diurnal variation: (*a*) values at cross-points of ten lines (computed weights are shown in parentheses); (*b*) adjusted values at the cross-points (normalised weights are shown in parentheses). From ref. 51.

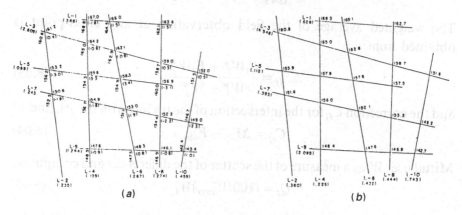

(a) (b)

assuming a linear diurnal variation in time, which for short times is quite acceptable. Weymouth and Lessard give a good description of the method (88,89). This technique is only reasonable for small surveys where the additional movement is not unreasonable. With the tie line method, a similar approach derived from geodetic practice is possible, whereby the differences are distributed linearly around the square loops in an iterative fashion. Another method, which is quite simply applied in practice, is due to Mittal (51). It was intended for gravity surveys, for which only instrument drift must be compensated, but it will also work for small magnetic surveys when only moderate diurnal variation plays a role. The method is not useful when rapid fluctuations due to magnetic storms, passing vehicles, etc. are present.

Assume that there are N survey lines which intersect with K_i intersections of the ith line with the others. Then:

$$\Delta F_{ij} = F_i - F_j \qquad (8.59)$$

where F_i and F_j are the field readings at the line crossing points. The method takes the normalised sum of the squares of these differences for assigning a weighting function to each line. This weight is:

$$W_i = K_i \bigg/ \sum_{j=1}^{K_i} (\Delta F_{ij})^2 \qquad (8.60)$$

The weights are then normalised to an average value of unity by taking:

$$W_i^* = W_i N \bigg/ \sum_{j=1}^{N} W_j \qquad (8.61)$$

since:

$$\left(\sum_{i=1}^{N} W_i^* \right) \bigg/ N = 1 \qquad (8.62)$$

The weighted average of the field observation at any cross-point is obtained from:

$$\Delta F_{ij} = \frac{F_i W_i^* + F_j W_j^*}{W_i^* + W_j^*} \qquad (8.63)$$

and the correction C_{ij} for the intersection of the ith line with the jth line is:

$$C_{ij} = \Delta F_{ij} - F_{\text{obs}} \qquad (8.64)$$

Mittal uses W_i as a measure of the scatter of the differences and computes:

$$Q_i = (100/W_{\text{max}}) W_i \qquad (8.65)$$

where Q_i is called the quality of the profile, and W_{max} is the maximum observed value of weights.

Least squares methods. Mittal's method although simple and quick does not distribute errors uniformly over the grid. For this purpose, least squares techniques have been preferred in marine and airborne geophysics, and a considerable number of papers have been devoted to the problem since the earliest construction of a magnetic map in the 1880s. A large bibliography is given by Ray (61) and a survey of work up to 1983 is presented by Green (27). The methods differ in their model assumptions. In one, due to Foster, Jines, and van der Weg (24) and Yarger, Robertson and Wentland (91), the diurnal variations are modelled by a polynomial power series which is used to adjust each line. Sander and Mrazek (67) use a Fourier series to model the error as a continuous time series for the entire survey, which reflects the idea that the diurnal changes are quite well approximated by such techniques when magnetic storms are not prevalent. Cloutier (19) used a discrete non-parametric spline technique. The coefficients for the series or splines in all the methods are determined by conventional least squares iterative techniques.

Ray suggests using observatory data for a preliminary correction before applying the models, which will help to eliminate rapid variations due to magnetic storms and micropulsations (61). He notes that correcting one line at a time using a series approximation will yield corrections which are dependent on the order of processing, and hence will not be optimal. A method which finds the minimum root mean square error by simultaneously adjusting all lines yields corrections which are not unique. Ray presents a method which is applicable to irregularly spaced surveys which do not have a pattern of traverses with equal numbers of intersections, as required by the method due to Foster. The price paid is the decomposition of a very large sparse matrix whose size is given by the number of intersections times the number of adjustment coefficients. For a large archaeological survey with 100 survey lines and 20 cross-traverses, the matrix for shift and tilt corrections will approach a size of nearly 500 000. For the aeromagnetic survey of North America, 2.5 million equations in 400 000 unknowns were handled. Progress in numerical analysis has made treating such large matrix problems feasible for large computers. Interested readers should consult Ray's fundamental paper for the details of the ridge regression techniques used based on the work of Hoerl (33,34). It seems reasonable to suppose that anyone who is able to write and run a program of the complexity described by Ray would also have access to two magnetometers for differential work!

Time recording and base station. Most modern commercial proton magnetometers incorporate a clock/calendar chip which allows the recording of the time and date of reading to high accuracy. If a second magnetometer of this type is used at a base station not more than 50 km from the site, then diurnal variation changes can be corrected for by linear interpolation and subtraction of the readings at both points. Some modern instruments provide a program internally for carrying out this correction when the two magnetometers are connected via a serial digital link which is hooked up upon return to the base station. Correction over distances greater than 50 km is not very reliable for short term fluctuations because of differences in induced telluric currents in deep-lying rock formations which will affect the local fields. Corrections relative to a remote observatory are not recommended except as a first step in a data reduction scheme based on a least squares technique.

Instrument drift and static field change correction
Modern proton magnetometers which have adequately temperature compensated quartz oscillators show no significant drift in readings. Optically pumped devices may have small thermally derived changes but since these are almost always used in a differential setup, both sondes will probably be affected in very much the same way. Fluxgate gradiometers used in archaeology have drifts of the order of 1 nT per hour, and periodic zeroing of the instrument is imperative if high accuracy is required. Since the drift is largely thermal and mechanical, it may not be constant in amplitude and direction, so that compensation by calibration and subsequent calculation is not practical. Considerable improvement through refinement of instrument design is possible here.

Reduction of heading error, decontamination of personnel
All magnetometers have some degree of heading error, that is, a change in the reading obtained when rotating the sonde. In a free precession proton sonde the effect is least when great care is taken in the choice of materials (aluminium instead of copper wire for the windings, no metal plugs or staffs in the immediate vicinity of the sonde, etc.). Heading

errors of well under 0.05 nT are readily obtained even with limited precautions. Optically pumped sondes of simple uncompensated types have an inherent heading error of the order of 0.5 nT. When used in a gradiometer, the total error between two sondes when rotated can be readily kept under 0.1 nT by careful alignment. Fluxgate gradiometers usually do not show a heading error and other drift effects predominate. But another source of heading error which occurs in practice is due to dirt on the sonde itself. Even a small magnetic soil particle which directly contaminates the sonde can have quite a catastrophic effect. Washing the sondes with clear water and a detergent before beginning the measurements each morning has proven to be a simple but effective remedy for this problem.

Heading error manifests itself in the evaluated reading as a series of alternating light and dark bands in the direction of the measurement lines if the sonde is rotated on alternate lines (see figure 8.14). This can be somewhat reduced by using a destriping algorithm (see chapter 4) during processing. Small random rotations during measurement effectively increase the variance of the readings and reduce the total accuracy of the measurements. There is no way to reduce this in the processing phase.

The person carrying a sonde is often a major source of heading error. Anomalies of up to 90 nT have been observed from sonde carriers who have dental appliances or other medical protheses. Clothing is also a problem, with zippers, rings, belt buckles, shoe nails, etc. being the principal offenders. Rubber boots or sandals, and a coverall made without any metal fasteners are easy remedies. A check with a metal detector for unseen items can be used. Great care must be taken with sonde carriers who smoke! The natural tendency to put one's cigarette lighter back into one's pocket after lighting up can produce dreadful results in the resulting measurements. After any interruption, the sonde carrier should be checked again for possible contamination.

A simple method for checking personnel with a differential or gradient magnetometer is to place the sonde on a staff in a fixed position over the ground at about 1 m height, then have the sonde carrier stand first north and then south of the sonde and as close as possible to it while making about two dozen readings. Two small histograms should result which must be absolutely identical. An appropriate subroutine for this type of check should be included in any microprocessor recording scheme with a Student T test made for the difference of the two mean readings if highest sensitivity is being used.

8.3.2. Position control

Survey grids and hand recording of data

Archaeologists usually lay out large excavations as a grid of squares. This organisation is eminently suitable for the layout of a geophysical survey grid as well since it can be used afterwards for an excavation. As opposed to automatic or semi-automatic positional recording schemes discussed below, the requirements are extremely simple and readily satisfied with negligible expenditure of technical effort.

If the area to be surveyed is relatively flat and free from obstacles, the easiest layout method is to use a set of six ranging poles, a steel measuring tape, and an optical square to establish an initial square of the desired size with two poles extending one side. First, four ranging poles are laid out in line by eye at three measured equal intervals. If desired, a level or a theodolite can be used for alignment, but alignment by eye is usually adequate. Using an optical square and measuring tape with at least twice the side length of the square at each of the two inner ranging poles in line, two further poles are positioned at right angles to the initial line at a distance equal to the square side length. The accuracy of the initial square is checked by measuring the diagonals, and minor adjustments are made until all side lengths and diagonals are correct. Now the two outer ranging poles can be replaced with wooden pegs, and then used to extend the first square to make a second square adjacent to the first. By sighting along the rows and columns of the squares and checking the diagonal sightings, a large field can be pegged out quickly in a very short time by two people. Accuracy over a large grid, figure 8.32, can readily be held to 20 cm or so, without the use of surveying instruments. If the terrain is not flat, then smaller squares must be used, and accuracy can be improved by the use of a theodolite for sighting.

The size of a measured square must depend on the planned measurement spacing, the flatness of the terrain, and logistic considerations related to cable lengths and type of magnetometer. Suitable values are either 10 or 20 m if position control for the measurements is to be by use of non-magnetic measuring tape, since longer tapes are subject to wind movement. During measurement, the non-magnetic tapes can be laid out according to the scheme shown in figure 8.32. Usually tape capsules are somewhat magnetic, so that they must be placed well beyond the working area. Tapes of 25 m length have proven satisfactory for a 20 m grid. If a measurement grid spacing of 1 m is wanted on bare ground, it is not necessary to place a tape at metre intervals. It has been observed that a sonde carrier can readily judge his position with adequate accuracy between two tapes laid at 2 m intervals, providing there are no obstacles to reading the tapes. A skilled sonde carrier can work with tapes placed at 4 m intervals and judge position to 1 m spacing, but this requires a totally bare and flat surface and is not to be recommended if anomalies close to the surface are expected whose shapes will be distorted by positioning errors. An alternative to the use of tapes which is useful in uneven or sloping terrain is a knotted grid of string (clothes line is useful), with the sonde placed systematically at the knots. This technique is only practical for very small squares because it is difficult to keep a larger string grid from becoming tangled when moving from square to square.

Fig. 8.32. An efficient arrangement for laying tapes for measurement of a square.

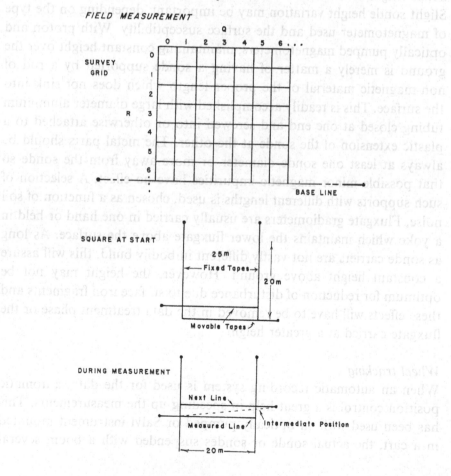

Many years of practice have shown that a major source of error in recording results is mistakes in the number of the square or its orientation relative to the grid when a large survey is underway. One good way to reduce this in the field is to use a set of prenumbered wooden pegs for the square corners. The numbers of the pegs are recorded along with the square numbers to provide redundancy which can be used for error correction during evaluation. It has been found that 30 cm long, 25 mm thick square oak sharpened pegs with the upper 10 cm painted with water resistant red luminescent paint and numbered in black on all four sides are easily made, can be reused for many years, and are readily visible. The pegs are placed in a random order during the layout of the grid, and the peg numbers recorded on a master job control sheet which also contains prerecorded square numbers.

Sonde height control

Slight sonde height variation may be important, depending on the type of magnetometer used and the surface susceptibility. With proton and optically pumped magnetometers, maintaining constant height over the ground is merely a matter of having a sonde supported by a rod of non-magnetic material of the proper length which does not sink into the surface. This is readily accomplished with large diameter aluminium tubing closed at one end and screwed into or otherwise attached to a plastic extension of the sonde at the other. The metal parts should be always at least one sonde diameter or more away from the sonde so that possible minor magnetic impurities have no effect. A selection of such supports with different lengths is used, chosen as a function of soil noise. Fluxgate gradiometers are usually carried in one hand or held in a yoke which maintains the lower fluxgate above the surface. As long as sonde carriers are not vastly different in bodily build, this will assure a constant height above ground. However, the height may not be optimum for reduction of disturbance due to surface iron fragments and these effects will have to be removed in the data treatment phase or the fluxgate carried at a greater height.

Wheel tracking

When an automatic recording system is used for the data, automatic position control is a great help in speeding up the measurements. This has been used with an optically pumped or Salvi instrument mounted in a cart, the actual sonde or sondes suspended with a boom several metres long to reduce the effects of magnetic materials in the supporting electronics. If the wheels of the cart are of large diameter, they will not slip appreciably when dragged over the measurement area. Bicycle wheels made of titanium and relatively non-magnetic stainless steel spokes have been used by Lemercier of Salvi's group in unpublished tests. The rotation of these wheels produced eddy currents in the closed metallic circuits with resultant magnetic effects, but these were much reduced by cutting the wheel rim and insulating the break to prevent such current flow. Wooden wheels have been used by Becker in Bavaria on a small four wheeled cart carrying an optically pumped gradiometer.

It is easy to arrange for an impulse to be generated and recorded at fractions of wheel rotation corresponding to a small distance travelled. Either a mechanical switch or a photodiode motion detector can be used. This technique is especially valuable with magnetometers which supply a nearly continuous stream of measurements. The cart must be positioned relative to a measured grid if accurate results are to be obtained. If only a rough survey is required, it may be sufficient to keep on a constant course by sighting on a pair of ranging poles appropriately placed at the end of each traverse. If the ground is bare, it may be possible to place the wheels in such a way that one always follows in the rut left by the other, so that comparatively constant traverse line spacing results. The cart technique will only work well over relatively smooth terrain, but it has the advantage of speeding up a measurement campaign considerably and is worth using when possible.

Wire guidance

The geophysics team of the Department of the Environment in Britain under Clark has used a wire guidance positioning scheme with a fluxgate gradiometer with success for many years (17,18) as shown in figure 8.33. In this arrangement, the fluxgate sonde is loosely attached to a long loop of non-magnetic wire which passes over pulleys mounted on tripods placed at each end of a traverse. One of the pulleys drives a potentiometer and an analogue signal is obtained which drives one axis of an x–y recorder. The other axis is driven by the fluxgate output. The result is a wiggle trace of the magnetic field gradient. The support tripods are displaced for each measurement traverse by a constant amount and the y recording axis stepped to scale. This technique provides exceptionally accurate positioning if the traverses are kept relatively short. The disadvantage of the wiggle trace as a display technique is discussed below. The wire guidance scheme could be retained and fluxgate output digitally recorded along with position obtained either from an analogue–digital converter connected to the potentiometer or from an absolute shaft encoder. This could be run simultaneously with the wiggle trace system, allowing a quick look at results in the field along with more sophisticated evaluation techniques applied later. Considerable care in construction would be required of the method if anything but a fluxgate gradiometer were to be used since sensitivity to various currents flowing in the system is higher. A wire guidance scheme was also used in an experiment with an Overhauser–Abragam magnetometer by Boyer and Ngoc (12).

Radio positioning

Perhaps the ultimate in positioning systems for archaeological magneto-meter surveys was proposed by Philpot and Evans at the Plessey Radar Research Centre in 1972 (22). The scheme was a true hyperbolic phase navigation system using two transmitters and two receivers for each axis for the prototype. Decca or Loran hyperbolic navigation systems have been used for many decades over distances of hundreds of kilometres,

Fig. 8.33 A wire guidance scheme used with a fluxgate gradiometer. From ref. 18.

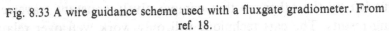

but the Plessey development was the first designed for archaeological site dimensions. At the end of two orthogonal axes there are four transmitters, and a fixed receiver is placed midway between them. The transmitters emit a constant signal at 24 MHz at slightly different frequencies. The fixed receiver picks up this channel and multiplies the received signals to produce a beat signal equal to the difference of the two transmitted frequencies. A movable receiver is attached to the measuring sonde. It also receives the two transmitters and produces a beat signal which is displaced in phase relative to the fixed receiver because of differences in arrival times of the signals. The phase difference in beat frequency from both receivers is proportional to the difference in distance between them. The difference gives the position of the moving receiver in hyperbolic coordinates centred on the transmitters. If the transmitters are far enough apart, the distances are nearly Cartesian. It would be possible actually to calculate hyperbolic coordinates in real time in the field today, although this was not done at the time of the experiments in the mid-1970s.

With the prototype, an accuracy of 10 cm over 90 m was achieved. In the preliminary proposal two transmitters and four receivers were suggested, with the moving transmitter also used to send back the magnetic field reading. This was dropped in favour of wire transmission of the fluxgate signal and use of two receivers and four transmitters instead in the test version. There is no theoretical difference when receivers and transmitters are interchanged, but there may be some practical advantages in one arrangement or the other. Choice of operating frequency was based on eliminating effects due to obstacles such as trees. Assuming trees of 50 cm diameter, the shortest useful wavelength was taken as 5 m or 60 MHz to avoid diffraction effects. The minimum frequency was based on the limits of phase measurement techniques then available, 10 MHz. The ultimate choice of frequency at 24 MHz was presumably dictated by available telemetry channels and interference considerations. Unfortunately, work was not continued on this promising approach for lack of funding, according to the internal Plessey report. Hadon-Reece of the Department of the Environment presented the results at an Archaeometry Symposium in Oxford in 1975. It is to be regretted that this interesting idea was never pursued further. Transmission of the magnetic field value back to the recording and fixed reference station, together with computation of position obtained from four receivers could be used for guidance information for the sonde carrier along with some kind of visual display to keep measurement lines reasonably straight. This arrangement would permit prospecting in otherwise inaccessible areas.

8.3.3. Data recording

Hand recording of data and position
Small scale surveys can be carried out using hand recording of data. If hand recording is used, it is advisable to write both the square and peg numbers on the sheet of gridded water resistant paper with centimetre rulings along with the data, the survey site name, and date. A scale of 1:10 or 1:20 depending on square size has proven to be a practical compromise between readability and ease of handling. A surveyor's clip board with leather strap is useful for holding the paper during recording. With 1:20, a 20 m square will fit nicely onto an A3 sheet of graph paper at 20 × 20 cm with room to write readings up to three digits at the centimetre crossing points. If a differential or gradient system is used, three digit plus sign recording will usually suffice for most sites unless the soil is unusually magnetic. If an absolute proton magnetometer is used, four digits may be necessary to allow for transitions at hundred nanotesla values. It may be difficult to write these cleanly at 1:20, and therefore 1:10 is to be preferred. Because of the necessity for frequent diurnal correction readings in this case, the large scale and smaller survey grid are to be recommended in any case.

Other data relative to the survey should be recorded on a master position control sheet which should also be used when an automatic recording scheme is available. Experience has shown that a log should be kept in a hard covered notebook with a fresh page taken for each day's work. In this log, the date, location, and numbers of measured squares should be noted, together with the tests for proper operation of the magnetometer and for the decontamination of the sonde carrier. The log should also be kept even when automatic recording is used to cross-check later for eventual errors. If an absolute magnetometer is used on magnetically very quiet days, it may be sufficient to record occasional base readings in the log book for later diurnal correction purposes.

Some conventions for position recording on a grid
Archaeologists sometimes number their excavation squares with letters

along one axis and numbers on the other. It is not a good idea to follow this scheme in magnetometry, especially if computer evaluation follows later. Square numbers are best recorded in standard matrix notation beginning with 0 in the upper left corner, for this simplifies putting things together in a program afterward. The corner of the grid nearest northwest should be chosen as zero, so that grid orientation is roughly north–south, east–west. This may simplify some problems of interpretation later. In the system developed at Scollar's laboratory during the early 1960s, a simple system of prefixes for a four digit recording system was developed, based on the idea that both hand and automatic recording techniques should use the same approach (71). Each day's work was given a code beginning with 4 and followed by a three digit number giving the day of the year, then followed a 3 and a three digit job number, and then the squares with a 2, a starting position indicator number of which more immediately, and for up to ten by ten squares, a two digit indicator in matrix form for row and column.

Since the magnetometer and the recording system usually have to be kept at a considerable distance from the working area, it may not be convenient to measure each square with the same orientation. Rather, to keep cable movement at a minimum, it is convenient to begin the measurement at any one of the four corners of a square. The starting corner is noted as the second digit of the square number. The traverse lines must be numbered if an automatic recording scheme is used, and numbers 1100–1120 were reserved for that purpose when using a 20 m grid with metre measurement intervals. This scheme was widely adopted by the archaeological prospecting community until automatic positioning schemes and microcomputer control became possible (72). With modern magnetometers incorporating internal processors and memory, the conventions used by the manufacturer must be accepted, although some of them are less convenient for archaeological purposes. Often these instruments are sufficiently non-magnetic that they can be carried close to the sonde.

Missing data and obstacles

Not infrequently, obstacles like bushes, haystacks, and trees are present in the survey area. If there is a considerable amount of surface iron, proton magnetometers may not be able to produce a measurement because signal is lost through high gradient. In both cases, a convention for recording a missing data item is required. With hand recording this is no problem, for the value can be left blank or better still, a cross made at the location(s) in question. With automatic recording a convention is required to indicate to evaluation programs that data has not been obtained. Here again, experience has shown that the simplest technique is to use the highest value which the recording system can accept as an indication of missing data. If desired special symbols which do not occur normally can be chosen. In a paper tape recording scheme used at Bonn for many years, 9999 was the missing data indicator. Later, when a microprocessor replaced the hardwired recording scheme, hexadecimal FFFF proved to be more convenient. The memory area for data is simply initialised to this value and then replaced with readings as they are made.

Interpolation of irregularly spaced data

An automatic positioning scheme which does not constrain a sonde carrier to a fixed course and regular stations produces irregularly spaced data. For nearly all evaluation techniques, data on a regular grid are required. In the case of irregularly spaced data at known geographic coordinates, the situation is analogous to the problem of constructing a digital terrain model in photogrammetry in order to obtain regularly interpolated readings (30). The interpolation algorithms discussed in chapter 5 on geometric processing of images apply to irregularly spaced magnetic data as well. However, there are many more magnetic readings than there are points in the photogrammetric problem, and simple interpolation, such as weighted distance averaging is usually adequate.

Magnetometers with built-in storage of data, position, and time

Many modern magnetometers and gradiometers have built-in micro-processors and a considerable amount of non-volatile memory for storing field or gradient readings, position, and time. The user can specify the line number at the beginning, a spacing increment along the line up to a maximum and a line number increment up to a maximum. Some instruments also provide for subgrids with finer position control. The reliability of each reading, given as a standard deviation along with the strength of the proton signal is sometimes recorded to aid in controlling the quality of each reading. A typical magnetometer of this type can store readings for about 2500 points, which is roughly half a day's work in a typical archaeological prospecting environment. Since computer memory costs have become negligible compared with that of the rest of the instrument, increasingly sophisticated internal software makes the use of such a commercial instrument even easier. Unfortunately manufacturers do not provide facilities which allow users to load codes suited to archaeological applications and a separately programmable microprocessor is much more flexible.

Microcomputer recording and control
Since the advent of cheap portable microcomputers, there is really no reason to use hand recording at all. A computerised recording scheme can be easily used for data with most modern magnetometers which do not have internal storage. Usually these provide a connection which looks to the computer like an attached terminal. Older instruments may require considerable modification to obtain this kind of operation. If a fluxgate gradiometer with analogue output only is used, then an analogue–digital converter will be required. In the past, it was useful to obtain differential operation essentially by constructing two instruments in one and then using a hard-wired subtraction scheme. With microcomputer control, it is much simpler just to use two absolute instruments triggered simultaneously and read out sequentially. Subtraction is done in the control program. The result can then be placed in a non-volatile storage medium for later readout. This takes less space than storing the absolute readings themselves, and no significant amount of time or programming is required.

Storage media used in the past include punched paper tape, magnetic tape cassettes, miniature floppy disks, and battery powered static CMOS memory which can hold readings for months. Field experience has shown that use of any device with moving parts, which can be contaminated by dust or dirt, leads to many breakdowns and erratic performance. In better modern portable personal computers, a sealed hard disk is available which does not suffer from dirt. A minor disadvantage is that power consumption is raised so that external batteries may have to be used if extended operation time is required.

Static CMOS memory has proven to be the most reliable of all, but this medium is not readily available for many commercial portables. A reasonable compromise, if one does not want to build or modify one's computer, is to use a $3\frac{1}{2}$ in disk with which many portables come equipped and to seal over the opening where the floppy is inserted with masking tape to keep out dirt. Sealing the keyboard with transparent flexible plastic is also to be recommended.

Organisation of the storage technique is media dependent. For floppies, the user is well advised to stick to whatever the operating system and a high level programming language run-time system offer. Any minifloppy offers more than enough storage for a very large prospecting campaign plus system and programs. With static CMOS, the most compact arrangement is to store two bytes per reading. Positional control information can be stored intermixed with measurements, if the first bit of the first byte is used as a flag to distinguish between the two types of data. This has the disadvantage of reducing the recordable dynamic range by half. This storage technique uses the static memory as if it were a magnetic tape, and having no inherent structure is the most compact. Alternatively, positional information can be stored in a separate memory area using a dynamic array of pointers containing the addresses of the data associated with a given position, or with the beginning and ends of a traverse. For software simplicity and compatibility with other storage media, the CMOS memory can be configured as a RAM disk and written to just like one with a removable medium. This mode of storage is not as compact as the others because of the overhead of the disk formatting and directory structures, but it is easy to implement with readily available RAM disk programs which are included with the operating system. They allow use of standard operating system calls and high level language software without modification as if writing to hard disk or removable media. Obviously, the RAM disk must be read out when returning to base, and for this purpose a file transfer program like KERMIT lends itself well to communicate between machines.

When data exchange and long term storage are of importance, the internal storage representation should be converted to ASCII which in the future almost any type of machine will be able to read. Painful experience has shown that raw ASCII data in digitalised form should always be archived since progress in evaluation techniques may be able to extract additional information which was not visible with less developed methods. In the course of transforming a data storage scheme to a standard type, no information should be removed or changed.

8.4. Treatment and display of archaeological magnetic data
8.4.1. Complexity of archaeological features

Archaeological features are frequently quite complex in shape and often have pronounced geometric forms, unlike the structures encountered in geological geophysical prospecting. Depth of burial may vary, but is seldom of interest. Sites are commonly close to the surface and the magnetic anomalies can delineate feature outlines well enough so that the archaeologist can identify them. He does this by mental comparison with excavated examples. A typical small site is shown in a low level aerial photograph in figure 8.34. Many dozens of sites of this type are known throughout northern Europe, and quite a few have been excavated. Geophysical measurement data from such a site must be recorded with a fine enough sampling interval to reproduce all of the features which may be of interest.

Fig. 8.34 A small Roman fort of the late third or early fourth century in the Rhineland.

8.4.2. Numerical range and accuracy of data

Recorded accuracy of measurement varies with the instruments chosen.
The dynamic range of magnetic field or field difference values also differs.
Requirements depend much on the local susceptibility contrasts between
archaeological features and undisturbed surroundings. Depending on
the soil, anomalies may be as low as 0.1 nT or as high as 500 nT. Usually,
an integer range of about ± 1000 is adequate for most sites.

With differential magnetometers, values may be negative or positive
integers established with reference to a sonde in a fixed position. The
use of signed integers with large dynamic range is one important way
in which magnetic data differs from scanned image data. This leads in
turn to some differences in methods of treatment. Other important
differences are: values at the lower end of the measurement range may
contain more significant information than those in the middle or high
end since these reflect weak anomalies produced by deep lying features;
sampling of the field is sparse compared with image data and every value
is of significance.

8.4.3. Spatial frequency content

The range of spatial frequencies in the data depends on the depth of
burial of the features, the susceptibility contrast between features and
surroundings, and the height of the measuring sonde above the surface.
In near-urban areas, surface iron fragments are likely to be prevalent,
producing isolated high values. Some sacrifice in sensitivity has to be
made either by raising the sensor height enough to reduce the effects of
such point-like disturbances or removing them by calculation (despiking)
before any other treatment is carried out on the raw data. High spatial
frequencies encountered in the data are significant, but often the
important high frequency geometric detail information is masked by
surface irregularities due to ploughing or small local variations in
susceptibility. Linear bandpass filtering techniques have proven helpful
to eliminate some of the near-surface and deeper geologically induced
effects. Use of a gradiometer with the lower sonde held some distance
above the ground provides implicit bandpass filtering so that further
high pass filtering during the data processing may not be necessary with
this type of instrument. On the contrary, it may be necessary to use
some form of low pass filtering to improve visibility of long wavelength
data.

8.4.4. Quantity of data for adequate sampling and site size

A typical archaeological survey in Europe, the Middle and Far East, will cover a hectare or more of ground, the actual prospected area depending very much on the type and period of the site which is of interest. The site shown in figure 8.34 required only an 80 × 80 m grid. However, some small sites can be adequately covered with a still smaller grid, and some large sites with strong anomalies require only mapping of larger features to delineate their extent. This results in a smaller quantity of data. Something between 2500 and 100 000 readings are typical for a site although one or two very large surveys have required nearly 1 000 000 readings (65,75).

Sites of great size are quite common in Europe where population densities in the past were much higher than in North America. The number of sites per square kilometre is also very much higher, especially in the major river valleys. For example, in the northern Rhineland, there are over 70 000 known archaeological sites in 16 000 km². Some are many hectares in extent, judging from surface scatters of finds or data from low level oblique aerial photography.

8.5. Treatment of magnetic data prior to display

8.5.1. Flagging and replacing missing data

Measuring the entire surface of a site is usually not possible. Roads, fences, houses, and other obstacles get in the way. Some type of indicator or flag for missing data must be recorded in these areas. Sometimes surface iron gives high readings which are rejected by the operator, usually by punching a missing data button on the recording control computer. If very weak anomalies are expected at the site, data are recorded at higher sensitivity. A decimal point is assumed before the last recorded digit or a binary point assumed in the middle of the low order byte. Missing input data are a vexing problem. The simplest approach is merely to set missing values to zero. If the display technique is not too sensitive and the number of missing values is not large, this zeroing will not be noticed. If missing data occur in areas of strong anomalies, they appear as medium gray spots on a black background. This is readily visible on an image processing display system. A better approach is to replace missing data with interpolated data from the surrounding area so that an abrupt readily visible change in gray value cannot occur. An iterative interpolation using a small separable convolution with near-

Gaussian weights produces a smooth result without high computational cost. The approach is rather like darning a hole in a sock, working in from the edges. If the holes are not large this approach is quite satisfactory.

The algorithm is quite simple:

If the central value in a 3 × 3 moving window is present, do nothing.

If less than five of the eight neighbours of the central value are present, do nothing.

If five or more are present, compute the missing central value as the weighted sum of the values present, such that the diagonally adjacent points have weights of 1, the immediately adjacent ones 2, normalised by dividing the sum of the weights of values actually present.

Repeat this process, passing over the data again and again until no central values are missing.

If there are large areas missing, then more elaborate techniques, such as filling with a two-dimensional spline surface, are more appropriate. The extreme values produced by surface iron must be removed before any filtering or filling operations are performed. One approach, which has proven computationally economical and robust, is thresholded median filtering or interquantile difference filtering as described in chapter 4. In this method, a value which is outside the 2.5% and 97.5% quantiles of all values in a moving window is replaced by the median. The window size is chosen interactively, with typical side lengths between 5 and 11 pixels. For windows larger than 19 there is no point in using median methods. Histogram techniques which search for outliers from a central distribution and flag them as missing data are similar in performance. For most magnetic data, a 5 × 5 window is adequate.

8.5.2. Linear filtering

As with image data, as discussed in chapter 4, magnetic data can be transformed from the spatial domain to the spatial frequency domain for further analysis and treatment. Compared with image data, magnetic measurements are not very numerous so that treatment in the spatial frequency domain is computationally quite economical. Fourier transform algorithms, which are available for nearly any computer in standard mathematical libraries, are used. Even more complex filtering techniques (5,58) are perfectly feasible for the relatively small amount of data compared with a typical image processing problem.

The real valued data are transformed to Fourier space by using a one-dimensional fast Fourier transform on the rows and columns. This results in a two-dimensional array of complex numbers. A function which has the desired properties is chosen and the transformed data multiplied by it, and then the inverse transform is applied. It differs from the direct transform only by a normalising factor. Since magnetic data have far fewer values than image data, it is usually possible to carry out the entire operation in fast storage, eliminating the need for the slow matrix transposition algorithms which are required when the data set has to be stored on disk to get at the rows and columns separately. The mathematics of a filtering operation carried out in the Fourier domain are discussed in chapter 4.

Gunn has shown (29) that many desirable operations can be considered as a general filtering operation which can be carried out in either the space domain by convolution or the Fourier domain by multiplication with a suitable function. For example, as Dean (21) showed, convolving a magnetic field with a filter whose frequency response is:

$$f(\omega) = \exp[h(u^2 + v^2)^{\frac{1}{2}}] \tag{8.66}$$

the field can be transformed to that which would be observed at a different level h above or below the level of measurement; h is positive for downward and negative for upward continuation of the field. Stability is always guaranteed for upward continuation, but not for downward, however. Downward and upward continuation have true physical meaning, but the results look very much like those obtained with much more arbitrary high and low pass filters when displayed as a grayscale image. With a filter response:

$$f(\omega) = [(1/n)(u^2 + v^2)^{\frac{1}{2}}]^n \tag{8.67}$$

the nth vertical derivative of the field can be obtained, a technique which is similar to one of the edge extraction techniques used in image processing, whereby the results closely follow the outlines of the field inflection points. These methods were first presented in considerable detail by Bhattacharyya (9,10,11). However, as described above, soil noise and archaeological signals have similar spatial frequency distributions so that symmetric filtering in the Fourier domain will not separate them very well.

8.5.3. Reduction to the pole

It is sometimes desired to filter the anomalies such that the peak lies directly over the centre of the feature producing it. This is especially

useful in low magnetic latitudes where distinct peaks do not correspond
to the archaeological features directly. The magnetic field is convolved
with a filter having a frequency response:

$$f(\omega) = \frac{(u^2 + v^2)^{\frac{1}{2}}}{D_1} \frac{(u^2 + v^2)^{\frac{1}{2}}}{D_2} \qquad (8.68)$$

where D_1 and $j = (-1)^{\frac{1}{2}}$:

$$D_1 = [jLu + jMv + N(u^2 + v^2)^{\frac{1}{2}}] \qquad (8.69)$$

is the factor accounting for the direction of magnetisation of the anomaly
and L, M, and N are the direction cosines of the magnetisation and D_2:

$$D_2 = [jlu + jmv + n(u^2 + v^2)^{\frac{1}{2}}] \qquad (8.70)$$

is the factor accounting for the direction of measurement with l, m, and
n being the direction cosines of the component of the field being
measured. The magnetic field values are transformed to those which
would have been observed if the measurement had been made at the
magnetic pole. Baranov and Naudy (6,7) were the first to introduce a
convolution operator which is the Fourier transform of the above
equation to accomplish the same result. Use of the Baranov and Naudy
convolution is computationally more intensive than use of the fast
Fourier transform. The Fourier equivalent of the operation can be
computed following Bhattacharyya. Reduction to the pole shifts the
anomaly so that it is located directly over the buried feature. This is of
interest when accurate prediction for excavation is required. The shift
in the spatial domain is obtained in the Baranov and Naudy method
by the use of an asymmetrical convolution which has phase shifting
properties in the spatial frequency domain. In the Bhattacharyya
technique deliberate phase shift in the transform domain is introduced
which is converted to a positional shift in the space domain upon
application of the inverse transform.

Both the Baranov and the Bhattacharyya methods suffer from
instabilities for measurements made nearer than about 15° from the
magnetic equator. Silva (78) introduced a different technique which is
quite stable at nearly all values of inclination. He formulates the problem
as one of generalised inverses and uses Tikhonov stabilising techniques
(84) developed for this case. In effect, the magnetisation distribution of
an equivalent magnetic dipole layer which accounts for the measured
field is computed. Then the inclination of the equivalent field and the
dipoles are changed to 90° and the field recalculated. The computational
load is considerably greater than that required for the Gunn method

which is good enough at high magnetic latitutdes. As discussed below, archaeologists often prefer the appearance of magnetic maps of shallow sites without reduction to pole so that the Silva method is mainly of interest for the accurate prediction of the location of deep lying anomalies at low magnetic latitudes.

8.6. Data display techniques
8.6.1. Contour plots

Two-dimensional data resulting from a survey have traditionally been presented in a variety of ways derived from conventional geophysical practice. Of these methods the most frequently used is contour plotting. It performs two functions simultaneously. It delineates the form of large, well-separated features, and allows one to see the numerical values of the data at the same time. The impression of darkness, lightness, and relief produced by the relative crowding or spreading of contours is easily interpreted by those practised in using them. However, a smoothing procedure must be used if continuous contours are to be obtained. Regardless of how this is actually carried out, it constitutes low pass filtering of the spatial information with the removal of fine features which may be of interest. Simple averaging over a moving window (convolution with equal weights) and convolution with variable weights characterise the internal algorithms of most contour plotting programs. For archaeological data, the results are not satisfactory if the detail sought is fine. This can be seen in the plot of figure 8.35 which shows the magnetic readings from the site photographed in figure 8.34. Perception of faint shapes is not properly supported.

8.6.2. Overprinting or symbol plots

One of the first surveys of a large archaeological site near Stanton Harcourt in the Thames Valley in 1959 by Linington and Aitken was recorded by plotting symbols of various degrees of blackness on graph paper with a pencil, the blackness corresponding to the strength of the reading. The approach was continued in the early 1960s by Linington (43,44) in Italy using a typewriter and at the same time by Scollar with a computer printer (69,70). A crude grayscale image of the data was thus obtained. A result of this kind for the data of figure 8.34 is shown in figure 8.36. If the symbols are chosen carefully, a range of about a dozen distinguishable gray levels is possible with commonly available symbol sets and overprinting (37,38,48). This approach quantises the input data into a limited number of intervals. It removes information since the number of bits required to represent a given value is less than before. This is tantamount to a reduction of signal to noise ratio. If the intervals are chosen non-linearly, with compression of higher value into fewer levels, then weaker anomalies can still be seen to some extent in the presence of strong ones. If the data are viewed at a distance, faint anomalies can be seen in the presence of strong ones. The limited resolving power of the eye apparently removes high frequency components and seems to smooth some of the quantisation noise. If the area surveyed is small, then the size of the symbols relative to the features becomes visually significant, and it is difficult to see fainter shapes.

8.6.3. Dot density plots

To overcome some of the limitations of contour presentation and symbol printing, 'dot density' display for a survey made in 1963 was introduced

Fig. 8.35 Contour plot of magnetic data obtained at the site shown in figure 8.34. From ref. 77.

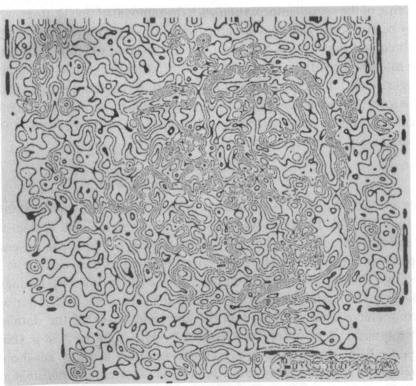

by Scollar (69). A computerised version using a flatbed plotter was published the following year (70). The method has been widely followed by the archaeological prospection community because it can be (tediously) carried out without a computer. Dots equal in number to the results after processing with small random displacements relative to the centre of each measured point are plotted on a square grid. The result resembles the stippled drawings often used to record excavation results and is acceptable to archaeologists. Viewed at a distance, the effect of a grayscale picture can readily be obtained, but quantisation noise due to the limited range of grayscale resolution of the technique remains. An example of a dot density plot is shown in figure 8.37 for the data of

Fig. 8.36 Symbol plot of magnetic data obtained at the site shown in figure 8.34. From ref. 77.

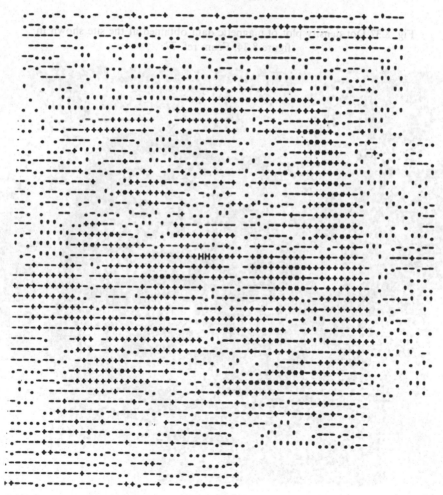

figure 8.34. An attempt to produce a true digital half-tone image was not attempted because the plotters of the epoch could not be used for this purpose with reasonable plot times. Modern half-tone generating algorithms (36) were not then known.

8.6.4. Isometric line traces

The need for viewing results immediately in the field and the advent of reliable instruments which gave continuous readings of the magnetic field differences (long base fluxgate gradiometers) in the late 1960s, along with availability of cheap battery operated x–y plotters, led naturally to the development of an isometric data display. A good result on a magnetically strong site with a quiet background is shown in figure 8.38.

Fig. 8.37 Dot density plot of magnetic data obtained at the site shown in figure 8.34. From ref. 77.

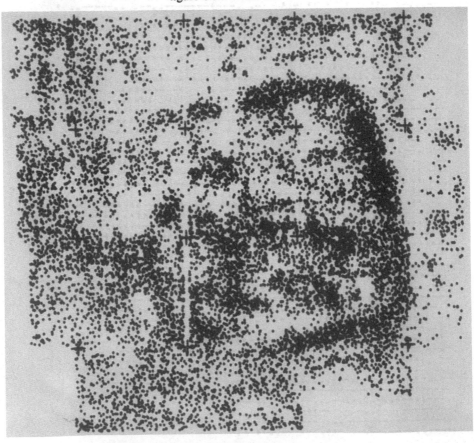

Fig. 8.38 Isometric data display of fluxgate gradiometer data, here showing Bronze Age circular ditches, Saxon sunken huts and other features at Radley, Oxfordshire. From ref. 18.

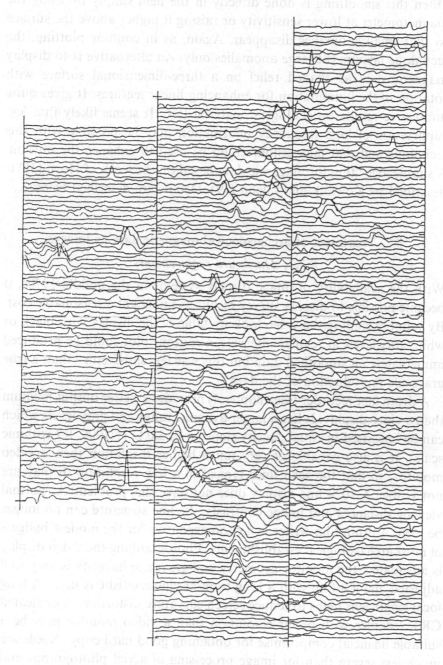

If one is to distinguish anything at all in this kind of a picture, the data must be heavily smoothed to remove small wiggles in the traces. Often this smoothing is done directly in the field simply by using the magnetometer at lower sensitivity or raising it higher above the surface so that small anomalies disappear. Again, as in contour plotting, the technique favours the large anomalies only. An alternative is to display magnetic data as shaded relief on a three-dimensional surface with rotation of the illumination for enhancing linear features. It gives quite impressive results on geological material (56). It seems likely that less suppression of high frequency information may be possible with this technique and that false colour could be used to enhance the illusion. A simple algorithm for magnetic data based on the work of Horn (35) is described by Kowalik and Glenn (39). Details are described below.

8.7. Treatment of archaeological magnetic data as images

8.7.1. Hardware requirements

With the availability of image display systems in the middle 1970s, it became possible to display survey results as images at a reasonable cost. By suitably loading the video look-up table of the display with black or white gray levels at regular intervals, a contour plot could be produced immediately on smoothed data but this was less satisfactory than a true grayscale reproduction of the data.

A drum film recorder (see chapter 4) provides a better output medium than a video display. It has a larger dynamic range and gray levels which can be linearised in equal density steps up to 3.0 D (1:1000, logarithmic scale). It is also capable of very much higher resolution than a video monitor, so that the scan lines typical of a low resolution display are not visible. A 1024 line display does not have this problem. At normal viewing distances and image contrasts, the line structure can no longer be seen. A drum film writer is far too expensive for the modest budgets of most archaeological establishments. Photographing the video display is adequate if the monitor geometry and grayscale linearity is very well adjusted and if a film with a long transfer characteristic is used. A long focal length lens helps to reduce screen curvature distortion. A dedicated CRT recorder with better resolution than a video monitor may be a suitable financial compromise for obtaining good hard copy. Needs are much less severe than for image processing of aerial photographs and great economies in hardware are therefore possible. Grayscale accuracy also does not have to be as good because the eye has no previous training in magnetic imagery and will tolerate more quantisation noise. False colour may also be appreciated.

8.7.2. Techniques of enlargement

Magnetic data almost never have the very large number of points which can be shown on a high resolution image display system. Most image processing systems provide some form of internal hardware expansion of imagery, so that the smaller amounts of data can be made to cover the entire screen. Having oversampled input data eases the requirements placed on the enlarging technique. Digital enlarging can be viewed as an interpolation process, whereby many output pixels are generated from few input values. The problem is mathematically related to the geometric transformation of images discussed in chapter 5, but much less severe since only a linear scale change over a limited range is required. The interpolator should have a spatial frequency response which produces as little aliasing as possible (55,79). Having oversampled input data helps. A grayscale image of the data measured at the site in figure 8.34, enlarged 12 times to nearly fill a 1000 line frame and subject to data compression as described below is shown in figure 8.39.

Pixel replication
The usual hardware approach to enlargement is by pixel replication, that is, each input data value is displayed using an integer number of pixels horizontally and vertically at some convenient ratio usually a power of two. If there is only a small amount of initial data this produces a very block-like appearance, for the output is vastly undersampled relative to the input and serious aliasing results.

Nearest neighbour interpolation
The coordinates of an output pixel are projected back in input image coordinates and the value of the output is taken as that of the nearest neighbour in input space. This produces somewhat less aliasing than does pixel replication and is rapidly computed. It is not satisfactory for degrees of magnification greater than about 2.

Bilinear interpolation
Considerable improvement in the enlargement technique can be obtained via bilinear interpolation working back from the calculated position of

a pixel in the output image. The brightness value is obtained as a distance weighted average of the four nearest neighbour pixels in the input image. In some more expensive image display systems, this function is available in hardware where it can also be used for rotation with less degradation of appearance than in nearest neighbour schemes. This interpolation technique, has a triangular transfer function in the space domain. In the frequency domain there is therefore a considerable loss of higher frequencies and the picture appears blurred at degree of enlargement larger than 4 or so.

Separable cubic spline sinc function approximation

The ideal interpolator, having least loss of high frequencies must have a rectangular transfer function in the frequency domain. The Fourier transform of this function is a $\sin(x)/x$ function in the space domain. The interpolator is a convolution with kernel weights which depend on the output pixel coordinates projected back into input coordinate space.

Fig. 8.39 The magnetic data from the site of figure 8.34 displayed as a grayscale image using arctangent compression. From ref. 77.

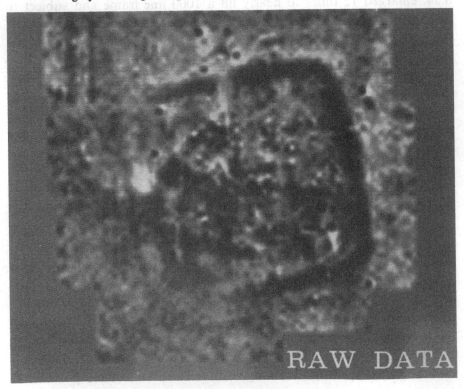

It is computationally too expensive for a large number of points for the small machines which are usually at our disposal. Typically an 11×11 convolution with a weight table look-up operation and a floating point multiply and add for each point in the kernel must be used for each output data point for the function to be well defined. For enlargement to 1000×1000 output points this is excessive. There are also large losses at the boundaries with such large convolutions used on small data sets.

As discussed in chapter 5, the limited dynamic range and coarse quantisation of most display devices makes it unnecessary to define the $\sin(x)/(x)$ function over many oscillations since the smaller additions to the pixel value go unnoticed. It is feasible to use a truncated approximation having only 16 (4×4) support points. The main positive and first negative oscillations of the $\sin(x)/x$ function is approximated by a separable cubic spline in the x and y directions which goes smoothly to zero at the outermost points. Thus only 16 multiplies, 16 adds and 16 table look-up operations for the weights are necessary for calculating the value of an output pixel. The separability of the spline approximation is an advantage. The function as discussed in chapter 5 is symmetric around zero and can be redefined as:

$$f(x) = \begin{cases} f_1(x) = (x-2)x^2 + 1, & 0 \le |x| < 1 \\ f_2(x) = [(-x+5)x-8]x+4, & 1 \le |x| < 2 \\ f_3(x) = 0, & 2 \le |x| \end{cases} \quad (8.71)$$

It is not necessary to calculate the coefficients anew for each data point. They are calculated once for 200 intervals and placed in a table. The values can be retrieved as needed, with considerable computational savings. The output image can be rotated while being enlarged if desired by including a rotational transformation in the back projection calculation. This is sometimes useful for putting together survey grids of different orientations from a large site. With a common orientation, they can be more readily merged.

The small size of the convolution kernel means that there is minimum loss of data at the boundaries of the input data. It has been found useful to limit the size of the output picture to something less than the total area available on the video monitor. For a typical 512×512 display, 400×400 points are suitable to allow for the annotation area around the data. The interpolator can produce a value which is negative or outside the range of single byte arithmetic. A maximum and minimum limit must be set to prevent negative results or results larger than a single byte can contain. Since differential or gradient data are signed integers, zero can be represented by an intermediate gray value or a colour in the middle of the spectrum if false colour is used.

The expression for an eight-bit output pixel using fixed point arithmetic is thus:

$$g(x, y) = \begin{cases} 0, & f(x, y) < 0 \\ [f(x, y)/2] \times 2, & 0 \le f(x, y) < 254 \\ 254, & f(x, y) \ge 254 \end{cases} \quad (8.72)$$

leaving the least significant bit 0. For displays which have pixels of less than eight bits, the expression must be appropriately modified.

It is vital that the data be despiked before enlargement or computational artefacts will appear in the result. Smooth enlargement to full screen size can readily be obtained up to a factor of 12 or so and appearance of fine detail is satisfactory. The result shown in figure 8.39 confirms this. Digital filtering and other enhancement techniques may be applied either before or after enlargement/rotation but the best order of operations is data dependent and should therefore be decided interactively.

8.7.3. Techniques of data range compression

The input data may have quite a large dynamic range, with features of interest both small and large simultaneously present. To prevent clipping due to the limited range of pixel representation on the screen or on film, a non-linear transformation of the input data is required which compresses the higher values. Empirically the arctangent has very satisfactory properties for this purpose as described by Scollar, Weidner, and Segeth (77). It goes smoothly and almost linearly through zero. It seems almost ideal for leaving the small anomalies around the zero level alone, but compressing the larger ones to a maximum of between ± 1.57. Since the arctangent can accept arguments of unlimited size, the input may be as large as required. After taking the arctangent, the result is multiplied by a constant which is chosen to keep the output pixel within the range of the display system monitor, and the result added to a similarly selected mid-range pixel value to give zero as a medium gray tone. For an eight bit display which can show values from 0 to 255, and for medium gray at 127 the multiplicative constant is 80.25. The input data which are to appear as an eight-bit output pixel are transformed in place by:

$$g(x, y) = \arctan[(f(x, y) - B) \cdot C] \cdot 80.25 + 127 \quad (8.73)$$

with *B*, *C* constants to be chosen for brightness and contrast. For other than eight bit pixel representation the constants must be modified appropriately. At least five bits should be available for a grayscale display, or contouring effects will be noticeable. For visibility of faint structures it has been found best to use white for maximum negative and black for maximum positive magnetic values. On sites in the higher latitudes of the northern hemisphere, positive anomalies south of a feature are stronger than negative ones north of it, due to the inclination of the field. True negative anomalies due to the absence or randomising of magnetic material in a more magnetic soil matrix typical of walls in non-magnetic stone are usually weaker than positive anomalies due to magnetically filled ditches and pits on the same site. Under these circumstances a slightly better dynamic display range may be obtained by setting the gray value of zero nearer to white than to black. The actual setting depends on the transfer characteristics of the video monitor used, and is best adjusted by experiment. Allowing zero to be represented by a value of about 85 is suitable. The whiter negative values south of the black positive anomalies (high latitude, northern hemisphere) give the results an appearance of relief which archaeologists find satisfying. This effect could be removed if desired by a reduction to the pole but since the archaeologists like it, we leave it. Figure 8.39 has been displayed and recorded using this technique.

Most modern display systems use the image buffer memory to index the video look-up table connected to the digital–analogue converters which drive the video monitor. This table allows a rapid global transformation of the image with a simple load operation. If desired, the look-up table can be loaded with an approximation to the arctangent transformation after the input values are normalised and made positive. This is very much faster than computing the arctangent for all the data points. It suffers from the disadvantage that very large input values may be clipped rather than compressed since the dynamic range of the display memory is sometimes smaller than that of the raw data.

It may be useful to reserve the least significant bit plane in a low cost display for superimposing crosses to delineate the survey grid and allow annotation. Higher priced display systems have one or more supplementary graphic bit planes. Using one of the grayscale bit planes reduces the available dynamic range. The cheapest display systems which contain only four bit planes are not very satisfactory but if annotation is written into the data after processing has been completed, then the result will be acceptable.

8.7.4. Compression by linear transformation

Input data scaling may be needed so that compression by the arctangent does not remove too much detail from larger anomalies. Simple linear transformation is good enough if the input data have had any trend removed by appropriate filtering, or if a gradiometer was used for measurement. The multiplication and shift factors in (8.73) can be determined by eye. Values are rapidly obtained by loading the look-up table appropriately, looking at the result and applying the best factors to the data if the dynamic range of the input data is small enough so that there is no clipping.

8.7.5. Mean/standard deviation normalisation

If the data have an overall Gaussian distribution, then a simple technique for automatic scaling is to subtract the global mean and divide by twice the standard deviation of the data values to make sure that no pixels go far into the arctangent compression range with resulting loss of detail.

8.7.6. Median/interquantile difference normalisation

Most data with a few strong and many weak anomalies do not have a Gaussian distribution. For such data, normalisation by subtraction of the median and division by the 5% interquantile difference is more robust than using the mean and standard deviation. Even if the data should be normally distributed, this method produces very much the same result as does the mean-standard deviation normalisation at little cost in calculation time when a fast median technique is used. A first pass through the data computes a histogram of the frequencies of all data values. From this histogram, the computation of any desired quantile is simple. The data of figure 8.39 were treated in this way.

8.7.7. Operations on the data after picture transformation

Box filtering and crispening
Once the data are in image form, they can be treated like any other picture data using one of the many algorithms available for this purpose (63) described in chapter 4. Crispening, box filtering, or separable Gaussian filtering (13,14) on the picture data after enlargement and arctangent compression gives considerable improvement in the appearance of the faint line-like anomalies which are generally of importance.

The previous compression and normalisation steps permit the filtering to be more effective in the middle gray range with resulting enhancement of weak anomalies. Crispening and box filtering (49) require the fewest number of operations per data point of any filtering technique. Separable Gaussian filtering can be carried out by repeating box filtering with windows of different sizes in one input/output pass with somewhat greater storage requirements, and with a slight increase in the number of operations. Figure 8.40 shows the data of figure 8.39 enhanced by crispening. Square windows produce quite satisfactory results if they are small relative to the total side length of the data after enlargement. Typical windows are from 5×5 to 21×21 pixels side length (1–5% of total image side length). Using the accumulator technique described in chapter 4 only four integer additions and subtractions per point are required to obtain the running mean of a window of any size in a single pass through the data. Using three window sizes, three sets of accumulator arrays for the sum of the window vertical sides, and six buffers for the single pass through the data. Directional filters can be easily implemented by using long rectangular instead of square windows and rotating the data by a small amount.

Fig. 8.40 Crispening applied to the image data of figure 8.39. From ref. 77.

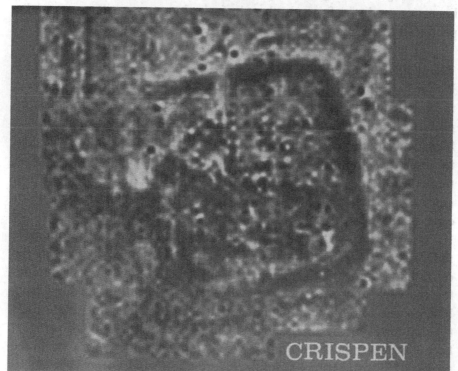

CRISPEN

topmost and bottom data line covered by the window, plus one output buffer, Gaussian filtering can be closely approximated for any window size with only 12 integer additions and subtractions. This technique is much faster than any operation using a Fourier approach in the frequency domain, requires little floating point arithmetic, no matrix transposes, and far less input/output activity.

The Gaussian low pass value can be used in place of the mean in the crispening operation. The Gaussian produces less banding (ringing) at the edges of strong anomalies with large high frequency components than does the primitive box filter. Selected bandpass filters can readily be constructed using a fourth window size and adding or subtracting various combinations of the accumulator array contents. If the resolution is high enough so that minor computing artefacts due to the square windows cannot be seen, then bandpass filtering can also be approximated simply by using the difference between two means computed from windows of two sizes. This requires only nine operations per point in a single pass through the data. Directional filters can be easily implemented by using long rectangular instead of square windows and rotating the data during the enlargement operation. This might be useful in reducing linear anomalies of non-archaeological character such as may be produced by ploughing.

The Wallis algorithm

If there are considerable variations in local contrast and brightness in the raw data, the Wallis algorithm (85) discussed in chapter 4 may be useful. This algorithm is less noise sensitive than is crispening or box filtering if large (greater than 5% of image side length) windows are used to obtain the standard deviation of the window. The data of figure 8.41 treated with the Wallis algorithm are shown in figure 8.42. The improvement over the same data displayed as a contour plot in figure 8.43 or even as a dot density plot as in figure 8.44 is spectacular.

The Wallis algorithm is somewhat more computationally intensive than box or separable Gaussian filtering, but is useful for equalising the overall appearance of the image. By using accumulator array techniques for the computation of means and standard deviations derived from box filtering, and by sampling larger data windows, it is computationally feasible for even the smallest machines. The window may be sampled because the standard deviation can be computed reliably for data in the range 0–255 with about 400 values, regardless of window size. The Wallis

Fig. 8.41 Magnetic data from part of the large scale (65 ha) survey of the Colonia Ulpia Trajana near Xanten in the Rhineland displayed as a grayscale image. From ref. 77.

Fig. 8.42 Enhancement of the data of figure 8.41 using the Wallis algorithm. From ref. 77.

algorithm should only be used with large windows (21×21 or more) to get stable standard deviations, otherwise noise will be introduced.

It is usually not useful to combine crispening, box filtering, and the Wallis algorithm. Each can be tried in turn on the data to see which gives the most pleasing result. The Wallis algorithm is most useful on data containing a very wide range of anomaly strengths, whereas box and Gaussian filtering appear to work best on sites with uniformly weak or uniformly strong anomalies. This is probably because the latter two methods do not include the measure of dispersion in the local window which the Wallis algorithm has. The Wallis algorithm does not enhance fine line structure as much as the other methods because the high pass

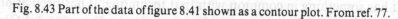

Fig. 8.43 Part of the data of figure 8.41 shown as a contour plot. From ref. 77.

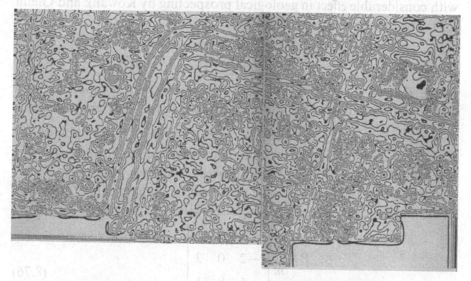

Fig. 8.44 The data of figure 8.41 shown as a dot density plot. From ref. 77.

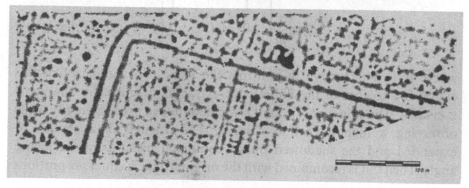

filtering component due to subtraction of the mean is usually restricted to a small fraction by choice of a low value of α. Data which have been filtered before transformation to image form usually require little further treatment. Strong filtering prior to enlargement which very much enhances the higher spatial frequency content of the data may result in computational artefacts because the response function of the enlarging interpolator is unduly stimulated. This should be avoided. Strong filtering after enlargement can bring up deficiencies in the enlarging process and is equally unacceptable.

Shadowing

Shadowing is a much better method than the use of isometric line traces, but it has something in common with that technique. It has been applied with considerable effect in geological prospecting by Kowalik and Glenn (39). The value of an output pixel, $g(x, y)$ is given by:

$$g(x, y) = \frac{(1 + P_0 P + Q_0 Q)}{(1 + P^2 + Q^2)^{\frac{1}{2}}(1 + P_0^2 + Q_0^2)^{\frac{1}{2}}} \tag{8.74}$$

where P_0 and Q_0 are the east–west and north–south components of a vector pointing toward the source of illumination. They are given by:

$$P_0 = -\tan(90 - \theta) \cos(\phi), \qquad Q_0 = -\tan(90 - \theta) \sin(\phi) \tag{8.75}$$

where θ is the elevation of the illumination above the horizon and ϕ is the azimuth of the illumination in degrees clockwise from north.

P and Q are the east–west and north–south components of the slope of the magnetic surface and are computed within a 3×3 window using:

$$P = \frac{1}{m} \begin{bmatrix} -1 & 0 & 1 \\ -2 & 0 & 2 \\ -1 & 0 & 1 \end{bmatrix} \tag{8.76}$$

$$Q = \frac{1}{m} \begin{bmatrix} 1 & 2 & 1 \\ 0 & 0 & 0 \\ -1 & -2 & -1 \end{bmatrix}$$

where m is a factor which controls the slope. The similarity of this expression to that of the Sobel edge extractor described in chapter 4 is evident. The values of θ, ϕ, and m are determined experimentally while observing a high resolution display. The original data should be expanded and the shadowed image computed using a low incidence angle. Then this is recombined with the original data in equal proportions and renormalised to produce the output picture. The idea of recombining with the original is to reduce the shadow contrast in a controlled way. Shadowing requires a locally smoothed image if it is to produce an improvement in appearance. If there is much local noise, then the shadowing is more irritating than helpful. If a very small area has been surveyed, then the natural smoothing provided by the interpolation expansion of the data to image size may be enough. In other cases, a Gaussian or box low pass filter should be applied after noise cleaning and possible Wallis equalising. Spread-out anomalies with slight amplitude or long linear features are the ones which profit most from this technique. The best direction and the angle of incidence for shadowing must be chosen interactively. Shadowing of a false colour image can produce a pleasing aesthetic appearance which laymen much appreciate. In favourable cases, faint detail which might otherwise have gone unnoticed becomes more evident. Because of the smoothing required, small point-like details will be reduced or eliminated.

8.8. Historical note: the development of magnetic prospecting in archaeology

The fact that burned clays acquire a higher magnetic susceptibility was known from Folgheraiter's work on pottery as early as 1896 (23). This effect was used to study the direction of the earth's magnetic field and published by Thellier in 1938. Enhancement of soil magnetism in the uppermost layer was studied and published by Le Borgne in 1955, when he conducted experiments in burning of surface scrub and measuring the enhanced susceptibility of the topsoil which resulted. References are given in chapter 7.

In the autumn of 1957 Belshé, then of the Department of Geodesy and Geophysics of Cambridge University, later director of archaeological rescue work with the US Army Corps of Engineers, first used magnetic field measurements in connection with a modern reconstruction and firing of an archaeological pottery oven. Scollar met with him and Aitken during the International Archaeometry Symposium in Washington DC on May 16, 1984 and asked them to recall the sequence of events which are reported here. Aitken published his version of the conversation afterwards in the Lerici-Linington memorial number of *Prospezioni Archeologiche* (Volume 10) and provided some additional colourful detail about early prospecting campaigns. In 1954 Belshé had used a fluxgate magnetometer in Washington. In 1956 he was in Cambridge and worked with Cook on kiln dating through measurement of fossil magnetic direction following Thellier, picking up the work from Creer who had preceded him. In the course of this work he met chemists and archaeologists from the British Museum through Cook. They had previously fired an experimental kiln. In August 1957, Eames and Bimson fired a kiln at Calke Wood, near Wattisfield Pottery, and Belshé used a marine proton magnetometer built at the Cambridge University Department of Geodesy and Geophysics and detected the rise in the anomaly on cooling. An Askania torsion balance was also used and the resultant observation of a 200 nT anomaly published. A fluxgate magnetometer was used to monitor the rise in susceptibility in the oven wall, and the external magnetometers were added more or less as an afterthought. Shortly thereafter in September 1957, using a grid of 1.5 m, a series of measurements were made with the Askania instrument in the neighbourhood of Kirstall Abbey near Leeds and three anomalies were detected which were thought to come from a forge. This can properly be thought of as the first application of magnetic prospecting in archaeology.

Belshé knew of Le Borgne's work from correspondence with Thellier and saw a possibility for detecting pit fillings as well as kilns. He met Webster during the report on the kiln firing given to the Society of Antiquaries in London in the winter of 1957. In 1958 Aitken obtained circuit diagrams of the Signals Research and Development Establishment proton magnetometer published by Francis and Waters from Belshé. Aitken and Hall made a first archaeological model using transistorised components, which was used in the field in March 1958 at Webster's Water Newton dig, as published that year in *Antiquity*. Other archaeologists involved were Hartley and Gillam on the 13 mile long Peterborough bypass rescue excavations, which included Water Newton.

A later survey was made at Dorchester near Oxford by one of Aitken's students, Linington, in March–May 1959. Scollar visited the site and saw the prototype magnetometer in operation for the first time. Both here and at Water Newton the results were recorded graphically with paper and pencil using a symbol plot. Considerable difficulty due to a mild magnetic storm was visible during his visit, and he decided that a differential magnetometer would be more appropriate. Upon returning to Bonn, a prototype was designed, built and tested in mid-1960. This was followed by a production version which was used for a series of large scale surveys beginning in 1963. Additional instruments of this type were constructed, one of which went to Linington at the Fondazione Lerici in Rome in 1967, and one to the Angel Mounds excavations under Black in Ohio in 1964.

An optically pumped magnetometer made by Varian was used for the first time in Italy in 1964 at Sybaris by Ralph of the University of Pennsylvania Museum. A fluxgate gradiometer was built by Alldred while at Aitken's laboratory, and the first surveys were carried out with that instrument in 1964. It was later produced commercially by Hall's company, Littlemore Scientific Engineering.

Computer evaluation of data began modestly with the acquisition of a second hand Siemens teletype machine by Scollar in 1959. The punched paper tapes were destined for the ER 56 Standard Elektrik Lorenz computer at Bonn University. By the time enough was understood to programme that unique machine, it had been exchanged for an IBM 1410. This had a paper tape reader using a different standard, and all the programs had to be redone. Automatic recording of archaeological magnetic data began with the purchase of a Teletype paper tape reperforator in 1961 by Scollar. A mechanical relay drive for a tape punch made by Serson was seen at the Ottawa Magnetic Observatory in late 1964. A design using bidirectional mechanical uniselector telephone switches inspired by this visit was constructed and used in the field in the Rhineland for the first time in 1966 in a joint campaign with Linington, by then at the Fondazione Lerici in Rome. It made about as much noise as a telephone exchange and was christened 'the clicking wonder' by those using it. It was replaced with a hardwired TTL logic scheme in 1970, which was in turn replaced by a design using one of the first commercial microprocessors in 1974.

Treatment of magnetic data from an archaeological site using a computer was first carried out by Scollar in 1963, following algorithms published by him for resistivity work in 1959. A prototype image processing system at Information International Inc. in Los Angeles was tested by Scollar in early 1971 based on methods previously developed for conventional computers. An image processing system using the first commercially available solid state high resolution colour display system together with film recorder and scanner was installed in late 1975 at Bonn and magnetic data treated as images evaluated with it on a production basis from early 1977 onwards.

Notes

(1) Aitken, M. J., Tite, M. S. 1962, A gradient magnetometer using proton free precession, *Journal of Scientific Instruments*, **39**, 625–9.

(2) Aitken, M. J., Alldred, J. C. 1965, Prediction of magnetic anomalies by means of a simulator, *Prospezioni Archeologiche*, **1**, 67–72.

(3) Alldred, J. C. 1964, A fluxgate gradiometer for archaeological surveying, *Archaeometry*, **7**, 14–19.

(4) Alldred, J. C., Aitken, M. J. 1965, A fluxgate gradiometer for archaeological surveying, *Prospezioni Archeologiche*, **1**, 53–60.

(5) Anderson, B. D. O., Moore, J. B. 1979, *Optimal Filtering*, Prentice-Hall, Englewood Cliffs, NJ.

(6) Baranov, V. 1957, A new method for interpretation of aeromagnetic maps: Pseudo-gravimetric anomalies, *Geophysics*, **22**, 359–83.

(7) Baranov, V., Naudy, H. 1964, Numerical calculation of the formula of reduction to the magnetic pole, *Geophysics*, **29**, 67–79.

(8) Bates, L. F. 1961, *Modern Magnetism*, 4th edn, Cambridge University Press, Cambridge.

(9) Bhattacharyya, B. K. 1965, Two dimensional harmonic analysis as a tool for magnetic interpretation, *Geophysics*, **30**, 829–57.

(10) Bhattacharyya, B. K., Navolio, M. E. 1976, A fast Fourier transform method for rapid computation of gravity and magnetic anomalies due to arbitrary bodies, *Geophysical Prospecting*, **24**, 633–49.

(11) Bhattacharyya, B. K. 1978, Computer modeling in gravity and magnetic interpretation, *Geophysics*, **43**, 912–29.

(12) Boyer, D., Ngoc, P. V. 1971, A continuous measurement technique for detailed magnetic prospecting. Application to an archaeological site, *Prospezioni Archeologiche*, **6**, 43–52.

(13) Burt, P. J. 1981, Fast filter transforms for image processing, *Computer Graphics and Image Processing*, **16**, 20–51.

(14) Burt, P. J. 1983, Fast algorithms for estimating local image properties, *Computer Vision, Graphics and Image Processing*, **21**, 368–82.

(15) Chapman, S., Bartels, J. 1962, *Geomagnetism*, Vols. I & II, The Clarendon Press, Oxford.

(16) Chu, D. C. 1974, Time interval averaging: theory, problems, and solutions, *Hewlett Packard Journal*, **25**, 12–15.

(17) Clark, A. J., Hadon-Reece, D. 1972, An automatic recording system using a Plessey fluxgate gradiometer, *Prospezioni Archeologiche*, **7**, 107–14.

(18) Clark, A. J. 1986, Archaeological geophysics in Britain, *Geophysics*, **51**, 1404–13.

(19) Cloutier, J. R. 1983, A technique for reducing low-frequency time dependent errors present in network-type surveys, *Journal of Geophysical Research*, **88**, 659–63.

(20) Cook, J. C., Carts, S. L. 1962, Magnetic effects and properties of typical topsoils, *Journal of Geophysical Research*, **67**, 815–28.

(21) Dean, W. C. 1958, Frequency analysis for gravity and magnetic interpretation, *Geophysics*, **23**, 97–127.

(22) Evans, D. M., Philpot, F. V. 1972, Hyperbolic Phase Navigation System, internal report, Plessey Radar Electronic Research Laboratory, The Plessey Co. Ltd, Martin Road, West Leigh, Havant, Hants.

(23) Folgheraiter, A. 1896, Determinazione sperimentale della direzione di un campo magnetico uniforme dalla orientazione del magnetismo da esso indotto, *Rendiconti Academia Lincei*, **5**, 127–35.

(24) Foster, M. R., Jines, W. R., van der Weg, K. 1970, Statistical estimation of systematic errors at intersections of lines of aeromagnetic survey data, *Journal of Geophysical Research*, **75**, 1507–11.

(25) Geyger, W. A. 1964, *Nonlinear Magnetic Control Devices*, McGraw-Hill, New York, NY.

(26) Graham, I. D. G., Scollar, I. 1976, Limitations on magnetic prospection in archaeology imposed by soil properties, *Archaeo-Physika*, **6**, 1–124.

(27) Green, A. A. 1983, A comparison of adjustment procedures for levelling aeromagnetic survey data, *Geophysics*, **48**, 745–53.

(28) Grivet, P. A., Malnar, L. 1967, Measurement of weak magnetic fields by magnetic resonance, *Advances in Electronics and Electron Physics*, **23**, 39–151.

(29) Gunn, P. J. 1975, Linear transformation of gravity and magnetic fields, *Geophysical Prospecting*, **23**, 300–12.

(30) Haas, A. G., Viallix, J. R. 1976, Krigeage applied to geophysics: The answer to the problem of estimates and contouring, *Geophysical Prospecting*, **24**, 49–69.

(31) Hardwick, C. D. 1984a, Important design considerations for inboard airborne magnetic gradiometers, *Geophysics*, **49**, 2009.

(32) Hardwick, C. D. 1984b, Non-oriented cesium sensors for airborne magnetometry and gradiometry, *Geophysics*, **49**, 2024–31.

(33) Hoerl, A. E., Kennard, R. W. 1970a, Ridge regression: biased estimation for nonorthogonal problems, *Technometrics*, **12**, 55–67.

(34) Hoerl, A. E. 1970b, Ridge regression: applications to nonorthogonal problems, *Technometrics*, **12**, 69–82.

(35) Horn, B. K. P. 1981, Hill shading and the reflectance map, *Proceedings of the IEEE*, **69**, 14–47.

(36) Hou, H. S. 1983, *Digital Document Processing*, John Wiley & Sons, New York, NY.

(37) Jarvis, J. F., Judice, C. N., Ninke, W. H. 1976, A survey of techniques for the display of continuous tone pictures on bilevel displays, *Computer Graphics and Image Processing*, **5**, 13–40.

(38) Knowlton, K., Harmon, L. 1972, Computer produced grey scales, *Computer Graphics and Image Processing*, **1**, 1–20.

(39) Kowalik, W. S., Glenn, W. E. 1987, Image processing of aeromagnetic data and integration with Landsat images for improved structural interpretation, *Geophysics*, **52**, 875–84.

(40) Köhler, K. 1958, Grundlagen für die Auswertung von magnetischen Anomalien, *Freiberger Forschungshefte C*, **41**, 1–128.

(41) Le Borgne, E., Le Mouel, J. 1964, Magnétomètres à protons, magnétomètres à pompage optique, Note No. 2, Institut de Physique du Globe, 1–124, Institut de Physique du Globe, Paris.

(42) Ledley, B. G. 1970, Magnetometers for space measurements over a wide range of field intensities, *Revue de Physique Appliquée*, **5**, 1, 164–8.

(43) Linington, R. E. 1968, The Rome computer system for treating archaeological survey results, part 1, *Prospezioni Archeologiche*, **3**, 19–36.

(44) Linington, R. E. 1969, The Rome computer system for treating archaeological survey results, part 2, *Prospezioni Archeologiche*, **4**, 9–58.

(45) Linington, R. E. 1972a, A summary of simple theory applicable to magnetic prospecting in archaeology, *Prospezioni Archeologiche*, **7**, 9–60.

(46) Linington, R. E. 1972b, Topographical and terrain effects in magnetic prospecting, *Prospezioni Archeologiche*, **7**, 61–84.

(47) Linington, R. E. 1974, The magnetic disturbances caused by DC electric railways, *Prospezioni Archeologiche*, **9**, 9–20.

(48) Macleod, I. D. G. 1970, Pictorial output with a line printer, *IEEE Transactions on Computers*, **19**, 160–2.

(49) McDonnell, M. J. 1981, Box filtering techniques, *Computer Graphics and Image Processing*, **17**, 65–70.

(50) Meilleroux, J. L. 1970, Progrès récents sur le magnétomètre à vapeur de césium type "asservi", *Revue de Physique Appliquée*, **5**, 19, 1–30.

(51) Mittal, P. K. 1984, Algorithm for error adjustment of potential field data along a survey network, *Geophysics*, **49**, 467–9.

(52) Montalenti, G. 1970, Le bruit de Barkhausen dans les matériaux ferromagnétiques, *Revue de Physique Appliquée*, **5**, 87–93.

(53) Mudie, J. D. 1962, A digital differential proton magnetometer, *Archaeometry*, **5**, 135–8.

(54) Mudie, J. D. 1965, The behaviour of the dipole moment of protons in a changing magnetic field, *Archaeo-Physika*, **1**, 93–108.

(55) Park, K., Schowengerdt, R. A. 1983, Image reconstruction by parametric cubic convolution, *Computer Vision, Graphics and Image Processing*, **23**, 258–72.

(56) Paterson, N. R., Reeves, C. V. 1985, Applications of gravity and magnetic surveys: The state of the art in 1985, *Geophysics*, **50**, 2558–94.

(57) Philpot, F. V. 1972, An improved fluxgate gradiometer for archaeological surveys, *Prospezioni Archeologiche*, **7**, 99–106.

(58) Pilkington, M., Crossley, D. J. 1987, A Kalman filter approach to susceptibility mapping, *Geophysics*, **52**, 655–64.

(59) Pratt, W. K. 1978, *Digital Image Processing*, Wiley-Interscience, New York, NY.

(60) Primdahl, F. 1979, The fluxgate magnetometer, *Journal of Physics E: Scientific Instruments*, **12**, 241–53.

(61) Ray, R. D. 1985, Correction of systematic error in magnetic surveys: An application of ridge regression and sparse matrix theory, *Geophysics*, **50**, 1721–31.

(62) Ripley, B. D. 1981, *Spatial Statistics*, John Wiley & Sons, New York, NY.

(63) Rosenfeld, A., Kak, A. C. 1982, *Digital Image Processing*, John Wiley & Sons, New York, NY.

(64) Roy, A. 1962, Ambiguity in geophysical interpretation, *Geophysics*, **27**, 90–9.

(65) Rüger, C. B., Scollar, I., Grewe, K. 1972, Neues zum Plan der Colonia Ulpia Trajana, *Bonner Jahrbuch*, **172**, 308–9.

(66) Salvi, A. 1970, Perfectionnements apportés aux magnétometres à résonance magnétique nucléaire à pompage électronique, *Revue de Physique Appliquée*, **5**, 131–4.

(67) Sander, E. L., Mrazek, C. P. 1982, Regression technique to remove temporal variation from geomagnetic survey data, *Geophysics*, **47**, 1437–43.

(68) Scollar, I. 1963, A proton precession magnetometer with diurnal variation correction, *Electronic Engineering*, **35**, 3, 177–9.

(69) Scollar, I. 1965, Recent developments in magnetic prospecting in the Rhineland, *Prospezioni Archeologiche*, **1**, 43–50.

(70) Scollar, I., Krueckeberg, F. 1966, Computer treatment of magnetic measurements from archaeological sites, *Archaeometry*, **9**, 61–71.

(71) Scollar, I. 1968a, A program package for the interpretation of magnetometer data, *Prospezioni Archeologiche*, **3**, 9–18.

(72) Scollar, I. 1968b, Automatic recording of magnetometer data in the field, *Prospezioni Archeologiche*, **3**, 105–10.

(73) Scollar, I. 1969, A program for the simulation of magnetic anomalies of archaeological origin in a computer, *Prospezioni Archeologiche*, **4**, 59–83.

(74) Scollar, I. 1970, Fourier transform methods for the evaluation of magnetic maps, *Prospezioni Archeologiche*, **5**, 9–41.

(75) Scollar, I. 1971, A magnetometer survey of the Colonia Ulpia Trajana near Xanten, West Germany, *Prospezioni Archeologiche*, **6**, 83–92.

(76) Scollar, I., Gubbins, D., Wisskirchen, P. 1971, Two dimensional digital filtering with Haar and Walsh transforms, *Annales de Géophysique*, **27**, 2, 85–104.

(77) Scollar, I., Weidner, B., Segeth, K. 1986, Display of archaeological magnetic data, *Geophysics*, **51**, 623–33.

(78) Silva, J. B. C. 1986, Reduction to the pole as an inverse problem and its applicaton to low-latitude anomalies, *Geophysics*, **51**, 369–82.

(79) Simon, K. W. 1975, Digital image reconstruction and resampling for geometric manipulation, reprinted in *Digital Image Processing for Remote Sensing*, R. Bernstein, ed. 1978, Proceedings IEEE Symposium Machine Processing Remotely Sensed Data, 3A1–3A11, Institute of Electrical and Electronic Engineers, New York, NY.

(80) Slocum, R. E. 1970, Advances in optically pumped He4 magnetometers: resonance and nonresonance techniques, *Revue de Physique Appliquée*, **5**, 109–12.

(81) Sternberg, R. S. 1987, Archaeomagnetism and magnetic anomalies in the American Southwest, *Geophysics*, **52**, 368–71.

(82) Strong, C. L. 1968, The Amateur Scientist: Building a sensitive magnetometer, *Scientific American*, **218**, 2, 124–8.

(83) Talwani, M. 1965, Computation with the help of a digital computer of magnetic anomalies caused by bodies of arbitrary shape, *Geophysics*, **30**, 797–817.

(84) Tikhonov, A. N., Arsenin, V. Y. 1977, *Solutions to Ill-posed Problems*, V. H. Winston & Sons, Washington, DC.

(85) Wallis, R. 1977, An approach to the space variant restoration and enhancement of images, in *Proceedings, Symposium on Current Mathematical Problems in Image Science*, Monterey, CA, Nov. 1976, reprint, *Image Science Mathematics*, C. O. Wilde, E. Barett, eds., Western Periodicals, North Hollywood, Ca.

(86) Waters, G. S., Phillips, G. 1956, A new method of measuring the Earth's magnetic field, *Geophysical Prospecting*, **4**, 1–9.

(87) Waters, G. S., Francis, P. D. 1958, A nuclear magnetometer, *Journal of Scientific Instruments*, **35**, 88–93.

(88) Weymouth, J. W. 1986, Archaeological site surveying program at the University of Nebraska, *Geophysics*, **51**, 538–52.

(89) Weymouth, J. W., Lessard, Y. A. 1986, Simulation studies of diurnal corrections for magnetic prospection, *Prospezioni Archeologiche*, **10**, 37–48.

(90) Yabuzaki, T., Ogawa, O. 1974, Frequency shift of self-oscillating magnetometer with cesium vapor, *Journal of Applied Physics*, **45**, 1349.

(91) Yarger, H. L., Robertson, R. R., Wentland, R. L. 1978, Diurnal drift removal from aeromagnetic data using least squares, *Geophysics*, **46**, 1148–56.

9

Electromagnetic prospecting

9.1. Definition and general concepts

The term 'electromagnetic' is used in prospecting as it is in physics, when use is made of phenomena varying in time which combine magnetic and electric fields. In contrast, electrical prospecting utilises the effects due solely to electric fields with the associated magnetic field being negligible, and magnetic prospecting uses only the earth's magnetic field, neglecting any effects produced by its temporal variation. All electromagnetic phenomena are describable using Maxwell's equations. They are the basis for defining the approximations which are needed in order to carry out the necessary theoretical calculations. In a linear, homogeneous, and isotropic medium which does not contain any sources, with conductivity σ, magnetic permeability μ, dielectric permittivity ε, and with $\mathbf{H}$ and $\mathbf{E}$ the magnetic and electric fields respectively, Maxwell's equations are:

$$\mathbf{\nabla} \cdot (\mu \mathbf{H}) = 0, \qquad \mathbf{\nabla} \cdot (\varepsilon \mathbf{E}) = q$$
$$\mathbf{\nabla} \times \mathbf{E} = -\mu \, \partial \mathbf{H}/\partial t, \qquad \mathbf{\nabla} \times \mathbf{H} = \sigma \mathbf{E} + \varepsilon \, \partial \mathbf{E}/\partial t \tag{9.1}$$

where q is the net volume density of electrostatic charge, which is usually zero at macroscopic levels because the numbers of negative and positive charges in a given volume are equal. The operator nabla, $\mathbf{\nabla}$ gives the local divergence, and $\mathbf{\nabla} \times$ the rotation or curl of the field.

A priori, all three properties, μ, σ, and ε are involved, and thus the situation is more complicated than it is in pure electrical or magnetic prospecting. Nevertheless there are approximations which simplify things considerably. The first of these comes from the comparison between $\sigma \mathbf{E}$, the conduction current, and $\varepsilon\,\partial E/\partial t$, the displacement current in the Maxwell–Ampère equation. In the case of a sinusoidal variation in time $\partial/\partial t = i\omega$, for a soil of conductivity $10^{-2}\,\text{S m}^{-1}$ and of relative dielectric permittivity $\varepsilon_r = 18$, σ and $|\varepsilon\omega|$ are equal at a frequency of 10 MHz. If the frequency is between 30 and 300 kHz or lower, $\sigma \gg \varepsilon\omega$, in which case the Maxwell–Ampère equation reduces to:

$$\mathbf{V} \times \mathbf{H} = \sigma \mathbf{E} \tag{9.2}$$

which is the case for induction; the partial differential equation which governs the electric (and similarly for the magnetic) field is therefore written:

$$\mathbf{V} \times \mathbf{V} \times \mathbf{E} = -\sigma\mu\frac{\partial \mathbf{E}}{\partial t} \tag{9.3}$$

or:

$$\mathbf{V}^2\mathbf{E} - \sigma\mu\frac{\partial \mathbf{E}}{\partial t} = 0 \tag{9.4}$$

This is a diffusion equation, analogous to the equation of heat. If one goes to very high frequencies, between 30 MHz and 3 GHz, the equation which governs the fields contains two terms dependent on σ and ε:

$$\mathbf{V}^2\mathbf{E} - \sigma\mu\frac{\partial \mathbf{E}}{\partial t} - \varepsilon\mu\frac{\partial^2 \mathbf{E}}{\partial t^2} = 0 \tag{9.5}$$

This is a propagation equation with damping.

One also ought to consider the case where the variation with time is very slow, and this is equivalent to writing:

$$\frac{\partial}{\partial t} \to 0, \qquad \mathbf{V} \times \mathbf{E} = 0 \tag{9.6}$$

In this quasistatic case, the laws which govern direct current electric phenomena in electric prospecting prevail. This approximation also applies when one uses a low frequency alternating current in electrical prospecting. Thus, there are two domains for electromagnetic prospecting: one, the low frequency region where only σ and μ intervene and where the phenomena are governed by a diffusion equation; the other, the high frequency region where dielectric permittivity intervenes and where the phenomena are governed by propagation laws. The first domain is utilised much more than the second, where only the soil radar methods to be discussed below are of importance.

Another approximation which is commonly used neglects the contrasts which may exist due to different magnetic properties of soil. The magnetic susceptibilities rarely exceed 300×10^{-5} SIu and the approximation $\mu = \mu_0(1 + \chi) \sim \mu_0$ is accurate to 3×10^{-3}. This approximation is much used in non-archaeological electromagnetic prospecting applications, but given the importance of magnetic susceptibility in archaeology, one wishes, on the contrary, to measure it.

9.1.1. Skin depth at low frequencies and sounding techniques

If one considers the simplest case, figure 9.1, with a homogeneous soil and a uniform electromagnetic field $(\mathbf{E}_{x_0}, \mathbf{H}_{y_0})$, the distribution of the fields in the soil obeys the equation:

$$\frac{d^2 E_x}{dz^2} - i\sigma\mu\omega E_x = 0 \tag{9.7}$$

in the low frequency domain for a sinusoidal variation with time of ω. The field E_x is in this case:

$$E_x(z) = E_{x_0} \exp[-(\sigma\mu\omega/2)^{\frac{1}{2}}z] \exp[-i(\sigma\mu\omega/2)^{\frac{1}{2}}z] \tag{9.8}$$

a field shifted in phase and attenuated as a function of depth. The skin depth is:

$$p = (2/\sigma\mu\omega)^{\frac{1}{2}} \tag{9.9}$$

where the amplitude of the field is equal to E_{x_0}/e (e being Neper's constant). This definition is purely theoretical, as is the notion of a homogeneous soil, but it is very useful for illustrating the fact that fields

Fig. 9.1 Coordinate system for a homogeneous soil and a uniform electro-magnetic field $\mathbf{E}_x, \mathbf{H}_y$.

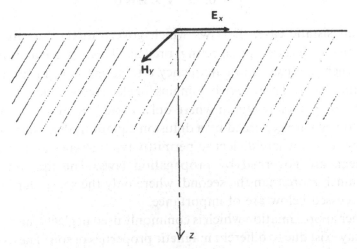

and the currents induced in them are concentrated in the neighbourhood of the surface, the so-called skin effect. This effect is unfavourable in terms of depth of investigation for electromagnetic techniques in general, but not for archaeological applications. The fact that *p* varies with frequency makes a sounding technique possible in which the terrain thickness taken into account in measurement is reduced by increasing *p*. There are thus three essentially different sounding techniques:

(1) At low frequencies, the depth of penetration is augmented by reducing the frequency.
(2) As in electrical prospecting, the depth is augmented by separating the source from the receiver.
(3) At high frequencies, the time required for a downward propagating signal to return from an interface is measured.

Of these three possibilities, only the last two find applications in archaeology as we shall see below.

9.1.2. Types of sources or signals utilised at low frequencies

An electromagnetic measuring system necessarily utilises a primary field source, and a receiver which measures one or more components of the total field. These are the sums of the primary field and a secondary field which results from the presence of the soil. One can imagine a whole variety of sources, but nevertheless in low frequency electromagnetic prospecting three kinds of sources are generally used:

(1) Sources located at a great distance which create a primary uniform field. These are, at less than 3 kHz, the natural variations of the earth's magnetic field. Above this frequency one must use radio transmitters located far from the prospecting area, low frequency broadcasting stations for example. The principal advantages of this type of source derive in part from the fact that they have a primary uniform field which is favourable for the depth of investigation and facilitates theoretical calculations. The very existence of such transmitters also permits economies in equipment and energy.
(2) A long wire which carries a current producing a primary field which decreases as $1/r$, where r is the distance to the wire, with current return either via the ground itself or by means of a large loop outside the surveyed area. The loop is not moved during the course of measurement.
(3) Electric or magnetic dipoles: electric dipoles are made up of two

electrodes which are forced into the ground between which a current is passed, whereas magnetic dipoles are loops or coils of wire. The primary field of these sources decreases as $1/r^3$, which is unfavourable for the depth of investigation, but magnetic dipoles have an advantage in that they are easily moved.

In archaeological prospecting, only sources at a great distance and magnetic dipoles have been widely used up to now. It is unlikely that this situation will change. The sources at a great distance which have been used are low frequency broadcast (150–250 kHz) stations in Europe. They have sufficient penetration (of the order of 5–10 m) so that the field is only slightly attenuated at the level of archaeological structures, and their strength is such that they can be used at up to 500 km from the transmitter. Their geographic distribution is reasonable in western Europe. In other parts of the world there are other transmitters whose signals can be received at low frequencies, but they may be too far away to be received solely via the ground wave.

Distant radio sources may be utilised in a number of different fashions. These are: measurement of one or more components of the magnetic or electric field; the inclination of the wave or the measurement simultaneously of both the electric and the magnetic field, the magneto-telluric technique of Cagniard (5) which allows the apparent resistivity of the earth to be obtained directly. In archaeological prospecting, distant radio sources have been used with measurement of the vertical component of the magnetic field for several transmitters at once. Magneto-telluric techniques have only been used experimentally because a satisfactory detector for the electric field was not available until the present time. Measurement of the electric field alone has not been used at all for the same reason.

Employing magnetic dipoles (loops or solenoids) as a transmitter offers the user a choice: one may transmit either a sinusoidal signal continuously or an impulse. The former favours the processing of the signal and allows the electronics to be much simpler, the latter permits much more powerful transmission with a wide frequency range. Magnetic dipoles with continuous sinusoidal emission associated with a receiver which also uses another dipole have given rise to the technique which was christened Slingram by the Swedish scientists who worked in mining geophysics during the 1930s. It is the most frequently used electromagnetic technique in archaeological prospecting. Utilisation of a pulsed emission allows magnetic viscosity effects in the soil to be detected, but it was developed primarily for finding metallic objects.

9.2. Research into applications of low frequencies in archaeology

The first applications of electromagnetic methods in archaeology came about at the end of the Second World War with the availability of surplus mine detectors. Little was published, but apart from the detection of metal objects, mention was made of responses obtained from pottery. These experiments were apparently unsuccessful in detecting buried structures. Research in the application of electromagnetic methods in archaeology began after 1960 and was dominated by one basic idea: to replace the electrical method which requires good contact between the ground and the electrode system, an important limitation and major source of problems. In 1962, Scollar (30) studied the problem theoretically in considering the response to be obtained from a homogeneous soil with a dipole–dipole Slingram arrangement of vertical coaxial coils on a horizontal axis, using the integrals of Wait (48). The conclusion drawn in this study was essentially negative in view of the complexity of the response, the dependence on the height of the coil system and the coil spacing. Unfortunately, Wait's calculations were made under an assumption which was later shown to be incorrect. Furthermore, (see below), it is now known that this coil arrangement is the most unfavourable possible for the determination of the resistivity of a homogeneous soil. Scollar selected the orientation following Wait's opinion that this arrangement produced the least error with changes in coil height.

Experiments were made from 1965 onwards in England using two devices, the soil conductivity meter (SCM) of Howell (18) and the Decco of Colani (7), the former was a Slingram device using coils with a perpendicular coil arrangement spaced 1 m apart using a sinusoidal frequency of 4 kHz, the latter a pulsed device with two horizontal concentric coils.

Utilisation of these two devices showed (8,9,19,23) that anomalies associated with archaeological structures were observed but that the results were correlated with the magnetic properties of the ground rather than with its electrical conductivity. Tite and Mullins (45,46) therefore began a series of experimental and theoretical studies with the aim of obtaining a better understanding of the physical phenomena involved in both devices. Their conclusions are very clear:

(1) Neither device measures the electrical conductivity. Rather the magnetic properties, susceptibility in the case of the SCM, magnetic viscosity in the case of the Decco, are involved, and the depth of investigation is less than with purely magnetic prospecting.

(2) In order to determine the conductivity, one must raise the frequency and in the case of an apparatus with a sinusoidal signal measure the quadrature component.

These conclusions were confirmed by tests made in France (34) with the EM15 made by Geonics, which has a frequency of 16 kHz, a coil separation of 0.83 m, measures the in-phase component only, and has a coil orientation of 35° to the vertical. This device permits rapid measurement of apparent soil magnetic susceptibility to a very limited depth. Two problems remain: soil conductivity, the target of the experiment, is not measured, and the depth of penetration is very low, much less than that available with purely magnetic or electrical techniques. In order to remove these obstacles, two paths were followed:

(1) Electromagnetic sources at a distance with a nearly uniform field at the depth of typical archaeological structures were utilised. This raised the depth of penetration which then depended only on the distance from the structure to the receiver. The higher frequencies reinforced the conductivity response. This path led to the development of the SGD technique of Tabbagh (33,34,36) and Nguyen Van Hoa, Jolivet and Tabbagh (25).

(2) The theoretical calculations were reconsidered, taking into account the magnetic susceptibility to determine at which frequencies and coil spacings the Slingram method could simultaneously permit the measurement of the magnetic susceptibility of the ground by the measurement of the in-phase component and the conductivity by measurement of the quadrature component of the secondary field. The theoretical calculations by Tabbagh (35) led to specifications for the optimum parameters. Their use has verified experimentally the possibility of obtaining the desired simultaneous measurements, as shown by Parchas and Tabbagh (27). These studies were subsequently extended to investigate the limits of the Slingram technique. At the same time, the middle of the 1970s saw the development of ground penetrating radar used principally on frozen soils in archaeology by Dolphin, Bollen and Oetzel (11) and Bevan and Kenyon (4).

9.3. Utilisation of sources at a great distance

9.3.1. Surface fields in a homogeneous soil created by a distant transmitter

Very low frequency transmitters have vertical antennae which create a primary field made up of vertical electric and horizontal magnetic components perpendicular to the direction of propagation. At a sufficiently great distance, of the order of ten times the wavelength in vacuum, this primary field can be considered as a plane wave with grazing incidence. In order to define a prospection technique, we will have to choose the component(s) to be measured, and hence we must describe the components present at the surface of a homogeneous soil.

Let the vertically polarised field be $(\mathbf{E}, \mathbf{H})$ with an angle of incidence θ, figure 9.2. The total field has three component E_z, E_x, H_y which can all be calculated from H_y alone. H_y in air satisfies the equation:

$$\nabla^2 H_y + k_0^2 H_y = 0 \qquad (9.10)$$

where $k_0^2 = \varepsilon_0 \mu_0 \omega^2$. In the ground it satisfies the equation:

$$\nabla^2 H_y - \gamma^2 H_y = 0 \qquad (9.11)$$

where $\gamma^2 = i\sigma\mu\omega$. At the air–ground interface there is continuity for H_y and E_x. The angle of incidence being θ, the dependence of H_y on x is:

$$\exp[-ik_0 \sin(\theta_x)]$$

Fig. 9.2 Components of a vertically polarised field $\mathbf{E}_i$, $\mathbf{H}_i$ with an angle of incidence θ.

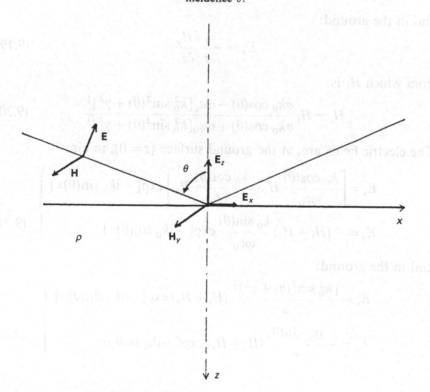

and therefore:

$$\partial^2 H_y/\partial x^2 = -k_0^2 \sin^2(\theta) H_y \tag{9.12}$$

On the other hand:

$$\partial^2 H_y/\partial y^2 = 0 \tag{9.13}$$

because we are dealing with a plane wave. Therefore in air:

$$\partial^2 H_y/\partial z^2 + k_0^2 \cos^2(\theta) H_y = 0 \tag{9.14}$$

and in the ground:

$$\partial^2 H_y/\partial z^2 - [\gamma^2 + k_0^2 \sin^2(\theta)] H_y = 0 \tag{9.15}$$

The field in air is therefore:

$$H_y = \{H_i \exp[-ik_0 \cos(\theta)z] + H_r \exp[ik_0 \cos(\theta)z]\} \exp[-ik_0 \sin(\theta)x] \tag{9.16}$$

where H_r is the reflected field. The field in the ground is therefore:

$$H_y = (H_i + H_r) \exp\{-[\gamma^2 + k_0^2 \sin^2(\theta)]^{\frac{1}{2}} z\} \exp[-ik_0 \sin(\theta)x] \tag{9.17}$$

The continuity of the electric field E_x permits calculation of the value of the modulus of the reflected field in air:

$$E_x = -\frac{1}{i\varepsilon_0 \omega} \frac{\partial H_y}{\partial z} \tag{9.18}$$

and in the ground:

$$E_x = -\frac{1}{\sigma} \frac{\partial H_y}{\partial z} \tag{9.19}$$

from which H_r is:

$$H_r = H_i \frac{\sigma k_0 \cos(\theta) - \omega \varepsilon_0 [k_0^2 \sin^2(\theta) + \gamma^2]^{\frac{1}{2}}}{\sigma k_0 \cos(\theta) + \omega \varepsilon_0 [k_0^2 \sin^2(\theta) + \gamma^2]^{\frac{1}{2}}} \tag{9.20}$$

The electric fields are, at the ground surface $(z = 0)$, in air:

$$\left. \begin{aligned} E_x &= \left[\frac{k_0 \cos(\theta)}{\omega \varepsilon_0} H_i - \frac{k_0 \cos(\theta)}{\omega \varepsilon_0} H_r \right] \exp[-ik_0 \sin(\theta)x] \\ E_z &= -(H_i + H_r) \frac{k_0 \sin(\theta)}{\omega \varepsilon_0} \exp[-ik_0 \sin(\theta)x] \end{aligned} \right\} \tag{9.21}$$

and in the ground:

$$\left. \begin{aligned} E_x &= \frac{[k_0^2 \sin^2(\theta) + \gamma^2]^{\frac{1}{2}}}{\sigma} (H_i + H_r) \exp[-ik_0 \sin(\theta)x] \\ E_z &= -\frac{ik_0 \sin(\theta)}{\sigma} (H_i + H_r) \exp[-ik_0 \sin(\theta)x] \end{aligned} \right\} \tag{9.22}$$

Using these expressions for the three components, one obtains simple expressions by calculating the wave impedance:

$$Z = E_x/H_y \qquad (9.23)$$

and the wave tilt:

$$W = E_x/E_z \qquad (9.24)$$

such that:

$$
\left.
\begin{aligned}
Z &= \frac{[k_0^2 \sin^2(\theta) + \gamma^2]^{\frac{1}{4}}}{\sigma} \\[2mm]
W &= -\frac{\omega \varepsilon_0}{k_0 \sin(\theta)} \frac{[k_0^2 \sin^2(\theta) + \gamma^2]^{\frac{1}{4}}}{\sigma}
\end{aligned}
\right\} \qquad (9.25)
$$

With the hypothesis that $\gamma^2 \gg k_0^2$, a logical assumption if the displacement currents in the ground are neglected, one has:

$$|Z| = \left(\frac{\mu\omega}{\sigma}\right)^{\frac{1}{2}}, \qquad E_z \to 0 \qquad (9.26)$$

in the ground and:

$$|W| = \left(\frac{\varepsilon_0 \omega}{\sigma}\right)^{\frac{1}{2}} \frac{1}{\sin(\theta)} \qquad (9.27)$$

By this means one has derived the classical equation of magneto-telluric resistivity prospecting:

$$\rho = \frac{1}{\sigma} = \frac{1}{\mu\omega}\left|\frac{E_x}{H_y}\right|^2 \qquad (9.28)$$

and for grazing incidence, one has a second method for measuring the resistivity of the soil:

$$\rho = \frac{1}{\varepsilon_0 \omega}\left|\frac{E_x}{E_z}\right|^2 \qquad (9.29)$$

These two methods for measuring the soil resistivity have the advantage that they are independent of variations in intensity of the source. Radio transmitters generally have an output power stability such that the value of E_x alone suffices for the determination of ρ. There is, however, one important difficulty in measuring E_x, namely that in air E_z is very much greater and that this creates difficulties in designing the detectors. For example, for $\sigma = 10^{-2}\ \mathrm{S\ m^{-1}}$ and $f = 200\ \mathrm{kHz}$, $|E_x/E_z| = 0.033$.

9.3.2. The case of a two- or three-dimensional buried structure

At the surface of a homogeneous or layered soil, there are three field components from a distant source E_z, E_x, H_y, and measurement of E_x

alone permits the determination of the soil resistivity. Archaeological structures, however, are inhomogeneities in the homogeneous or layered ground. It is necessary to examine the effects produced by simple geometric structures. Since the primary field is uniform, long cylindrical structures require only the solution of a two-dimensional problem, whereas compact structures which do not have any significant elongation in the horizontal direction require a solution in three dimensions.

Two-dimensional horizontal structures

Let $k_0^2 \ll \gamma^2$, that is, one examines the variations in resistivity or field strength over a region small compared with the wavelength, variations which can only be due to a buried structure. Since it is always possible to decompose the horizontal field into one at right angles and one parallel to the direction of the structure, one can reduce each case to two subcases, one where the structure is parallel to E_x and one where it is parallel to H_y.

For a structure parallel to E_x where $\partial/\partial x = 0$, the two coupled equations:

$$\nabla \times \mathbf{E} = -i\omega\mu\mathbf{H}, \qquad \nabla \times \mathbf{H} = \sigma\mathbf{E} \qquad (9.30)$$

permit one to show that there is only one mode which exists in the ground corresponding to E_x, H_y and H_z. A new component appears in this case namely H_z. For a structure parallel to H_y, the same line of reasoning allows us to show that there is only a single mode in the ground corresponding to the components E_x, H_y and E_z. The calculation of the effect due to a cylindrical structure is not possible analytically, and we must use a numerical method. The method of finite elements is the most powerful of these. The results will be presented below. The method is described in section 9.8.

Three-dimensional structures

The modal decomposition which can be used for a two-dimensional structure does not exist for a three-dimensional body, and the existence of such a feature causes variations in all six field components, which furthermore are cross-coupled. Evidently the calculation for a three-dimensional body is much more complex than for a two-dimensional one, and the use of a method like that of finite elements with a grid placed over the entire domain is very difficult. But the method of moments permits the treatment of a body immersed in a homogeneous or layered earth using a collection of equivalent dipole sources. The principle will be described below in the section devoted to the Slingram method. One important point must be noted: the variations in E_x, H_z and H_y are much greater than those for E_y, H_x and E_z.

The two techniques, finite elements and the method of moments, permit one to introduce the difference in magnetic susceptibility between the buried structure and its environment as well as the differences in resistivity. Therefore one can verify that for a large class of models, the magnetic properties have a negligible effect on the response obtained. At the frequencies under consideration (around 200 kHz), the responses due to the conductivity have amplitudes which are close both in phase and in quadrature referred to the primary horizontal magnetic field. The component of H_z relative to H_0 in order to have more readily identifiable for the anomalies obtained.

9.3.3. The SGD method

Principle

In this technique the vertical component of the magnetic field is measured. This component only exists in the presence of inhomogeneities. It permits them to be determined easily with satisfactory sensitivity without compensation by subtraction of an average value for the field. It is easy to orient a detector vertically, whereas this is not the case horizontally. The field intensity stability of the transmitters is very good, and therefore it is not necessary to measure the ratio relative to another component. Nonetheless in the theoretical calculation, it is useful to express the anomaly obtained in terms of the ratio between the measured field H_z and the horizontal magnetic field H_0 which would exist in a homogeneous or layered ground in the absence of the anomaly. It has also proven interesting to measure both the in-phase and quadrature component of H_z relative to H_0 in order to have more readily identifiable anomalies, that is, to measure H_0 at a fixed point in the prospecting area in order to obtain a phase reference. In figure 9.3, the anomalies obtained for compact structures, both more and less conductive than the surrounding environment, are shown as well as those for elongated structures with the same contrasts. The in-phase or quadrature anomaly always has an S-shaped form, the corresponding anomaly of the absolute value of the field has two peaks with zero value over the centre of the buried structure. The in-phase anomalies are more readily identified. On the one hand they are simpler, and on the other, as one can determine experimentally in the case where local inhomogeneity is superimposed

Fig. 9.3 Calculated ratios of H_z/H_0 for conductive and resistive compact and elongated structures with various contrasts at 200 kHz. From ref. 44.

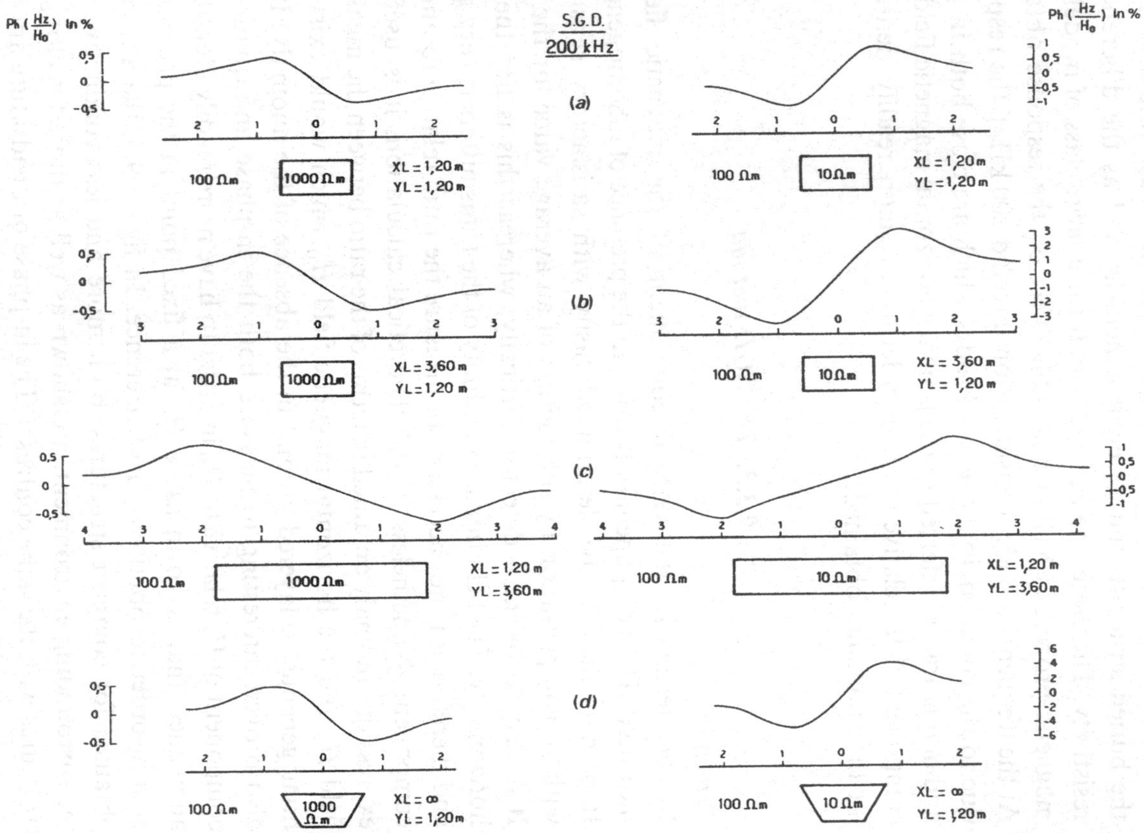

on a slow regional variation, one obtains the varied appearances shown
in figure 9.4 due to the displacement of the zero level.

One important feature of this method lies in the fact that it allows
the determination of the orientation of long structures crossed by a
measurement profile by comparing the amplitude of the anomalies
obtained from several transmitters in different directions. This allows
widely spaced profiles. Let α and β be the angles between the direction
of the buried feature and the apparent directions of the transmitters. If
we decompose the horizontal magnetic fields of each transmitter into
two components, one perpendicular to, and the other parallel to the
structure, only the first will give rise to anomalies in the vertical field
and one has:

$$(\Delta H_z)_1 = k H_{01} \cos(\alpha)$$
$$(\Delta H_z)_2 = k H_{02} \cos(\beta) \tag{9.31}$$

The coefficient k is related to the structure itself and to the surrounding
soil and is the same for the two transmitters, allowing for a small

Fig. 9.4 Different appearance of an anomaly of the absolute value of the
field when a local inhomogeneity is superimposed on a slow regional
variation. From ref. 36.

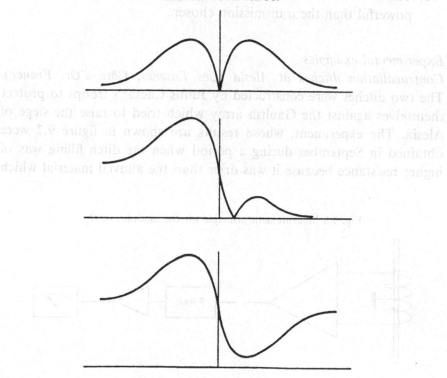

correction factor for the difference in frequency. Therefore one can deduce the value of the ratio $p = \cos\alpha/\cos\beta$, the acute angle $\beta - \alpha = q$ being known, by searching for a null with the horizontal detector, and it follows that:

$$\beta = \tan^{-1}[-\cot(q) + p/\sin(q)] + n\pi \qquad (9.32)$$

Measuring equipment

The equipment needed for this method must be able to measure either the absolute value of H_z, or the ratio H_z/H_0. In the first case, the device is very simple, figure 9.5. It has a vertical detector, filter, and amplifier. In the second case one needs a horizontal and a vertical detector with the same electrical properties, a filtering and synchronous detection system which does not introduce any phase shift between the two channels, and a divider which computes the ratio H_z/H_0, figure 9.6. Two points have to be treated with care in the technical solution:

(1) The detector has to be totally insensitive to the influence of the electric field, and for this purpose one uses two short coils mounted symmetrically.

(2) The filter must eliminate nearby transmissions which may be more powerful than the transmission chosen.

Experimental examples

Contravallation ditches at Alesia (Les Laumes, Côte d'Or, France). The two ditches were constructed by Julius Caesar's troops to protect themselves against the Gaulish army which tried to raise the siege of Alesia. The experiment, whose results are shown in figure 9.7 were obtained in September during a period when the ditch filling was of higher resistance because it was drier than the alluvial material which

Fig. 9.5 Measurement principle for the modulus of H_z.

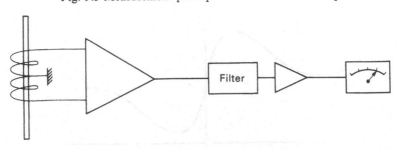

constitutes the Laumes plain. The conventional resistivity prospection shown here was carried out with a Wenner configuration using an electrode spacing of 1 m. The profile *A–B* was measured using the absolute value method on two sources, one at Allouis 163.84 kHz, the other Europe 1 (in the Saarland) at 180 kHz. The ratio of relative anomalies gives a value of $p = 0.587$ and the angle between the two transmitters is $+38°$, with the orientation of the ditches of $-18°$, corresponding well with the direction of the resistivity anomalies measured conventionally.

Gallo–Roman building at Aligny, Nièvre, France. The anomaly obtained for this small structure, *ca* 17×15 m in extent, has two peaks which correspond to the bulk of the construction material. The walls, known through excavation, are not individually detectable, figure 9.8. The appearance of the anomalies is similar for both of the sources utilised. The measurements were made on a double grid of profiles spaced 5 m apart with readings taken every 50 cm.

Levelled barrow at St. Martin sur Nohain, Nièvre, France. The barrow at Les Breuillards appears as a stony mass of 20 m diameter which is more resistant than the surrounding soil. The blackish border which is visible from the air and on the ground suggests the existence of a ditch which is not visible in the apparent resistivity map of the site, figure 9.9. The variations of the vertical field in phase with the horizontal field

Fig. 9.6 Measurement principle for the ratio H_z/H_0.

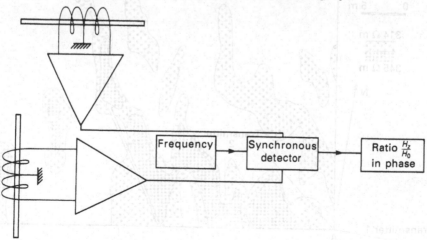

show the association of a maximum and a minimum separated by a line which is approximately oriented toward each transmitter. The anomaly obtained is caused by the mass of the barrow, and it is not possible to see any internal structure.

Gallo–Roman theatre at Entrains, Nièvre, France. This theatre is a semicircle with a diameter of 110 m. The southern and northeastern

Fig. 9.7 SGD experiment on the contravallation ditches at Alesia giving (*a*) the absolute value of two sources at 163.84 and 180.0 kHz and (*b*) measured resistivities. From ref. 36.

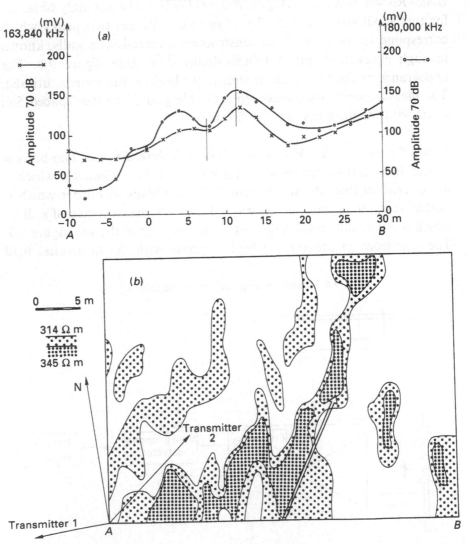

portions are relatively well preserved. SO_2 prospecting was carried out using two series of parallel profiles space 1.10 m apart. The map of the anomalies of $T_1^c T_2^e$ for the in-phase signal is shown in Fig. 9.7(a) and 9.11 for each survey. Figure 9.12 shows the results of the orientation calculation carried out each time one of the profiles displayed an identifiable S-shaped anomaly. The orientation of the wall of the feature is readily visible, with the inclinations of the anomaly varying with direction.

Limits and precautions of the S-B method

The method has two good features: detecting variations in soil resistivity and its good spatial penetration. It has three additional advantages: measuring direct detection of the orientation of...

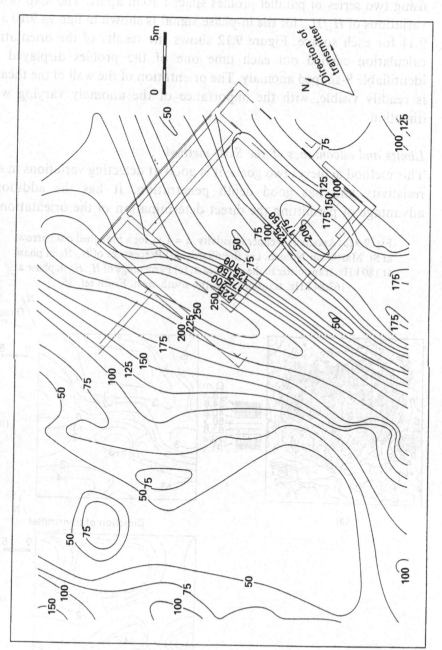

Fig. 9.8 Contoured anomalies with two peaks for a Gallo–Roman building at Aligny. From ref. 36.

portions are relatively well preserved. SGD prospecting was carried out using two series of parallel profiles spaced 10 m apart. The map of the variations of H_z/H_0, for the in-phase signal is shown in figures 9.10 and 9.11 for each source. Figure 9.12 shows the results of the orientation calculation carried out each time one of the profiles displayed an identifiable S-shaped anomaly. The orientation of the wall of the theatre is readily visible, with the importance of the anomaly varying with direction.

Limits and advantages of the SGD method
This method achieves two goals: it is good at detecting variations in soil resistivity and has good depth penetration. It has the additional advantage of permitting the direct determination of the orientation of

Fig. 9.9 (a) Apparent Wenner resistivity, $a = 1$ m, for a ploughed out barrow, at St. Martin sur Nohain, Les Breuillards. (b) Percentage of H_z/H_0 in phase at 180 kHz, transmitter to the north-east. (c) Percentage of H_z/H_0 in phase at 163.84 kHz, transmitter to the south-west. From ref. 36.

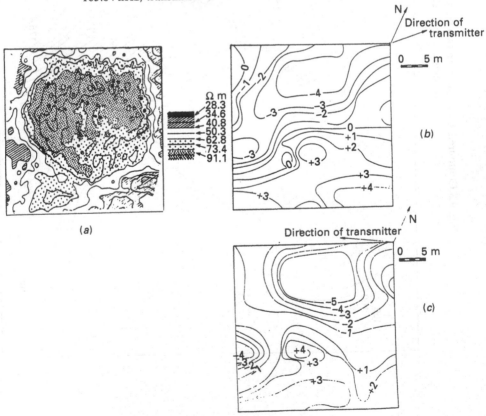

Fig. 9.10 Gallo–Roman theatre at Entrains, in-phase variations of H_z/H_0 in per cent for 180 kHz. From ref. 24.

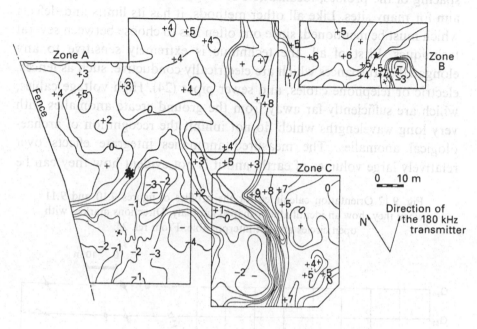

Fig. 9.11 Gallo–Roman theatre at Entrains, in-phase variations of H_z/H_0 for 163.84 kHz.

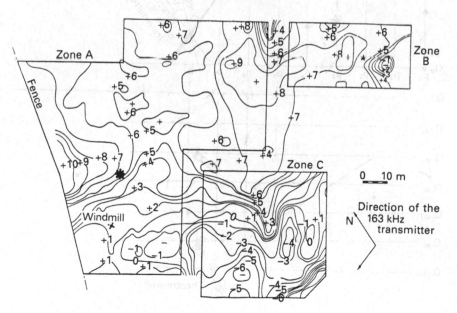

long structures which are cut by a single profile, and hence permits wider spacing of the profiles, localisation of long structures being a principal aim for many sites. Like all other methods, it has its limits and defects which must be mentioned, since one often has to choose between several techniques. First of all, the technique is extremely sensitive to any elongated disturbances which are electrically conductive, such as fences, electric or telephone cables, and sewer pipes (24). High voltage cables, which are sufficiently far away from the ground create anomalies with very long wavelengths which do not inhibit the recognition of archae-ological anomalies. The measured anomalies integrate effects over relatively large volumes of earth, and it is not evident how they can be

Fig. 9.12 Orientation calculated for the profiles of figures 9.10 and 9.11 when they show an identifiable S-shaped anomaly. Directions marked with open circles are not interpretable. From ref. 24.

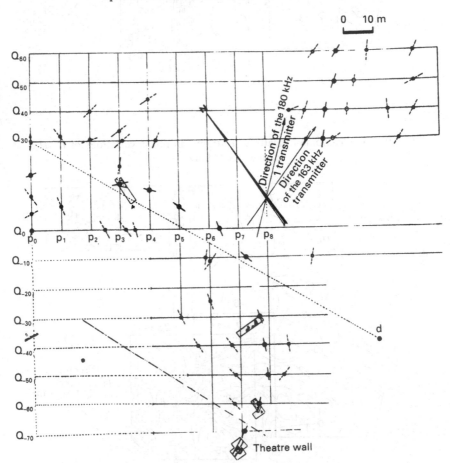

used to detect relatively small structures. They do not give a measure of apparent resistivity of the soil directly, but simply permit the localisation of lateral variations in resistivity. All the experiments carried out up to now have been made point by point. The technique is not suited for continuous measurements since the mechanics of keeping the detector absolutely vertical when moving over a rugged terrain are complex.

9.3.4. The magneto-telluric method

As described above, the existence of the E_x field component allows us to envisage the measurement and mapping of apparent resistivities. Of the three possible kinds of measurement, E_x alone, Z, or W, Z, the magneto-telluric approach appears to be the most interesting. Detectors for the magnetic field components are much better understood than those for the electric field, and the measurement of a magnetic component perpendicular to E permits great tolerance in the orientation of the detector for E_x.

The theoretical calculations which one can carry out on models in order to appreciate the potential of the method show well-defined anomalies centred on the buried bodies, but they vary considerably with the orientation of the E_x field, figures 9.13 and 9.14. Bodies parallel to the magnetic field induce discontinuities in the electric field which produce very marked anomalies. This relationship is not as simple as that in H_z utilised in the SGD technique, and it cannot serve as the basis for the calculation of the orientations of structures. The method is worthy of use if one can solve the problem of a detector for the horizontal electric field. Effectively, one must have a detector whose dimensions are compatible with the spatial variations in the field which one wishes to measure. A detector made up of two small electrodes is realisable, but it necessitates inserting them in the ground as in resistivity prospecting. The solution developed by Guineau (17) for targets which are deeper, utilising two open capacitor plates lying on the ground has not given rise to realisations free of E_z influence when the spacing between the plates is small.

9.4. Utilisation of magnetic dipoles: the Slingram method
9.4.1. The response of a homogeneous soil in the presence of a magnetic dipole

When one places a magnetic dipole on the surface of a homogeneous soil of electrical conductivity σ and magnetic permeability $\mu = \mu_0(1 + \chi)$,

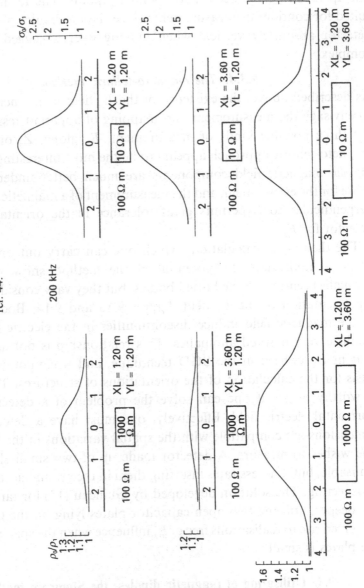

Fig. 9.13 Theoretical calculation of three-dimensional magneto-telluric anomalies with different contrasts, cross-section and length of rectangular cross-section, XL and YL being the length and width of the feature. From ref. 44.

differing from that of a vacuum, one creates induced (Foucault) currents in the ground if the moment of the dipole varies with time, and a magnetisation proportional to the magnetic susceptibility χ. Initial experiments showed that the magnetic properties of the soil are significant, this means that these cannot be neglected as is done habitually in mining geophysics. A mathematical solution to the problem of the response of the ground to a magnetic or electric dipole with vertical or horizontal orientation was proposed by Sommerfeld (32) and it has been frequently used since. It is presented here for the case of a vertical magnetic dipole. It is more complicated for a horizontal magnetic dipole, but the principle remains the same.

Let, as in figure 9.15, a vertical magnetic dipole of moment M and volume v be located above a homogeneous earth. Because of the rotational symmetry of the dipole axis, a cylindrical coordinate system in (r, θ, z) is used with the z axis directed downward, in which the surface of the ground corresponds to the plane $z = 0$ and the dipole is located at height $z = -h$. The rotational symmetry corresponds to $\partial/\partial\theta = 0$ and the coupled equations:

$$\mathbf{V} \times \mathbf{E} = i\omega\mu\mathbf{H}, \qquad \mathbf{V} \times \mathbf{E} = \sigma\mathbf{E} \tag{9.33}$$

permit us to show that there are two independent modes, of which one only, namely (E_θ, H_z, H_r) can be generated by a vertical magnetic dipole. Since $\mathbf{V} \cdot \mathbf{E} = 0$, one may introduce a potential vector $\mathbf{F}$, which satisfies:

$$\mathbf{E} = -\mathbf{V} \times \mathbf{F}, \qquad \mathbf{H} = \frac{1}{i\omega\mu}\mathbf{V} \times \mathbf{V} \times \mathbf{F} \tag{9.34}$$

Fig. 9.14 Theoretical calculation of two-dimensional magneto-telluric anomalies with different contrasts, cross-section and length of trapezoidal cross-section, $\overline{XL}$ being the mean value of the width. From ref. 44.

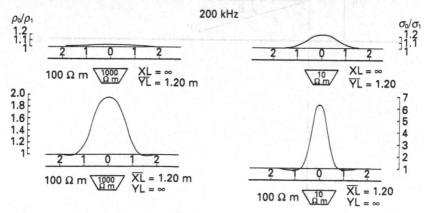

This potential has only one component, F_z, because it is sufficient for the determination of:

$$E_\theta = \frac{\partial F_z}{\partial r}, \qquad H_r = \frac{1}{i\omega\mu}\frac{\partial^2 F_z}{\partial r\,\partial z}, \qquad H_z = \left(-\frac{\partial^2 F_z}{\partial r^2} - \frac{1}{r}\frac{\partial F_z}{\partial r}\right)\frac{1}{i\omega\mu} \tag{9.35}$$

This potential satisfies the equation:

$$\nabla^2 F_z = -i\omega M/v \tag{9.36}$$

in the air and:

$$\nabla^2 F_{z_1} - \gamma^2 F_{z_1} = 0 \tag{9.37}$$

where $\gamma^2 = i\sigma\mu\omega$ in the ground. By introducing the solution in the air corresponding to the primary potential, that of the dipole in a vacuum, F_z can be written in the form:

$$F_z = \frac{i\omega\mu_0 M}{4\pi}\frac{1}{(r^2 + z^2)^{\frac{1}{2}}} + F_{z_s} \tag{9.38}$$

where F_{z_s} is the solution of $\nabla^2 F_{z_s} = 0$.

Using the method of separation of variables, or by means of a Hankel transform, we obtain the solutions of the two equations as:

$$\left.\begin{aligned}
F_{z_s} &= \int_0^\infty \alpha(\lambda)\,\exp(\lambda z)J_0(\lambda r)\,\mathrm{d}\lambda \\[2mm]
F_{z_1} &= \int_0^\infty \beta(\lambda)\,\exp(-uz)J_0(\lambda r)\,\mathrm{d}\lambda
\end{aligned}\right\} \tag{9.39}$$

Fig. 9.15 A vertical loop above a homogeneous soil, coordinate notation.

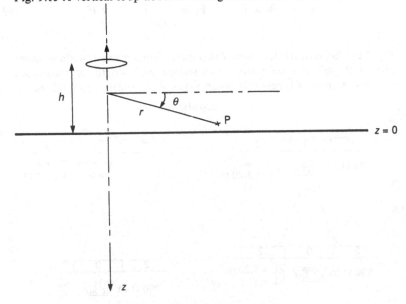

where J_0 is a Bessel function of the first kind and order zero and where $u = (\lambda^2 + \gamma^2)^{\frac{1}{2}}$.

The conditions at the boundary, requiring continuity in E_θ and H_r, allow us to calculate:

$$\alpha(\lambda) = \frac{i\omega\mu_0 M}{4\pi} \exp(-\lambda h)\left(\frac{\lambda/\mu_0 - u/\mu}{\lambda/\mu_0 + u/\mu}\right) \tag{9.40}$$

By introducing the change of variables proposed by Wait (48):

$$g = \lambda p, \qquad B = r/p, \qquad A = (h + |z|)/p \tag{9.41}$$

where:

$$p = (2/\sigma\mu\omega)^{\frac{1}{2}} \tag{9.42}$$

is the depth of penetration and the integrals:

$$T_0(A, B) = \int_0^\infty R(g)g^2 \exp(-gA)J_0(gB)\, dg \tag{9.43}$$

$$T_1(A, B) = \int_0^\infty R(g)g^2 \exp(-gA)J_1(gB)\, dg \tag{9.44}$$

where:

$$R(g) = \frac{g(1 + \chi) - (g^2 + 2i)^{\frac{1}{2}}}{g(1 + \chi) + (g^2 + 2i)^{\frac{1}{2}}} \tag{9.45}$$

one obtains for the secondary magnetic fields in air:

$$H_r = -\frac{M}{4\pi}\frac{1}{p^3} T_1(A, B) \tag{9.46}$$

$$H_z = \frac{M}{4\pi}\frac{1}{p^3} T_0(A, B) \tag{9.47}$$

Calculation of these integrals is only possible numerically, either by direct integration or by transforming the integral into a convolution and calculating the convolution according to Anderson (2). The amplitude and phase of each component depends on the distance of the source dipole, r, the frequency, f, and the properties σ and χ of the soil. In order to have a response which is independent of the value of the magnetic moment, one expresses the measured fields by their ratio to the primary field at the point under consideration. Thus, for example, for a detector with a vertical axis measuring H_z at point r and at height $-h$, the primary field is:

$$H_p = -\frac{M}{4\pi}\frac{1}{r^3} \tag{9.48}$$

and the ratio of the fields is:

$$\frac{H_s}{H_p} = B^3 T_0(A, B) \tag{9.49}$$

this ratio being expressed either in per cent or in parts per million. Each field component is a linear combination of the integrals $T_1(A, B)$, $T_0(A, B)$ defined above and:

$$T_2(A, B) = \int_0^\infty R(g) \exp(-gA) g J_1(gB) \, dg \tag{9.50}$$

9.4.2. The case of a buried structure

To calculate the response obtained in the presence of an inhomogeneity, one has to resolve a problem in three dimensions, because even if the buried body is long, the source is point-like. The method of moments permits us to account for the presence of the inhomogeneity by replacing it with a series of equivalent electric and magnetic sources. The moments of the sources are defined below, following Tabbagh (41).

In a soil where there is no true source and where the properties of the ground are σ and μ, and where there is a structure with properties σ_s and μ_s, we call $\mathbf{E}$ and $\mathbf{H}$ the total fields and $\mathbf{E}_p$ and $\mathbf{H}_p$ the primary fields which exist when $\sigma_s = \sigma$ and $\mu_s = \mu$. These fields therefore satisfy the following equations within the volume of the inhomogeneity:

$$\nabla \times \mathbf{E} = -i\omega\mu_s\mathbf{H}, \qquad \nabla \times \mathbf{H} = \sigma_s\mathbf{E} \tag{9.51}$$

for the total fields and

$$\nabla \times \mathbf{E}_p = -i\omega\mu\mathbf{H}_p, \qquad \nabla \times \mathbf{H} = \sigma\mathbf{E}_p \tag{9.52}$$

for the primary fields. By subtraction one obtains:

$$\nabla \times (\mathbf{E} - \mathbf{E}_p) = -i\omega\mu(\mathbf{H} - \mathbf{H}_p) - i\omega(\mu_s - \mu)\mathbf{H} \tag{9.53}$$

$$\nabla \times (\mathbf{H} - \mathbf{H}_p) = \sigma(\mathbf{E} - \mathbf{E}_p) + (\sigma_s - \sigma)\mathbf{E} \tag{9.54}$$

This shows that the existence of the structure is equivalent to that of magnetic sources of magnetisation $M_s = (\mu_s - \mu)\mathbf{H}$ and electric sources of current density $J_s = (\sigma_s - \sigma)\mathbf{E}$, sources which generate an anomalous field $(\mathbf{E} - \mathbf{E}_p)$ and $(\mathbf{H} - \mathbf{H}_p)$ which adds to the primary field $\mathbf{E}_p$, $\mathbf{H}_p$. Since the intensity of each source depends on the total field at the point considered, which itself depends on the other sources, it is necessary to cut the volume of the structure up into N cells in which it may be assumed that the field is uniform and where the six unknowns (the three components each of the electric and magnetic fields) can be determined.

In writing each of these components as the sum of the primary field and the field created by all the sources, one obtains a system of linear equations in $6 \times N$ unknowns. Knowing the intensity of each source its contribution to the measured field is then calculated. The complete calculation of the effect of an inhomogeneity is difficult because the number of unknowns is large and because the primary field has to be recalculated and the system of linear equations solved again for each position of the source dipole. The case of an object or group of metallic objects can be treated more simply to the extent that the object or group can be made equivalent to a single magnetic dipole with negative moment if the metal is a good conductor and non-magnetic (bronze) and positive with phase shift if the metal is magnetic and a poor conductor (iron). One difficulty is that isolated objects are never isotropic and one does not know the orientation of the moment. This can completely modify the appearance of the anomaly obtained.

9.4.3. *Defining optimal characteristics of a prospecting device*

This definition depends on what one wants to know about the subsoil, taking into consideration all the potentialities of the method. One wishes to have an instrument which permits:

(1) measurement of the electrical conductivity;
(2) measurement of magnetic susceptibility;
(3) detection and identification of metallic objects;
(4) all of these with good penetration depth, good lateral resolution, simple anomaly shape (no change of sign for an anomaly and the centre of the anomaly over the centre of the buried structure)

If the choice is limited to a single transmitting and a single receiver coil and one frequency of operation, the parameters which must be chosen are: the frequency, f, the spacing between the coils, L, the height of the instrument above the ground, h, and the orientation of the coils to each other and to the ground. In order not to restrict the depth of investigation, it is necessary to avoid closely spaced coils. Wide spacing which will reduce lateral resolution and augment apparent anisotropy excessively must also be avoided. An operating frequency of not more than 100 kHz and a coil spacing not greater than 2 m should be chosen. In this case, no matter what coil orientation is chosen, a response proportional to the magnetic susceptibility for the in-phase component independent of frequency, and a quadrature response proportional to the conductivity rising with frequency are obtained. The quadrature response to conductivity neglects magnetic viscosity. Figures 9.16 and 9.17 show the result when using two coils parallel to each other and at an angle of 35° to the vertical, with a spacing of 1.50 m. The 35° angle is one of two possible angles for minimising direct coupling between two coils, the other being 90°. For the range of resistivities between 20 and 2000 Ω m encountered in typical soils and for a spacing of 1.50 m, the in-phase response is practically independent of the conductivity if frequencies less than 10 kHz are used. The quadrature response, on the other hand, gives the conductivity directly. The determination of conductivity from the quadrature response is always simple because the condition of low induction number $i\sigma\mu\omega L^2 < 1$ always exists in archaeological prospecting since one cannot space the coils too widely. The choice of frequency permits simple measurement of susceptibility or not, at will. The adoption of a frequency which is slightly less than 10 kHz has two practical non-negligible advantages. There are no radio transmitters below 10 kHz, and the instrument can be used everywhere without having to

Fig. 9.16 H_s/H_p in parts per million for two coils parallel to each other and at an angle of 35° to the vertical with spacing of 1.50 m, resistivity 100 Ω m, susceptibility $\chi = 30 \times 10^{-5}$ SIu, phase and quadrature response as a function of frequency.

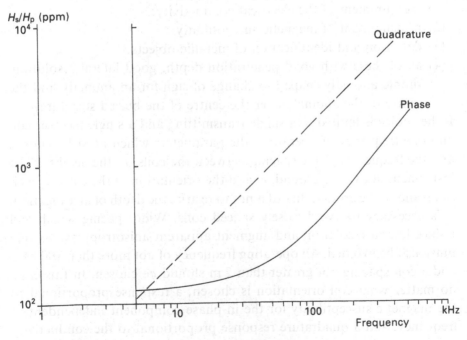

take such transmitters into account. For common values of conductivity and susceptibility, the responses in-phase and in-quadrature are of the same order of magnitude permitting the use of simple electronics for separating them.

For detection of metal objects, the most important parameter is the coil spacing. The detection of a metal object requires identification of the response without ambiguity, a response which is clearly different from that which is produced by a variation in soil susceptibility (38). Thus, an instrument with coil spacing of 50 cm is capable of identifying all objects with a depth of 30 cm, one with spacing of 1.50 m can only identify groups of objects with several kilograms of metal at the order of a metre depth, and an instrument with coil spacing greater than 3 m can only detect very voluminous objects (another Vix cauldron!). The detection of metal objects is of interest in archaeological prospecting

Fig. 9.17 H_s/H_p in parts per million for two coils parallel to each other and at an angle of 35° to the vertical with spacing of 1.50 m, phase and quadrature response at 8 kHz for susceptibilities of (a) 0, (b) 50×10^{-5} and (c) 100×10^{-5} SIu as a function of resistivity in ohm metres. From ref. 42.

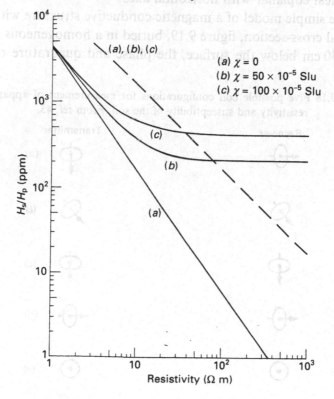

and during an excavation. Conventional metal detectors giving precise locations suffice. Slingram devices with 1.50 m coil spacings which can detect a group of objects at greater depth provide a specialised advantage which is particularly useful.

The choice of coil orientation is a delicate one. Experimental comparison among all the possibilities is long and costly. A theoretical calculation based on a three-dimensional model is necessary in order to determine the appearance of anomalies over buried bodies of simple geometric shapes, and to judge the depth of investigation which can be expected (43). Since it is desired to measure both the apparent conductivity and susceptibility of the soil, the choice is reduced to five configurations which give a response which is sufficiently strong on a homogeneous or layered soil. These five configurations are shown in figure 9.18.

These are coils:

(a) perpendicular to each other, the transmitter axis vertical;
(b) both at 35° to the vertical;
(c) horizontal coplanar, both axes vertical;
(d) vertical coaxial on a horizontal axis;
(e) vertical coplanar with horizontal axes.

For the simple model of a magnetic conductive structure with square horizontal cross-section, figure 9.19, buried in a homogeneous soil with the top 40 cm below the surface, the phase and quadrature responses

Fig. 9.18 Five possible coil configurations for measurement of apparent resistivity and susceptibility of the soil. From ref. 43.

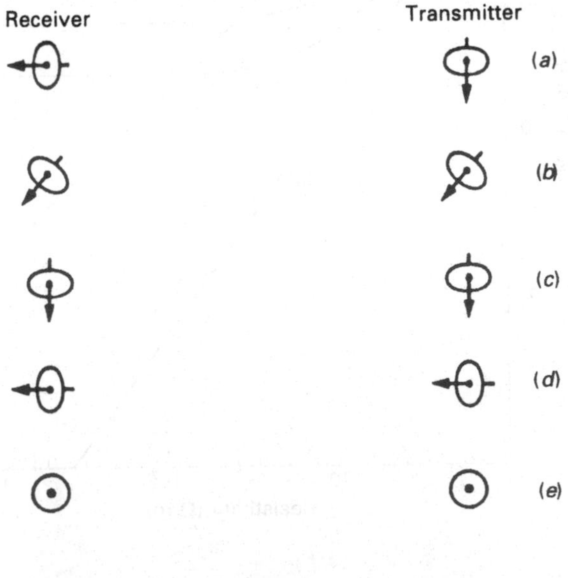

presented in figures 9.20, 9.21, and 9.22 are obtained for spacing of 1.5 m, 8 kHz, and 15 cm height. The anomalies are expressed in terms of the ratio of apparent susceptibility to that of a homogeneous soil for the in-phase component, and the resistivity for the quadrature component. The overall result for the depth of investigation is given in figure 9.23 which shows a decreasing amplitude of the anomaly as a function of the depth to the top of elongated and square structures. The general conclusion is that the most sensitive arrangement is that of (*a*) above, and that structures which are conducting are much more readily detectable than features which are more resistive than the surrounding soil. An analogous asymmetry also exists for methods which use sources at a great distance, and this is one of the weaker points for electromagnetic prospecting at low frequencies as opposed to purely electrical methods. In summary, an arrangement with mutually perpendicular coils, spaced 1.50 m at a frequency slightly less than 10 kHz used at 15 cm above the surface with measurement of in-phase and quadrature signals is optimum.

9.4.4. *Examples of the use of the method*

The theoretically optimum instrument is unavailable at the present time; three other devices have been reported in the literature:
  (1) The SCM of Howell (18), an instrument with coil orientation (*a*) above, with a spacing of 1 m using a frequency of 4 kHz, carried

Fig. 9.19 Coordinate system for a magnetic conductive structure of square horizontal cross-section buried in a homogeneous soil. From ref. 43.

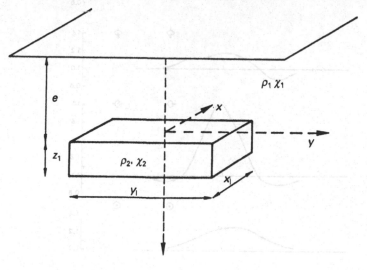

at 3–8 cm above the soil. It is hand carried and does not have spacers which allow precise definition of the working height. It measures the absolute value of the soil response where the in-phase component and hence the magnetic response predominates according to Tite and Mullins (46).

Fig. 9.20 The in-phase response for the conductive body of figure 9.19 for the five coil configurations of figure 9.18 at 15 cm height above the surface for a conductive body and typical susceptibility contrast. From ref. 43.

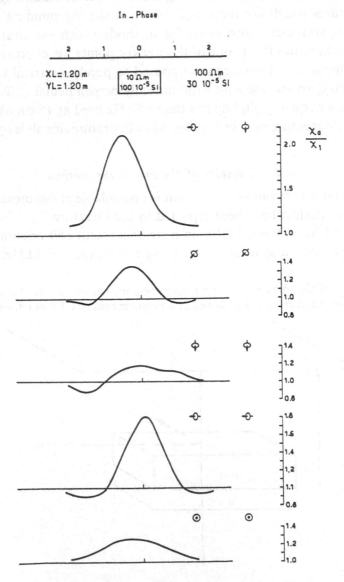

(2) The EM15 made by Geonics Ltd, of Toronto has coils at 35° to
the vertical, spaced 83 cm and a working frequency of 16 kHz. It
measures the in-phase component only, and hence the magnetic
susceptibility (38).

Fig. 9.21 The quadrature response for the body of figure 9.19 for the five
coil configurations of figure 9.18 at 15 cm height above the surface for
a conductive body and typical susceptibility contrast. From ref. 43.

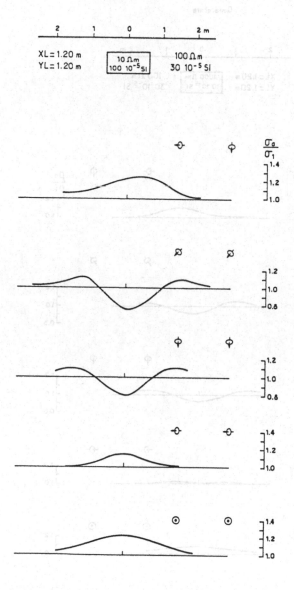

(3) The SH3 constructed at the CRG Garchy for archaeological prospection and specified in ref. 35 for the simultaneous measurement of conductivity and susceptibility. The frequency is 8 kHz, coil spacing 1.50 m and orientation 35°. The theoretical calculations above had not been carried out at the time of its construction. The height utilised is 15 cm.

Fig. 9.22 The quadrature response for the body of figure 9.19 for the five coil configurations of figure 9.18 at 15 cm height above the surface for a resistive body and typical susceptibility contrast. From ref. 43.

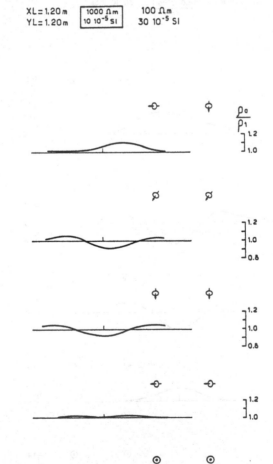

Fig. 9.23 The depth of investigation for the five coil configurations and for various resistivity and susceptibility contrasts. From ref. 43.

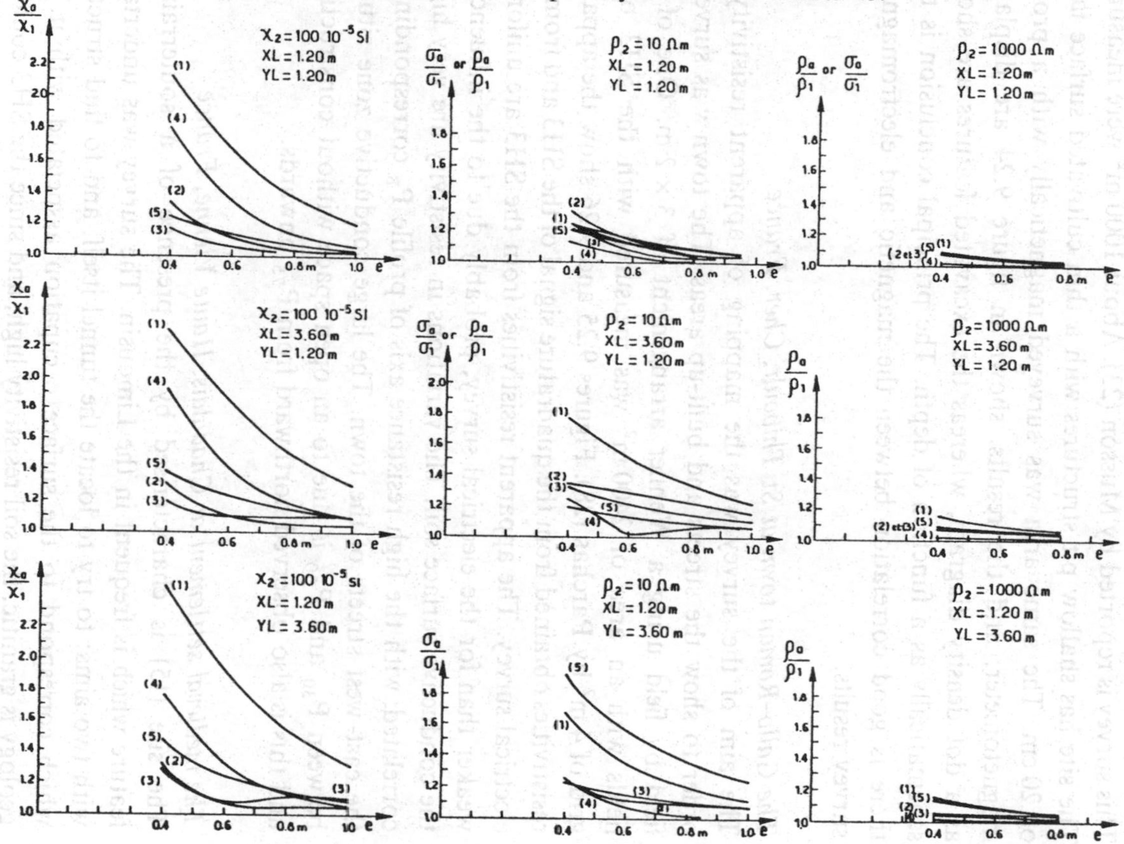

The examples given below were carried out with one or two of these devices.

The Iron Age hill fort at South Cadbury Castle, Somerset, England

This survey is reported by Musson (23). About 1000 m² were measured. The site has shallow pit structures with a thin cultivated surface layer of 20 cm. The same area was surveyed magnetically with a proton magnetometer, and the results, shown in figure 9.24 are displayed as a dot density diagram, whereas the excavated features are shown schematically as a function of depth. The principal conclusion is that there is good correlation between the magnetic and electromagnetic survey results.

The Gallo–Roman town at St Thibault, Cher, France

The aim of the survey was the mapping of apparent resistivity in order to show the streets and built-up areas. The town was surveyed field by field using a Wenner arrangement of 3×2 m. One of the fields with an area of 2400 m² was measured with the SH3 on a grid of 4 m² by Parchas (28). Figures 9.25 and 9.26 show the apparent resistivities obtained from the quadrature signal of the SH3 and from the electrical survey. The apparent resistivities from the SH3 are uniformly weaker than for the electrical survey, probably due to the influence of the conductive surface soil. The variations in resistivity are very highly correlated, with the high resistance axis of profile P_8 corresponding to the east–west streets of the town. The large conductive zone situated between P_{30} and P_{60} is due to an open space without constructions, and this is also observed northward from P_{75} onwards.

The medieval settlement at Chadalais, Haute Vienne, France

The site (15) is characterised by the presence of a souterrain, a feature which is frequent in the Limousin. The survey was undertaken with two aims: to try to locate the tunnel itself, and to find structures which correspond to the surface occupation associated with it. The geology is granitic, the soil resistivity high, and since the SH3 could not furnish a correct measurement of such high values, it was necessary to use normal resistivity prospecting to try to detect the cavities of the tunnel and SH3 to find the surface structures which have a more magnetic fill, such as pits, silos, etc. The results obtained are shown in figure 9.27. The resistivity map shows conductive zones with resistance less than

1300 Ω m which correspond to clay veins and weathering in the granite. These conductive structures mask the resistive structure of the souterrain. The map of magnetic susceptibility (SH3 in-phase component) shows pockets of higher values which on excavation proved to be archaeological structures: ditches, rubbish pits or silos.

Fig. 9.24 Proton magnetometer and SCM surveys at South Cadbury Castle, with excavated features. From ref. 23.

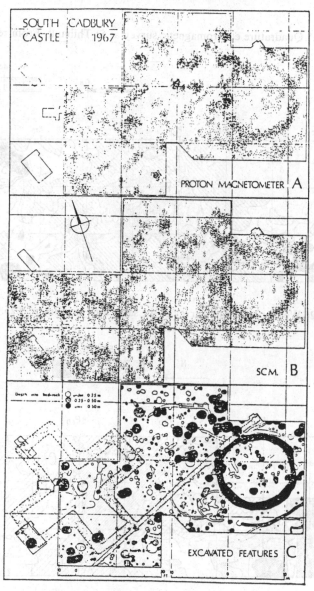

The Gallo–Roman pottery workshop at Jaulges-Villiers-Vineux, Yonne, France

This workshop has long been known for its abundant production of luxury pottery, Terra Sigillata, and Varnished Ware. Usually a site of this type is surveyed magnetically, but its size, more than 10 ha, would have made this very time consuming, and the archaeologists merely wanted to know its extent and rough organisation, rather than the location of every kiln. Therefore a survey on a wide grid (50 m^2) was

Fig. 9.25 Quadrature electromagnetic survey at St Thibault. From ref. 28.

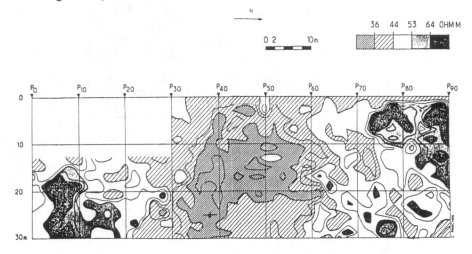

Fig. 9.26 Wenner resistivity survey of the same area as in figure 9.25. From ref. 28.

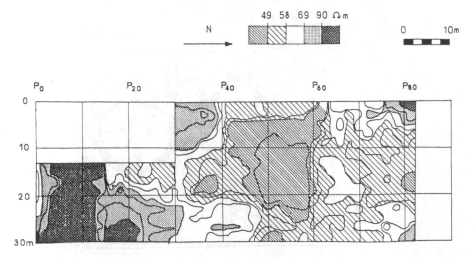

carried out with the EM15 over an area of 9 ha after having treated a
1 ha area using various methods. In a survey with widely spaced
measurements, the sampling interval does not fulfil the conditions of
Shannon's Theorem and there is a risk of aliasing. This imposes two
constraints: (1) The quantity measured must be significant in itself, that
is, it can be expressed as an apparent property of the soil. This excludes
normal magnetic prospecting where an isolated measurement has no
significance. (2) There must be a technique for checking afterwards to
see if aliasing has taken place or not. This can be done by using a
variogram $\gamma(h)$, a graph showing the mean standard deviation between
two distant measurements in h permitting determination of the limiting
value of $\gamma(h)$ when h tends to zero. If $\gamma(h)$ also tends to zero, there is no
aliasing, whereas if $\gamma(h)$ tends to a non-zero limit, there is a so-called
'nugget' effect which can correspond either to aliasing or to random
errors in the measurements. Figure 9.28 shows the variogram $\gamma(h)$ and the
covariance $K(h)$ of the measurements made with the EM15. There is
nearly no nugget effect. The map of apparent magnetic susceptibility is
shown in figure 9.29. Because of the small spacing between the coils of
the EM15, the measurements only show the upper 30 cm of the ground,
that is, they are limited mainly to the cultivation layer. A comparison
carried out on the test zone to the southwest of the site shows that the

Fig. 9.27 (a) Wenner resistivities $a = 2$ m at Chadalais Les Abouts. (b)
Apparent magnetic susceptibilities (in-phase response of the SH3). From
ref. 15.

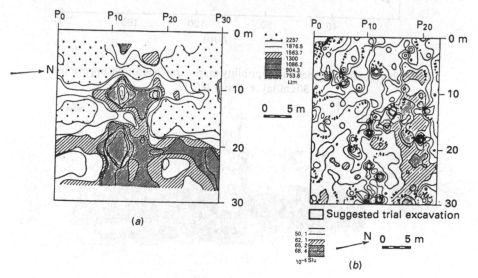

susceptibility of the cultivation layer is representation of the subsurface soil as well. The site extends in a northwest–southwest direction along the Auxerre–Autun highway, and the 60×10^{-5} SIu contour can be interpreted as the limit of the site, with the 120×10^{-5} SIu contour showing the production units which associate kilns and waste pits.

The Bronze Age site at Marchesieux, Manche, France

In 1961, some late Bronze Age socketed axes were discovered in the St Clair marsh at Marchesieux during clearing of drainage ditches. The

Fig. 9.28 Variogram and covariance of measurements made with an EM15 at Jaulges over a 50 m² grid. From ref. 42.

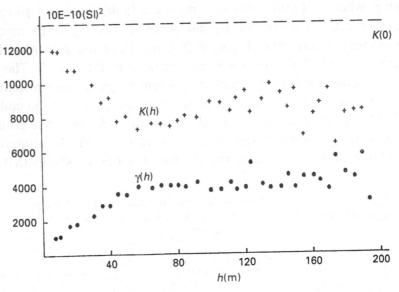

Fig. 9.29 Apparent magnetic susceptibility map at Jaulges for the uppermost 30 cm layer. From ref. 42.

exceptional drought of 1976 which dried out the ditches completely permitted the excavation of two hoards D_I and D_{II} (47). The question which arises at all such sites where hoards are found was posed: are there more hoards or archaeological structures associated with those already excavated which would aid in understanding their significance? Actually, Bronze Age hoards are quite common in Europe, with over 1200 known in France alone, but those which have been excavated *in situ* are rare, and their purpose is unknown.

Therefore, an electromagnetic survey was undertaken at Marchesieux using the EM15 and the SH3 with the aim of localising other hoards and any archaeological structure which might exist in their vicinity. The survey covered approximately 2.5 ha and resulted in the discovery of six new hoards, D_A–D_F in figure 9.30, but no archaeological structure corresponding to permanent occupation in the neighbourhood of the hoards was found. At this state there were eight hoards spread out over 80 m, located along a line which perhaps corresponds to an old trackway through the bog, distributed in groups of 2–4–2. The magnetic susceptibility was not mapped because this is, in general, very weak. The peat of the bog and the clay or silt below it have susceptibilities less than 10×10^5 SIu. At three points, shown in black in figure 9.30, the susceptibilities were very high, and they corresponded to a soil with a yellow-orange colour, attributed to the remains of fires. The excavation, the position of these points at the limits of contemporary fields, and the testimony of the farmers proved that these traces of fire correspond to contemporary weed burning. This is very important because it shows that fire leaves very strong traces in this soil. We can thus state that there was no permanent habitation there in the Bronze Age, since no other trace of this kind was found.

The map of apparent resistivity is shown in figure 9.31 as well as the results of nine associated electrical soundings. The limit of the bog appears well in the 29 Ω m contour, the loess of the neighbouring plateau having a resistivity slightly above 30 Ω m and the peat a resistivity between 20 and 25 Ω m. The more conductive zone is due to the presence of a layer of marine clay, with 12 Ω m, located on the loess and below the peat. No anomaly suggests archaeological structures with transported materials which would be much more resistant than those naturally present. The most probable hypothesis is that the hoards were put into the bog far from any human occupation. The results agree with those obtained at another hoard site (39).

9.4.5. *Interest and limits of the Slingram method*

Compared with the electromagnetic methods previously described, and with conventional magnetic and electrical prospecting, the Slingram technique has two unique properties: the simultaneous measurement of conductivity and magnetic susceptibility, and the capability to detect metallic masses made of bronze. It corresponds to the objective pursued at the beginning of research into electromagnetic methods, namely the measurement of resistivity without having to insert electrodes. But this does not imply that it can substitute completely for the other methods, and comparison with them must be made. The choice of criteria for such a comparison is very open, and it is useful to describe the limitations of

Fig. 9.30 Hoards discovered at Marchesieux (Marais de St Clair) using the EM15 and the SH3. From ref. 39.

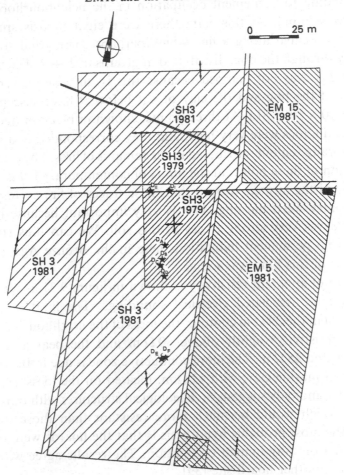

each technique which arise from the physical principles involved, rather than from the technical details of a particular instrument which may be improved in the future. Several publications (40,44) have developed this subject, and the main conclusions are presented here.

Fig. 9.31 Apparent resistivities at Marchesieux (quadrature response) and nine electrical soundings. From ref. 39.

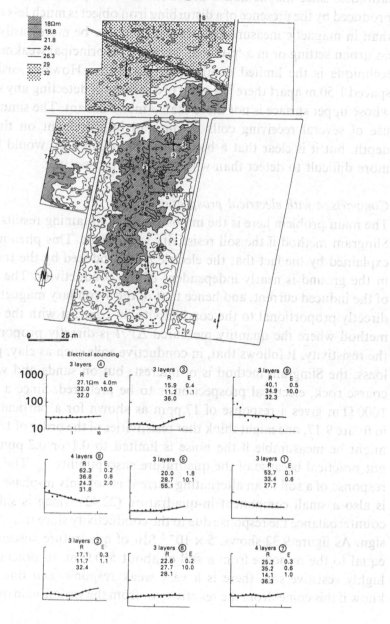

Comparison with magnetic prospection

The principal advantage of the Slingram method is that it measures the magnetic susceptibility of the soil directly. In a layered earth it allows one to follow slow variations in thickness and to observe lenticular structures. It also permits measurements on a widely spaced grid. There are two minor additional advantages: there are no problems at low latitudes; since the method is active rather than passive, the blind spot produced by the presence of a disturbing iron object is much less extensive than in magnetic measurement. It can therefore be more easily used in an urban setting or in a 'polluted' milieu. The principal weakness of the technique is the limited depth of investigation. However, using coils spaced 1.50 m apart there is a good probability of detecting any structure whose upper surface is not more than a metre distant. The simultaneous use of several receiving coils permits an improvement on the search depth, but it is clear that a body more than 2 m deep would be much more difficult to detect than with magnetic prospecting.

Comparison with electrical prospection

The main problem here is the impossibility of obtaining results with the Slingram method if the soil resistivity is too high. This phenomenon is explained by the fact that the electric field produced by the transmitter in the ground is nearly independent of the conductivity. The intensity of the induced current, and hence that of the secondary magnetic field is directly proportional to the conductivity, in contrast with the electrical method where the quantity measured $\Delta V/I$ is directly proportional to the resistivity. It follows that, in conductive soils such as clay, peat, and loess, the Slingram method is of interest, but on sands and weathered coarse rock, electrical prospection is to be preferred. Since a soil with 1000 Ω m gives a response of 17 ppm as shown for a particular device in figure 9.17, one might think that resistivities of the order of 10 000 Ω m might be measurable if the noise is limited to 0.1 or 0.2 ppm. This is not practical because of the quadrature susceptibility χ_q. The magnetic response of a soil to an alternating current is not only in-phase but there is also a small component in-quadrature (22,42) which is sufficient to counterbalance the response due to the conductivity since it is of opposite sign. As figure 9.32 shows, 5×10^{-5} SIu of quadrature susceptibility is equal to the response from a soil of about 500 Ω m. In practice, over a highly resistive soil, there is a very weak response and one does not know if this comes from the resistivity or from the variation in quadrature susceptibility. A solution exists in principle for separating the two responses. It consists in utilising two frequencies, since the response in conductivity is directly proportional to frequency and that of the quadrature susceptibility independent of it. One may conclude that the difference in response at the two frequencies is due entirely to the conductivity. The second weakness of the Slingram method is shared with other electromagnetic techniques whereby a conductive target is much easier to detect than a resistive one. In archaeological prospecting it is precisely the resistive anomalies due to a stone construction which are often sought.

Possibilities for improvement of the Slingram technique

One possibility is the utilisation of two or more frequencies in order to improve performance on resistive soils. Another consists in using several receiver or transmitter coils. There are two ways of doing this:

(1) Favouring the response from deep structures as compared with superficial structures by taking into account the differences between the fields measured at two receiving coils.

(2) Using the sounding concept, which utilises the variation in response with the spacing of the coils or the joint use of sounding and

Fig. 9.32 The quadrature susceptibility which counterbalances resistivity response posing a practical limit to electromagnetic measurement of high resistivities for single frequency measurements. From ref. 42.

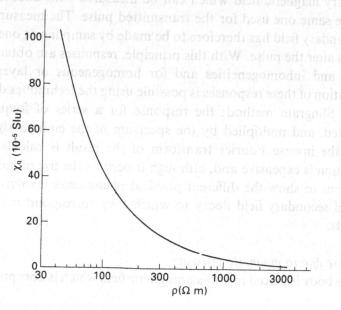

profiling to study both depth and horizontal changes in soil properties.

The Slingram method does not lend itself to a sounding approach using a variable frequency if one wishes to remain in the range of low frequencies. This is true even though at 1 MHz, the depth of penetration is still relatively good. But it does lend itself perfectly to an approach whereby the transmitter and receiver are progressively separated, at least for the quadrature response. The in-phase response poses mechanical problems for this kind of operation. The use of a network of receivers might be envisaged, as is done in the measurement of potential in electrical prospecting.

9.5. Emission of pulse signals

9.5.1. *Measurement principles*

In all of the methods discussed up to now, the signal delivered by the source is sinusoidal and the transmission continuous. A well-defined working frequency is used. However, there is another way to induce Foucault currents in a conductor which consists of transmitting a transient and working in the time domain. Figure 9.33 shows an abrupt step change in current in the transmitting coil. This change carries with it an important change in the magnetic field in the conductor and the creation of a counter voltage and current. This, in turn creates a secondary magnetic field which can be measured with another coil or with the same one used for the transmitted pulse. The measurement of the secondary field has therefore to be made by sampling at one or more instants after the pulse. With this principle, responses are obtained from objects and inhomogeneities and for homogeneous or layered soils. Calculation of these responses is possible using the techniques developed for the Slingram method: the response for a series of frequencies is computed, and multiplied by the spectrum of the emitted signal, and finally the inverse Fourier transform of the result is calculated. This calculation is expensive and, although it permits the use of models, it is important to show the different physical phenomena involved and the types of secondary field decay to which they correspond with simpler methods.

Response due to magnetic viscosity
When a body is placed in a weak magnetic field which is abruptly reduced

to zero, the induced magnetisation vanishes suddenly, but a residual viscous remanent magnetisation remains. This does not disappear quickly, rather it decays regularly with log(t) after the field is removed. In an electromagnetic impulse system, the disappearance of the induction corresponds to a variation in flux which is the derivative of the falling current in the transmitter, which goes quickly to zero. The viscous magnetisation gives a variation of flux as $1/t$ at the receiver. Experiments using this method (8,9) show that this latter phenomenon is of considerable importance in instruments used in archaeology. This reasoning is valid for an object, a structure, or the subsoil itself. It is correct only for an object in a vacuum because it does not consider the influence of the other soil properties, σ and μ, on the diffusion of the signals from the transmitter and the buried object.

Response of a conductive object
If one equates all conductive objects to a resistive–inductive circuit, the response of such a circuit placed in a vacuum will produce a field which

Fig. 9.33 A step function of the current in a transmitting coil, the primary field, and the secondary field which results from the presence of a buried conductive object. From ref. 34.

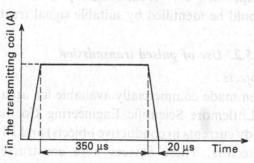

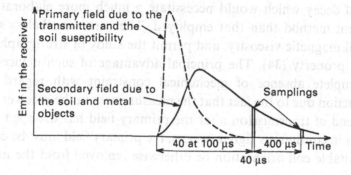

decays as $\exp(-\alpha t)$ with $\alpha = R/L$, (34) and the more conductive the object, the slower the decay. The variation of the response with the depth z of the object is governed by the function $r^4/(z^2 + r^2)^3$ where r is the radius of the transmitting and receiving coil. This has a great influence on the depth of investigation, but it is independent of the type of emitted signal and is the same for a continuous sinusoidal signal.

Response of a homogeneous earth

The response v of a homogeneous earth to a single turn coil used as the receiver and transmitter and driven by an impulse has been studied by several authors (10,26). Expressed in terms of the variation of flux after the abrupt cessation of current as a Heaviside function, it depends on the time and on the properties of the soil as given by the expression:

$$v = (\mu\sigma)^{\frac{3}{2}}/t^{\frac{5}{2}} \tag{9.55}$$

where μ is considered to be a property independent of time, the viscosity being ignored. It depends on the radius of the coil as r^4 (r^2 for transmission, r^2 for reception), and is considerably reinforced by any augmentation of the radius. As seen above, the response of an object drops off with r. Therefore there are different types of signal decay at the receiver: t^{-1}, $\exp(-\alpha t)$, $t^{-5/2}$, each having a different cause. Theoretically these could be identified by suitable signal treatment.

9.5.2. Use of pulsed transmission

Detection of metal objects

Instruments have been made commercially available for archaeological prospecting by the Littlemore Scientific Engineering Co: namely the DECCO (decay of eddy currents in conductive objects) and PIM (pulsed induction metal) detectors. These devices were constructed for the detection of metal objects. They do not permit separation of the different types of decay which would necessitate a much more elaborate signal treatment method than that employed. They are nonetheless sensitive to a soil magnetic viscosity, and permit the study of stratigraphy based on this property (34). The principal advantage of such devices lies in the complete absence of mechanical constraint with regard to coil construction due to the fact that the measurement in the receiver is made at the end of transmission after the primary field has dropped to zero, whereas in continuous signal devices, the primary field must be cancelled by a suitable coil orientation or otherwise removed from the measured field. It is therefore relatively simple in impulse instruments to use coils of a size which are adapted to the kind of analysis desired. One can readily change coils during the measurement in order to improve the localisation of an object and vary the relative positions of the receiver and transmitter at will. Experimental results are reported below.

Possibilities for use apart from metal detection
The ease with which it is possible to modify the coil system permits the use of these devices for the measurement of soil magnetic viscosity in the laboratory (16,22), but their utilisation in the field for this purpose has been restrained by the limited depth of investigation with available coil systems (45). To measure conductivity and viscosity with a pulsed system, two possibilities exist: using a Slingram technique with sufficiently well-separated dipoles or using transmitting/receiving coils of larger radius. The first solution has little chance of giving better results than that obtained with the Slingram method where the separation of in-phase and quadrature signals permits a simultaneous measurement of conductivity and susceptibility in a much simpler manner than the same measurement made in a pulse system through identification of different types of decay. The second solution has not been tried up to now in archaeological prospecting. There ought to be no problem, inasmuch as the systems developed for mining prospecting have more than sufficient power and do not have any restrictions on mechanical rigidity, as is the case when continuous signals are employed.

Two lines of research can be imagined:

(1) The use of a transmitter/receiver of modest size, say 1 to 2 m in diameter which is moved over the surface to be investigated and which would permit the measurement of the conductivity and the apparent viscosity of the ground.

(2) The use of a very large loop as a transmitter surrounding the entire area to be investigated, which would give a nearly uniform vertical field. For a uniform primary vertical field, the response obtained on a homogeneous earth is zero for all components because the hypotheses $\partial/\partial x = 0$ and $\partial/\partial y = 0$ imply the fact from the equations:

$$\mathbf{V} \times \mathbf{E} = -\mu \, \partial\mathbf{H}/\partial t, \qquad \mathbf{V} \times \mathbf{H} = \sigma\mathbf{E} \qquad (9.56)$$

that the secondary vertical field is also zero. Therefore it is not possible with this arrangement to determine the conductivity or apparent viscosity of a layered medium, but the presence of an inhomogeneity could be detected with great sensitivity and depth. The characteristics of the SGD method are approached but it is not possible to determine the orientation of long structures as can be done by methods using horizontal primary fields.

9.6. Detection of metal objects

9.6.1. General principles

There are many metal detectors, and their legal or illegal use for archaeological aims is very significant. At the present state of the art, they are all electromagnetic devices working at low frequencies which exploit the electrical conductivity via induced currents or the magnetic permeability of metals. They all use coils for emission and reception and thus belong to the Slingram class of devices, most of them being of continuous transmission types. The numerous classifications which one may find in the literature produced by the manufacturers are based on the electronic solutions adopted for the identification of the response of an object and therefore a function of the cost of the device. We are interested here in three general principles which may allow an improvement in their operation.

The first of these lies in the focusing of the field. This may be explained as follows: instead of using a single transmitting coil, several are used such that their additive or subtractive effects limit the volume of the significant primary field. An improvement in lateral resolution is sought together with an increased depth of penetration. This is the same technique which is used in electromagnetic well logging or in the Slingram method when using several transmitting or receiving coils. It can be used for pulsed transmission as well as for a continuous signal.

The second one is one of discrimination: it consists in trying to identify the metal by the phase of the response. At very low frequencies, and for an isotropic object, iron gives a response displaced in phase by $0°$ compared with the primary field, whereas a good metallic conductor such as bronze gives a response dephased by $180°$ by Lenz's law. By raising the frequency, and for metals which are poor conductors, a quadrature response appears. Nevertheless, it must be emphasised that the discrimination between magnetic and conductive bodies assumes that the object is isotropic. This is never the case and the sign of the response may be exactly the contrary to that expected for objects of complex shape such as horseshoes or those with large plane surfaces.

Discrimination is currently employed for continuous signal devices, but more advanced pulse instruments may do so in the future.

The third principle consists in annulling the response due to the physical properties of the soil itself in order to enhance the anomalies from metal objects. This can only be carried out once, and since the soil response varies during the course of a line of measurements, the compensation cannot be perfect.

9.6.2. Detection of metal objects and archaeological prospection

Metal detectors have been used on excavations for a long time in order to separate the corroded remains of an object from the surrounding soil. This prevents the loss of valuable information for possible restoration. They also aid in finding small objects during excavation or in the sieves used on spoil heaps. These uses will not be discussed here.

The use of metal detectors in archaeological prospecting raises several methodological problems: Why look for metal objects? Is their frequency on the site greater than outside its limits? What specific information can such objects give us with regard to the date or type of occupation? Do we have to look for objects which have migrated to the topsoil layer or for objects *in situ* and hence more difficult to detect? The answers to all of these questions lead to a long debate and some examples are discussed below. The first objective is a search for things *in situ*, either because these objects carry an important amount of archaeological information, as is the case for Bronze Age hoards (Marchesieux above), or because the metal objects permit the characterisation and mapping of important structures. This is the case presented below when searching for nails from a 'murus gallicus'. The second aim is to find all the objects in the topsoil and therefore to gain as much information as possible about the remains before destruction through ploughing. The third goal is to deal with metal objects in much the same way as one deals with fragments of pottery or flint in surface scatters, that is, to utilise them for statistical estimates of the position of a site or its internal structures, and perhaps to gain dating information.

Site at Reuilly, Cher, France

In a field aptly named the 'Field of Treasure', a bronze hoard was found accidentally through ploughing in 1967. Archaeologists were able to recover all of the objects and to excavate a small area of 25 m² around the point of discovery. No archaeological structure accompanied the hoard. The field name suggests that this was not the first hoard discovered here and a campaign was undertaken in 1971 using a DECCO 684 over a surface of 3200 m². The EM15 was used afterwards to see if there were any non-magnetic objects. No bronze objects were found, but the map obtained, figure 9.34, shows the number of important metal responses and their surface distribution. The density is nearly one object for every 2 m². It is difficult to explain the fluctuations over the surface, which may reflect an older field system or preferential zones for depositing organic wastes used earlier as fertiliser.

Site of Norville, Seine-Maritime, France

Four heeled axes of the middle Bronze Age were found here by chance during the winter of 1981–2. From the approximate positions given by the discoverers, it was possible to carry out a survey with a metal detector in order to locate more axes as well as with the SH3 to locate other possible hoards. Twelve other axes were found with the detector, eleven in one small area, the twelfth at the entry to the field. In figure 9.35 only the axes are marked. Excavation showed that nothing was left *in situ* below the cultivated topsoil. The hoard was entirely within the topsoil,

Fig. 9.34 Surface distribution of metal responses at Reuilly, Champ de Trésor. From ref. 34.

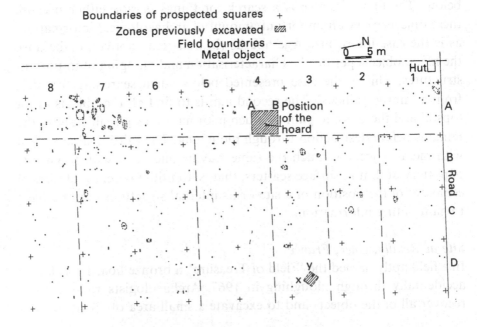

and the centre of the area where the eleven axes were found can be considered to be the original position of the hoard with an uncertainty of the order of a metre.

Sutton Hoo, Suffolk, England

The cemetery of Sutton Hoo, widely known through the discovery of a funeral boat of one of the first Anglo-Saxon kings in 1938, was the subject of an intense prospecting campaign carried out by a group from the University of Birmingham (6). Among the methods used was a complete coverage of the site with a metal detector in order to locate important groups as well as isolated objects. The figure 9.36 shows the result of this survey by Royle and Copp. Metal objects are very abundant and the interpretation is difficult inasmuch as the area was used as a training camp during the Second World War. The prewar excavation most certainly resulted in a lot of metal objects being dropped on the

Fig. 9.35 Survey area, electromagnetic anomalies (open circles) and middle Bronze Age axes (closed circles) at Norville. From ref. 39.

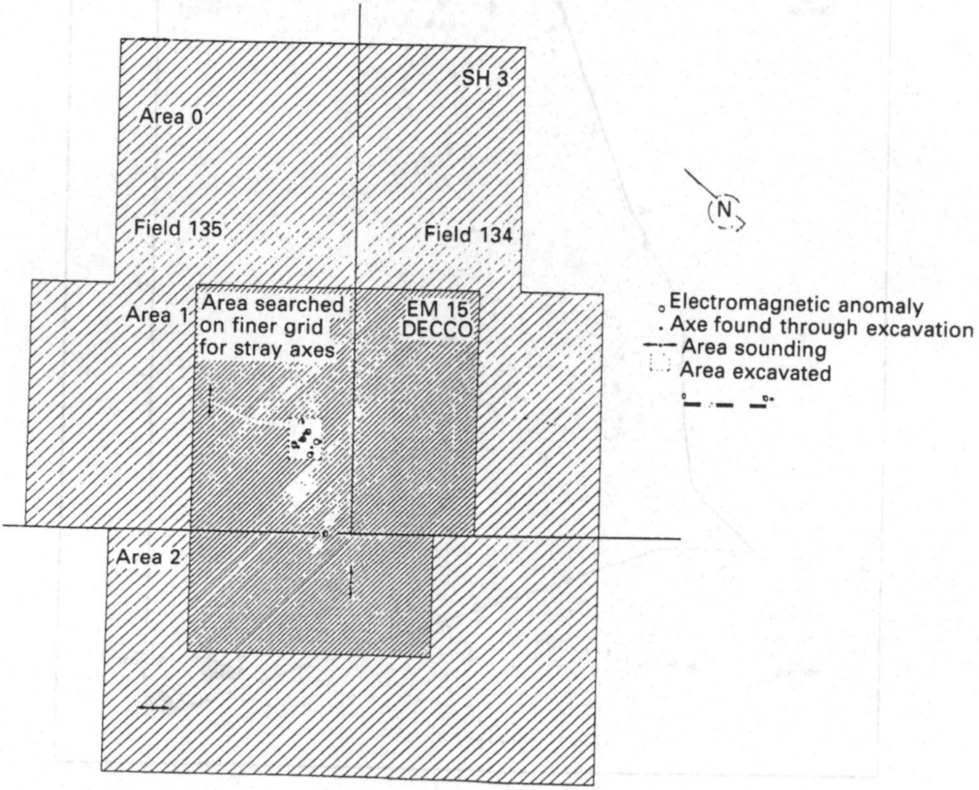

site. The main interest in this survey lies in the quantity of information obtained for one of the most difficult cases of the systematic use of metal detectors and in the fact that seven different methods were used in anticipation of future test excavations.

Oppidum of Murcens, Lot, France
This hill fort which was occupied at the end of the Iron Age and which probably played an important role in the Gallic wars, is surrounded by

Fig. 9.36 Sutton Hoo metal detector survey. From ref. 6.

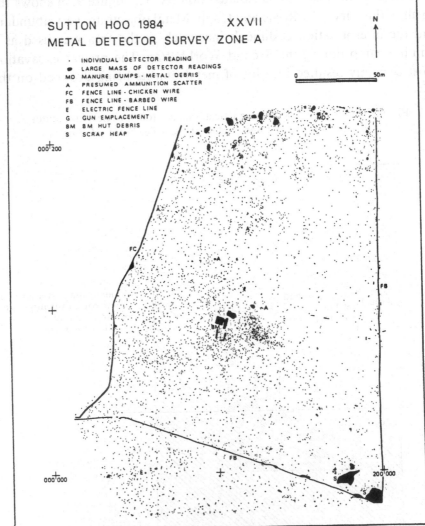

a 'murus gallicus'. The wooden beams of the wall were joined with 30 cm long iron nails. These nails are sufficiently large to be detected *in situ*, and their anomalies are large enough so that no confusion with surface iron fragments is possible. Hesse carried out a survey to map the course of the wall by identification of the nails. The result is shown in figure 9.37.

9.7. Soil radar

No low frequency method satisfactorily solves the sounding problem, a vertical description of the variations of the properties of the subsurface soil at a given point. The Slingram and electrical methods permit a geometric sounding by spacing the coils or electrodes, but this method is slow and implies a considerable increase in lateral effects. No frequency dependent sounding is possible because the conductivity and permittivity vary. That is why the idea of utilising the propagation of a high frequency

Fig. 9.37 The 30 cm iron nails found in cutting at the oppidum of Murcens.

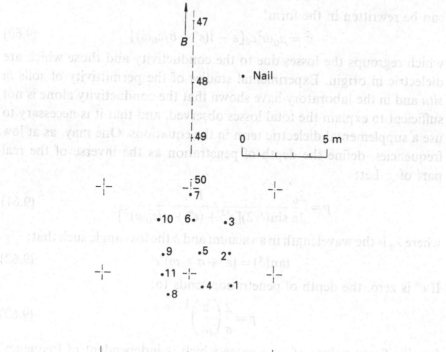

signal downward and observing its reflection at various interfaces is *a priori* seductive. It is similar to the seismic reflection technique, although much more rapid, since one point needs only 100 ns, and nothing impedes the movement of an antenna over the soil surface. One difficulty arises, however, namely the strong absorption of a high frequency signal in the ground.

9.7.1. The dielectric permittivity of soils and the attenuation factor

The behaviour of the electric field is governed by the equation:

$$\nabla^2 \mathbf{E} - \sigma\mu \, \partial\mathbf{E}/\partial t - \varepsilon\mu \, \partial^2\mathbf{E}/\partial t^2 = 0 \tag{9.57}$$

For harmonic variation in time this equation can be written:

$$\nabla^2 \mathbf{E} - (i\sigma\mu\omega - \varepsilon\mu\omega^2)\mathbf{E} = 0 \tag{9.58}$$

and $\sigma < \varepsilon\omega$ prevails. One may not, however, neglect σ relative to $\varepsilon\omega$ inasmuch as σ causes damping even if this occurs at distances greater than a wavelength. The propagation factor γ such that:

$$\gamma^2 = i\sigma\mu\omega - \varepsilon\mu\omega^2 \tag{9.59}$$

can be rewritten in the form:

$$\gamma^2 = \mu_0\omega^2\varepsilon_0[\varepsilon' - i(\varepsilon'' + \sigma/\varepsilon_0\omega)] \tag{9.60}$$

which regroups the losses due to the conductivity and those which are dielectric in origin. Experimental studies of the permittivity of soils *in situ* and in the laboratory have shown that the conductivity alone is not sufficient to explain the total losses observed, and that it is necessary to use a supplemental dielectric term in the equations. One may, as at low frequencies, define the depth of penetration as the inverse of the real part of γ. Let:

$$p = \frac{\lambda_0}{2\pi} \frac{1}{\sin(\delta/2)[\varepsilon'^2 + (\varepsilon'' + \sigma/\varepsilon_0\omega)^2]^{\frac{1}{4}}} \tag{9.61}$$

where λ_0 is the wavelength in a vacuum and δ the loss angle such that:

$$\tan(\delta) = (\varepsilon'' + \sigma/\varepsilon_0\omega)/\varepsilon' \tag{9.62}$$

If ε'' is zero, the depth of penetration tends to:

$$p = \frac{2}{\sigma}\left(\frac{\varepsilon}{\mu_0}\right)^{\frac{1}{2}} \tag{9.63}$$

for significant values of ω, a value which is independent of frequency. This relationship is not confirmed experimentally. Figure 9.38 shows the variation in the depth of penetration against frequency for two very different soil types, one a humid well-textured agricultural soil with $\rho = 40\ \Omega$ m, $\varepsilon' = 2.77$, and the other a sand with $\rho = 700\ \Omega$ m, $\varepsilon' = 0.67$ at 400 MHz. At low frequencies, up to 1 MHz, p decreases regularly as $1/\omega^{\frac{1}{2}}$ due to conductivity. This diminishes between 1 and 100 MHz and above 100 MHz increases to approach the slope in $1/\omega$. The utilisation of high frequencies in archaeology is therefore limited to 1 GHz at best, a frequency for which the penetration is well under a metre. It is customary to express the attenuation with an amplitude coefficient in dB m^{-1} called α, and the relation between α and p is $\alpha = 8.68/p$. Two other parameters are important for understanding the behaviour of the electromagnetic signal, the phase velocity and the wavelength, which are derived from the imaginary part of the propagation factor:

$$\lambda = 2\pi/I(\gamma) \tag{9.64}$$

$$v = \omega/I(\gamma) \tag{9.65}$$

where:

$$\lambda = \lambda_0 \frac{1}{[\varepsilon'^2 + (\varepsilon'' + \sigma/\varepsilon_0\omega)^2]^{\frac{1}{4}} \cos(\delta/2)} \tag{9.66}$$

$$v = c \frac{1}{[\varepsilon'^2 + (\varepsilon'' + \sigma/\varepsilon_0\omega)^2]^{\frac{1}{4}} \cos(\delta/2)} \tag{9.67}$$

in which c is the phase velocity of electromagnetic waves in a vacuum. If σ and ε'' are very small, v tends to $c/\varepsilon^{\frac{1}{2}}$ and λ towards $\lambda_0/\varepsilon'^{\frac{1}{2}}$. For the

Fig. 9.38 Depth of penetration of electromagnetic waves as a function of frequency in (a) a humid agricultural soil and (b) a dry sand. From ref. 37.

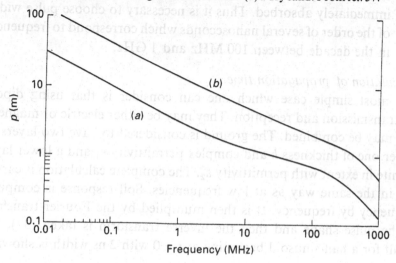

two soils used experimentally above, it equals: for the humid soil at 10 MHz $v = 0.15c$, and $\lambda = 4.6$ m, at 100 MHz $v = 0.18c$ and $\lambda = 0.55$ m and for the sand at 10 MHz $v = 0.3c$, $\lambda = 9$ m and at 100 MHz $v = 0.3c$ and $\lambda = 0.9$ m. There is weak dispersion, since velocity varies little. This reflects the invariability of ε' with frequency below 1 GHz as noted by Ruault and Tabbagh (29).

9.7.2. *Methods of measurement and interpretation of results*

Given the weak variation with frequency of the depth of penetration, a frequency sounding requires the use of three decades between 1 MHz and 1 GHz. This meets with two difficulties:

(1) The antennae or coil systems for the magnetic or electric fields have a relatively narrow bandwidth, a decade at most, which would necessitate using several sets of systems.

(2) Due to the preponderant effect of the dielectric permittivity above 10 MHz, the absolute value of the received signal varies as the sine of the frequency and its interpretation is very difficult when the transmitting antenna is close to the surface (37). It is therefore preferable to use a time measurement, in which a pulse is transmitted and the arrival of reflections from the interfaces present in the ground is observed. The pulse width must be such that it permits separation between two interfaces situated at two depths. For example, one may choose to show a difference of 15 cm, but the pulse must not be absorbed too much. It is therefore useless to transmit a spectrum which is rich in high frequencies if these are immediately absorbed. Thus it is necessary to choose pulse widths of the order of several nanoseconds which correspond to frequencies in the decade between 100 MHz and 1 GHz.

Calculation of propagation time

The most simple case which one can consider is that using dipoles for transmission and reception. They may be either electric or magnetic, and may be combined. The ground is considered to have two layers, an upper one of thickness h and complex permittivity ε_1 and a lower layer, infinite in extent with permittivity ε_2. The complete calculation is carried out in the same way as at low frequencies. Soil response is computed frequency by frequency. It is then multiplied by the Fourier transform of the pulse shape and then the inverse transform is taken (37). The result for a half-sinusoid beginning at $t = 0$ with 2 ns width is shown in figure 9.39(a) for different values of the parameters ε_1, ε_2 and h for a dipole spacing of 1 m and a height of 0.5 m. Two responses are observed: the first due to the air–ground interface, the second to the interface between the layers. The responses are S-shaped with the first proportional to $(\varepsilon_1^{\frac{1}{2}} - 1)/(\varepsilon_1^{\frac{1}{2}} + 1)$ while the second is very damped, with 30 dB difference between them. In the calculations $\varepsilon'' = \varepsilon'/10$ and the sign of the second response depends on the sign of $(\varepsilon_2^{\frac{1}{2}} - \varepsilon_1^{\frac{1}{2}})/(\varepsilon_2^{\frac{1}{2}} + \varepsilon_1^{\frac{1}{2}})$.

Fig. 9.39 Calculated soil response for a half-sinusoid for different permittivity contrasts and first layer thickness. From ref. 37.

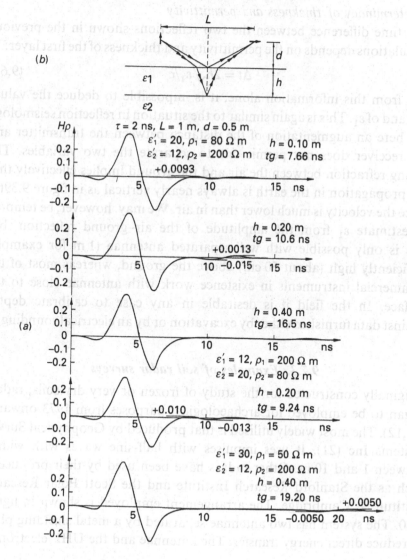

But the most important conclusion to be drawn from this calculation is the good correspondence between propagation time given by a ray approximation denoted as *tg* in the figure and the real time calculated. This result is surprising because the form of the different wave surfaces is far from plane and one is close to the source. Here we have a phenomenon which is well known in seismic prospecting. It is therefore possible to analyse the arrival times in terms of ray theory and to reconstruct the position of the interfaces with satisfactory precision. This phenomenon is linked to the fact that the dispersion is weak.

Indeterminacy of thickness and permittivity

The time difference between the two reflections shown in the previous calculations depends on the permittivity and thickness of the first layer:

$$\Delta t = 2h\sqrt{\varepsilon_1}/c \tag{9.68}$$

and from this information alone, it is impossible to deduce the values of h and of ε_1. This is again similar to the situation in reflection seismology but here an augmentation of the distance between the transmitter and the receiver does not permit one to separate the two variables. The strong refraction between the air and the ground implies effectively that the propagation in the earth is always nearly vertical as in figure 9.39(*b*) since the velocity is much lower than in air. We may, however, be tempted to estimate ε_1 from the amplitude of the air–ground reflection, but this is only possible with well-separated antennae (1 m for example) sufficiently high (about 50 cm) above the ground, whereas most of the commercial instruments in existence work with antennae close to the surface. In the field it is desirable in any case to calibrate depths against data furnished either by excavation or by an electrical sounding.

9.7.3. *Examples of soil radar surveys*

Originally constructed for the study of frozen or very dry soils, radars began to be employed for archaeological purposes from 1973 onwards (11,12). The most widely utilised is that produced by Geophysical Survey Systems Inc (21). It uses impulses with half-sine waves with widths between 1 and 10 ns. Other radars have been used by their producers, such as the Stanford Research Institute and the Scott Polar Research Institute at Cambridge. The arrangement employed is shown in figure 9.40. The system has two antennae separated by a metal shielding plate to reduce direct energy transfer. The antennae and the UHF electronics

are pulled along a profile on the ground surface and the power supply
and signal conditioning circuitry are placed either at a point near the
profile or in the vehicle pulling the antennae, if used. Displacement is
at the rate of $1-2$ m s^{-1}. A pulse repetition frequency of 50 kHz is used.
The results are displayed in the form of a section in time with
displacement on the abscissa and time in ordinate downward. The
signal can be recorded for later processing or recorded directly on
paper. The blackening of the recording paper is proportional to
the intensity of the received signal above a threshold, and the results
may also be shown on a CRT. A horizontal reflector, for example,
the passage from soil to a rock, gives a series of parallel black bands
in the display.

Fig. 9.40 (*a*) Soil radar arrangement and (*b*) waveforms.

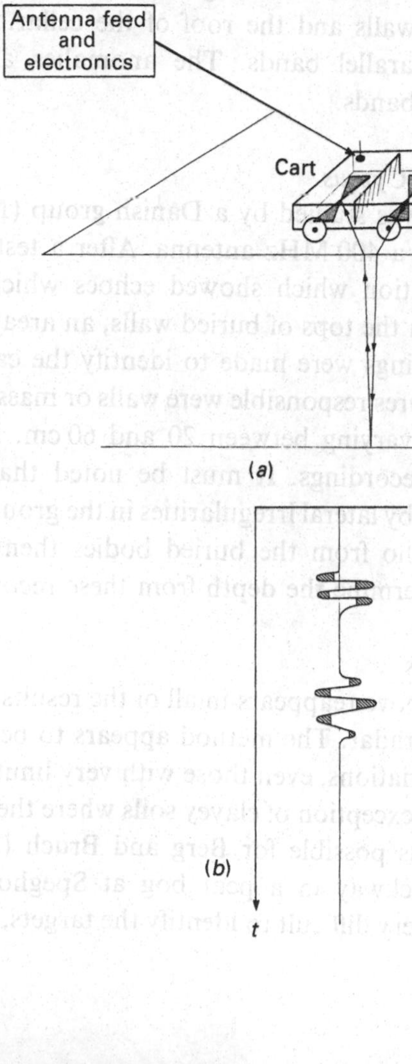

Antenna feed and electronics

Cart

(*a*)

(*b*)

t

Gain increases with time after the pulse so as to reinforce the most delayed signals in order to favour deep reflectors and compensate attenuation. The maximum gain is of the order of 90 dB, but the dynamic range on the thermal recording paper is only about 20 dB, and most of the recording systems only use one gray level. Better results would probably be obtained if a digital storage scheme which preserved the dynamic range of the data were used, followed by subsequent image processing to bring out the weak echoes in the presence of strong ones.

The historical site of Stenton Mansion, Pennsylvania, USA

The survey using a GSSI radar by Bevan and Kenyon (4) aimed at detecting cellars and substructures in a park covered by a lawn. The pulse used was of the order of 15–30 ns and the results displayed based on an estimated velocity of 7.6 cm ns^{-1}. Figure 9.41 shows examples of results obtained over the walls and the roof of the cellar. Each echo has three or five dark parallel bands. The anomalies are seen as displacements of the dark bands.

Site at Hala Sultan Tekke, Cyprus

This late Bronze Age site was studied by a Danish group (14) with the aid of a GSSI radar using a 400 MHz antenna. After a test on a strip at the edge of the excavation which showed echoes which could be interpreted as coming from the tops of buried walls, an area of 5000 m^2 was surveyed. Twelve cuttings were made to identify the causes of the clearest echoes. The structures responsible were walls or masses of stones with a top surface depth varying between 20 and 60 cm. Figure 9.42 shows two examples of recordings. It must be noted that here too, structures are made visible by lateral irregularities in the ground response rather than by a clear echo from the buried bodies themselves. The authors did not try to determine the depth from these recordings.

Limitations and advantages

The approach discussed above reappears in all of the results which have been published using soil radar. The method appears to be an efficient technique for detecting variations, even those with very limited volumes or weak contrast, with the exception of clayey soils where the absorption is very strong. It was thus possible for Berg and Bruch (3) to detect remains of a wooden trackway in a peat bog at Speghoje Mose in Denmark. However, it is very difficult to identify the targets, their depth, volume, and the material of which they are made. It is only possible to map the points or zones which are 'abnormal' and which must afterwards be elucidated by excavation. Thus soil radar appears to be complementary to the other prospecting techniques which always give indications concerning the contrast of properties and the volumes of structures.

Fig. 9.41 GSSI soil radar responses at Stenton Mansion, Pa. From ref. 4.

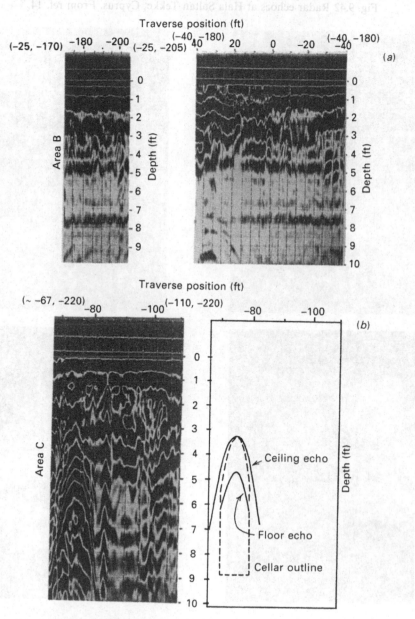

Radar contributes a richness of detail when the signals are not absorbed too much which the other methods do not furnish. On very dry volcanic soils, radar can yield results which are comparable to those obtainable with a resistivity survey. Examples of prospecting under these conditions are given by Imai, Sakayama, and Kanemori (20).

Fig. 9.42 Radar echoes at Hala Sultan Tekke, Cyprus. From ref. 14.

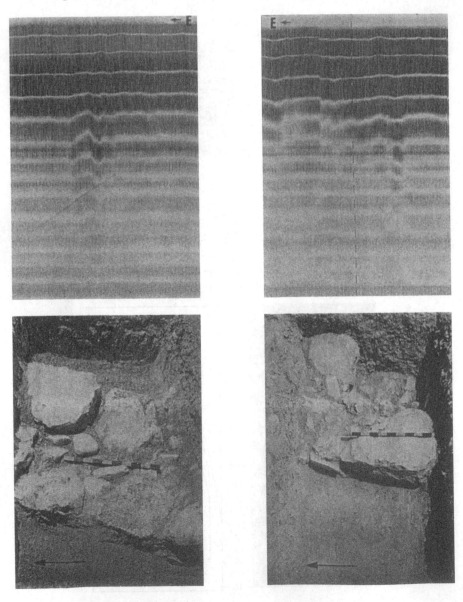

9.7.4. Airborne radar

In principle, radar does not require contact with or close proximity to the ground. There is no reason for not using it in an airborne installation. But this application raises three questions:

(1) Can one obtain sufficient resolution relative to sought-for archaeological structures?

(2) Since reflection from the surface is very strong can one obtain information relative to the subsurface?

(3) What is the effect of surface vegetation?

The answer to the first question lies in using a synthetic aperture antenna. In a system of this type described for example by Elachi, Bicknell, and Rolando (13), the radar illuminates the ground to the side of the aircraft and at right angles to the flight track. It emits short bursts of waves which reflect from the surface. On the surface, points whose echoes return at the same time are located on the circumference of a circle centred around the nadir point of the aircraft. Points having the same Doppler delay are located on hyperbolas. Therefore it is possible to determine the point on the ground from which the echo returns for a given delay and Doppler shift. Under these conditions the ground resolution depends only on the precision with which these two measurements can be carried out, and in particular it does not depend on the height of the aircraft. In order to remove all ambiguity due to the movement of the aircraft, it is possible to sum the different echoes received from each point during the course of time by utilising the phase which the reflected signal must have and taking into account the movement of the aircraft with respect to them. The signal treatment in such a system is very complex and cannot be discussed in detail here. As an example of what can be accomplished, it may be noted that the synthetic aperture radar in the SEASAT satellite has a ground resolution of 25 m.

Interest in such radars appeared at the beginning of the 70s and they have been utilised for mapping of regions of the globe where conventional aerial photography is impossible due to the presence of a continuous cloud cover. An interesting archaeological experiment was conducted in Mexico using an airborne radar belonging to the Jet Propulsion Laboratory. It operated in the L band, at 1.225 GHz and had a resolution of 15 m laterally and 8 m longitudinally. The experiment, reported by Adams, Brown, and Culbert (1), aimed at proving the existence of and mapping a set of drainage canals constructed by the Maya in the Yucatan which permitted intensive agriculture supporting a dense population.

The tests, which were carried out between 1978 and 1980, effectively showed the existence of the network and thus demonstrated the presence of intensive agriculture.

This experiment is unfortunately unique, for on the one hand it is not easy to obtain the use of equipment of the type available to the Jet Propulsion Laboratory and on the other hand there are few problems in archaeology which require the detection of fossil relief with a resolution of the order of tens of metres. This situation can probably be improved to some extent. One of the major aims of this experiment was to show that it is possible to measure through heavy vegetation and not only through cloud using L band radar. If the ground resolution could be improved, synthetic aperture radar would offer the possibility of studying fossil relief under forest or brush cover. This type of cover usually preserves old relief better than cultivated areas where agricultural practice destroys them. The use of this technique could furnish interesting information in the forested regions of Europe where other prospecting techniques cannot be readily employed because of the difficulty in moving measuring equipment or the impenetrability of the evergreen tree canopy for conventional aerial photography.

9.8. Appendix: The finite element method

The finite element method allows the solution of partial differential equations by transforming them into a set of simultaneous linear equations with suitable conditions. The idea is that a continuous variable can be approximated by a set of piecewise simple functions. A major advantage of the method is that once the problem has been set up, all problems of that class can be analysed such that the technique is independent of the geometry in that class. The method (31,49) is one of the numerical techniques which offer a solution to the problem of calculating the approximate values of a physical quantity $f(x, y, z, t)$ inside a domain Ω, where f satisfies a partial differential equation $L(f) = 0$, with boundary conditions such that f satisfies the limiting conditions $\partial f / \partial n = 0$ or $f = h(x, y, z, t)$.

To accomplish this, the domain is divided up into elements, segments, triangles, tetrahedra, etc. depending on the number of dimensions of the problem. Within each element, an interpolation function is chosen to represent f. Thus, for example, in a two-dimensional problem in x and z, and for triangular elements, $f(x, z) = \alpha_n x + \beta_n z + \gamma_n$ with the three coefficients linearly dependent on the three values of f at the three corners of the triangle f_i, f_j, and f_k. The method consists in calculating all of the values of f at the nodes of the domain by solving a system of linear equations which contains as many unknowns as there are points. For this, a quantity $\mathscr{F}(f, x, y, z, t, h)$ is defined such that $\partial\mathscr{F}/\partial f_i = 0$ at every node, that is, equivalent to its being globally minimum.

To define $\mathscr{F}$, the usual method consists in stating it as a functional which corresponds to the Euler partial differential equation. In problems in electromagnetism such a functional physically represents the total electromagnetic energy contained in the domain. In thermal problems this is not the case because the temperature represents the stored energy, and the functional has to be simply considered as a purely mathematical aid.

Mathematically, given a functional defined over a domain Ω for a function f and its derivatives in two dimensions:

$$\mathscr{F} = \iint_\Omega F(x, y, f, \partial f/\partial x, \partial f/\partial y)\, dx\, dy \qquad (9.69)$$

It is strictly equivalent to find the free extrema of the functional and to look for functions which satisfy the Euler equation:

$$\frac{\partial F}{\partial f} - \frac{\partial}{\partial x}\left[\frac{\partial F}{\partial(\partial f/\partial x)}\right] - \frac{\partial}{\partial y}\left[\frac{\partial F}{\partial(\partial f/\partial y)}\right] = 0 \qquad (9.70)$$

The partial differential equations which are needed in archaeological geophysical prospecting are few and simple. These are the Laplace equation $\Delta V = 0$ in electrical or magnetic prospecting, and the Helmholtz equation in electromagnetic prospecting with a single frequency $\Delta E - \gamma E = 0$. For example, in magnetic prospecting, the field created by the presence of an inhomogeneity derives from a potential V, $\mathbf{H} = -\nabla \cdot V$ which itself satisfies the Laplace equation over the whole of the domain. If one considers a two-dimensional problem and a cylindrical inhomogeneity for which $\partial/\partial y = 0$, the Laplace equation becomes:

$$\partial^2 V/\partial x^2 + \partial^2 V/\partial z^2 = 0 \qquad (9.71)$$

This equation is the Euler equation of a functional:

$$\mathscr{F} = \iint_\Omega \tfrac{1}{2}[(\partial V/\partial x)^2 + (\partial V/\partial z)^2]\, dx\, dz \qquad (9.72)$$

where effectively:

$$\frac{\partial F}{\partial V} = 0, \qquad \frac{\partial}{\partial x}\left(\frac{\partial F}{\partial(\partial V/\partial x)}\right) = \frac{\partial^2 V}{\partial x^2} \qquad (9.73)$$

The domain is divided up into triangular elements. Inside these V is approximated by a linear function. This is not difficult because only the first derivatives intervene in F and the calculation of $\partial \mathcal{F} / \partial V_i$ does not require integration over the whole domain but simply for the six triangles for which V_i is a vertex. For each of these triangles we write:

$$\frac{2 \iint \frac{1}{2}((\partial V / \partial x)^2 + (\partial V / \partial z)^2) \, dx \, dz}{\partial V_i} = \iint \left(\frac{\partial V}{\partial x} \frac{\partial(\partial V / \partial x)}{\partial V_i} + \frac{\partial}{\partial z} \frac{\partial(\partial V / \partial z)}{\partial V_i} \right) dx \, dz$$

(9.74)

and this expression is linear! Adding the six expressions gives us a linear expression which links the values in V_i to those of the six neighbouring nodes. By doing the same for all the nodes one obtains the desired system of equations. The matrix corresponding to this system is sparse and banded with the non-zero coefficients on the diagonal and in six lines parallel to it. The solution of such a system is relatively simple when the number of nodes is less than a thousand.

Notes

(1) Adams, R. E. W., Brown, W. E., Culbert, T. P. 1981, Radar mapping archaeology and ancient Maya land use, *Science*, **213**, 1457–63.

(2) Anderson, W. L. 1979, Computer program numerical integration of related Hankel transform of order 0 and 1 by adaptive digital filtering, *Geophysics*, **44**, 1287–305.

(3) Berg, F., Bruch, H. 1982, Georadar: Archaeological interpretation of soil radar data, *PACT News*, **7**, 285–94.

(4) Bevan, B., Kenyon, J. 1975, Ground penetrating radar for historical archaeology, *Masca News Letter 2*, **11**, 2–7.

(5) Cagniard, L. 1953, Basic theory of the magneto-telluric method of geophysical prospecting, *Geophysics*, **18**, 605–35.

(6) Carver, M. O. H. 1985, *Bulletin of the Sutton Hoo Research Committee*, **3**, 31.

(7) Colani, C. 1966, A new type of locating device – I. The instruments, *Archaeometry*, **9**, 3–8.

(8) Colani, C., Aitken, M. J. 1966a, A new type of locating device – II. Field trials, *Archaeometry*, **9**, 9–19.

(9) Colani, C., Aitken, M. J. 1966b, Utilisation of magnetic viscosity effects in soils for archaeological prospection, *Nature*, **212**, 1446.

(10) Colani, C., Grau, G. 1967, Die Grenzreichweite von Metallsuchgeräten in schwach leitenden Medien, *Frequenz*, **21**, 209–14.

(11) Dolphin, L. T., Bollen, R. L., Oetzel, G. N. 1974, An underground electromagnetic sounder experiment, *Geophysics*, **39**, 49–55.

(12) Dolphin, L. T., Barakat, N. 1975, *Electromagnetic Sounder Experiments at the Pyramid of Giza*, Stanford Research Institute, Stanford, Ca.

(13) Elachi, C., Bicknell, T., Rolando, L. J., Chialin, Wu 1982, Space borne synthetic aperture imaging radars, *Proceedings of the IEEE*, **70**, 1174–209.

(14) Fisher, P. M., Follin, S. G. W., Ulriksen, P. 1980, Subsurface interface radar survey at Hala Sultan Tekke, Cyprus, *Studies in Mediterranean Archaeology*, **63**, 48–51.

(15) Gauthier, F., Conte, P. 1986, Apports de la prospection géophysique à l'étude d'un ensemble de structures de surface lié à une cavité aménagée, *Revue d'Archéometrie*, **10**, 1–10.

(16) Graham, I. D. G., Scollar, I. 1976, Limitations on magnetic prospection in archaeology imposed by soil properties, *Archaeo-Physika*, **6**, 1–124.

(17) Guineau, B. 1975, Exemple d'application de la méthode magnéto-tellurique de prospection géoph. à l'étude de structures ou de formations géologiques situées sous un très faible recouvrement, *Geophysical Prospecting*, **23**, 104–24.

(18) Howell, M. I. 1966, A soil conductivity meter, *Archaeometry*, **9**, 20–3.

(19) Howell, M. I. 1968, The soil conductivity anomaly detector (SCM) in archaeological prospection, *Prospezioni Archeologiche*, **3**, 101–4.

(20) Imai, T., Sakayama, T., Kanemori, T. 1987, Use of ground-probing radar and resistivity surveys for archaeological investigations, *Geophysics*, **52**, 137–50.

(21) Morey, R. M. 1974, Continuous subsurface profiling by impulse radar, in *Proceedings: Conference on Subsurface Exploration*, American Society of Civil Engineers, 213–32.

(22) Mullins, C. E. 1974, The magnetic properties of soil and their application to archaeological prospecting, *Archaeo-Physika*, **5**, 143–348.

(23) Musson, C. R. 1968, A geophysical survey at South Cadbury Castle, Somerset, using the Howell soil conductivity anomaly detector (SCM), *Prospezioni Archeologiche*, **3**, 115–21.

(24) Nguyen Van Hoa 1977, Utilisation de la méthode électromagnétique, SGD pour la prospection de structures archéologiques, Thesis, Université de Paris 6, Paris.

(25) Nguyen Van Hoa, Jolivet, A., Tabbagh, A. 1977, Utilisation de la phase dans la méthode de prospection électromagnétique S.G.D., *Revue d'Archéometrie*, **1**, 115–26.

(26) Ott, R. H., Wait, J. R. 1972, On calculating transient EM fields of a small current carrying loop over a homogeneous earth, *Pure and Applied Geophysics*, **95**, 157–62.

(27) Parchas, C., Tabbagh, A. 1978, Simultaneous measurements of electrical conductivity and magnetic susceptibility of the ground in electromagnetic prospecting, *Archaeo-Physika*, **10**, 682–91.

(28) Parchas, C. 1979, Mesures simultanées de la conductivité électrique et de la susceptibilité magnétique du sol, Thesis, Université de Paris 6, Paris.

(29) Ruault, P., Tabbagh, A. 1980, Etude expérimentale de la permittivité diélectrique des sols dans la gamma de fréquence 100 MHz–1 GHz en vue d'une application à la télédetection de l'humidité des sols, *Annales de Géophysique*, **36**, 477–90.

(30) Scollar, I. 1962, Electromagnetic prospecting methods in archaeology, *Archaeometry*, **5**, 146–53.

(31) Segerlind, L. J. 1976, *Applied Finite Element Analysis*, John Wiley & Sons, New York, NY.

(32) Sommerfeld, A. 1926, Über die Ausbreitung der Wellen in der drahtlosen Telegraphie, *Annalen der Physik*, **81**, 1135–43.

(33) Tabbagh, A. 1973, Méthode de prospection électromagnetique S.G.D. Utilisation de deux sources, *Prospezioni Archeologiche*, **7**, 125–33.

(34) Tabbagh, A. 1974, Méthodes de prospection électromagnétique applicables aux problèmes archéologiques, *Archaeo-Physika*, **5**, 351–437.

(35) Tabbagh, A. 1974b, Définition des charactéristiques d'un appareil électromagnetique classique adapté à la prospection archéologique, *Prospezioni Archeologiche*, **9**, 21–33.

(36) Tabbagh, A. 1977, Deux nouvelles méthodes géophysiques de prospection archéologique, Thesis, Université de Paris 6, Paris.

(37) Tabbagh, A. 1979, Sur l'utilisation des hautes et très hautes fréquences électromagnétiques en prospection archéologique, *Revue d'Archéometrie*, **3**, 97–113.

(38) Tabbagh, A. 1982, L'interprétation des données en prospection électromagnétique avec les appareils SH3 et EM15, *Revue d'Archéometrie*, **6**, 1–11.

(39) Tabbagh, A., Verron, G. 1983, Etude par prospection électromagnétique de trois sites à dépôts de l'Age du Bronze, *Bulletin Société Prehistorique Française*, **80**, 375–89.

(40) Tabbagh, A. 1984, On the comparison between magnetic and electromagnetic prospecting methods for magnetic feature detection, *Archaeometry*, **26**, 171–82.

(41) Tabbagh, A. 1985, The response of a three dimensional magnetic and conductive body in shallow depth electromagnetic prospecting, *Geophysical Journal of the Royal Astronomical Society*, **81**, 215–30.

(42) Tabbagh, A. 1986a, Applications and advantages of the Slingram electromagnetic method for archaeological prospecting, *Geophysics*, **51**, 576–84.

(43) Tabbagh, A. 1986b, What is the best coil orientation in the Slingram electromagnetic prospecting method?, *Archaeometry*, **28**, 185–96.

(44) Tabbagh, A. 1986c, Sur la comparaison entre la prospection électrique et trois méthodes de prospection électromagnetique pour la detection des contrastes de réssistivité des structures archéologiques, *Prospezioni Archeologiche*, **10**, 49–64.

(45) Tite, M. S., Mullins, C. E. 1969, Electromagnetic prospecting, a preliminary investigation, *Prospezioni Archeologiche*, **4**, 95–102.

(46) Tite, M. S., Mullins, C. E. 1970, Electromagnetic prospecting on archaeological sites using a soil conductivity meter, *Archaeometry*, **12**, 1, 97–104.

(47) Verron, G. 1977, Informations archéologiques, site de Marchésieux, *Gallia Préhistoire*, **20**, 370–4.

(48) Wait, J. R. 1955, Mutual electromagnetic coupling of loops over a homogeneous ground, *Geophysics*, **20**, 630–8.

(49) Zienkiewicz, O. C. 1971, *The Finite Element Method in Engineering Science*, p. 521, McGraw-Hill, New York, NY.

10

Thermal prospecting

10.1 Introduction

The temperature at the surface of the soil is one of the physical parameters upon which an archaeological prospecting method may be based. It has been observed that variations of surface temperature reveal subsurface structures. These effects have been noted both on the ground and in aerial archaeology (see chapter 1) during the melting of snow or frost. Such observations are difficult to analyse systematically because it is necessary that favourable meteorological conditions prevail. Such visible traces are extremely evanescent. In southern Germany where snowfall is frequent in winter, this phenomenon has been systematically exploited by Braasch as is shown in figure 10.1 which displays the ditches of a Roman road through delayed melting of the covering snow (1). One might leave it at that, and limit oneself to the indirect effects of temperature anomalies, but this parameter has an important property: a measurement using a radiometer may be made very rapidly at a distance. Therefore one has a prospecting technique using temperature measurement which can be carried out from an aircraft, and the existence of an airborne method allows the gap to be filled in the panoply of techniques which neither electromagnetic nor the magnetic airborne methods can fill because of the limited size of the structures sought.

The measurement of temperature at ground level is a possible option. The first tests made by Hesse (4) as well as a group of tests made to check calculations have demonstrated the fact. Ground measurements pose difficult problems in the correction of temperature variation with time, and they do not have any significant advantage compared with other surface measurement techniques. The virtue of an airborne method which allows coverage of very large areas in a short time constitutes the principal reason for interest in thermal prospecting in archaeology and justifies a deeper study of the subject.

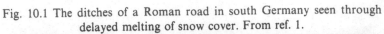

Fig. 10.1 The ditches of a Roman road in south Germany seen through delayed melting of snow cover. From ref. 1.

10.2. Soil temperature

10.2.1. Surface temperature, temperature profile, and temporal change

Anyone may observe that the temperature of the ground varies with time, and knows that this variation is determined by meteorological conditions. It is one of the parameters regularly measured in climatic studies and in weather forecasting. The simplest experiment for describing the temperature in the subsoil and analysing its temporal evolution consists in recording the changes with time in a vertical profile at several points. Figure 10.2 shows the variations obtained for two examples, one

Fig. 10.2 The variation in temperature at five depths in soil: (*a*) over a four day period at Compiegne (Oise); (*b*) over a seven day period at Boissy Chatel (Seine et Marne) France. From ref. 8.

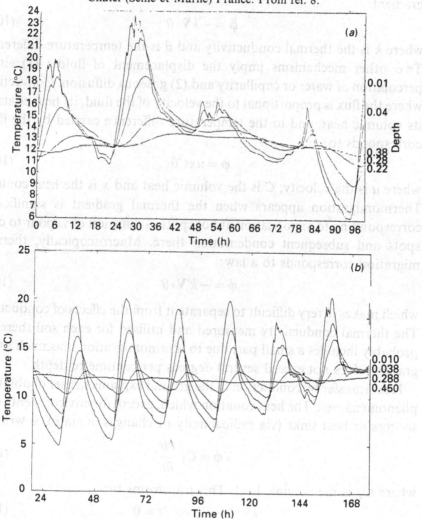

a sequence of four days at Compiegne (Oise), the other one of seven days at Boissy Chatel (Seine et Marne). In both cases one can see very clear diurnal variations which decreases with depth upon which is superimposed a number of rapid variations which are only observed at the near-surface (1–4 cm), and slow changes, especially at Compiegne, which are felt at all depths. The variation is not sinusoidal, and the second harmonic is important.

Several heat transfer mechanisms in the ground can be considered *a priori*: conduction is one whereby heat is transmitted without movement of material principally via the solid soil matrix, or via fluids present in the pores. The flux ϕ is proportional to the inverse of the temperature gradient:

$$\phi = -k\nabla \cdot \theta \tag{10.1}$$

where k is the thermal conductivity and θ is the temperature difference. Two other mechanisms imply the displacement of fluids: (1) either percolation of water or capillarity and (2) gaseous diffusion. Convection, where the flux is proportional to the velocity of the fluid, its heat content, its volumic heat, and to the temperature difference carried by the fluid corresponds to:

$$\phi = uxC\theta \tag{10.2}$$

where u is the velocity, C is the volumic heat and x is the heat content. Thermomigration appears when the thermal gradient is significant, corresponding to vaporisation at hot spots, a diffusion of vapour to cold spots and subsequent condensation there. Macroscopically, thermo-migration corresponds to a law:

$$\phi = -k'\nabla \cdot \theta \tag{10.3}$$

which makes it very difficult to separate it from the effects of conduction. The thermal conductivity measured and utilised for each soil therefore probably includes a small part due to thermomigration inasmuch as the gradient does not exceed several degrees per centimetre depth.

Heat transfer by conduction in the ground explains the set of observed phenomena well. The heat equation which, except for environments with sources or heat sinks (via radioactivity or changes of state), is written:

$$\nabla\phi = C_V \frac{\partial\theta}{\partial t} \tag{10.4}$$

where C_V is the volumic heat. This transforms to:

$$k\nabla^2\theta - C_V\, \partial\theta/\partial t = 0 \tag{10.5}$$

or:

$$\nabla^2\theta - \frac{1}{\Gamma}\frac{\partial\theta}{\partial t} = 0 \tag{10.6}$$

when the diffusivity:

$$\Gamma = k/C_V \tag{10.7}$$

is introduced. In the ground, the temperature θ satisfies this equation with the limiting condition at the surface:

$$-k\frac{\partial\theta}{\partial z}\bigg|_{z=0} = q(t) \tag{10.8}$$

where q is the flux exchanged either between the ground and the atmosphere or by radiation. To account for the experimental results shown in figure 10.2 one must use a Fourier transform where $F(\theta)$ is the transform of θ and $F(q)$ that of q, and study each component of θ which satisfies the equation:

$$\nabla^2 F(\theta) - (i\omega/\Gamma)F(\theta) = 0 \tag{10.9}$$

for which the solution:

$$F(\theta)(\omega) = F(q)(\omega)\frac{\exp(-i\pi/4)\Gamma^{\frac{1}{2}}}{\omega^{\frac{1}{2}}k}\exp[-(\omega/2\Gamma)^{\frac{1}{2}}z]\exp[-i(\omega/2\Gamma)^{\frac{1}{2}}z] \tag{10.10}$$

in an infinite half-space accounts perfectly for the damping and phase shift with depth for each component of the variation ω. These take into account the fact that the surface temperature is shifted in phase relative to the apparent movement of the sun; the diurnal temperature maximum occurring around three o'clock in the afternoon and the annual peak in the northern hemisphere at the beginning of August. This expression also leads to the introduction of thermal inertia:

$$P = k/\Gamma^{\frac{1}{2}} \tag{10.11}$$

which permits us to account globally for the reaction of the ground to a quantity of heat. Damping with depth can be characterised, as in electromagnetism by a skin effect:

$$p = (2\Gamma/\omega)^{\frac{1}{2}} \tag{10.12}$$

for which the values are close to 15 cm for the diurnal variation and 3 m for the annual variation depending on the value of diffusivity Γ for a soil. Recording temperatures at different depths allows determination of diffusivity via either the damping or the phase shift.

Analysis of temperature variation in the ground therefore leads to two conclusions:

(1) The main transfer mechanism is conduction.
(2) Variations in temperature are attributable to a quantity of heat applied to the surface $q(t)$ which varies with time, and which is worthy of study.

In all the discussion which follows, the notation θ is used, representing not the absolute temperature at a given point, but rather its difference with respect to the average air temperature T_M, so that:

$$T_s = T_M + \theta \qquad (10.13)$$

(where the subscript s denotes 'soil').

10.2.2. Surface energy balance

At the ground surface, energy may exist in different forms: radiation, latent heat, convective heat. Conservation of energy requires that their sum be zero, and decomposing the whole into four terms gives:

$$R_n + LE + H + Q = 0 \qquad (10.14)$$

where all energy gained by the ground is considered positive. In this equation R_n is the net radiative gain, LE is, the heat absorbed or given off by change of phase of water, H is the convective flux in the atmosphere (in a gas the conductive flux is usually negligible), and Q is the heat flux in the ground. The net radiation R_n itself contains three terms:

(1) Solar radiation, R_g, diffuse and direct, which reaches the ground in the short wavelengths between 0.4 and 1.1 μm. This radiation is reflected by the ground in the proportion a, a being the albedo, and absorbed as $(1 - a)$. Solar radiation depends on the latitude, the time of day, the slope of the ground, and the transmission of the atmosphere.
(2) Thermal radiation of the atmosphere R_a, which is at long wavelengths as a consequence of its temperature, and it is absorbed by the ground as a function of its emissivity ε_s.
(3) Vertical thermal radiation of the ground R_s.

One has therefore:

$$R_n = (1 - a)R_g + \varepsilon_s R_a + R_s \qquad (10.15)$$

with $R_a = \sigma T_a^4$ (Stefan–Boltzmann Law) and $R_s = -\varepsilon_s \sigma_s T_s^4$. Therefore:

$$R_n = (1 - a)R_g + \varepsilon_s(T_a^4 - T_s^4) \qquad (10.16)$$

by setting $\theta = T_a - T_{ground}$ (T_a is the atmospheric temperature), that is, assigning T_a to the average reference temperature:

$$(T_a^4 - T_s^4) \approx 4T_a^3 \theta \tag{10.17}$$

is approached and this gives a linear dependence of the radiation relative to the surface temperature difference.

The convective heat flux H in the atmosphere depends on the turbulent diffusivity which is in turn affected by the wind and the roughness of the ground surface. This flux is proportional to the difference in temperature between the air and the ground. The latent heat flux LE which corresponds to a displacement of water vapour also depends on the turbulent diffusivity but is proportional to differences in water vapour content between the air and the ground surface. The flux in the ground Q is the most important term for archaeologists, inasmuch as it determines the surface temperature directly. The magnitude of its instantaneous value is much lower than those of the other three terms.

The surface energy balance is very complex, and it is out of the question to be able to measure all the terms during a survey. The convective and latent heat fluxes are largely influenced by the vegetation, which is directly responsible for the surface roughness and for evapotranspiration. Therefore we are led to consider separately the cases of bare soils with a sparse inactive vegetation, and soils with active plant cover. This is tantamount to distinguish conditions in autumn or winter from those in spring or summer in temperate cultivated regions. A temperature anomaly at the ground surface can have as its cause one of the terms of the temperature balance or an abnormal reaction which corresponds to the presence of an inhomogeneity in the subsoil where the thermal properties vary with respect to the surrounding soil.

10.2.3. Heat flow in the ground

Periodic and transient variations in heat flux

As figure 10.2 and a large collection of published studies show, the heat flux shows periodic variations, among which the two most evident are daily and yearly, plus unpredictable transient variations which are essentially linked to air mass movements. The formation of an anticyclone corresponds to a negative flux in winter and a positive one in summer. The arrival of a depression leads to a negative flux in summer and usually to a positive one in winter. These periodic variations have the advantage of being predictable, but they are of little interest in archaeological prospecting.

The annual variation in the heat flux, which is of the order of $10 \, \text{W m}^{-2}$ in the temperate zones (2), penetrates the ground to a satisfactory extent, but the perturbations which are imposed by changes in the surface through ploughing and changes in humidity, as well as the existence of transient variations render its use difficult. In order to see anomalies which are linked to the annual variation, one would have to remeasure the temperature at the same place a large number of times in order to average out the effects of the transient variations. This is, in fact, impractical. The daily variation is much stronger, $100 \ \mathrm{W \ m^{-2}}$, but much more capricious. In rainy periods with cloud cover it is nearly zero. The slight penetration depth which it offers, being damped by a factor of nearly 10 in the plough layer alone, makes its direct use unlikely for anomalies which correspond to archaeological structures at depth. It is not completely useless, however, inasmuch as the properties of the surface layer often reflect contrasts which are much deeper. Rapid transient variations are without interest, but slow transient variations which exist over several days give a sufficient depth of penetration, and if intense enough may give anomalies which reflect buried archaeological structures.

Determination of slow transient flux variations
Variations in heat flux at the surface of the ground can be measured in three ways: via a fluxmeter, direct measurement of gradient, or by recording the evolution with time of the temperature in a vertical profile. A fluxmeter consists of a small plate of material of known thermal conductivity. The temperature gradient is measured between two of its faces. The measurement is concentrated at one point, and it cannot be made at the surface of the soil. The thermal conductivity must be identical to that of the surrounding medium, and this is difficult to realise. Direct measurement of the vertical gradient is by application of the formula:

$$q = -k \ \partial\theta/\partial z \qquad (10.18)$$

and this is only possible if the difference ∂z between two measured points is very small compared to the depths of penetration of the variations under consideration. This condition is also difficult to realise. The measurement of the evolution of the temperature at several points in a vertical profile has the advantage of being integrating in effect because it includes a thickness of soil which may be important. Practically, it permits one to utilise the measurements of temperature made by meteorological stations. Several methods may be utilised to calculate the flux from the variations in temperature at several depths (9). That which gives the best results consists in decomposing the surface variations in flux into a set of Heaviside functions $U(t)$. Its principle is described here, and it can be applied even if one has only a single depth.

The temperature at a point in a homogeneous soil for a variation in flux $Q_0 U(t)$ is given by:

$$\theta(z, t) = 2\frac{Q_0}{P} t^{\frac{1}{2}} \text{ ierfc}\left[\frac{z}{2(\Gamma t)^{\frac{1}{2}}}\right] \quad (10.19)$$

where ierfc is the integral of the complementary error function:

$$\text{erfc}(x) = (2/\pi^{\frac{1}{2}}) \int_x^\infty \exp(-u^2)\, du \quad (10.20)$$

Any variation $Q(t)$ can be expressed as a discrete series of functions $U(t - i\Delta t)$ beginning at different multiples of a time step Δt. By introducing the function:

$$S(z, n) = (n\Delta t)^{\frac{1}{2}} \text{ ierfc}[z/2(\Gamma n\Delta t)^{\frac{1}{2}}] \quad (10.21)$$

one may write, for the instant in time $t_i = (i - 1)\Delta t$:

$$\theta(z, t_i) = \frac{2}{P}\left[Q_1 S(z, i - 1) + \sum_{m=2}^{i-1} (Q_m - Q_{m-1})S(z, i - m)\right] \quad (10.22)$$

Therefore, from the temperature differences:

$$Q_1 = [\theta(z, \Delta t) - \theta(z, 0)]\frac{P}{2}\frac{1}{S(z, 1)} \quad (10.23)$$

$$Q_2 = \frac{[\theta(z, 2\Delta t) - \theta(z, \Delta t)](P/2) - Q_1[S(z, 2) - S(z, 1)]}{S(z, 1)} \quad (10.24)$$

and so on.

If there are several recording depths, one can use the method of least squares in calculating step by step the successive values of Q_i such that the sum of the squares of the differences between the theoretical expression for the differences of temperature and the observed values of this difference at all depths are minimised. An example of the results obtained with this method of calculation is shown in figure 10.3 for a period of a month at three meteorological stations in the Paris basin where measurements were made at depths of 10, 20, 50, and 100 cm. These results show the good correlation which exists between the slow transient variations in flux between points separated over a distance of about 100 km. This fact has been verified systematically for data from meteorological stations as well as for stations set up at surveyed sites as Perisset has shown (8). It allows us to use the values obtained from the meteorological stations to interpret the results of prospecting campaigns carried out in the same region without having to implant measuring devices in the soil of the site studied.

The weak point of this technique, like all the methods which only calculate the variation in time, is that there is no information about the stationary flux value upon which these variations are superimposed. One does not know what the true zero value is, in other words, what the flux is at time $t = 0$. To overcome this difficulty, it is convenient either to begin the calculation on a day where the vertical temperature profile is uniform, or to calculate the flux at $t = 0$ from:

$$-k \, \partial\theta/\partial z$$

possibly by using the average of several pairs of depths, or by utilising a least squares method, or by calculating the average flux during the recording period using the average of temperatures at two depths and calculating the mean flux by:

$$\bar{\phi} = -k \, \partial\bar{\theta}/\partial z \qquad\qquad (10.25)$$

Fig. 10.3 Thermal flux variations at three meteorological stations in the Paris basin over a month.

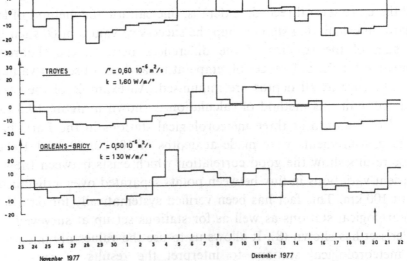

FLUX OF HEAT IN THE GROUND (in W/m²)

AUXERRE $\Gamma = 0,70 \; 10^{-6} \mathrm{m}^2/\mathrm{s}$
           $k = 1,80 \; \mathrm{W/m/}°$

TROYES $\Gamma = 0,60 \; 10^{-6} \mathrm{m}^2/\mathrm{s}$
           $k = 1,60 \; \mathrm{W/m/}°$

ORLEANS-BRICY $\Gamma = 0,50 \; 10^{-6} \mathrm{m}^2/\mathrm{s}$
                $k = 1\,30 \; \mathrm{W/m/}°$

November 1977   December 1977

10.3. Thermal anomalies and choice of a favourable measurement time for bare soil

10.3.1. Effects due to microrelief

The primary cause of a temperature anomaly at the ground surface lies in a local variation in one or more of the terms of the heat balance equation. We can neglect the fact that corresponding to every temperature anomaly, there is an augmentation of ground radiation which tends to reduce it, since this is a second order effect. The term of the heat balance equation which changes over short distances is $(1 - a)R_g$ in the presence of relief, microrelief, a shadow from a tree, a hedge. LE changes abruptly in the presence of a large quantity of available surface water, or transpiration from vegetation. The convective heat flow into the atmosphere is only important at times of strong wind, and it creates a tail of heat or cold as a function of obstacles such as houses or trees, depending on air temperature. If one considers a uniform field with constant roughness, no vegetation and where there are no puddles, the anomalies due to microrelief are primarily to be taken into consideration. They are characterised by the association of warm and cool zones, the latter on the less favourably exposed side. When incident radiation is sufficiently far from vertical, they can even be observed at noon, but the morning is better. The only important precondition is that there should be no cloud cover. The passage of small clouds in a nearly clear sky is not serious because they pass too rapidly to leave a thermal shadow on the ground.

10.3.2. Effects due to inhomogeneous subsoil

The second possible cause for localised anomalies at the surface lies in the presence of an heterogeneous region in the subsoil which modifies its reaction to the heat flux, the ease with which it can remove heat downwards or restore it upwards depending on the situation. The prediction and interpretation of the effect of such an inhomogeneous zone requires a model calculation in which the flux at the surface is considered horizontally uniform and variable in time.

Choice of model and method of calculation
The model to be considered must be realistic and at the same time as simple as possible. For this purpose we consider a two-dimensional rather than a three-dimensional case. It must contain at least three different zones, a structure, a surrounding undisturbed soil, and a

covering layer representing the arable zone which plays an important role since it retards and attenuates the variations of flux and temperature. Such a model is shown in figure 10.4. It is symmetric and therefore the calculation can be carried out over half of the figure. If one considers the harmonics of $Q(t)$, the problem to be resolved is two-dimensional in x and z. If slow transient variations are taken into account, the problem has three dimensions x, z and t. In both cases a numerical method must be used. The finite element method (see section 9.8) is employed to solve:

$$k\left(\frac{\partial^2\theta}{\partial x^2}+\frac{\partial^2\theta}{\partial z^2}\right)-C_V\frac{\partial\theta}{\partial t}=0 \tag{10.26}$$

or:

$$k\left(\frac{\partial^2\theta}{\partial x^2}+\frac{\partial^2\theta}{\partial z^2}\right)-i\omega C_V\theta=0 \tag{10.27}$$

with boundary conditions:

$$-k\,\partial\theta/\partial z=q(t),\qquad \text{in } z=0 \tag{10.28}$$

For the harmonic variations one must solve a system of linear equations which can be written in matrix form as:

$$([K]+i[C])[\theta]=[Q] \tag{10.29}$$

where $[Q]$ is the vector arising from the conditions at the boundaries, $[K]$ and $[C]$ are the matrices derived from the heat equations and $[\theta]$ is the vector representing the temperature at each support point in the domain. For the transient variations, a recursive calculation is necessary (13). For a time step Δt, the initial vector $[\theta_0]$ being known, one calculates $[\theta_1]$ by applying the formula:

$$\left(\frac{2[K]}{3}+\frac{[C]}{\delta t}\right)[\theta_1]+\left(\frac{[K]}{3}-\frac{[C]}{\delta t}\right)[\theta_0]=\frac{[Q_0]}{3}+2\frac{[Q_1]}{3} \tag{10.30}$$

Fig. 10.4 A model for a buried structure with a covering layer: 1 = arable surface soil; 2 = archaeological structure; 3 = subsoil. From ref. 10.

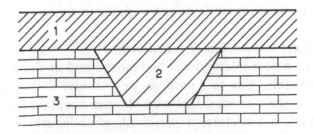

A grid using equilateral triangles, as shown in figure 10.5, is adopted. At the surface, the boundary condition of equation (10.27) is imposed on the axis of symmetry:

$$\partial\theta/\partial x = 0 \tag{10.31}$$

and at the lateral limit of the domain and at depth one has the choice of either:

$$\partial\theta/\partial n = 0 \tag{10.32}$$

(the Neuman condition) or that θ has the values which it would have for a two layer soil. The latter has been adopted.

Inhomogeneous subsoil and the definition of a coefficient of contrast
If one considers the simple case shown in figure 10.4 the number of parameters which can determine the sign of an anomaly *a priori* is seven: the thickness of the arable layer and two properties for each of the three environments, arable, subsoil, and feature. The amplitude of the anomaly will depend mostly on the volume of the structure and in the case of transient variation of temperature, the past history of the flux will play a role. It is therefore difficult initially to appreciate the influence of each

Fig. 10.5 The grid of equilateral triangles adopted for the finite element calculation of heat flux. From ref. 13.

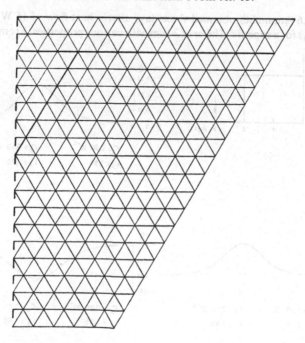

parameter. The numerical calculation does not provide any analytic expression which would allow us to offer a qualitative explanation at the cost of several approximations. The only method at our disposal is let the programme grind out results. It is prudent, in terms of computing time, to look at the analytic expressions governing a layered earth as a guide in order to appreciate how inhomogeneities affect things.

In an analytic calculation for a layered earth, each layer takes into account coefficients of thermal inertia: $(P_2 - P_1)/(P_2 + P_1)$, for example, for the second layer of a two layer soil (11). Transposing this in the case of a three layer soil, we can introduce a contrast coefficient:

$$C = (P_3 - P_2)/(P_3 + P_2) \qquad (10.33)$$

3 being the index for the surrounding medium and 2 the index for the buried structure. The results of the numerical calculation systematically give an anomaly increasing with C and for which the sign is that of C in the case where the soil gains heat. This is illustrated by figure 10.6 which shows anomalies obtained at $t = 3$ days after a flow of heat of 15 W m^{-2} during two days for a superficial layer of 20 cm with the properties:

$$\Gamma_1 = 0.6 \times 10^{-6} \text{ m}^2 \text{ s}^{-1} \qquad \text{and} \qquad k_1 = 1.00 \text{ W m}^{-1} \,^\circ\text{C}^{-1}$$

and a structure 60 cm wide and 30 cm deep for three cases. In the first

Fig. 10.6 Anomalies obtained three days after a heat flow of 15 W during two days for a superficial layer of 20 cm with various properties. From ref. 10.

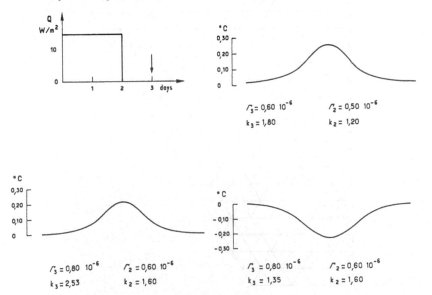

case:

$\Gamma_2 = 0.5 \times 10^{-6}\ m^2\ s^{-1}$, $k_2 = 1.2\ W\ m^{-1}\ °C^{-1}$, $\Gamma_3 = 0.6 \times 10^{-6}\ m^2\ s^{-1}$

and $k_3 = 1.8\ W\ m^{-1}\ °C^{-1}$ so that $C = 0.156$; in the second:

$\Gamma_2 = 0.6 \times 10^{-6}\ m^2\ s^{-1}$, $k_2 = 1.6\ W\ m^{-1}\ °C^{-1}$, $\Gamma_3 = 0.8 \times 10^{-6}\ m^2\ s^{-1}$

and $k_3 = 1.35\ W\ m^{-1}\ °C^{-1}$ so that $C = -0.156$ and in the third:

$\Gamma_2 = 0.6 \times 10^{-6}\ m^2\ s^{-1}$, $k_2 = 1.6\ W\ m^{-1}\ °C^{-1}$, $\Gamma_3 = 0.81 \times 10^{-6}\ m^2\ s^{-1}$

and $k_3 = 2.53\ W\ m^{-1}\ °C^{-1}$ so that $C = 0.156$. In the first case, the anomaly obtained is $+0.25$, in the second -0.23 and in the third $+0.22°C$. The sign of the anomaly follows that of C, but its amplitude is not strictly proportional to it. When the values of Γ and k are stronger as in the third case, there is a better distribution of heat and the anomalies are weaker. The three cases considered here correspond to those encountered experimentally: the first is an earth structure cut into a hard calcareous rock, the second to structures in a poorly conducting gravel, and the third to a soil rich in sand in a sandy subsoil. The contrast coefficient thus defined allows us to interpret the effect of an inhomogeneity simply. It is analogous to the coefficient:

$$C = (\rho_3 - \rho_2)/(\rho_3 + \rho_2)$$

which is encountered in electrical prospecting or to differences in susceptibilities as seen in magnetic prospecting.

The role of the surface layer

Homogeneous surface layer. The primary effect of the surface layer when it is homogeneous is to damp diurnal and short term variations completely. Conclusions drawn from a large number of calculations (9,13) show a number of effects due to the three parameters, thickness, thermal diffusivity, and thermal conductivity. An increase in thickness decreases the amplitude of the anomalies and delays their appearance. A rise in conductivity k_1 reduces the anomalies without influence on the time lag. A reduction of diffusivity has the same effect as an increase in thickness. Consequently, thin surface layers are favourable for thermal prospecting. Agricultural practices which aerate the soil such as ploughing or harrowing reduce the conductivity but have little effect on the diffusivity, and thus give rise to rather favourable situations which happily coincide with a complete absence of surface vegetation. Soils with high conductivity like alluvium, which is rich in fine sand, are relatively unfavourable.

Inhomogeneous surface layer. The presence of inhomogeneities in the surface layer is common, due either to changes in field boundaries followed by different crops, variations in the subsurface rock, or because of the presence of archaeological material. In such cases the detection of anomalies does not pose a problem, because shallow inhomogeneities induce strong anomalies which follow the diurnal rhythm. Paradoxically, the main difficulty is to prevent these anomalies from being so strong that they mask those of deeper origin. This risk is greatest when the effect of diurnal variation is maximum and it is important to know this. If one considers an inhomogeneity in the surface layer whose depth may vary within this layer, figure 10.7, calculation shows (13) that the moment when the anomaly due to diurnal variation is maximum occurs between 12.00 and 19.00 hours depending on thickness and the position in the superficial layer. If the inhomogeneity is at the surface, then the maximum is produced in the early afternoon and when not, it is delayed until the late afternoon.

The role of an inhomogeneity in the surface layer is illustrated in figure 10.8 which shows results obtained at three different times on the same day (survey GDTA of 31 March, 1977). The anomalies which were observable during the morning were completely eclipsed by those due to modern field boundaries and this effect faded during the course of the afternoon.

Inversion of transient heat flux. When heat flow has a constant direction and sign, the rules of interpretation are simple. The sign of an anomaly is that of the contrast coefficient and by taking the geological or pedological context into account, a structure which is detected can be interpreted. When the flux changes sign, the situation is more complex because one must also take into account those characteristics of the surface layer, thickness, and diffusivity, which determines the time which is needed for the anomalies to invert. Four examples are shown in figure

Fig. 10.7 An inhomogeneity in the surface layer whose depth varies with this layer. From ref. 13.

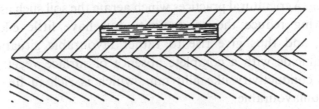

Fig. 10.8 Results obtained at three different times on 31 March, 1977 showing the role of an inhomogeneity in the surface layer: (*a*) 07 h 47, (*b*) 13 h 32, (*c*) 17 h 08.

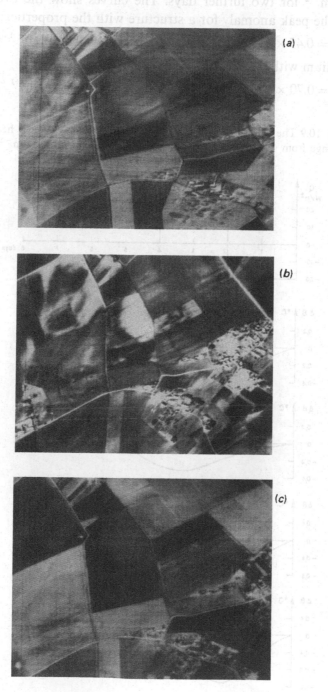

10.9. They are given for different surface layers. It is assumed that the heat flux changes first by $-20\ \mathrm{W\ m^{-2}}$ for two days running, then $+20\ \mathrm{W\ m^{-2}}$ for two further days. The curves show the evolution in time of the peak anomaly for a structure with the properties:

$$\Gamma_2 = 0.45 \times 10^{-6}\ \mathrm{m^2\ s^{-1}} \qquad \text{and} \qquad k_2 = 0.80\ \mathrm{W\ m^{-1}\,^{\circ}C^{-1}}$$

in a medium with:

$$\Gamma_3 = 0.70 \times 10^{-6}\ \mathrm{m^2\ s^{-1}} \qquad \text{and} \qquad k_3 = 1.80\ \mathrm{W\ m^{-1}\,^{\circ}C^{-1}}$$

Fig. 10.9 The evolution with time of the peak anomaly after a heat flux change from $-20\ \mathrm{W\ m^{-2}}$ for two days, followed by $+20\ \mathrm{W\ m^{-2}}$ for a contrast coefficient of 0.29. From ref. 10.

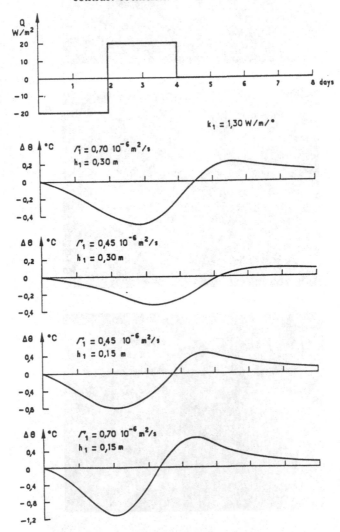

The contrast coefficient is thus 0.29. It may be noted that the anomaly reverses more readily if the surface diffusivity is large and its thickness low. For measurements made during the fifth day, the anomaly may be either negative or positive, depending on the characteristics of the surface soil. It is preferable not to measure after a heat flux inversion which has a globally zero value.

10.3.3. Choice of the optimal measurement time

Since the heat flux which creates the anomalies and the anomalies themselves vary with time, it is the object of this section to establish the rules defining the most favourable moments which furnish information about archaeological structures. These rules are, in non-hierarchical order:

(1) It is preferable to make measurements without clouds in order to record anomalies due to microrelief. Navigation is at the same time more exact and flying conditions more enjoyable.

(2) It is better to take readings during the morning so that the effects of diurnal variation on surface inhomogeneities do not mask deeper weak anomalies. A flight during the first half of the night might be of interest, but the microrelief will not appear and the quality of navigation will be degraded.

(3) It is better to fly after a sufficiently well-marked long variation in transient flux. Calculations of slow variations in transient flux using data from meteorological stations allow verification of whether or not this condition has been fulfilled.

10.4. Use of a scanning radiometer

10.4.1. Radiometric measurement of temperature

Every body having an absolute temperature T which is different from zero emits radiation whose intensity is a function of temperature. For a black body, that is one whose coefficients of absorption of incident radiation and emission are equal to 1, the luminance is a function of temperature T and the wavelength emitted follows Planck's law:

$$L(\lambda) = \frac{2\hbar c^2}{\lambda^5} \frac{1}{[\exp(\hbar c/kT\lambda) - 1]} \tag{10.34}$$

where k is Boltzmann's constant, $\hbar$ is Planck's constant and c is the velocity of light. Planck's law is shown graphically in figure 10.10. The luminance as a function of λ passes through a maximum which depends on the temperature and for which the value is expressed by Wien's law:

$$\lambda_{max} = 2897/T \qquad (10.35)$$

where λ is in micrometres. For typical soil temperatures, the wavelength of radiation is mostly in the infra-red. Real bodies, and in particular, the surface of the ground are not black bodies, and the emissivity $\varepsilon(\lambda)$, the ratio of the real luminance to that of a black body, is used.

Measurement is made through the atmosphere and since this also has bands of infra-red absorption, a wavelength window must be chosen which allows measurement of the ground temperature and not that of the air. There are two such windows in the infrared, 3–5 μm and 8–14 μm. The choice of wavelength to be utilised depends on several factors. The 8–14 μm window corresponds to the maximum of the Planck curve and the luminance of the ground is stronger. The 3–5 μm window corresponds to a wavelength where the sensitivity peaks inasmuch as the quantity

$$\frac{1}{L(\lambda)}\frac{\partial L(\lambda)}{\partial T}$$

is stronger, which allows detection of warm points more readily. But

Fig. 10.10 Planck's law giving the luminance as a function of temperature and wavelength. From ref. 8.

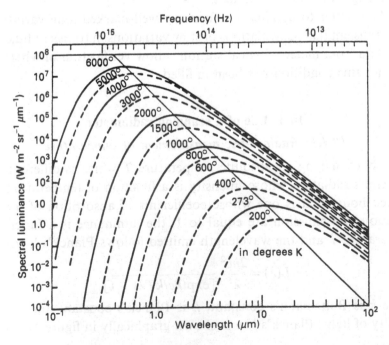

during the day there is reflection of solar radiation which masks that emitted by the ground. Therefore, for daytime measurement the 8–14 μm band is used. Multispectral scanners for the infra-red exist which allow simultaneous measurement in several wavebands. This type of device allows identification of different types of bare rock given the variations in emissivity at different wavelengths (5). Nonetheless, experience on humid soils in temperate climates has shown that the emissivities are uniform (10). Humidity increases emissivity in the whole of the thermal infra-red and the fractionation of rocks into small grains of soil increases uniformity.

10.4.2. The scanning radiometer

Principle of operation

The radiometer is made up of a photovoltaic detector or a photo-conductor which is sensitive to a wide band of radiation. A filter is placed in front of it to define the wavelengths utilised precisely. To record temperatures in different parts of scenes, two solutions exist: an array of detectors with electronic scanning, or a single detector with opto-mechanical scanning. For airborne radiometers, the instrument used up to now has a single detector and a rotating mirror whose axis is parallel to that of the aircraft. The arrangement is shown in figure 10.11.

Fig. 10.11 An airborne radiometer with rotating mirror and single detector. From ref. 8.

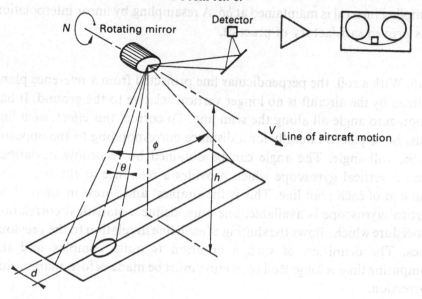

Radiation emitted by the ground over a small conic angle θ (instantaneous field of view, IFOV) reflects from the mirror and is directed at the detector. The rotation of the mirror provides the scanning of a line which is recorded over an angle ϕ. The movement of the aircraft gives the displacement from one line to the next. The data is digitised either before or after recording with an angular step sampling of δ which, in order to satisfy the Shannon sampling theorem, must have a value which is less than or equal to $\theta/2$. In this type of radiometer, the spacing between the lines of the image d is equal to the ratio between the speed of the aircraft and that of the rotation of the mirror $d = V/N$ for a regularly sampled image. Therefore it is necessary that the spacing between the points of a line at the nadir point directly under the aircraft $h\delta$ be equal to d, and the flight altitude relative to the ground must be proportional to ground speed.

Image distortions and their correction
Anamorphic distortion. Since sampling takes place at constant angular steps, the viewed point on the ground corresponding to each sample is circular, of diameter $h\delta$ at the nadir, becoming elliptical when the angle of view is no longer vertical. The length of its major axis is $h\delta/\cos^2(\alpha)$, where α is the angle between the angle of view and the vertical. This deformation which is equivalent to the projection of a plane on a cylinder must be corrected by resampling of each line such that the sampling interval is maintained at $h\delta$. A resampling by linear interpolation has proven satisfactory in practice.

Roll. With a roll, the perpendicular line projected from a reference plane defined by the aircraft is no longer vertical relative to the ground. It has a non-zero angle all along the scan line. To correct this effect, each line must be displaced laterally for a distance corresponding to the opposite of the roll angle. The angle can be obtained by recording deviations from a vertical gyroscope which provides a reference at the beginning and end of each scan line. This is the simplest and surest method. If no vertical gyroscope is available, one must define a statistical correlation procedure which allows the shifting of each line in relation to the previous lines. The definition of such a criterion is quite complex and the computing time is long. Roll correction must be made before anamorphic correction.

Pitch. When the aircraft pitches upwards, lines relative to complete coverage of the ground may be missing, and similarly may be doubled when it dives. If one has a vertical gyroscope, the angle of pitch is known and the correction consists of resampling to create new lines of data at regular intervals based on the lines actually recorded. If there is no gyroscope, it is very difficult to correct for pitch. The distortion can be readily distinguished from that due to roll because it is symmetrical with respect to the median line of the image. In practice, in airplanes currently used, there is much less pitch than roll.

Global distortion of the image. After the three corrections mentioned above have been carried out, there are usually two remaining types of distortion present in the image. The first has high spatial frequencies and is essentially due to the remaining errors in the correction of roll, linked to the imperfections in the gyroscope. This type of distortion only appears when flying conditions are really bad, and therefore we shall not consider it any further. Distortion which has low spatial frequency properties is due to slow variations in the altitude and speed of the aircraft during the scan. Examples are shown in figure 10.12. They can be corrected if the aircraft has a radar altimeter and a Doppler ground speed radar whose readings are regularly recorded along with the data. A complete inertial navigation system also provides the necessary data but all these systems are extremely expensive and they do not permit correction of deformation of the image due to ground relief. That is why it is just as simple to correct the image *a posteriori* by comparing it with a topographic map. This correction is carried out in two steps (15). Using control points which are well defined in both the map and the image which has been corrected for roll, pitch, and anamorphism, a

Fig. 10.12 Examples of global image distortion: (a) theoretical ground image; (b) variation of orientation; (c) yaw; (d) variation of height; (e) variation of flight speed. From ref. 15.

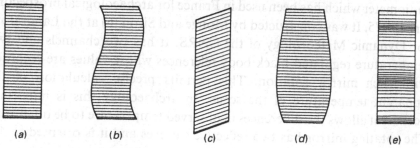

(a) (b) (c) (d) (e)

function is established which gives the correspondence between the coordinates of points in the image and points in the map. In practice, this is a bicubic spline defined on many patches, a patch being about 1000×1000 pixels (15). One calculates the gray value of each pixel in the corrected image from the neighbours of the corresponding point in the initial image. Practically, one can take the nearest neighbour (see chapter 5). Figures 10.13 and 10.14 show the localisation of the control points in the initial image and in the map. Figure 10.15 shows the corrected image, and for comparison figure 10.16 shows the original image without any correction whatsoever.

Enhancement of the image. It is possible to improve the contrast of most images. As for aerial photographs, discussed in chapter 4, we utilise a Wallis equation (4.16) to equalise the mean over a moving window while at the same time augmenting the standard deviation:

$$V_{(i,j)} = (V_{(i,j)} - M_0)\frac{A\sigma_1}{A\sigma_0 + \sigma_1} + \alpha M_1 + (1 - \alpha)M_0 \qquad (10.36)$$

where α is a factor which allows sudden changes of mean value to be passed and A limits the maximum multiplication factor (gain) when the standard deviation σ is too low. In the equation M_0 is the observed mean in the moving window, M_1 is the desired mean, and σ_1 is the standard deviation desired. For thermal images, large windows of 100×100 pixels have proven desirable. Radiometer images acquired line by line sometimes show line structure which is removed by 'destriping'. This operation is carried out by equalising the mean value of each line by reference to the mean of the surrounding lines. Typically 200 lines are used for computing the reference.

10.4.3. The ARIES radiometer

As an example, it is interesting to give the specifications for the ARIES radiometer which has been used in France for archaeological prospecting since 1975. It was constructed by Monge and Sirou (6) at the Laboratory for Dynamic Meteorology of the CNRS. It has two channels and two temperature regulated black body references whose values are recorded with each mirror rotation. This permits precise calculation of the apparent temperature of the scene. In archaeology this is important because it allows the differences in observed temperature to be quantised. The rotating mirror has two reflective surfaces and it is oriented to 45°

Fig. 10.14 Spline corrected image based on ground control points, from ref. 15.

Fig. 10.15 Image corrected for roll, pitch, and aspect ratio, and based on of control at successive pixels. From ref. 15.

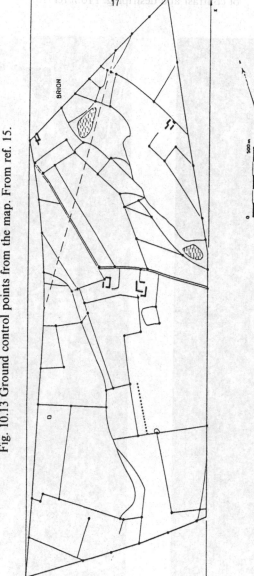

Fig. 10.13 Ground control points from the map. From ref. 15.

Fig. 10.14 Spline corrected image based on ground control points. From
ref. 15.

Fig. 10.15 Image corrected for roll, pitch, and anamorphism with amelioration
of contrast and destriping. From ref. 15.

Fig. 10.16 Uncorrected image. From ref. 15.

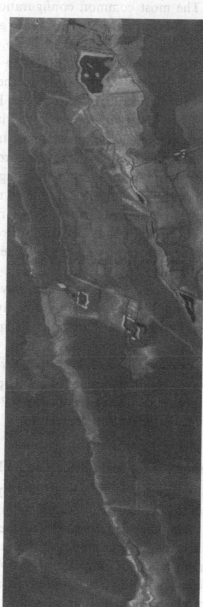

to its rotational axis, so that the two channels are recorded sequentially. Measurements are quantised to eight bits. Depending on the gain of the electronics, the least significant bit corresponds to a temperature difference between 0.04 and 0.25°C. Each channel may be utilised in the thermal infra-red, the near infra-red, or the visible region, depending on the choice of filter. The most common configuration sets up the first channel for the visible and near infra-red in order to acquire information about the state of the soil surface and the albedo. On bare soils the albedo is strongly correlated with humidity. The second channel is utilised for the measurement of temperature in the band from 10.5–12.5 μm, as defined by an interference filter. It uses an Hg–Cd–Te detector cooled by liquid nitrogen. The values are recorded in the aircraft using pulse code modulation of a signal on an analogue tape recorder.

The mirror is rotated with a motor which is synchronised to the quartz oscillator which controls the recording and the digitising. It has four possible rotational speeds, 4.55, 9.1, 18.2, and 36.4 rps. The IFOV θ is 2.8 mrad and the angular sampling step δ, 1.74 mrad. The total scan angle is 90° and therefore 900 points per line are recorded. A vertical gyroscope which furnishes the roll and pitch angle is attached to the mounting of the radiometer.

In archaeological prospecting one wishes to have the best possible resolution at ground level. For that reason a slow-flying aircraft is used which is capable of 80 knots or 41.2 m s^{-1}. The spacing between the scan lines is thus 1.13 m which therefore requires a flight altitude of 650 m above the ground. The width of the strip covered is about 1200 m. After anamorphic correction, the number of points per line is 1144, but the two ends are truncated to reduce the line to 1024 points so that the image has straight sides after the roll correction.

10.5. Example of thermal prospecting on bare soils

With the ARIES radiometer, the surface covered by a flight line has a width of about 1 km and a length of several kilometres. Therefore a number of parallel flight lines can be used and the surface covered is very large. For such areas, a large number of anomalies can be seen which may be of man-made, geological, or pedological origin. The explanation of all the anomalies seen represents an enormous task, inasmuch as one rarely has other sources which give a comparable density of information. For a complete interpretation it is necessary to correlate knowledge from various sources ranging from geology to rural history of the modern period and the past century. One is therefore in the situation where only a small part of the observable anomalies are interpretable. The cases where one has a sufficient amount of control information concerning the thermal anomalies are of great interest, and among those a certain number of examples are presented here. These are, a bit arbitrarily, divided into localised structures and old field patterns.

10.5.1. Localised structures

The ditch of the neolithic enclosure at Noyen sur Seine (Seine et Marne)
This enclosure is made up of an interrupted ditch abutting on an old arm of the Seine. It is one of the types of structures characteristic of the late middle neolithic in the Paris basin. This site was discovered by aerial photography and later, as a result of a deep ploughing which brought up a considerable quantity of material, it was excavated (7) before being destroyed entirely by gravel quarrying. Thermal prospecting was carried out on the 4 May, 1976 at 08 h 56 when the main field was bare, having just been planted with maize. In figure 10.17 one can see a warm white band where the temperature of the soil is more than 22°C. The subsoil is silty in this region. There are two cold black zones where the temperature is less than 21°C and here the subsoil is gravel. The ditch only shows up as a warm anomaly of 0.5°C where it is cut into the gravel. It does not show up in the silt. The depth and width of the ditch are known from the excavation and it was possible to measure the thermal properties of the fill and the surface layer. Therefore, a model can be constructed for which the surface layer has a thickness of 24 cm, a conductivity k_1 of 1.35 W m^{-1} °C^{-1} and a diffusivity Γ_1 of 0.72 $\times$ 10^{-6} m^2 s^{-1}. The ditch has a width at the top of 2 m, sides at 60° to the vertical, a depth of 0.5 m and thermal properties $\Gamma_2 = 0.5 \times 10^{-6}$ m^2 s^{-1} and $k_2 = 1.25$ W m^{-1} °C^{-1}. The surrounding gravel has the properties $\Gamma_3 = 0.9 \times 10^{-6}$ m^2 s^{-1} and $k_3 = 2.40$ W m^{-1} °C^{-1}. Therefore the contrast coefficient is $C = 0.18$. The transient variation in thermal flux can be estimated from the data of the meteorological station at Troyes which allows calculation of the evolution of the anomaly with time from the 25 April onwards. This is shown in figure 10.18. It reached 0.45°C on the morning of the 4 May.

A small Gallo–Roman structure at Leyment (Ain)

Large scale earth moving undertaken during the construction of the Autoroute A42, Lyon–Geneva, permitted verification of information acquired by thermal prospecting executed shortly beforehand. This was the case for the foundations of a small Gallo–Roman structure which appeared very clearly in the thermal scan of the 25 March, 1981 at 07 h 58, figure 10.19, which does not correspond to any superficial anomaly inasmuch as the visible and near infra-red image does not show anything. This structure appears as a cold anomaly agreeing perfectly with the fact that the flight was carried out during a period of increasing warmth as shown by the heat flux values measured at the nearby meteorological station at Amberieu, figure 10.20.

Deserted village at St Vincent (l'Hostellerie de Flée, Maine et Loire)
The two preceding examples are those of simple structures in a relatively homogeneous environment. This assists identifying them. The deserted

Fig. 10.17 Neolithic enclosure at Noyen sur Seine (Seine et Marne), thermal image of 4 May, 1976. From A. Tabbagh, Thesis, 1977, University of Paris 6.

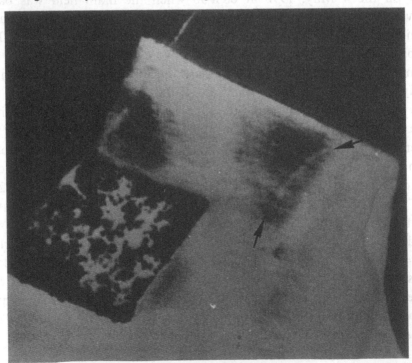

village which was sought at St Vincent could not correspond to a simple anomaly or group of anomalies. The possible variety of structures (walls, ditches, pits, accumulation of debris) presents a collection of thermal anomalies which are difficult to interpret in detail and from which only the large rectilinear anomalies shown in figure 10.21 emerge. The identification of the village is based in this case on two criteria. The area appears disturbed relative to calm surroundings. Furthermore, its position corresponds well to the indications given by medieval texts. The cold anomalies in black in figure 10.21 dominate. They are interpreted as being connected with a larger amount of stone, since the flight took place on the 27 March, 1981 at 07 h 57 after a marked heat flow as shown in figure 10.22.

Gallic and Gallo–Roman complex at Maillys (Côte d'Or)
The thermal survey carried out at this site on the 26 March, 1982 at 09 h 11 shows a complex set of cold marks corresponding to a Roman road and camp, and a rural settlement of the iron age, figure 10.23.

Fig. 10.18 Evolution of an anomaly with transient flux variation measured at Troyes from the 25 April–4 May, 1976. From A. Tabbagh, Thesis, 1977.

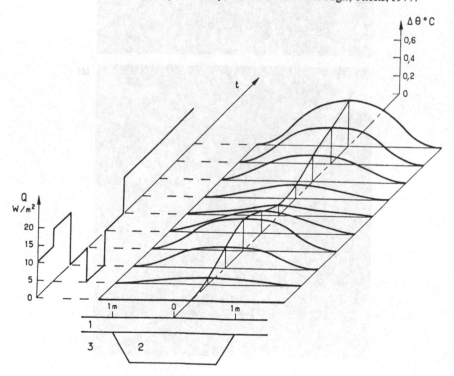

Some of the remains had previously been recorded by conventional
aerial photography. As figure 10.24 shows, there was no clear transient
variation in heat flux before the flight. This obliges us to suspect that
the cause of the observed anomaly lies in a variation of the thermal

Fig. 10.19 A Gallo–Roman structure near Leyment (Ain), scan of 25 March,
1981: (*a*) thermal, (*b*) visible.

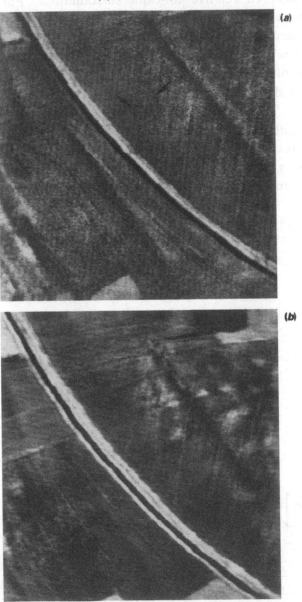

(*a*)

(*b*)

properties of the surface layer itself. The slow heat flux variation shown in figure 10.24 gives an anomaly of the order of $-0.05°C$ for a long structure of reasonable volume under a surface cover of 25 cm. Unfortunately the flight took place at the beginning of the rise in heat flux of the 24 and 25 March. This interpretation is corroborated by the fact that some of the traces are observed by conventional photography on bare soil.

10.5.2. Old field boundaries

Because thermal scanning allows us to cover a very large area, it is well suited to the study of the traces of older field systems. Two types of characteristic anomalies may be used. Microrelief which is particularly evident in the whole of the centre of the Paris basin where it appears to be correlated with a long tradition of grain farming is the first. The second is the traces which correspond to a soil which is more compacted or richer in stones compared with the lower thermal inertia of the surrounding soil, or to old ditches with a fill having a lower thermal inertia.

Microrelief
Figure 10.25 shows part of the thermal scan made on the 8 March, 1979 at 09 h 40 near the south-west part of the commune of Chatenay sur Seine (Seine et Marne). Microrelief is particularly abundant here. As figure 10.26 shows, this microrelief corresponds to field boundaries which are extant in the cadastral maps of the beginning of the nineteenth century. The arrows in the figure show the direction of ploughing the strip fields. This shows that microrelief appears preferentially at the ends

Fig. 10.20 Heat flux at the Amberieu meteorological station which is near Leyment.

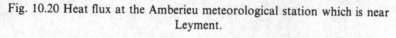

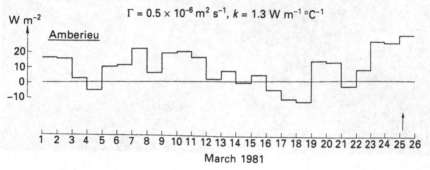

$\Gamma = 0.5 \times 10^{-6} \, m^2 \, s^{-1}, \, k = 1.3 \, W \, m^{-1} \, °C^{-1}$

March 1981

of the fields at right angles to the ploughing direction. The same phenomenon was observed at Lion en Beauce (3) which corroborates the traditional explanation of this microrelief as ploughing hillocks. These occur due to an accumulation of earth related to the cleaning effect on the plough at each half turn at the end of the long field. The

Fig. 10.21 Deserted village at St Vincent (Maine et Loire). From the *Proceedings of IGARSS Symposium Strasbourg*. ESA SP215 (published by ESA scientific and technical publications branch, August 1984).

Fig. 10.22 Heat flux prior to the flight over the village at St Vincent on the 27 March, 1981.

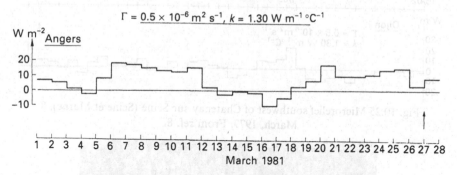

$$\Gamma = \underline{0.5 \times 10^{-6}\,m^2\,s^{-1}},\ k = 1.30\ W\,m^{-1}\,{}^{\circ}C^{-1}$$

Fig. 10.23 Gallic and Gallo–Roman complex at Maillys (Côte d'Or). From *Proceedings of IGARSS Symposium, Strasbourg.* ESA SP215 (published by ESA scientific and technical publications branch, August 1984).

Fig. 10.24 Heat flux prior to the flight over Maillys on the 26 March, 1981.

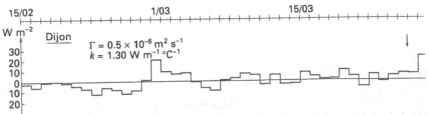

Fig. 10.25 Microrelief southwest of Chatenay sur Seine (Seine et Marne), 8
March, 1979. From ref. 8.

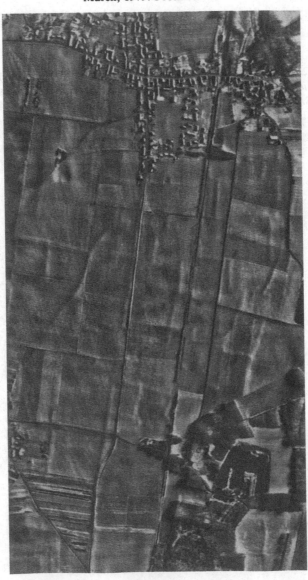

same type of microrelief has been found at Montcresson (Loiret), figure 10.27. Here it is interesting to note that the effect disappears towards the west side of the village, and this may confirm the fact that the microrelief is associated with a particular type of agriculture.

Fig. 10.26 The field boundaries at Chatenay sur Seine in the cadastral map of 1823. From ref. 8.

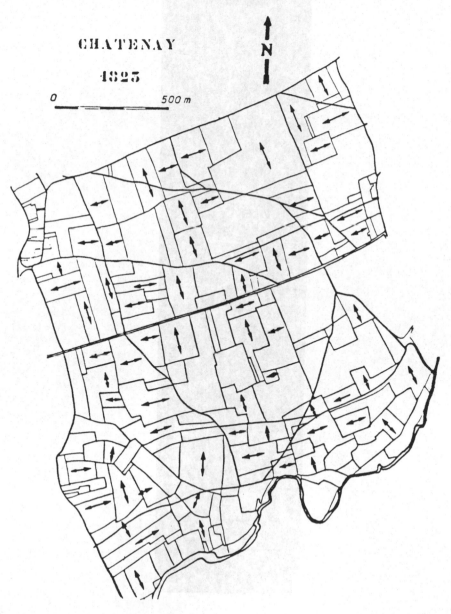

Fig. 10.27 Microrelief at Montcresson (Loiret).

Heterogeneous linear marks

Figure 10.28 shows results obtained at Senneçay and Levet (Cher) where microrelief does not exist but where, on the contrary, traces corresponding to linear features with strong thermal inertia are abundant. In this scan made late in the morning (11 h 18) one can easily see the modern

Fig. 10.28 Linear markings at Senneçay and Levet (Cher). From ref. 10.

field boundaries with cold black marks corresponding to an older field system. The flight took place after a period of increased heat flux. Figure 10.29 shows the older field boundaries as dotted lines corresponding to the cadastral plan of the beginning of the nineteenth century. This accounts for nearly all the marks.

10.6. Thermal anomalies from vegetation

When they are dead or inactive, plants essentially act as a badly conducting screen since the vegetable fibres and the water which they contain are poor heat conductors. The roughness of the surface, which they influence directly, may also play a role in reducing the convective exchange between the surface and the atmosphere. When the plants are alive and active they regulate their temperature by transpiration, and

Fig. 10.29 The older field boundaries at Senneçay and Levet as dotted lines corresponding to the cadastral map of the beginning of the 19th century. From ref. 10.

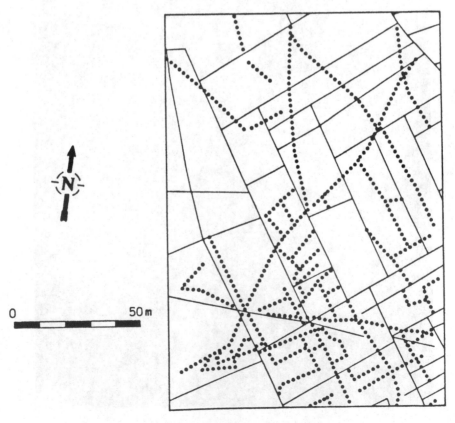

this phenomenon plays a major role in the surface energy balance. The phenomenon of transpiration by plants has been intensively studied by bioclimatologists since the survival of the vegetation depends on this directly. If a plant lacks water, it cannot prevent its temperature from rising, and it wilts and dies. Irrigation has the main function of maintaining transpiration. Therefore it is not surprising that during a period of growth one observes well-marked cold anomalies in plants. Nevertheless, since anomalies associated with buried archaeological structures appear, it is necessary to understand why some of these favour plant transpiration. Is this because they offer more water and nutrients for the plant which permits more rapid maturation, or is it tied to the presence of weeds in greater numbers or a particular type? These questions cannot be answered at present because it would be necessary to carry out a study of the physiological parameters of the vegetation which may be modified by the presence of one or another type of structure. However it is important to note that the effect exists, if only empirically known. Utilisation of this effect may allow the acquisition of important information about certain archaeological structures. The two cases presented below arise from measurements made at the beginning of May in the Paris basin after a short rainy period which left the upper soil humid. The winter crops were wheat or barley.

10.6.1. A site at Villeneuve La Guyard (Yonne)

The flight over the site of Prepoux at Villeneuve La Guyard was carried out on the 4 May, 1976 at 09 h 08. This site was already known through conventional aerial photography. It is an ensemble of proto-historic square or circular enclosures. In three of the fields at the site, two were covered with vegetation, one in wheat having an average temperature of about 20°C (14), and the other, in barley, having a bimodal temperature distribution, which corresponds to different subsoils, one with an average temperature of 15.5°C to a silt, the other with an average of 17°C to a gravel. The third field which was bare had an average temperature of 26°C. The scan is shown in figure 10.30 and the interpretation in figure 10.31. The complex of enclosures already known appears clearly, despite the fact that the image was not corrected for roll.

10.6.2. A site at Maisy (Aisne)

Several thermal surveys were carried out in the Aisne valley in northeast France. The double ditch shown in figure 10.32 located at a point called

'le passage de Maisy' is a large quadrangular enclosure, probably Gallo–
Roman in date. It appears in winter grain in the scan made on the 5
May, 1976 at 08.15 hours, but was not seen in two other flights made
over bare soils in March 1979 and March 1984. It was first observed in
conventional aerial photographs in August 1984, although the Aisne
valley has been systematically flown since 1966. During the test of May
1976 the vegetation had a temperature of 16°C, and that of the bare
ground was between 22 and 26°C depending on the kind of soil. This
led to an image where the bare soil pixels were saturated, thus requiring
mean equalisation treatment.

In the two examples shown, the thermal effect of vegetation cover is
important and there are differences of several degrees compared with

Fig. 10.30 Scan of prehistoric complex at Villeneuve La Guyard (Yonne),
4 May, 1976. From A. Tabbagh, Thesis, 1977.

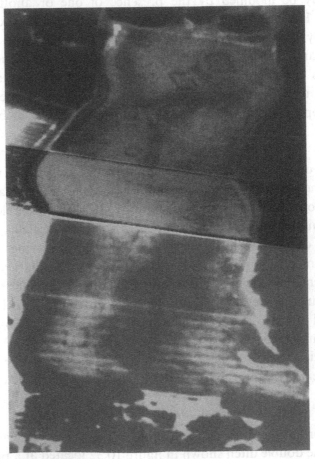

bare soils. Archaeologial structures are visible, and it would be useful to develop the empirical approach by flying during growth periods and studying the relationship between buried archaeological structures and transpiration of the plants growing over them.

Notes

(1) Braasch, O. 1983, *Luftbildarchäologie in Süddeutschland*, Gesellschaft für Vor- und Frühgeschichte in Wüttemberg, Aalen.

(2) Carson, J. F., Moses, H. 1963, The annual and diurnal heat exchange cycles in upper layers of soil, *Journal of Applied Meteorology*, **2**, 397–406.

(3) Fourteau, A. M., Tabbagh, A. 1979, Parcellaire fossile et prospection thermique, resultats des recherches à Lion en Beauce (Loiret), *Revue d'Archéometrie*, **3**, 115–23.

(4) Hesse, A. 1970, La prospection géophysique en archéologie, Problèmes, résultats et perspectives, *Mises à Jour*, **5**, 369–90.

Fig. 10.31 The interpretation of the scan of figure 10.30. From A. Tabbagh, Thesis, 1977.

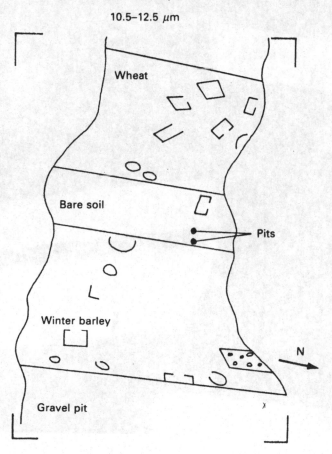

(5) Kahle, A. B., Goetz, A. F. H. 1983, Mineralogic information from a new airborne thermal infra-red multispectral scanner, *Science*, **222**, (4619) 24–7.

(6) Monge, J. L., Sirou, F. 1975, ARIES: un radiomètre multispectral à balayage, paper presented at, 5ème journée d'Optique Spatiale, Marseille.

(7) Mordant, D., Mordant, C. 1977, Noyen sur Seine, habitat néolithique de fond de vallée alluviale, *Gallia Préhistoire*, **20**, 229–69.

(8) Perisset, M. C. 1980, Prospection thermique de subsurface – application à l'archéologie. Thesis, Université de Paris 6, Paris.

Fig. 10.32 A double ditch enclosure in the Aisne valley (Maisy) in winter grain on 5 May, 1976.

(9) Perisset, M. C., Tabbagh, A. 1981a, Calcul des variations transitoires lentes du flux de chaleur à la surface du sol en vue de l'interprétation des résultats obtenus en prospection thermique de subsurface, *La Météorologie*, **6**, 39–45.

(10) Perisset, M. C., Tabbagh, A. 1981b, Interpretation of thermal prospection on bare soils, *Archaeometry*, **23**, 169–87.

(11) Tabbagh, A. 1973, Essai sur les conditions d'application des mesures thermiques à la prospection archéologique, *Annales de Géophysique*, **29**, 179–88.

(12) Tabbagh, A. 1976, Les propriétés thermiques des sols, *Archaeo-Physika*, **6**, 128–48.

(13) Tabbagh, A. 1977, Sur la détermination du moment de mesure favorable et l'interprétation des résultats en prospection thermique archéologique, *Annales de Géophysique*, **33**, 243–54.

(14) Tabbagh, A. 1979, Prospection thermique aéroportée du site de Prépoux (Villeneuve la Guyard, Yonne), *Revue Archéologique de l'Est*, **30**, 101–4.

(15) Tabbagh, J. 1983, Correction géometrique globale d'une image obtenue par un radiometrie à balayage embarqué, *Revue d'Archéometrie*, **7**, 11–25.

(16) Wallis, R. 1977, An approach to the space variant restoration and enhancement of images, in *Proceedings, Symposium on Current Mathematical Problems in Image Science*, Monterey, Ca, Nov. 1976, reprint, *Image Science Mathematics*, C. O. Wilde, E. Barett, eds., Western Periodicals, North Hollywood, Ca.

INDEX